Recent Trends in Computational Intelligence and its Application

About the Conference

The role of IT in every nook and cranny is a remarkable accomplishment, right from agriculture sector to exploration of planetary-science. This conference takes on a journey to all the stakeholders as a treat to establish a captivating knowledge for the forthcoming student community.

The increase in computing power and sensor data has driven Information Technology on end devices, such as smart phones or automobiles bringing together the practitioners of IT much delightful. The conference bundled with technical talks, is aimed to educate and vitalize the audience with eminent speakers sharing their experiences of oneself and business use cases which also examines and discuss various other interdisciplinary researches that could accelerate Information Technology.

This conference would be a unique conflux of networking, recreation, education, and practice to achieve global advantage. This conference improves work flow with a better understanding of how to build intelligent applications, learn and observe usage of novel intelligent techniques etc.

The significance of Information Technology and its impact on the future of mankind has increased productivity level exponentially with increase in the performance and functionality of all sorts of products and services.

Brands that integrate Information technology clinch a crucial edge over their competitors. Companies adopting IT could exploit their Big Data programs to harvest maximum potential out of their investments.

The widespread application of IT across the globe includes manufacturing, engineering, retail, e-commerce, health care, education, financial services, banking, space exploration, politics (to help predict the sentiments of voter demographics), etc.

IT professionals will be in strong demand for the future due to their widespread benefits in a variety of fields.

Recent Trends in Computational Intelligence and its Application

Proceedings of the 1st International Conference on Recent Trends in Information Technology and its Application (ICRTITA, 22)

Edited by

Sugumaran D.

Associate Professor, Department of Information Technology,
Vel Tech Rangarajan Dr. Sagunthala R&D Institute of
Science and Technology, Chennai, India

Souvik Pal

Associate Professor, Department of Computer Science and Engineering,
Sister Nivedita University (Techno India Group), Kolkata, India

Dac-Nhuong Le

Associate Professor, Faculty of Information Technology,
Haiphong University, Haiphong, Vietnam

Noor Zaman Jhanjhi

Associate Professor, School of Computer Science,
Taylor's University, Subang Jaya, Selangor Darul Ehsan, Malaysia

CRC Press
Taylor & Francis Group
Boca Raton London New York

CRC Press is an imprint of the
Taylor & Francis Group, an **informa** business

First edition published 2023
by CRC Press
4 Park Square, Milton Park, Abingdon, Oxon, OX14 4RN

and by CRC Press
6000 Broken Sound Parkway NW, Suite 300, Boca Raton, FL 33487-2742

© 2023 Vel Tech Rangarajan Dr. Sagunthala R&D Institute of Science and Technology

CRC Press is an imprint of Informa UK Limited

ISBN: 978-1-032-48410-5 (pbk)
ISBN: 978-1-003-38891-3 (ebk)

DOI: 10.1201/9781003388913

Typeset in Times LT Std
by Aditiinfosystems

Table of Contents

Recent Trends in Computational Intelligence and its Application – Sugumaran D. et al. (eds)
© 2023 Taylor & Francis Group, London, ISBN 978-1-032-48410-5

List of Figures

List of Tables

Foreword

Dear Delegates,

It makes me exceptionally contented that the Department of Information Technology is conducting International Conference on Recent Trends in Information Technology and its Application 2022 (ICRTITA) on August 29 & 30, 2022.

Regardless of lack of familiarity, Artificial Intelligence (AI) is a technology that offers a variety of tools to reassess ways to combine information, evaluate the data, and apply the ensuing insights to enhance decision-making. Smart Computers employed with artificial intelligence and machine learning exemplify massive ability of machines to exhibit intelligent behaviour. The learning, thinking and self-rectification of computers had given this era astounding experiences to human in various fields which could automatically improve many applications on robotics, data mining, navigation, bioinformatics etc. Smart systems in vehicles, buildings and farms save time, money and lives to enjoy a more customised business environment.

I personally believe that this conference will give the participants a chance to share their extensive knowledge, learn more about solution-based principles, and have a wonderful experience.

Prof. Dr. S. Salivahanan
Vice Chancellor

Recent Trends in Computational Intelligence and its Application – Sugumaran D. et al. (eds)
© 2023 Taylor & Francis Group, London, ISBN 978-1-032-48410-5

Preface

We feel greatly honoured to have been assigned the job of organising the International Conference on Recent Trends in Information Technology and its Application 2022 (ICRTITA '22) on August 29 & 30, 2022, at Veltech Rangarajan Dr Sagunthala R & D Institute of Science and Technology.

The International Conference's proceedings published in a disciplined manner for the use of the researchers is a great privilege for us. The goal of the conference is to facilitate better interaction among researchers, scientists, engineers from various parts of the globe at one place together with the discussion of challenges encountered and the solutions adopted in the fields of Artificial Intelligence, Deep learning etc. Out of 210 papers submitted from different nations, around 70 have been accepted, which is highly impressive. The conference has been divided into major sessions like Augmented and Virtual Reality, Block chain Technology, Cloud Computing, Computational Intelligence and Computer Vision, Deep Learning and Artificial Intelligence, Human Computer Interaction, Remote Sensing and GIS etc.

We intend to utilize this occasion to extend our gratitude to Col. Prof. Dr. Vel. R. Rangarajan and Founderess Dr. Sagunthala Rangarajan from the bottom of our hearts for their trust in us to organise this International Conference on Recent Trends in Information Technology and its Application 2022 (ICRTITA '22). We sincerely thank our beloved Chairperson & Managing Trustee, Mrs. Rangarajan Mahalakshmi Kishore for her continuous support and guidance in organising the conference. We really need to convey our sincere gratitude to the Vice Chancellor Prof. Dr. S. Salivahanan and Head of Information technology Dr. C. Mahesh of Veltech Rangarajan Dr Sagunthala R & D Institute of Science and Technology for their significant support in making this conference successful.

With all our due respect, we would thank our chief guest for his presence at the inauguration ceremony for delivering motivational and inspiring speech even in your busy schedule.

This event would not have been possible without the cooperation of all the paper authors and conference speakers. All papers which are being selected and approved will be published in Conference Proceedings Series on Intelligent Systems Data Engineering bookseries of Taylor and Francis (Scopus Indexed).

About the Editors

D. Sugumaran working as an Associate Professor in the Department of Information Technology, Vel Tech Rangarajan Dr. Sagunthala R & D Institute of Science and Technology, Chennai, Tamil Nadu, India, since 2013. He is having 18 years of teaching experience and 3 years of industrial experience. He has Published numbers of research articles in various Journals. Apart from teaching He has been actively involving in software designing and development activities. In his education career he has received Bachelor of Computer Science degree in 1997 from AVC College, Mayiladuthurai affiliated to Bharathidasan University, Trichy, Tamil Nadu, India. He completed and received his Master of Computer Application degree in 2000 from S.K.S.S Arts College, Bharathidasan University, Trichy, Tamil Nadu, India and Master of Engineering (Computer Science) degree in 2009 from Vel Tech Multi Tech Engineering College, Anna University, Chennai, Tamil Nadu, India He completed doctorate of philosophy in computer science and engineering in Vel Tech Rangarajan Dr. Sagunthala R & D Institute of Science and Technology, Chennai, Tamil Nadu, India.

Souvik Pal is an Associate Professor in the Department of Computer Science and Engineering at Sister Nivedita University (Techno India Group), Kolkata, India. Prior to that, he was associated with Global Institute of Management and Technology; Brainware University, Kolkata; JIS College of Engineering, Nadia; Elitte College of Engineering, Kolkata; and Nalanda Institute of Technology, Bhubaneswar, India. Dr. Pal received his MTech, and PhD degrees in the field of Computer Science and Engineering from KIIT University, Bhubaneswar, India. He has more than a decade of academic experience. He is author or co-editor of more than 15 books from reputed publishers, including Elsevier, Springer, CRC Press, and Wiley, and he holds three patents. He is serving as a Series Editor for "Advances in Learning Analytics for Intelligent Cloud-IoT Systems", published by Scrivener-Wiley Publishing (Scopus-indexed); "Internet of Things: Data-Centric Intelligent Computing, Informatics, and Communication", published CRC Press, Taylor & Francis Group, USA; "Conference Proceedings Series on Intelligent Systems, Data Engineering, and Optimization", published CRC Press, Taylor & Francis Group, USA; Dr. Pal has published a number of research papers in Scopus/SCI/SCIE Journals and conferences. He is the organizing chair of RICE 2019, Vietnam; RICE 2020 Vietnam; ICICIT 2019, Tunisia. He has been invited as a keynote speaker at ICICCT 2019, Turkey, and ICTIDS 2019, 2021 Malaysia. He has also served as Proceedings Editor of ICICCT 2019, 2020; ICMMCS 2020, 2021; ICWSNUCA 2021, India. His professional activities include roles as Associate Editor, Guest Editor, and Editorial Board member for more than 100+ international journals and conferences of high repute and impact. His research area includes cloud computing, big data, internet of things, wireless sensor network, and data analytics. He is a member of many professional organizations, including MIEEE; MCSI; MCSTA/ACM, USA; MIAENG, Hong Kong; MIRED, USA; MACEEE, New Delhi; MIACSIT, Singapore; and MAASCIT, USA.

Dac-Nhuong Le (Lê Đắc Nhường) has an M.Sc. and Ph.D. in computer science from Vietnam National University, Vietnam in 2009 and 2015, respectively. He is an Associate Professor on Computer Science, Deputy Head of Faculty of Information Technology, Haiphong University,

Vietnam. He has a total academic teaching experience of 15+ years with many publications in reputed international conferences, journals, and online book chapter contributions (Indexed By: SCI, SCIE, SSCI, Scopus, ACM, DBLP). His area of research includes Soft computing, Network communication, security and vulnerability, network performance analysis and simulation, cloud computing, IoT, and Image processing in biomedical. His core work in network security, soft computing and IoT, and image processing in biomedical. Recently, he has been on the technique program committee, the technique reviews, the track chair for international conferences under Springer-ASIC/LNAI Series. Presently, he is serving on the editorial board of international journals and he authored/edited 20+ computer science books by Springer, Wiley, CRC Press**.**

Noor Zaman Jhanjhi is currently working as Associate Professor | Director Center for Smart society 5.0 [CSS5], and Cluster Head for Cybersecurity cluster, at School of Computer Science and Engineering, Faculty of Innovation and Technology, Taylor's University, Malaysia. He is supervising a great number of Postgraduate students mainly in cybersecurity for Data Science. CSS5 has extensive research collaboration globally with several institutions and professionals. Dr. Jhanjhi is Associate Editor and Editorial Board for several reputable journals including PeerJ Computer Science, Frontier Journals, etc. PC member for several IEEE conferences worldwide, and guest editor for reputed indexed journals. Active reviewer for a series of Q1 journals has been awarded globally as a top 1% reviewer by Publons (Web of Science). He has been awarded an outstanding Associate Editor by IEEE Access for the year 2020. He has high indexed publications in WoS/ISI/SCI/Scopus, and his collective research Impact factor is more than 700 points as of the first half of 2022. He has international Patents on his account and edited/authored more than 40 research books published by world-class publishers. Including Springer, CRC Press, IGI Global USA, Wiley, Intech Open Science, Eliva Press, etc. He has great experience in supervising and co-supervising postgraduate students, and an ample number of 20 plus PhD and Master's students graduated under his supervision. He is an external PhD/Master thesis examiner/evaluator for several universities globally. He has completed more than 25 international funded research grants successfully. He has served as a Keynote/Invited speaker for more than 40 international conferences, presented several Webinars worldwide, and chaired international conference sessions. His research areas include Cybersecurity, IoT security, Wireless security, Data Science, Software Engineering, and UAVs.

Dr. Jhanjhi has an intensive background of academic quality accreditation in higher education besides scientific research activities, he had worked for academic accreditation for more than a decade and earned ABET accreditation twice for three programs at the college of computer sciences and IT, King Faisal University Saudi Arabia. He also worked for National Commission for Academic Accreditation and Assessment (NCAAA), Education Evaluation Commission Higher Education Sector (EECHES) formerly NCAAA Saudi Arabia, for institutional level accreditation. He also worked for the National Computing Education Accreditation Council NCEAC) Pakistan. He has experience in teaching advanced era technological courses including, Mobile Application Development (Android), Cloud Computing, Mobile Computing, Wireless Networks and Security, Secure Software Systems, Computer Security, .Net Framework programming, etc. besides other undergraduate and postgraduate courses. Postgraduation supervision, graduation projects, internships, and thesis supervision.

Recent Trends in Computational Intelligence and its Application – Sugumaran D. et al. (eds)
© 2023 Taylor & Francis Group, London, ISBN 978-1-032-48410-5

Introduction

Department of information technology, Vel Tech Rangarajan Dr. Sagunthala R&D Institute of Science and Technology, India is organizing 1st International Conference on Recent Trends in Information Technology and its Application 2022 (ICRTITA '22) on August 29 & 30, 2022, in hybrid mode (online as well as on-site, due to the COVID-19 pandemic) The ICRTITA '22 aims to bring together leading academicians, scientists, researcher scholars, and UG/PG graduates across the globe to exchange and share their research outcomes. It targets to provide a state-of-the-art platform to discuss all aspects (current and future) of Recent Trends in Information Technology and its Application. This will enable the participating researchers to exchange their ideas about applying existing methods in these areas to solve real-world problems. The proceedings of conference will be submitted for publication in Conference Proceedings Series on Intelligent Systems Data Engineering book-series of Taylor and Francis. [SCOPUS Indexed].

Recent Trends in Computational Intelligence and its Application – Sugumaran D. et al. (eds)
© 2023 Taylor & Francis Group, London, ISBN 978-1-032-48410-5

Details of Programme Committee

Chief Patrons

Col. Prof. Dr. Vel. R. Rangarajan, Chancellor & Founder President.

Dr. Sagunthala Rangarajan, Foundress President.

Mrs. Rangarajan Mahalakshmi Kishore, Chairperson & Managing Trustee.

Patrons

Prof. Dr. S. Salivahanan, Vice Chancellor.

Prof. Dr. E. Kannan, Registrar.

General Chair

Dr V. Srinivasa Rao, Vel Tech, Chennai, India.

Dr. C.Mahesh, Vel Tech, Chennai, India.

Convenor

Dr. C.Mahesh, Vel Tech, Chennai, India.

Co-Convenor

Mr.R.Hariharan, Vel Tech, Chennai.

Program Chairs

Mr.Karthikayan.P.N., Vel Tech, Chennai

Mrs.Dhilshat Fathima, Vel Tech, Chennai.

Keynote Speakers

Prof. Dr. Hee Yong Youn, Honor Professor Department of Computer Engineering, College of Information and Communication Engineering, Sungkyunkwan University, South Korea.

Mr. Souvik Pal, Associate Professor Department of Computer Science and Engineering, Sister Nivedita University, Kolkata, India.

Dr. P. Arish, Data Scientist, Entropik Technologies Pvt. Ltd., Chennai, India

Dr. G.Umarani Srikanth, Professor, Bharath Institute of Higher Education and Research Chennai, India.

Mr. Balaji Rajakesari.M, Scientist/Engineer Project, Manager, ISRO.

Prof. Dr. R. Logeswaran, Professor & Dean, Faculty of Information Technology, City University, Malaysia.

Prof. Dr. Saravanan Muthaiyah, Professor, Multimedia University, Malaysia.

Dr.Surendiran, Assistant Professor, Department of CSE, National Institute of Technology, Puducherry

Dr. Varalakshmi Perumal, Professor, Anna University, Chennai.

Dr. P. Suresh, Professor Dean – International Relations, Vel Tech Rangarajan Dr.Sagunthala R&D

Institute of Science and Technology, Avadi, Chennai

Publicity Chair

Dr. Sugumaran.D, Vel Tech, Chennai

Mr Durairaj.K, Vel Tech, Chennai Mr Karthikayan.P.N, Vel Tech, Chennai Mrs Ramya.D, Vel Tech, Chennai

Publication Chair

Mrs Dhilsath Fathima.M, Vel Tech, Chennai

Dr. N.Noor Alleema, Vel Tech, Chennai

Mrs Jayanthi. K, Vel Tech, Chennai

Mrs Sakunthala Prabha. K.S, Vel Tech, Chennai

Mr Pradeepkumar.S, Vel Tech, Chennai

Finance Chair

Mr.R.Hariharan, Vel Tech, Chennai

Mr.Karthikayan.P.N, Vel Tech, Chennai

Organizing Committee Members

Dr.N.Noor Alleema, Vel Tech, Chennai

Dr.Nithyanandam, Vel Tech, Chennai

Dr. Sugumaran D, Vel Tech, Chennai

Mr Karthikayan.P.N, Vel Tech, Chennai

Mr R Hariharan, Vel Tech, Chennai

Mrs. Dhilsath Fathima. M, Vel Tech, Chennai

Mr Pushpakumar.R, Vel Tech, Chennai

Mrs Ramya.D, Vel Tech, Chennai

Mrs Deepa.J, Vel Tech, Chennai

Mrs Sakunthala Prabha. K.S, Vel Tech, Chennai

Mrs Lijetha C Jaffrin, Vel Tech, Chenna

Mrs Jayanthi.K, Vel Tech, Chennai

Mr Pradeepkumar.S, Vel Tech, Chennai

Mr Durairaj.K, Vel Tech, Chennai

Mrs. Vijayalakshmi V, Vel Tech, Chennai

Technical Reviewer Board

Prof. Dr. Selwyn Piramuthu, Professor of Information Systems, University of Florida, USA

Prof. Dr. Michael Opoku Agyeman, Professor of Computer Engineering, University of Northampton, UK

Prof. Dr. Suzanne Kieffer, Associate Professor, Université Catholique De Louvain, Belgium

Prof. Dr. Biju Bajracharya, Assistant Professor of Cyber Security at East Tennessee State University, USA

Prof. Dr. Dawit Mengistu, Senior Lecturer, Kristianstad University, Sweden

Prof. Dr. Goran Dambic, Head of Software Engineering Department, Algebra University, Croatia

Prof. Dr. Naveen Chilamkurti, Acting Head of Department of Computer Science and Engineering, La Trobe University, Australia

International/National Advisory Committee

Dr. G.Umarani Srikanth, Professor/CSE Bharath Institute of Higher Education and Research

Dr.N.Kanya, Addl Dean E&T.HoD , DR.M.G.R Educational and Research Institute, Chennai

Dr. S.T.Selvamani, Professor, Chennai Institute of Technology, Chennai

Dr. Ilavarasan, Professor, Pondicherry Engineering College

Dr. M.Anand Kumar, Asst Prof, NIT Surathkal

Dr. Jagadesh Kakarla, Asst Professor, IITDM Kancheepuram, Chennai

Dr. M.Sridevi, Asst Prof. National Institute of Technology, Tiruchirappalli

Dr. J.Jayashree, Asst Prof, VIT .Vellore

Dr. S.K.Manigandan, Asst Prof, Vel Tech High Tech Dr RR and Dr SR Engineering college, Chennai

Dr. P.Arish, Data Scientist, Entropik Technologies Pvt. Ltd., Chennai.

Dr. Dahlia Sam, Associate Professor, DR.M.G.R DR.M.G.R Educational and Research Institute, Chennai

Dr. Karthikeyan.R, Professor, Vardhaman College of Engineering, Hyderabad

Dr.P. Sakthivel, Professor, GEC, Chennai

Dr. Valarmathi, Asst Prof, VIT Chennai

Dr. S.Poonkuntran, Professor and Dean, VIT Bhopal University, Kothrikalan, Madhya Pradesh

Dr. P.Velvizhy, Asst Prof, College of Engineering, Guindy, Anna University, Chennai

Dr. J.Vijayashree, Asst Prof, NIT Vellore

Dr. S.Amutha, Prof and Head/IT, Saveetha Engineering College, Chennai

Dr. Dharbaneshwer.S.J, Senior Data Scientist, Entropik Technologies Pvt. Ltd., Chennai

Dr. K.Arthi, Associate Professor, SRM Institute of Science and Technology, Kattankulathur Campus.

Dr. Varalakshmi, Professor, MIT, Chennai

Dr.R.Balamurugan, Asst Prof., VIT Vellore

Dr. Gunasekaran, Professor, Ahalia School of Engineering and Technology. Kerala

Dr. Arun Anoop.M, Associate Prof /CSE, Alva's Institute of Engg & Technology, Mangalore

Dr. E.Shanmugapriya, Asst Prof College of Engineering, Guindy. Anna University, Chennai

Dr. C.Lakshmi, Professor, SRM Institute of Science and Technology

Dr. N.Snehalatha, Associate Prof., SRM Institute of Science and Technology, Kattankulathur Campus

Dr. V.Uma Rani, Associate Prof, Saveetha Engineering College, Tandalam, Chennai

Dr. T.Ravi, Professor, Shadan Women's college of Engg and Technology, Hyderabad

Dr.Shankar Ram.N, Professor/CSE, Dr.NGP College of Technology, Coimbatore

Dr. J.Banumathi, HOD- IT, University College of Engineering, Nagercoil

Dr. J.Vijila, Assistant Professor, College of Engineering, Nagercoil

Dr. Muneeswari, Professor, VIT, Hyderabad

Suma Priyadarshani, Technology lead lnfosys, Bangalore

Dr P.Arish, Data Scientist, Entropik Technologies Pvt. Ltd., Chennai.

Dr.P.Venkata Krishna, Professor, Shri Padmavathi Mahila University, Thirupathi

Dr.Venningston, Professor/CSE, NIT Srinagar, Kashmir

Dr.J.Janet, Professor, Sri Krishna College of Engineering, Coimbatore

Dr.Surendiran, Asst.Prof., NIT, Puducherry

Dr.G.Rohit, Asst Prof, VIT, Chennai

Recent Trends in Computational Intelligence and its Application – Sugumaran D. et al. (eds)
© 2023 Taylor & Francis Group, London, ISBN 978-1-032-48410-5

1

S-Café—A Smart Ubiquitous Chatbot-based Café for Human Motional Analysis and Coffee Brewing

S. Arun[1], SaiSucharitha K.[2], Mohamed Anees F.[3], K. Kalaivani*

Department of Computer Science & Engineering,
Vels Institute of Science Technology & Advanced Studies, Chennai, India

Abstract—A Smart Ubiquitous Chatbot based Café for Human Emotional Analysis and Coffee Brewing is characterized by persistent sadness and lack of interest or pleasure. Depression is a mood disorder and it is observed that more than 264 million people (of all ages) around the world have suffered. Depression is different from usual mood fluctuations and short-lived emotional responses to challenges in everyday life and may become a serious health condition when it lasts long with moderate or severe intensity. On the other hand, Coffee – a brewed drink prepared from roasted coffee beans, the seeds of berries from certain coffee species is a famous stress buster. It is observed that chlorogenic acid, ferulic acid and caffeic acid have the ability to reduce the inflammation of nerve cells that takes place in the brains of people with depression, thus relieving some of the discomfort and distress due to depression. Hence, this study focuses on developing a Café that utilizes a smart ubiquitous chatbot to understand emotions (stress, depression, happiness, etc.) and brew a perfect cup of coffee for them. Natural language processing techniques and rule-based algorithms were applied for making that perfect cup of coffee.

Keywords—Café, Chatbot, Human emotional analysis, Coffee brewing

I. Introduction

Humans have changed the world in many aspects in terms of technology through the advancement of computers. Smart chatbots can be a better example of the evolution of technology that humans have made today. Chatbots are simple computer programs that can be used as a chat assistant. These computer programs have the ability to respond to humans as they do. Some chatbots has the ability to respond not only through texts but through voice as well. But this paper concentrates only on the text responses by the chatbots. The chatbots can be developed in the form of a web page, system software, etc.

The human–machine interaction in a chatbot may be in the form of a question-answer session [9]. The replies will be given to the users based on the queries and the match words

*Corresponding Author: kalai.se@velsuniv.ac.in,
[1]arun.se@velsuniv.ac.in, [2]sucharithakandepi2000@gmail.com, [3]mohammedanees182002@gmail.com

DOI: 10.1201/9781003388913-1

that are being stored in the knowledge base. If no matching word is found, a similar response will be given.

Here the chatbot is trained to reply to a few complex questions as well as queries regarding the coffee type, and give suggestions according to the user's taste and need so that a perfect cup of coffee could be brewed for the user. The approximate input sentences are stored in a Relational Data Base Management System (RDBMS). The important keywords can be extracted from the user texts and similar sentences that match those keywords can be found. These chatbots would coordinate and generate responses based on those input sentences. The sentence similarity is found by the application of TF-IDF (term frequency and inverse document frequency), cosine similarity and N-gram. The user interface (UI) could be designed with various frameworks such as Django, flask, etc. In this modern world, chatbots are used in various fields so that much time can be saved and immediate responses can be received [4].

II. Literature Survey

In our case, brewing a perfect cup of coffee for the user and understanding their emotions such as stress, depression, happiness and so on, is vital in our case. We need a strong foundation in our natural language processing (NLP) algorithms to work along with understanding the user, as well as our machine learning (ML) model which will figure out which coffee it's supposed to brew based on the emotions that are being shared and calculated by the chatbot [9]. By understanding large-scale matrix inversion which is done by the spark parallel computing platform [1], the proposed algorithm shall be a better foundation for designing a high-performance linear algebra library on Spark. Learning about Xatkit framework which is an open source and its modular architecture facilitates the separate evolution of language processing in a unique way [3]. These systems pretty much allow natural language generation

(NLG) and sentiment analysis which is a key to our chatbot development. Using the three-stage NLP module [4], first obtaining sentences, texts and detection of affirmative, negative and interrogative sentences and with the inclusion of subject and other elements, we orthographically narrow it down using the algorithms to realize using morphology and orthography [2]. With further research, it is witnessed that the NLG module could be used to avoid monotonous responses from the users. Nowadays, chatbots are being used by educational sites as well so that students can interact with the machine and increase their knowledge [10]. Processing a huge amount of data could be a bit challenging while dealing with the users' queries, knowledge base supports and multilingual aspects. To overcome these challenges, the chatbot architecture should be supplemented with the implementation of artificial intelligence (AI) strategies like NLP, which intends to understand the user texts and ML techniques that generate the queries based upon the user responses.

III. Existing System with Limitations

Existing systems in this domain are limited to practical difficulties. There aren't many systems to accurately predict the stress levels of any individual and even if they did, they do not produce a solution to reduce it [6]. Two-dimensional string arrays are used to create a database in which its request and questions are stored in even rows and responses and replies from the users are stored in odd rows. Columns hold the queries that can be answered by the chatbot. This database has a row that has a default response, which is displayed when the machines could not recognize the user's text. The chatbots embedded with a smart AI domain that identifies the sentences can generate responses accordingly. From the input sentences, a score will be evaluated. The more similar reference sentences are matched based on the score, similar to the ranking system.

A bigram is used to calculate the similarities among the sentences. Bigrams are a couple of words together being separated as different input sentences. These bigram words are stored in a database that lays a foundation for the replies of the chatbot. Here, the data are stored in the RDBMS and the interpreter figures out the pattern and chooses the apt response.

IV. Proposed System

This paper is a study of developing a smart chatbot powered by various domains of AI. There are modules in Python that can be used to design a convenient UI as well as for database connectivity. With a combination of an interface, database connectivity and networking, a chatbot can be designed and an easier web service can be enabled. Many chatbots are created on online platforms that are available across the world. As the Internet connects to a wide range of users, more queries and responses can be updated in the database, and users can raise more queries which can further enhance the smartness of the chatbot to the next level. Application programming interface (API) for a chatbot can be created by using various cascading style sheets, java scripts are employed in designing and Python can handle the backend connectivity. The chatbot contains various algorithms that are embedded in it.

Pre-processing of dataset

Here, the dataset has the collection of raw data that is to be preprocessed before further consideration. The data processing has three phases:

1. Tokenization
2. Removal of stop words
3. Lemmatization

A. Tokenization

Tokens are the basic building blocks of a sentence. The process of breaking down a sentence or paragraph into phrases, words and lexical units is called tokenization. These tokens are further processes for text mining or parsing which is useful in linguistics, text summarization and word processing which is a part of lexical analysis. The definition of a word or token relies upon simple heuristics. This excludes the whitespace characters.

B. Removal of Stop Word

Stop words are nothing but the most commonly used word in a sentence. These include articles, prepositions, pronouns, adjectives, etc. These words are widely used and may have low esteem but have to be removed from a sentence entirely so that matching words could be sorted out easily. This process can be termed as the removal of stop words. The recurring stop words are organized according to the terms of collection frequency and removed from the document.

C. Lemmatization

The term Lemmatization refers to the grouping of various inflected words together and these words can be considered as a single item. The process of lemmatization can be associated with the term stemming, which refers to the reduction of a word into its simplest form called a lemma. Lemma refers to the root word, that can't be simplified further. These stemmers can be used to process the datasets, which may increase the accuracy of the processed datasets [6].

V. Architecture

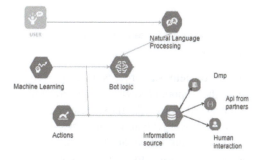

Fig. 1.1 Architecture diagram

Source: https://cyfuture.com/blog/a-comprehensive-guide-to-chatbot-machine-learning/

- Users can interact through a device on a messaging platform, these messages are processed through NLP.
- Then the chatbot can generate a response from the database and present it to the user through API.
- The bot then trains itself on how to respond to various users on receiving various datasets, this process is called ML. As humans help the bot to improve itself, it is called supervised learning.

The outline of the architecture and the system architecture of this chatbot application is shown in Fig. 1.1 and Fig. 1.2. The user enters the question in the UI in the form of text and is considered as input. The UI receives the user query and it is exported to the chatbot API. Here, the data processing is carried out which involves tokenization. During this process, the removal of stop words is done and the feature extraction depends on various algorithms like n-gram, TF-IDF and cosine likeness. The queries and responses are stored in a database for further retrieval. Tokenization separates the words in a sentence for increased processing. At some point, it meets with the words stored in the knowledge database and during this process, the functions are disposed of.

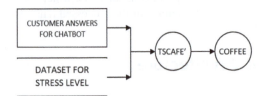

Fig. 1.2 System architecture

Source: Author

Stop words removal process is employed to extract the important keywords and to remove the unwanted words like most frequently used words, and words that do not possess any specific meaning like a, an, of, thus, etc. This step is mainly used to reduce computational complexity and processing time. The feature extraction based on n-gram and TF-IDF is characterized to decrease the process in the document. This ranks the lemma as per the documents so that the speed and adequacy of the document are upgraded [11].

TF-IDF stands for term frequency and inverse document frequency. This is used to calculate the weight of each term in the sentence [11]. This term frequency is used to check the number of occurrences in a particular sentence.

$$TF = TFI$$

The above formula can be used to calculate the term frequency.

VI. Functional Diagram

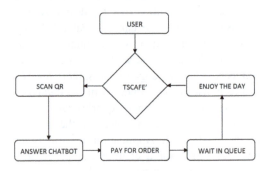

Fig. 1.3 Functional diagram

Source: Author

Figure 1.3 shows the functional architecture of this proposed framework. It demonstrates the complete flow of the process with every minute detail. This architecture also highlights different necessary algorithms which we have used. First the user logs in to the café's home page then scan the QR code for the questionnaire and the chatbot will interact with the customer (user). After the conversation, the stress level of the customer will be analyzed with the help of the answers given by the customer [9]. The bot will respond based on the answers given and will confirm the order with the customer before the billing and then the customer will pay for the order. After the payment confirmation, they should wait for the order to get ready.

VII. Algorithm

The proposed framework utilizes the features of TF-IDF and NLP to recognize the actions to be performed by the smart chatbot for coffee brewing. In information retrieval and ML, TF-IDF is widely used by researchers to represent the significance of string/collection of strings, which are commonly referred to as corpus [12]. Term frequency refers to the occurrences of a particular term in a document. Frequency can be defined in n number of ways: (i) raw count, (ii) logarithmic-scaled frequency and (iii) Boolean frequency [11].

Inverse document frequency checks whether a word is common or uncommon in the document. IDF is calculated as follows:

$$idf(t, D) = \log\left(\frac{N}{\text{count}(d \in D : t \in d)}\right)$$

Where,

t = the term for which the commonness is measured.

d = document

N = number of documents

D = corpus

The reason behind using IDF is to identify frequently used words, such as a, if, as, etc., the weighting can be minimized by taking IDF of the recent term.

TF represents the information on how often a term appears in a document and IDF depicts information about the relative rarity of a term in the collection of documents. By multiplying these values, the final TF-IDF can be computed as follows:

$$tfidf(t, d, D) = tf(t, d) . idf(t, D)$$

Thus, TF-IDF is widely used in information retrieval, ML, keyword extraction and text summarization [13].

Natural language processing

Natural language processing is a domain of AI that deals with the input streams like text and voice as humans do [8]. This enables human–machine interactions. It is the ability of the machine to understand the human language in both text and speech formats. This part of AI is used to handle a large amount of natural data. It is a subfield of linguistics and information technology.

Natural language processing uses AI to take real-world input, processes it and changes that into a machine-understandable form, irrespective of whether the language is spoken or written [7]. Humans have eyes to see and ears to hear, machines have a camera to see and mics to recognize speech. The component of a machine in which inputs can be produced is called sensors. NLP converts the input into code and processes it. NLP-based bot details

- Accurate interpretation.
- Lesser training data inputs to achieve natural language capability.
- Resolve conflicts using statistical modeling.
- Comprehensive communication for user responses.
- Identification of user inputs.

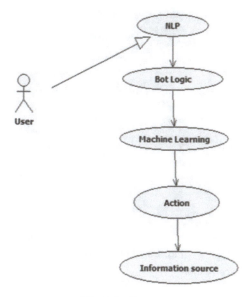

Fig. 1.4 Process

Source: Chatbot

- Learn faster.
- Ability of re-formatting the input

VIII. Experimental Results

A. Testing

Type equation here. It is a pre-final step before completing any product and it ensures that the product can be completed a way far better than it was before. This involves testing the final product in all possible aspects which helps the development team improve the product. There are various stages of testing software that undergo different exercises. If there is any error or defect in a product, it is sent to the development team to reduce the error. This process is done repeatedly until an error-free product is developed and it meets all the requirements given by the client. In our case, the chatbot is being tested with different complex queries, and many datasets are fed to the database in order to make it a better product.

Fig. 1.5 Final output

Source: Author

IX. Conclusion

A smart chatbot is a better application for a human–machine conversation that uses TF-IDF and NLP techniques. With the application of these techniques, a smart chatbot can be designed to determine the user's emotions with comparatively higher accuracy and thus makes human–machine interaction easier. Thus, this chatbot can be utilized in cafés as customers' chat partners for a while and suggest a suitable cup of coffee according to their emotions based upon the responses from the chat.

References

1. J. Liu, Y. Liang and N. Ansari, "Spark-Based Large-Scale Matrix Inversion for Big Data Processing," in IEEE Access, vol. 4, pp. 2166-2176, 2016, doi: 10.1109/ACCESS.2016.2546544.
2. A. Ait-Mlouk and L. Jiang, "KBot: A Knowledge Graph Based ChatBot for Natural Language Understanding Over Linked Data," in *IEEE Access*, vol. 8, pp. 149220-149230, 2020, doi: 10.1109/ACCESS.2020.3016142.
3. G. Daniel, J. Cabot, L. Deruelle and M. Derras, "Xatkit: A Multimodal Low-Code Chatbot Development Framework," in IEEE Access, vol. 8, pp. 15332-15346, 2020, doi: 10.1109/ACCESS.2020.2966919.
4. M. Polignano, F. Narducci, A. Iovine, C. Musto, M. De Gemmis and G. Semeraro, "HealthAssistantBot: A Personal Health Assistant for the Italian Language," in IEEE Access, vol. 8, pp. 107479-107497, 2020, doi: 10.1109/ACCESS.2020.3000815.
5. S. García-Méndez, F. De Arriba-Pérez, F. J. González-Castaño, J. A. Regueiro-Janeiro and F. Gil-Castiñeira, "Entertainment Chatbot for the Digital Inclusion of Elderly People Without Abstraction Capabilities," in IEEE Access, vol. 9, pp. 75878-75891, 2021, doi: 10.1109/ACCESS.2021.3080837.
6. Li, Y.; Su, H.; Shen, X.; Li, W.; Cao, Z.; Niu, S. DailyDialog: A Manually Labelled Multi-turn Dialogue Dataset. In Proceedings of the

Eighth International Joint Conference on Natural Language Processing (Volume 1: Long Papers), Taipei, Taiwan, 27 November–1 December 2017; Asian Federation of Natural Language Processing: Taipei, Taiwan, 2017; pp. 986–995.

7. Sojasingarayar, A. Seq2Seq AI Chatbot with Attention Mechanism. Master's Thesis, Department of Artificial Intelligence, IA School/University-GEMA Group, Boulogne-Billancourt, France, 2020.

8. Weizenbaum, J. ELIZA–A Computer Program for the Study of Natural Language Communication between Man and Machine. Commun. ACM 1966, 9, 36–45. [CrossRef]

9. Kim, J.; Oh, S.; Kwon, O.W.; Kim, H. Multi-Turn Chatbot Based on Query-Context Attentions and Dual Wasserstein Generative Adversarial Networks. Appl. Sci. 2019, 9, 3908. [CrossRef]

10. Winkler, R., Söllner, M. (2018): Unleashing the Potential of Chatbots in Education: A State-Of-The-Art Analysis. In: Academy of Management Annual Meeting (AOM). Chicago, USA.

11. Havrlant L, Kreinovich V (2017) A simple probabilistic explanation of term frequency-inverse document frequency (TF-IDF) heuristic (and variations motivated by this explanation). Int J Gen Syst 46(1): 27–36.

12. Bafna P, Pramod D, Vaidya A (2016) Document clustering: TF-IDF approach. In: IEEE int. conf. on electrical, electronics, and optimization techniques (ICEEOT). pp 61–66

13. Baker K, Bhandari A, Thotakura R (2009) An interactive automatic document classification prototype. In: Proc. of the third workshop on human-computer interaction and information retrieval. pp. 30–33.

Recent Trends in Computational Intelligence and its Application – Sugumaran D. et al. (eds)
© 2023 Taylor & Francis Group, London, ISBN 978-1-032-48410-5

2

A Deep Hybrid Neural Network Model to Detect Diabetic Retinopathy from Eye Fundus Images

Guntaka Rama Mounika[1]
Post-Graduation Student, VFSTR University

Kamepalli Sujatha*
Associate Professor, VFSTR University

Abstract—Diabetes is becoming a common disease almost in all the ages of people and is impacting a variety of human organs such as eyes, teeth, kidney, heart, skin, etc. This disease is causing various diseases in different organs of human body. Eyes are one of the important organs in human body through which one can see the beauty of the world. But, because of diabetes, many people are facing eye retinopathy problems. Due to these issues, many people are losing their eyesight and also going complete blind sometimes if it is not identified and treated at early stages. Damage to the retinal blood vessel obstructs the light passing through the optical nerves, rendering the diabetic patient blind. There are some manual procedures for detecting diabetic retinopathy (DR), one of which is screening, but it requires a professional ophthalmologist and a great deal of time. So, there is an immediate need of automation in detecting the diabetic retinopathy. By using convolutional neural network (CNN), we were able to identify multiple phases of DR severity. The main objective of this research work is to identify the diabetic retinopathy stages by analyzing the eye fundus images. Thus, we developed a hybrid deep neural network model to automate the diagnosis process of diabetic retinopathy using the transfer learning models with prior training of ResNet and DenseNet. The model is effective in detecting various stages of diabetic retinopathy.

Keywords—Diabetic retinopathy, Blindness, Screening, Eye fundus images, ResNet, DenseNet

I. Introduction

Human eye is a noticeable organ in the human body, which provides better visualization of the world to the human. This organ can be affected by many of the general diseases such as diabetes. Diabetic retinopathy (DR) occurs when there is a severe damage in the retina vessels; due to which many people are losing their eye sight. Diabetic retinopathy is characterized by a change that takes place in the retinal blood vessels. Vision loss can be caused

*Corresponding Author: sujatha_kamepalli@yahoo.com
[1]ramamounika2000@gmail.com

DOI: 10.1201/9781003388913-2

when retinal blood vessels of certain diabetic people enlarge and leak fluid which leads to diabetic retinopathy.

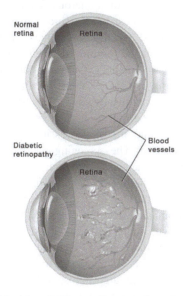

Fig. 2.1   Visualization of normal retina and diabetic retina

Source: https://www.mayoclinic.org/diseases-conditions/diabetic-retinopathy/symptoms-causes/syc-20371611

Figure 2.1 represents the difference between normal retina and diabetic retina. Retinopathy is a medical term which caused due to Changes in the blood vessels of the retina and the aberrant growth of new blood vessels on the retinal surface. This occurs in some cases and may result in blindness in the long run. If a person has diabetes for a long time, they are more likely to acquire DR at some point in their lives. It is possible to detect diabetic retinopathy at one of five stages: "mild, moderate, severe, proliferative or no disease" at all. There is a possibility to reduce the damage to one's vision by earlier detection and improved treatment. Medical specialists with the necessary expertise are required in order to detect diabetic retinopathy at its early stages. It is possible that automating the process of detecting diabetic retinopathy from eye photographs may be proven useful for medical practitioners in detecting the disease at an early stage and providing appropriate therapy at the appropriate time.

The automatic learning ability of the systems, due to "ML" and "DL" models (Sujatha & Srinivasa Rao, 2019) (Sujatha et al., 2021), can easily automate the process of DR disease detection by processing the eye fundus images. Convolutional networks, also known as convolutional neural networks (CNN), have a great deal of attraction in image, video and audio processing (Krizhevsky et al., 2012) (Zeiler & Fergus, 2013)(Sermanet et al., 2014) (Simonyan & Zisserman, 2015). These can be used for image segmentation, object detection, anomaly detection in images, etc. This has been made possible by the existence of a lot of benchmark image datasets such as ImageNet and other datasets in various data science communities (Jia Deng et al., 2009)(Dean et al., 2012). In particular, the "ImageNet Large-Scale Visual Recognition Challenge (ILSVRC)" (Russakovsky et al., 2015) played a key role in the development of various deep learning architecture. The success of CNNs in these competitions was the impetus for the decision to use CNN for this piece of study. A CNN is made up of one or more convolutional layers (typically followed by a subsampling step), which are then followed by one or more fully connected layers, just like a regular multilayer neural network. A CNN's architecture is meant to take advantage of the 2D hierarchical structure of an input image, so it can process that image more effectively (or other 2D input such as a speech signal). This is accomplished through the use of local connections and tied weights, and is then followed up with a form of pooling that produces features that are translation invariant. Another advantage of CNNs is that they are simpler to train and have a significantly reduced number of parameters in comparison to fully connected networks that have the same number of hidden units. CNNs train themselves by stacking numerous trainable stages one on top of the other. This allows them to take into account the hierarchical representation of images. In this research, we considered deep learning models ResNet and DenseNet, we developed a hybrid model to detect and forecast diabetic retinopathy using photographs of the eye's fundus.

The key objectives of this research work are

- Understand the causes, symptoms of diabetic retinopathy disease and need of automation process in detecting the disease.
- Development of deep hybrid neural network by considering pre-trained transfer learning models.
- Training and validating the model to detect the diabetic retinopathy disease from eye fundus images.
- Testing the model by giving random fundus images to detect the stage of diabetic retinopathy disease.

The paper is structured as follows:

In the second section, we explored the recent works done in the same area. The third section explains the proposed model, in the fourth section, dataset descriptions and results were discussed by comparing them with the existing models and finally, in the fifth section we concluded the paper.

II. Literature Survey

(Prabhu et al., 2019) Detection of diabetic retinopathy can be done by looking for the characteristic that demonstrates the disease's symptoms. Fundus pictures are used to extract the features from the eye fundus images, which show the illness symptoms. Hierarchical classification is used to identify distinct stages of disease based on the collected data. Due to the increased number of patients and a decrease in the number of ophthalmologists available to treat them, there was an underline need for a detection method, and the system has produced excellent results in terms of sensitivity and specificity (Sudha et al., 2020). Various image

processing algorithms were discussed to identify healthy and unhealthy images to find issues with the detection system. Pre-processing and identification of numerous features were employed. Low-cost, high-accuracy diagnostics were proposed by the researchers, and the technique can be employed even on low-performance computers (Qureshi et al., 2019). The analysis of CAD systems for diagnosing DR disease was discussed. CAD systems are used for diverse purposes, such as computing intelligence and image processing have also been discussed. The screening methods used in numerous detection research publications were also studied, as were the difficulties and outcomes of those studies. Some of the issues and automated disaster recovery methods were shown, along with suggested solutions. (Kumaran & Patil, 2018) Since the raw fundus images might be challenging to analyze by machine learning algorithms, a brief survey of several pre-processing and segmentation strategies has been presented here. They have provided a short overview to help others learn about the most recent advancements and research in their field. Researchers in the subject can benefit from this as well, as it can aid in spotting trends and patterns. (Panchal et al., 2018) Diabetic retinopathy detection methods have been thoroughly studied. Classifiers such as those for blood vessels, the optic disc, exudates (EXs) and hemorrhages are among the many tools available for sorting out the various aspects (HMs). Fundus photos are employed in a variety of ways in the survey reports. They have also included accuracy in terms of specificity and sensitivity. Kamaladevi et al. (2018) have put forth a methodology for automatically detecting diabetic retinopathy. Retinal pictures of affected patients have been analyzed and the features were retrieved and recognized. It uses classifiers such as GradientBoost, RandomForest and Voting for feature extraction and classification in the system. More accurate prediction by RandomForest continues to reign over others

(Gupta & Chhikara, 2018). The experimental findings of the various ML algorithms have been compared to each other. Sensitivity, specificity, AUC (Area Under the Curve) and accuracy are just a few of the metrics under consideration. DR is on its way to segmenting blood arteries and identifying lesions, as shown in the review. Deep neural network outcomes were also compared. The optimum method is presented from the numerous analyses. It has also proved to be extremely effective in identifying the necessary characteristics (Sayed et al., 2017). The diagnosis of diabetic eye disease in fundus images has been evaluated between probabilistic neural network (PNN) and support vector machines (SVM). Pre-processing and then machine learning techniques are used to analyze the data. Both strategies are then contrasted and analyzed in terms of the results they produce. SVM outperformed PNN in terms of detection accuracy (Kaur et al., 2019). Pre-processing, segmentation and classification are the three key phases of diabetic retinopathy detection proposed by the authors. The model was built in MATLAB and the results were analyzed based on several factors. For the purpose of detection, an NN classification strategy is presented in this paper. A comparison and evaluation of the suggested model and the SVM classification is provided. Compared to a NN-optimized result, this one is substantially more efficient. Kaur et al. (2019) suggested a technique for detecting and classifying various phases of DR disease. Using compactness analysis and bounding box approaches, the retinal vasculature was segmented. In the end, blood vessels and hemorrhage were characterized using the RandomForests technique.

III. Methodology

The architecture of the method that is being discussed can be seen in Fig. 2.2. The DenseNet-100 framework initially receives an input sample along with the bounding box as part of the initial data transmission. After this step, the Custom ResNet is trained to be able

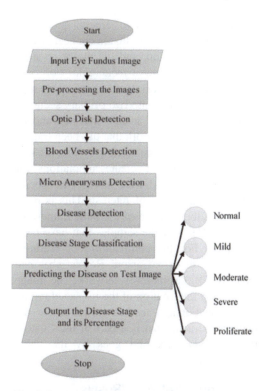

Fig. 2.2 Architecture of the presented methodology

Source: Author

to classify the locations. Finally, an accuracy estimate is provided for each unit based on the metrics that are currently being applied in the field of computer vision.

The ResNet is able to identify the regions of interest in the sample eye fundus images. After this step, the Custom Center Net receives training on how to classify the places that have been located. Finally, an accuracy estimate is provided for each unit based on the metrics that are currently being applied in the field of computer vision. It is essential to precisely identify the location of the afflicted region based on the input retinal samples in order to make sure that the training process goes as smoothly as possible. In order to accomplish this goal, we made use of the labeling software to construct the sample annotations. An XML file is used to contain the created annotations.

Within this file are two crucial pieces of information: (i) the class that is connected with each impacted region, and (ii) the box values that are used to draw a rectangular box over the detected. The work that is being presented can be broken down into two primary sections. "Dataset preparation" is the first step and an upgraded ResNet network that has been trained for eye disease classification is the second. In the first step, we create annotations for photos that include disease in order to specify the particular region of interest. This is done while the other portion of the built framework trains ResNet over the samples that have been annotated. For the computation of features, we made use of a Custom ResNet that had DenseNet-100 as its basis network. The image sample and the location of the impacted region in the input image are both required for the features extractor that is a part of the Custom ResNet framework and goes by the name DenseNet-100.

A. Residual Network (ResNet)

He et al. (2015) introduced "Residual Network (ResNet)" in their paper, making it one of the most well-known DL models. ResNet is one of the most standard and effective DL models to

date. Figure 2.3 is the residual block of ResNet model.

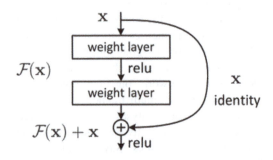

Fig. 2.3 Block of residuals in the ResNet model

Source: https://miro.medium.com/max/1140/1*D0F3UitQ2I5Q0Ak-tjEdJg.png

Figure 2.4 describes the architecture of ResNet Model.

IV. Experimental Results

Here we focused on the results that were obtained from the application of the deep hybrid network on the dataset of eye fundus images.

A. Dataset Description

Aravind Eye Hospital in India provided us with the data. Diabetic retinopathy prevention

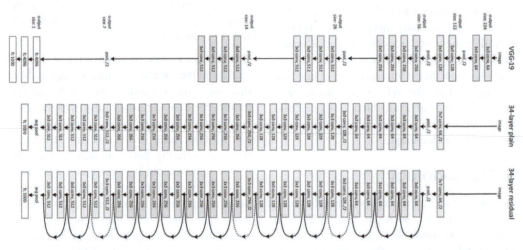

Fig. 2.4 Deep residual network architecture (Model considered for feature extraction) (He et al., 2015)

and early detection in rural populations is very difficult as the medical screening in these areas is very challenging. We acquired two public datasets on Kaggle, one published in 2019, and the other published in 2015. Those data have been pre-processed to reduce their size, since the original images were large and hard to manipulate, the processed train datasets are of 7GB. Both train datasets were labelled by clinicians on a 0–4 scale, showing the severity of the disease symptoms from "0 as Normal, 1as Mild, 2 as Moderate, 3 as Severe, and 4 as Proliferate." Figure 2.5 indicates the eye fundus images corresponding to different levels of disease.

B. Results and Discussions

Experimentation was carried in Google Colab with 32GB RAM, on a DELL laptop with internal configuration of 16GB RAM and 1TB hard disk. Figure 2.6 illustrates the model summary of proposed hybrid neural network by combining ResNet and Densenet.

We developed a webpage that asks for uploading an image to predict the diabetic retinopathy stage. Figure 2.7 shows the home page to upload an eye fundus image for testing.

After uploading the image, by clicking on test, the proposed model was implemented on

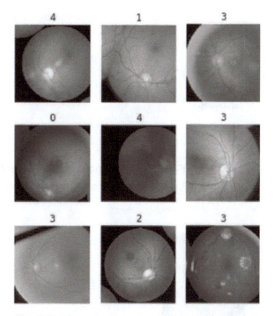

Fig. 2.5 Sample eye fundus images from 5 class labels

Source: Kaggle data science community

the test image and displays the severity of DR disease. Figure 2.8 shows that by implementing the proposed hybrid model on the test image it was predicted having mild symptoms of the disease with less% and severe symptoms with high%, that is, the uploaded image has severe diabetic retinopathy image with 74%.

Layer (type)	Output Shape	Param #
model_1 (Model)	(None, 204800)	23587712
dense_2 (Dense)	(None, 512)	104858112
dropout_1 (Dropout)	(None, 512)	0
dense_3 (Dense)	(None, 512)	262656
dropout_2 (Dropout)	(None, 512)	0
dense_4 (Dense)	(None, 1)	513

```
Total params: 128,708,993
Trainable params: 105,121,281
Non-trainable params: 23,587,712
```

Fig. 2.6 Proposed model summary

Source: Model summary from the developed model (Implementation)

Diabetic Retinopathy

Fig. 2.7 Eye Fundus image uploading to detect the retinopathy stage

Source: Results obtained from the implementation of proposed method

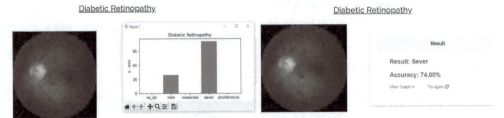

Fig. 2.8 Predicted diabetic retinopathy stage [Mild (less %), severe (high % 74%)]

Source: Results obtained from the implementation of proposed method

Figure 2.9 shows that by implementing the proposed hybrid model on the test image it was predicted that it is having no diabetic retinopathy with 100%.

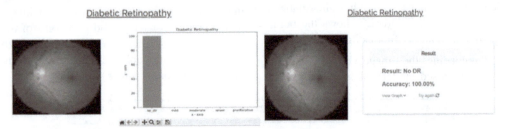

Fig. 2.9 Predicted diabetic retinopathy stage [No diabetic retinopathy (high%) 100%]

Source: Results obtained from the implementation of proposed method

Figure 2.10 shows that by implementing the proposed hybrid model on the test image it was predicted having no diabetic retinopathy with 99%.

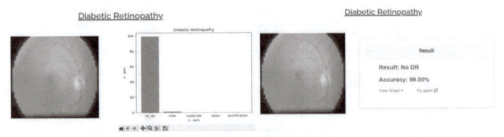

Fig. 2.10 Predicted diabetic retinopathy stage [No diabetic retinopathy (high%) 99%]

Source: Results obtained from the implementation of proposed method

The existing models work with an accuracy of more than 90% but they are unable to find the percentage if that particular image belongs to the predicted class. When it comes to diagnosing the various stages of diabetic retinopathy, the proposed hybrid model performs well. This model not only detects the stage of disease but it also finds the percentage of accuracy of the test image belongs to the identified disease.

V. Conclusion

In this study, we constructed a deep hybrid neural network to detect diabetic retinopathy and also to determine the stage of the disease based on images of the fundus of the eye. When developing the model, we took into account both the ResNet and DenseNet models that had been pre-trained. For the computation of features, we uses Custom ResNet that had DenseNet-100 as its basis network. Because of this research, we are able to swiftly diagnose diabetic retinopathy with a high level of accuracy using our trained neutral networks, and our method will assist to limit the damage that is caused by diabetic retinopathy at an early stage. The performance of the proposed hybrid model is quite satisfactory in detecting all stages of diabetic retinopathy, ranging from the absence of diabetic retinopathy to the proliferative stage. This model is not only capable of identifying the stage of the disease, but it can also determine the percentage of accuracy with which the test image belongs to the disease that has been recognized. Our system can be improved further by training our neutral networks model on various eye disorders; as a result, a person would be able to obtain a solution to all eye diseases from a single location, that is, not only diabetic retinopathy but also other eye diseases can be detected by training the developed model with the corresponding samples of eye images.

References

1. Dean, R., Van Kan, J. A. L., Pretorius, Z. A., Hammond-Kosack, K. E., Di Pietro, A., Spanu, P. D., Rudd, J. J., Dickman, M., Kahmann, R., Ellis, J., & Foster, G. D. (2012). The Top 10 fungal pathogens in molecular plant pathology. *Molecular Plant Pathology*, *13*(4), 414–430. https://doi.org/10.1111/j.1364-3703.2011.00783.x

2. Gupta, A., & Chhikara, R. (2018). Diabetic Retinopathy: Present and Past. *International Conference on Computational Intelligence and Data Science (ICCIDS 2018)*, *132*, 1432–1440. https://doi.org/10.1016/j.procs.2018.05.074

3. He, K., Zhang, X., Ren, S., & Sun, J. (2015). Deep residual learning for image recognition. *IEEE Computer Society Conference on Computer Vision and Pattern Recognition*, 1–12. https://doi.org/10.1109/CVPR.2016.90

4. Jia Deng, Wei Dong, Socher, R., Li-Jia Li, Kai Li, & Li Fei-Fei. (2009). ImageNet: A large-scale hierarchical image database. *IEEEXplore*, 248–255. https://doi.org/10.1109/cvprw.2009.5206848

5. Kaur, P., Chatterjee, S., & Singh, D. (2019). Neural network technique for diabetic retinopathy detection. *International Journal of Engineering and Advanced Technology*, *8*(6), 440–445. https://doi.org/10.35940/ijeat.E7835.088619

6. Krizhevsky, B. A., Sutskever, I., & Hinton, G. E. (2012). ImageNet Classification with Deep Convolutional Neural Networks. *Communications of the ACM*, *60*(6), 84–90.

7. Kumaran, Y., & Patil, C. M. (2018). A brief review of the detection of diabetic retinopathy in human eyes using pre-processing & segmentation techniques. *International Journal of Recent Technology and Engineering*, *7*(4), 310–320.

8. M. Kamaladevi, S. Sneha Rupa, & T. Sowmya. (2018). Automatic detection of diabetic retinopathy in large scale retinal images. *International Journal of Pure and Applied Mathematics*, *119*(12), 14181–14189.

9. Panchal, N. A., Thakore, D. D. G., & Pawar, D. T. D. (2018). Detection of Diabetic Retinopathy: A Survey. *International Journal for Research in Applied Science and Engineering Technology*, *6*(4), 2852–2857. https://doi.org/10.22214/ijraset.2018.4475

10. Perronnin, F., Sánchez, J., & Mensink, T. (2010). Improving the Fisher kernel for large-scale image classification. *Lecture Notes in Computer Science (Including Subseries Lecture Notes in Artificial Intelligence and Lecture Notes in Bioinformatics)*, *6314 LNCS*, 143–156. https://doi.org/10.1007/978-3-642-15561-1_11

11. Prabhu, N., Bhoir, D., & Shanbhag, N. (2019). Diabetic retinopathy screening using machine learning for hierarchical classification. *International Journal of Innovative Technology and Exploring Engineering*, *8*(10), 1943–1948. https://doi.org/10.35940/ijitee.J9277.0881019

12. Qureshi, I., Ma, J., & Abbas, Q. (2019). Recent development on detection methods for the diagnosis of diabetic retinopathy. *Symmetry*, *11*(749), 1–34. https://doi.org/10.3390/sym11060749

13. Russakovsky, O., Deng, J., Su, H., Krause, J., Satheesh, S., Ma, S., Huang, Z., Karpathy, A., Khosla, A., Bernstein, M., Berg, A. C., & Fei-Fei, L. (2015). ImageNet Large Scale Visual Recognition Challenge. *International Journal of Computer Vision*, *115*(3), 211–252. https://doi.org/10.1007/s11263-015-0816-y

14. Sayed, S., Inamdar, D. V., & Kapre, S. (2017). Detection of diabetic retinopathy using image processing. *International Journal of Innovative Research in Science, Engineering and Technology*, *6*(1), 99–107. https://doi.org/10.1007/978-3-319-70016-8_22

15. Sermanet, P., Eigen, D., Zhang, X., Mathieu, M., Fergus, R., & LeCun, Y. (2014). Overfeat: Integrated recognition, localization and detection using convolutional networks. *2nd International Conference on Learning Representations, ICLR 2014*, 1–16.

16. Simonyan, K., & Zisserman, A. (2015). Very deep convolutional networks for large-scale image recognition. *3rd International Conference on Learning Representations, ICLR 2015*, 1–14.

17. Sudha, V., Priyanka, K., Kannathal, T. S., & Monisha, S. (2020). Diabetic Retinopathy Detection. *International Journal of Engineering and Advanced Technology*, *9*(4), 1022–1026. https://doi.org/10.35940/ijeat.d7786.049420

18. Sujatha, K., & Srinivasa Rao, B. (2019). Recent Applications of Machine Learning : A Survey. *International Journal of Innovative Technology and Exploring Engineering (IJITEE)*, *8*(6), 263–267. https://doi.org/Retrieval Number: F10510486C219 /19©BEIESP

19. Sujatha, K., Venkata Krishna Kishore, K., & Srinivasa Rao, B. (2021). Animal Breed Classification and Prediction Using Convolutional Neural Network Primates as a Case Study. *2021 Fourth International Conference on Electrical, Computer and Communication Technologies (ICECCT) | 978-1-6654-1480-7/21/$31.00 ©2021 IEEE | DOI: 10.1109/ICECCT52121.2021.9616928*, 1–7.

20. Zeiler, M. D., & Fergus, R. (2013). Visualizing and understanding convolutional networks. *Lecture Notes in Computer Science (Including Subseries Lecture Notes in Artificial Intelligence and Lecture Notes in Bioinformatics)*, *8689 LNCS*(PART 1), 818–833. https://doi.org/10.1007/978-3-319-10590-1_53.

Recent Trends in Computational Intelligence and its Application – Sugumaran D. et al. (eds)
© 2023 Taylor & Francis Group, London, ISBN 978-1-032-48410-5

3

Optimization of Power Generation Costs Using Support Vector Machines

A. R. Danila Shirly[1]
Assistant Professor of Electrical and Electronics Engineering,
Loyola-ICAM College of Engineering and Technology, Chennai

M. V. Suganyadevi*
Associate Professor of Electrical and Electronics Engineering,
Saranathan College of Engineering, Trichy

A. Shunmugalatha[2]
Professor of Electrical and Electronics Engineering,
Velammal College of Engineering and Technology, Madurai

S. Rochan Rohan[3], P. S. Shakkthi Naresh[4], S. Reehan Fazil[5], B. Rithesh[6]
UG Student of Electrical and Electronics Engineering,
Saranathan College of Engineering, Trichy

Abstract—Effective load dispatch is considered to be an optimization effort taken on routine basis essential for the grid's functionality. In order to meet demand for electricity while minimizing operational costs and maintaining system equality and difference constraints, economic load dispatch (ELD), a method used in electricity generation, schedules dedicated power system yields. The generation schedule must be capable of directly accommodating the variable and erratic load demand at various buses without noticeably raising operational expenses. This paper tries to address very difficult ELD problems with transmission loss and varied cost curves using a new regression approach called support vector machine (SVM). This method's effectiveness has been proven on Indian 181 and IEEE 118 bus systems. The results of this proposed method show that this SVM algorithm is capable of finding significantly more cost-effective load dispatch solutions than particle swarm optimization (PSO), hybrid particle swarm optimization (HPSO), multiagent system (MAS), multiagent particle swarm optimization (MAPSO), hybrid multi agent based particle swarm optimization (HMAPSO) and extreme learning machine (ELM). Furthermore, when compared to other approaches, the calculation time is comparatively short.

Keywords—Power system optimization, Support vector machines, Economic load dispatch, IEEE bus system, Transmission loss, Regression

*Corresponding author: mvsuganyadevi@gmail.com
[1]ardanilashirly@gmail.com, [2]apmraja@gmail.com, [3]rocherohan2001@gmail.com, [4]shakkthinaresh01@gmail.com, [5]reeehanfazil@gmail.com, [6]rithishig1@gmail.com

DOI: 10.1201/9781003388913-3

I. Introduction

A vital daily optimization effort for a grid's operation is effective load dispatch. Economic load dispatch (ELD) in the generation of electricity has as its major objective of scheduling devoted power system yields to meet the demand for the least amount of money possible in terms of operational costs while also maintaining systemic equality and difference restrictions. The generation schedule must be able to directly accommodate the fluctuating and impulsive load demand at different buses without considerably increasing operating costs. [1]. The best generating schedules with non-convex fuel pricing capabilities are presented by conventional search methods while taking into account load sample models that are unable to find the most efficient solution on a global scale. Various solutions were created to address ELD difficulties.

Recently, evolutionary solutions to the ED issue have been put up. With more global probing capacity at the beginning of the execution and a local seek near the finish, the particle swarm optimization (PSO) algorithm [2, 3] can produce exceedingly fine solutions in a short amount of time. Breeding and subpopulation approaches are two aspects that an hybrid particle swarm optimization (HPSO) [5] brings to PSO. Different aspects of the distribution network are impacted by DG, and their ideal placement and size are essential to plummet the power losses and refine the voltage stability [6]. The following are the primary contributions of this paper: The Economic Dispatch problem of two test systems is solved using PSO, HPSO, MAS, MAPSO, ELM, HMAPSO and SVM algorithms. Performance evaluation of the PSO, HPSO, MAS, MAPSO, ELM and HMAPSO algorithms with SVM is done for the purpose of lowering generating costs. Using the suggested method, the most desirable price of generation [7] for the Indian 181 bus structure and the IEEE 118 bus structure are estimated. SVMs have two special abilities: they can increase the margin for good generalization and they can improve the effectiveness of the kernel technique for learning nonlinear functions [8]. SVM, also known as a support vector machine (SVM), is an ideal hyperplane that may be written by way of rectilinear function of subcategory of training data to solve a quadratic programming (QP) problem that is linearly constrained [9–11]. Additionally, the SVM has been enhanced for satisfying the nonlinear regression prediction model known as the SVM used in the regression process, which also includes Vapnik's -insensitive loss function. [12]

II. Mathematical Formulation

Reducing the entire cost of fuel for generating, as illustrated below, is the main criterion.

$$\text{Minimize } F = \sum_{j=1}^{n} F_{cj}(P_j) \qquad (1)$$

where, in accordance with power stability or voltage stability requirements, P_j represents the real power provided by unit j and $F_{cj}(P_j)$ reflects the energy price function of unit j.

$$\sum_{j=1}^{n} P_j = P_D + P_L \qquad (2)$$

where actual power demand and transmission line loss are represented by PD and PL, respectively, and P_j represents the active power produced by unit j. The restriction on generational capacity is described as

$$P_{j\min} \leq P_j \leq P_{j\max} \text{ for } j = 1, 2, 3, \ldots, n \qquad (3)$$

where $P_{j\min}$ signifies the generating unit j's minimal real power output and $P_{j\max}$ denotes the generating unit j's greatest active power output. The producing unit j's energy price function is specified as a means of

$$F_{cj}(P_j) = a_i P_j^2 + b_i P_j + c_i \qquad (4)$$

A. Particle Swarm Optimization (PSO)

The search space, or solution region, is not one-dimensional but rather multidimensional in nature, and as the name implies, the particles or components swarm or move around in it. Let x_d and v_d represent the position and speed of a particle in a search space. The value of p_{best} represents the particle's prior best position, which was recorded. G_{best} is thought to be the best global particle index even while particles from the group are present [9]. The positions and velocities of individual particles are premeditated as follows:

$$v_{d+1} = m * (w * v_d + m_1 \cdot \text{rand}() * (p_{\text{best}} - x_d)) \\ + m^2 \cdot \text{rand}() * (g_{\text{best}} - x_d) \quad (5)$$

$$x_{d+1} = x_d + v_{d+1} \quad (6)$$

where x_d represents its present location, v_d represents its speed and d is its creation pointer. The following formulas can be utilized to determine the weight, W

$$w = w_{\max} - \frac{w_{\max} - w_{\min}}{iter_{\max}} X_{iter} \quad (7)$$

The highest number of generations is known as $iter_{\max}$, whereas the count of generations in the present state is known as *iter*. The previous method uses a maximum value called $v_{\max}$ to control the component flow of motion. A particle may fly through a good solution if $v_{\max}$ is set too high. Particles might not be able to look past local solutions if $v_{\max}$ is set too low. In several PSO tests, $v_{\max}$ was adjusted at 0.1 to 0.2 of each dimension's variable's dynamic range.

B. Hybrid Particle Swarm Optimization

This algorithm uses both the PSO and the genetic algorithm as natural selection mechanisms (GA). The search region is confined by p_{best} and g_{best} since PSO's technique is strongly dependent on them. PSO iteratively modifies the existing searching locations [10]. The selection is used to replace agent positions with moderate evaluation values with those with strong evaluation values. On the other hand, HPSO uses the selection approach to focus its search, notably in the current effective zone, by sending low-evaluated particles there.

C. Multiagent System (MAS)

For many years, considerable research on a computation using agents have been accomplished in the machine learning area [11]. It is a technique for interacting with numerous agents to accomplish a single objective.

1. Agents are present in a particular setting and carry out particular functions.
2. Agents can communicate with other agents in the region by being aware of their local surroundings.
3. Agents make an effort to accomplish or achieve predetermined goals or duties.
4. Because agents are capable of learning, they can react quickly to changes in themselves.

MAS can solve and optimize a wide range of complex issues. In this research, combination of PSO and MAS are done to develop a new optimal algorithm, which is then compared against SVM. The mentioned particulars are to be defined in general when utilizing this to achieve solution:

1. Each agent's meaning and purpose;
2. The place in which all the agents reside;
3. Description of a local environment; and
4. Many behavioral rules are present to govern how agents interact with their surroundings. They are the agent of universe' laws.

D. MultiAgent Particle Swarm Optimization (MAPSO)

In this study, this method is used to address agents used in MAPSO and PSO to devise action plans or results for optimization problems. To begin, a lattice-like environment is created, with individual agents assigned to one of the lattice points.

1. *Agent Work:* A particle that symbolizes a suggested remedy for optimum challenge is an agent in this context. Selecting the fitness value of the agent is an optimum challenge.

2. *Environment:* Every agent in MAS resides in a certain setting. A lattice-like structure, as shown in Fig. 3.1, organizes the environs. A lattice-point-fixed circle serves as the representation for each actor in the environment. $L_{size} \times L_{size}$, where L_{size} is an integer, equals L. The total number of particles affects the size of PSO.

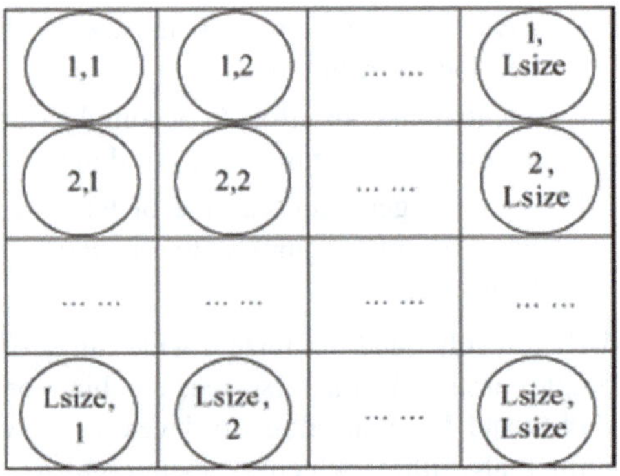

Fig. 3.1 Assembly of surrounding

Source: Author

3. *Local Environment:* Since each MAS agent can only see what is right in front of them, the definition is essential. The agent and its size are represented as (i, j) and $i, j, i, j = 1, 2, \ldots, L_{size}$, respectively. $N_i, j = i_1, j, I j_1, i_2, j, I j_2$.

   where

   $$i^1 = \begin{cases} i-1 & i \neq 1 \\ L_{size} & i = 1 \end{cases}, j^1 = \begin{cases} j-1 & j \neq 1 \\ L_{size} & j = 1 \end{cases}$$

   $$i^2 = \begin{cases} i+1 & i \neq L_{size} \\ 1 & i = L_{size} \end{cases}, j^2 = \begin{cases} j+1 & j \neq L_{size} \\ 1 & j = L_{size} \end{cases}$$

   $$(8)$$

   From Eq. (11), each agent is surrounded by 4 fellow surroundings. They generate a limited local environment that it will be able to perceive only.

4. *Agent Behavioral Strategies:* Every agent contests and collaborates with its neighbors to disseminate important information over the available space, and it can also leverage the PSO's knowledge.

Based on these behaviors, three operators are created for the agents.

Competition and Cooperation Operator: Assume this operation is used at the point Ij), and that the vector represented by $Ij = (1, 2, \ldots, n)$ on the search space has real values. If agent i, j fits condition (12), unless it is a loser, it is a winner. Assuming that agent $m = \min K_{i,j} = (n_1, n_2, \ldots, n_i)$ has the lowest suitability value among K_i, neighbors, j's agents, that is, $m \in K_{i,j}$ and $K_{i,j}$.

$$f(\alpha_{i,j}) \leq f(m) \qquad (9)$$

Whenever $\alpha_{i,j}$ triumphs the agent lattice preserves its location without moving around in the space of possible pursue solutions. A new era New_{ij} will take its lattice-point if it loses, causing it to expire.

PSO Operator: The proposed MAPSO technique quickly and precisely finds the best solution by combining the competition and cooperation operators with the PSO evolution mechanism. In the event of a violation, (8) and (9) renew each agent's positions, which are then finalised at the minimum or maximum bound of the solution environment.

E. Particle Swarm Optimization based on Hybrid Multi-Agent Systems (HMAPSO)

Hybrid multi agent based particle swarm optimization (HMAPSO) is a newfangled method that combines PSO and MAS [14]. GA methods have been utilized to resolve challenging optimization issues. There are a number of downsides to GA performance, according to recent studies. PSO's search capacity increases early in the run and declines as it progresses. If the particle's current location overlaps the PSO's global ideal location and its weight and previous velocity are both greater than 0, it will be ejected. PSO does not offer best alternative for the stagnation crisis. Instead, a more contemporary method termed as HPSO, which includes the breeding and sub-population processes. After neighborhood operators, rivalry and cooperation, the agent is provided to PSO in MAPSO as a particle. It has been demonstrated that PSO and its hybrid multi-agent algorithms produce more optimal results than conventional optimization methods. HMAPSO combines the benefits of MAPSO and HPSO.

F. Support Vector Machine

The SVM is becoming more and more popular due to its many appealing qualities and promising empirical performance. SVM can implement the structure risk minimization (SRM) concept, which is more effective than the typical empirical risk minimization method employed by the majority of neural networks. Compared to ERM, SRM minimizes an upper bound on the Vapnik-Chernoverkis dimension generalization error. SVM can perform well in generalization, which is the purpose of learning issues, because of this distinction. SVM is theoretically sound, allowing it to identify the optimum global solution while avoiding local minimization. It can tackle high-dimensional issues while avoiding "dimension disaster" as per Kernel Hilbert Spaces Theory.

The SVM classifier handles the following basic challenge, which is described as an optimization problem, to design this optimal separation hyperplane as shown in Fig. 3.2.

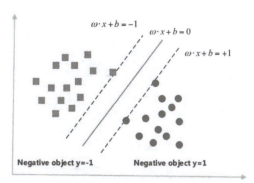

Fig. 3.2 Highest margin and the optimum hyperplane

Source: Author

The penalty parameter C is proportional to the degree of constraint violation, ξ_i is the

slack variable, the "kernel" function $\varphi(.)$ is the "kernel," and the threshold is b. Because of the kernel mapping technique, SVM models can perform separations even with very intricate borders. There are numerous kernel mapping functions available. For sigma fluctuations, MSE remains constant in linear kernel. Similarly, the MSE value of poly kernel is not less than that of RBF type. As a result, the RBF kernel type is employed to define the loadability margin in the SVM training model. When utilizing the RBF kernel, the penalty parameter C and the RBF kernel parameter Sigma are two parameters that are coupled to the SVM model.

G. Extended Learning Machine (ELM)

This algorithm is a high-efficiency learning technique. It addresses issues with sluggish training speed and over-fitting as compared to the current neural network learning method. This was created as a generalized single hidden layer feed forward neural network that does not require a hidden layer that mimics neurons. In contrast to other back propagation neural networks, ELM's hidden nodes are produced at cluster sampling, Activation processes in neurons, however, are nonlinear piecewise continuous processes. The ELM single layer architecture is shown in Fig. 3.3.

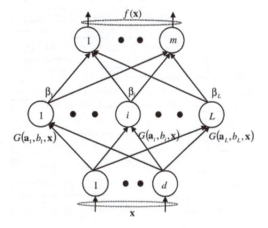

Fig. 3.3 ELM single layer architecture

Source: Author

The practical Indian 181 bus structure and IEEE118 bus system are considered for simulation studies. Each technique is validated with 2 test systems having nonlinear characteristics. The initial iteration, producing expenditure, and implementation time averages from 50 trials are shown in this study. Figures 3.4 and 3.5, respectively, depict the single line diagrams of the IEEE 118 bus and Indian 181 bus systems. Despite facts state that the electricity supply plan in India is very different from the analysis conducted in this work, the goal of the paper is to investigate the ELD efficiently to calculate the optimal cost of power generation.

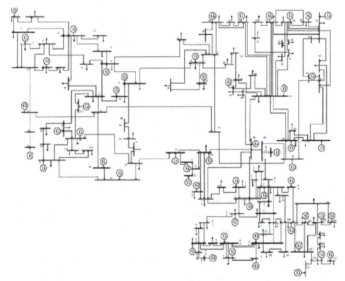

Fig. 3.4   One line diagram of IEEE 118 bus test system

Source: Author

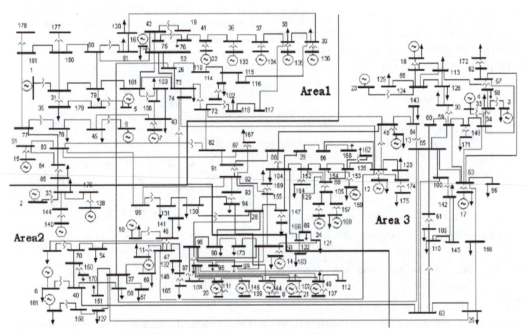

Fig. 3.5   One line diagram of Indian 181 bus test system

Source: Author

Table 3.1 IEEE 118 Bus/Indian 181 bus systems parameter for SVM and ELM for optimal cost of generation

	Regression Methods	Training Data Sets	Testing Data Sets	Activation Function	Training Time In Sec	Testing Time In Sec	MSE
Indian 181 Bus System	**SVM**	**28656**	**7184**	**Sigmoid**	**11.712**	**0.089**	**5.244e^{-005}**
				RBF	**9.506**	**0.055**	**2.333e^{-006}**
	ELM	28656	7184	Sigmoid	17.462	0.086	1.251e^{-005}
				RBF	10.578	0.047	1.115e^{-006}
IEEE 118 Bus Test System	**SVM**	**29360**	**7300**	**Sigmoid**	**14.755**	**0.013**	**3.234e^{-005}**
				RBF	**9.830**	**0.011**	**2.855e^{-006}**
	ELM	29360	7300	Sigmoid	19.636	0.013	1.925e^{-005}
				RBF	17.791	0.010	1.028e^{-006}

Source: Author

The MATLAB-based computer programme runs on a Pentium-IV with 128 MB of RAM and a 1.6 GHz clock speed. Table 3.1 displays the parameter values for SVM and ELM for the IEEE 118 Bus/Indian 181 Bus System. A total of 36,660 data samples were obtained by widely altering demand on specific bus and supply. Of 36,660 samples, 80% (29,360) of which were used for training and remaining 20% (7030) were used for testing.

III. Test System and Discussion

Considering the simulation outcomes displayed in Table 3.2, using PSO, HPSO, MAS, MAPSO, HMAPSO, SVM and ELM, it is possible to obtain the most affordable power system generation costs for the Indian 181 bus system and IEEE 118 bus systems. Figure 3.6(a) and (b) illustrate the representations of IEEE118 test bus and Indian 181 bus system and its optimum cost of electricity generation utilizing various methods.

Figures 3.7(a) and (b) illustrate representations of IEEE118 bus and Indian 181 bus system respectively and its computational time utilizing various methods. These results demonstrate that SVM and HPSO can quickly reach the lowest cost of generating electricity. The outcomes of the various algorithms demonstrate their usefulness and computational efficiency. The

Table 3.2 The price and implementational period of generation for IEEE 118 bus systems & Indian 181 bus systems using different algorithms

Methods	Indian 181 Bus System		IEEE 118 Bus System	
	Price in Rs/MW	Period in Sec	Price in Rs/MW	Period in Sec
PSO	224850.64	0.8455	1,99,851	0.5877
HPSO	219863.17	0.7236	1,98,641	1.8324
MAS	212756.09	0.8952	2,00,566	0.4551
MAPSO	213658.95	2.33178	2,53,217	1.4527
HMAPSO	222540.12	1.227336	2,25,341	0.6012
SVM	**201851.78**	**0.536576**	**1,57,354**	**0.3652**
ELM	224853.77	0.7066	2,69,501	0.6354

Source: Author

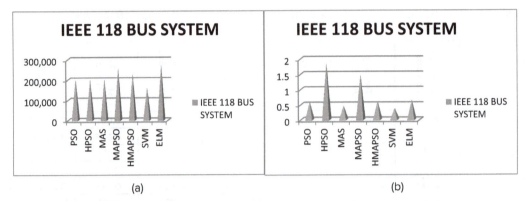

Fig. 3.6 IEEE 118 bus system generation cost and time for various techniques

Source: Author

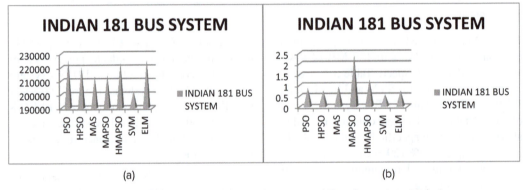

Fig. 3.7 Indian 181 bus system generation cost and time for various techniques

Source: Author

HPSO algorithm is practical, easy to use and effective at solving non-linear optimization issues. SVM results in improvement of cost of generation decrease. With decreased average execution time, it converges quickly and produces accurate results. The outcomes show that the suggested strategy is workable and that the SVM algorithm compared to other algorithms, is superior in efficacy and applicability for any optimization assignment.

IV. Conclusion

By accounting for varying load demand, ELD can be resolved. This research investigated the usage of a SVM as well as a number of other strategies. The suggested system can produce lower generation costs while also dramatically lowering transmission line losses, according to modeling data. Furthermore, the implementation time is roughly consistent and shorter for the 2 test systems under consideration when compared to the PSO, HPSO, MAS, MAPSO, HMAPSO and ELM approaches. Thus, the ELD's extremely nonlinear discontinuous cost functions issue seems to be amenable to solution by the SVM technique, which also provides a globally enhanced optimum solution.

References

1. C. Wang and S. M. Shahidehpour, "Effects of ramp-rate limits on unit commitment and economic dispatch," *IEEE Transactions on Power Systems*, vol. 8, no. 3, pp. 1341-1349, August 1993.

2. Abido M. A. Optimal power flow using particle swarm optimization. Electric Power Energy Syst 2002; 24:563–71.

3. Ahmed F. Ali. A hybrid particle swarm optimization and genetic algorithm with population partitioning for large scale optimization problems; Ain Shams Engineering Journal Volume 8, Issue 2, June 2017, Pages 191–206

4. Chau KW. A split-step particle swarm optimization algorithm in river stage forecasting. J Hydrol 2007; 346(3–4): 131–5.

5. Esmin AAA, Lambert-Torres G, Zambroni de souza AC. A hybrid particle swarm optimization applied to loss power minimization. IEEE Trans Power Syst 2005; 20(2): 859–66.

6. Gaing Zwe-Lee. Particle swam optimization to solving the economic dispatch considering the generator constraints. IEEE Trans Power Syst 2003; 18(3): 1187–95.

7. Xu Zhang, "Hybrid Particle Swarm Optimization Algorithm for Process Planning," Mathematics, 2020.

8. D. B. Fogel, "Hybrid Algorithm of Particle Swarm Optimization and Grey Wolf Optimizer for Improving Concerger Performance" Journal of Appilied Mathematics, Volume 2017.

9. Vedran Kirinčić, Ervin Čeperić, Saša Vlahinić & Jonatan Lerga (2019) Support Vector Machine State Estimation, Taylor & Francis, Applied Artificial Intelligence, 33: 6, 517–530, DOI: 10.1080/08839514.2019.1583452.

10. Faith Ahmeth Senel, " A novel Hybrid PSO-GWO Algorithm for Optimization Problems" Springer, Engineering with Computers, pp 1359–1373 (2019).

11. Suganyadevi, M.V., Babulal, C.K. (2013). Online Voltage Stability Assessment of Power System by Comparing Voltage Stability Indices and Extreme Learning Machine. In: Panigrahi, B.K., Suganthan, P. N., Das, S., Dash, S. S. (eds) Swarm, Evolutionary, and Memetic Computing. SEMCCO 2013. Lecture Notes in Computer Science, vol 8297. Springer, Cham. https://doi.org/10.1007/978-3-319-03753-0_63.

12. M. Tripathy and S. Mishra, "Bacteria Foraging-Based solution to optimize both real power loss and voltage stability limit," IEEE Trans. Power Systems, vol. 22, no. 1, pp. 240–248, Feb. 2007.

13. Piotr Dziwinski "A New Hybrid particle swarm optimization and Genetic Algorithm Optimization Method Controlled by Fuzzy Logic", IEEE Transactions on Fuzzy Systems, pp 1140–1154, Dec 2019.

14. Rajesh Kumar, Devendra Sharma, Anupam Kumar, "A new hybrid multi-agent-based particle swarm optimisation technique" , International Journal of Bio-Inspired Computation 2009 – Vol. 1, No.4 pp. 259–269.

15. Zimmerman R, Gan D. MATPOWER: A Matlab Power System Simulation Package, Ithaca. NY: Cornell University Press; 1997.

16. Suganyadevi, M. V., Babulal, C. K. & Kalyani, S. Assessment of voltage stability margin by comparing various support vector regression models. Soft Comput 20, 807–818 (2016). https://doi.org/10.1007/s00500-014-1544-x.

17. Zhang. X, Support Vector Machines, Encyclopedia of Machine Learning and Data Mining, Springer, Boston, MA, 2017.

4

Masked Face Recognition by Reconstructing Face Using Image Inpainting with GAN

Jinal Bhagat*

Assistant Professor, Department of Information Technology,
ADIT, V. V. Nagar, India

Mahesh Goyani[1]

Assistant Professor, Department of Computer Engineering,
GEC, Modasa, India

Abstract—The security systems nowadays rely more on biometric features such as IRIS, fingerprint and face, out of which face recognition has proved to be the most secure and reliable medium to identify a person. The Severe Acute Respiratory Syndrome Coronavirus 2 (SARS- CoV2) has presented greater challenges not just to human health, but security systems as well. In this pandemic situation, people wear masks, which hides the features useful in the recognition of face. So, the performance of face identity verification methods gets compromised, hampering accurate distinguishing between masked individuals. For masked face recognition, many methods have been presented recently. Still only a few are accurate and give satisfactory performance. The main reason behind this performance drop of face recognition methods is that the reference images are unmasked. This research introduces a method to generate a masked dataset and using state-of-the-art techniques of deep learning and feature extraction methods to identify a face. We have proposed the use of generative adversarial networks and multiple face recognition methods in our research. Three benchmark datasets, CelebFaces Attributes (CelebA) Dataset, Labeled Faces in Wild (LFW) and Flickr-Faces-HQ Dataset (FFHQ) are used for our research. For evaluation of our recognition algorithm, Confusion matrix is used from which f1 score, precision, and accuracy is calculated. We have obtained a satisfactory accuracy level up to 98% for our proposed algorithm.

Keywords—Masked face recognition, Deep learning, Generative adversarial networks, Image inpainting, Face reconstruction

I. Introduction

Face recognition techniques used today have been actively using facial features like nose, eyes and lips. In situations like working in laboratories and people with various medical conditions covering their faces with masks becomes essential. One of the recent

*Corresponding author: jinalrana93@gmail.com
[1]mgoyani@gmail.com

DOI: 10.1201/9781003388913-4

situations faced across the globe was the 2019 SARS CoV2. It has highly impacted the face recognition techniques and further, the security systems relying on it. We have seen large performance drop in face recognition algorithms while considering masked face probes. [1] [2] In return, the present security systems tend to become ineffective and unusable. Hence, new research and development is required to establish a method, where security systems work with face masks, without having to remove them. Since other biometric systems like fingerprint identification rely on physical contact, the urge to find an alternative face recognition technique with masks on, has received a great demand.

The active research done in face recognition has taken a hit because of the presence of face mask. New challenges have come up and more precise research is now required to develop an algorithm that is able to recognize both, masked and unmasked faces. We have built a masked faces dataset and introduced a precise algorithm to reconstruct the occluded mask area of the face region, and then fed them to classification algorithm to recognize face among the given probes. Combining generative adversarial networks (GAN) with image inpainting and using right parameters and weights has helped in improving the accuracy of masked face recognition. The main objective is to increase the performance of masked face recognition using deep learning Approaches, thus helping biometric security systems.

II. Related Work

One of the most researched topics in the field of biometric studies nowadays is facial recognition as it provides fairly better characteristics for identification than any other methods, more precisely, occluded or masked face recognition. Several previous research papers [3], [4] targetted general mediums of face occlusion, for example, wearing hat, scarf, a sunglasses, etc. The facial mask as occlusion is not mentioned before pandemic situation.

After the pandemic, the study presented by Naser et al. [5] states the adverse effect of face mask on the performance of face recognition algorithm. The NIST[6] also presented the negative effects of mask on the performance of face recognition algorithms.

Therefore in Ref. [7], they proposed a solution to improve the performance of masked face recognition algorithms, the novel function self-restrained triplet (SRT) loss is also presented in research. On other hand, Ref. [8] presented novel training method to maximize the features information extracted from facial images using semi-siamese network. And also studied the NIR-VIS masked face recognition. In Ref. [9], they proposed a feature extraction network called ResSaNet. This ResSaNet network is a combined version of self-attention network and residual block from convolutional neural network.

To highlight the problem of occluded faces, Weitao Wan and Jiansheng Chen [10] proposed the trainable module named MaskNet, which can be included in existing CNN architecture. Due to less availability of masked face data, Hongxia et al. developed an algorithm that can produce masked face photos. They have done it by extracting all important key features from facial regions, and also presented algorithm based on large margin cosine loss.[11].

In his research, Walid Hariri [12] applied networks like AlexNet,VGG-16 and ResNet, to extract deep features from the facial regions, for example, eyes and forehead and then applied the Bag-of-features paradigm. In their research, Hoai et al. [13] proposed combined method to identify masked faces. They have used deep learning techniques with local binary pattern(LBP) whereas in Ref. [14] they used ResNet 50 architecture.

This paper [15] considers the problem of person re-identification. It solves the problem by coalescing masked face and probe face images. Mengyue et al. [16] contributed new dataset in their research named as MFSR. To address the face recognition problem with mask they used a novel network known as IAMGAN (identity aware mask GAN). It applies multi-level identity module for recognization and evaluation purposes. The author Naser et al. presented unique research in their paper [17], they presented comparison of state-of-the-art face recognition algorithms with that of performance of human ability to verify the face. In their paper [18], they first proposed their created dataset named as FGNET-MASK then proposed different deep learning techniques.

III. Research Methodology

Benchmark datasets Labeled Faces in wild (LFW) [19], CelebFaces Attributes Dataset (CelebA) [20] and Flickr-Faces-HQ Dataset (FFHQ) [21] are used in this research. Each dataset contains images of human faces in different conditions. LFW contains images that are captured in wild, not in any artificial setup. celebA contains the images of only celebrities.

A. Face Reconstruction

The most concerned challenge in face recognition is to cop up with the situations when in all the scenarios the subjects are non-cooperative. In masked face recognition, around 70% of the features of the face are occluded, thus only 30% of the face (eyes and forehead) can be uses for feature extraction, but with such less features we increase the rate of false positive recognition of the face. Thus, to improve accuracy and reduce the matching time, the occluded part needs to be re-constructed. For which we used the combination of Generative Adversarial Networks (GAN) [22] and Image In-Painting Technique to remove mask and

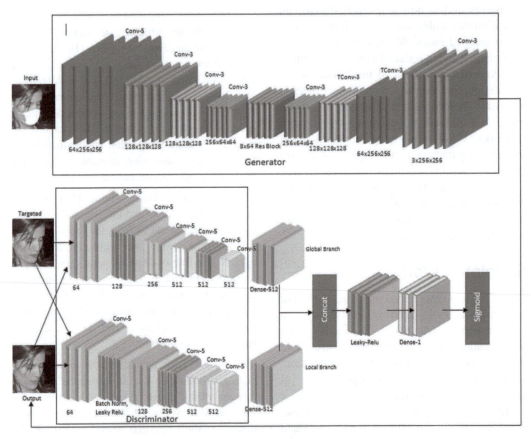

Fig. 4.1 GAN architecture

generate whole face from the image that best matches the features of ground truth images.

The detailed architecture of generative adversarial network is shown in Fig. 4.1 where generator and discriminator architecture are shown with all convolutional layers. The masked image is given to generator which produces the image output and this output image is given to discriminator to evaluate if it is fake or real. This process runs until discriminator labels output image as real.

Some of the face reconstruction using Image inpainting and GAN is shown in Table 4.1.

Table 4.1 Images generated by GAN

Original Image	Inpainted Image

B. Proposed Method

As shown in Fig. 4.2, this section represents how MFR system is developed step by step.

First, from the collection of images with the corresponding ground truth the face and mask is detected. This is usually done to verify if the basic object is actually present in an image or not as there can be multiple faces or no faces at all. The images can be taken in completely wild environment, from which the face is hard to detect. For example, it is hard to find the face from the video or image taken by the surveillance camera, because that is crowded places, so multiple faces could be present and also the background is continuous moving. If the face or face mask is not found from the image, we would discard that image, as it is not worth to run the whole algorithm on random non-face images.

If there is no mask in the image, a simulated mask is generated on the face. A dilib-based model is used here that detects all facial landmarks with face tilt. As per the facial tilt and all the recognized landmarks, a suitable mask template is chosen. Necessary alteration like resizing and reshaping is done in the face mask depending on facial landmarks. The block diagram in its entirety may be found in the image below. MaskTheFace [23] offers a variety of masks to choose from. Mask dataset collection is problematic in a variety of situations. We have generated a masked face dataset using MaskTheFace.

After this step the masked image is sent to detectron2 algorithm to generate mask image that will be used to feed in inpainting technique. Two images are required while deploying image inpainting technique.

1. The sampled image or masked image required for inpainting and restoration.
2. The mask image, depicting the restoration areas required to inpaint over. Ideally both the images should have the same dimensions. Here the mask image is divided into two types of pixels, white pixels represents the areas that need to be restored, while black pixels require no action to be taken.

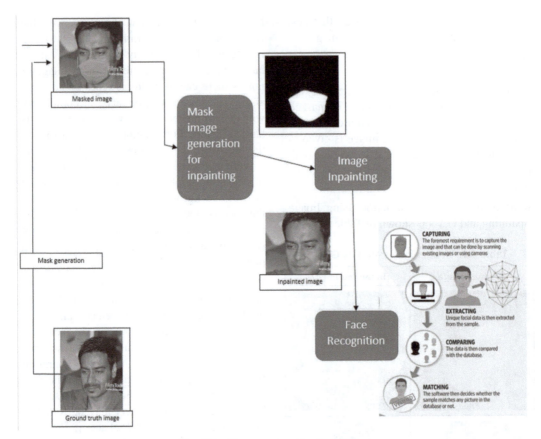

Fig. 4.2 Proposed method workflow

After generating mask image, the input image and mask image is feed into DMFN [24] network. In which the generative networks generate the occluded part of the face. The network creates pixels in empty or zero pixels. This is how the mask from the face is removed. As shown in Fig. 4.3, the result of an inpainted image is visible.

Fig. 4.3 Mask image with input Image of same dimensions

So now when we have obtained our image restored it is then supplied to face recognition algorithm. From which the features are extracted from full face. Then the image is embedded and compared with already available dataset. The distance is found between this inpainted image and all the images available in the dataset. The inpainted image is assigned in the class whose distance is minimum than threshold value (0.4).

All of the face recognition deep learning algorithms used in research are: DeepFace, DeepID, VGGFace, FaceNet, ArcFace, OpenFace and Dlib. The cosine distance calculation is performed in this classification work. For each face/person, we construct embeddings, which define the individual in quantitative data and then classification is been carried out.

IV. Results

Proposed model is trained on LFW dataset first. From various deep learning techniques, we have used Image inpainting technique which uses generative adversarial networks to regenerate image patches. Generative adversarial networks are used to generate accurate and similar look-alike face images. In GAN, VGG layers are used for feature extraction from the images. With the help of generative network and discriminator network similar face image can be generated, and this pixel generation is then set to occluded mask region of the face. The l1 loss, self-guided regression loss, adversarial loss, fm-vgg, fm-dis and alignment losses are calculated to evaluate performance of GANs. Here we have set the learning rate of neural network as 2.000e-04. Here semantic structure prevention issue is addressed. We used self-guided regression loss to correct the image semantic levels. Here the discrepancy between generated image and ground truth image is calculated to navigate the similarity measure of feature map. Following equation is calculated to compute regression loss:

$$\mathcal{L}_{self-guided} = \sum_{t=1}^{2} \omega^t \frac{M_{guidance}^t \odot (\psi_{I_{gt}}^t - \psi_{I_{output}}^t)_1}{N_{\psi_{I_{gt}}^t}} \quad (1)$$

A. Performance Analysis of Generative Adversarial Networks (GAN)

To evaluate the performance of generative networks, we calculated structural similarity index (SSIM) and PSNR index between generated image and original image.

1. Structural Similarity Index (SSIM)

SSIM [25] is used to check how similar two images are. SSIM works around three basic features of an image: luminance, contrast and structure. The comparison is carried out on the base of these features.

The SSIM is calculated as

$$SSIM(x, y) = [l(x, y)]^\alpha \cdot [c(x, y)]^\beta \cdot [s(x, y)]^\psi \quad (2)$$

2. Peak Signal-to-Noise Ratio (PSNR)

PSNR is the ratio of signal's power to noise. PSNR is popularly used to evaluate the quality of an image as well as the video. It is calculated using mean squared error. If a given image I has noise free mxn approach, the mse is calculated as follows:

$$MSE = \frac{1}{mn} \sum_{i=0}^{m-1} \sum_{j=0}^{n-1} [I(i, j) - K(i, j)]^2 \quad (3)$$

From which PSNR is defined as

$$PSNR = 10 \cdot \log_{10} \left(\frac{MAX_I^2}{MSE} \right) \quad (4)$$

B. Results of Face Recognition Algorithms

In this research, seven different face recognition models are used for classification of the images and we used 10 cross-fold validation technique. The dataset of ground truth images is divided in 10 folds each comprised 200 images.

All the evaluation parameters obtained are mentioned with confusion matrix in Table 4.2 and Fig. 4.4

5. Conclusion

The aim of this research was to develop a system capable of detecting masked faces. We presented our efforts towards masked face recognition. We collected different datasets and completed it by generating synthetic mask on face images. In this research, we recognized masked faces by reconstructing occluded part of face. We used image inpainting method using generative adversarial networks. After removing occlusion from faces we classified all images using several face recognition algorithms. We got 98% accuracy in recognizing masked faces.

Table 4.2 Result analysis

	Result analysis for Positive (1) class				Result analysis for Negative (0) class			
	Precision	Recall	F1-score	Accuracy	Precision	Recall	F1-score	Accuracy
VGG-Face	0.49	0.94	0.64	0.98	1.00	0.98	0.99	0.98
Arc-Face	0.58	0.92	0.71	0.98	1.00	0.99	0.99	0.98
Deep-Face	0.15	0.89	0.25	0.89	1.00	0.89	0.94	0.89
Deep ID	0.31	0.87	0.45	0.96	1.00	0.96	0.98	0.96
Dlib	0.31	0.91	0.46	0.96	1.00	0.96	0.98	0.96
Face-Net	0.91	0.86	0.88	0.99	1.00	1.00	1.00	0.99
Open-Face	0.73	0.53	0.62	0.99	1.00	0.99	0.99	0.99

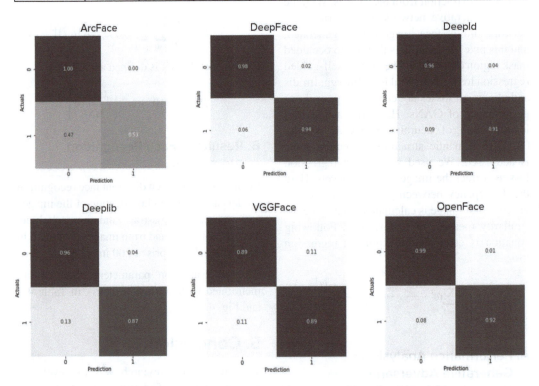

Fig. 4.4 Confusion matrix for all face recognition algorithms

References

1. N. Damer, J. H. Grebe, C. Chen, F. Boutros, F. Kirchbuchner, and A. Kuijper, "The Effect of Wearing a Mask on Face Recognition Performance: an Exploratory Study." arXiv, Aug. 20, 2020. Accessed: May 27, 2022. [Online]. Available: http://arxiv.org/abs/2007.13521

2. N. Damer, F. Boutros, M. Süßmilch, F. Kirchbuchner, and A. Kuijper, "Extended evaluation of the effect of real and simulated masks on face recognition performance," *IET Biom.*, vol. 10, no. 5, pp. 548–561, Sep. 2021, doi: 10.1049/bme2.12044.

3. L. Song, D. Gong, Z. Li, C. Liu, and W. Liu, "Occlusion Robust Face Recognition Based on Mask Learning With Pairwise Differential

Siamese Network," in *2019 IEEE/CVF International Conference on Computer Vision (ICCV)*, Seoul, Korea (South), Oct. 2019, pp. 773–782. doi: 10.1109/ICCV.2019.00086.

4. R. Biswas, V. González-Castro, E. Fidalgo, and E. Alegre, "A new perceptual hashing method for verification and identity classification of occluded faces," *Image Vis. Comput.*, vol. 113, p. 104245, Sep. 2021, doi: 10.1016/j.imavis.2021.104245.

5. B. BIOSIG *et al.*, "The Effect of Wearing a Mask on Face Recognition Performance: an Exploratory Study." 2020.

6. Ngan, Mei L., Patrick J. Grother, and Kayee K. Hanaoka., "Ongoing face recognition vendor test (FRVT) Part 6A: Face recognition accuracy with masks using pre-COVID-19 algorithms.," 2020.

7. F. Boutros, N. Damer, F. Kirchbuchner, and A. Kuijper, "Self-restrained Triplet Loss for Accurate Masked Face Recognition," *ArXiv210301716 Cs*, Dec. 2021, Accessed: Dec. 20, 2021. [Online]. Available: http://arxiv.org/abs/2103.01716

8. H. Du, H. Shi, Y. Liu, D. Zeng, and T. Mei, "Towards NIR-VIS Masked Face Recognition," *IEEE Signal Process. Lett.*, vol. 28, pp. 768–772, 2021, doi: 10.1109/LSP.2021.3071663.

9. W.-Y. Chang, M.-Y. Tsai, and S.-C. Lo, "ResSaNet: A Hybrid Backbone of Residual Block and Self-Attention Module for Masked Face Recognition," in *2021 IEEE/CVF International Conference on Computer Vision Workshops (ICCVW)*, Montreal, BC, Canada, Oct. 2021, pp. 1468–1476. doi: 10.1109/ICCVW54120.2021.00170.

10. W. Wan and J. Chen, "Occlusion robust face recognition based on mask learning," in *2017 IEEE International Conference on Image Processing (ICIP)*, Beijing, Sep. 2017, pp. 3795–3799. doi: 10.1109/ICIP.2017.8296992.

11. H. Deng, Z. Feng, G. Qian, X. Lv, H. Li, and G. Li, "MFCosface: A Masked-Face Recognition Algorithm Based on Large Margin Cosine Loss," *Appl. Sci.*, vol. 11, no. 16, p. 7310, Aug. 2021, doi: 10.3390/app11167310.

12. W. Hariri, "Efficient Masked Face Recognition Method during the COVID-19 Pandemic," *ArXiv210503026 Cs*, May 2021, Accessed: Dec. 20, 2021. [Online]. Available: http://arxiv.org/abs/2105.03026

13. H. N. Vu, M. H. Nguyen, and C. Pham, "Masked face recognition with convolutional neural networks and local binary patterns," *Appl. Intell.*, Aug. 2021, doi: 10.1007/s10489-021-02728-1.

14. B. Mandal, A. Okeukwu, and Y. Theis, "Masked Face Recognition using ResNet-50," *ArXiv210408997 Cs*, Apr. 2021, Accessed: Dec. 20, 2021. [Online]. Available: http://arxiv.org/abs/2104.08997

15. Q. Hong, Z. Wang, Z. He, N. Wang, X. Tian, and T. Lu, "Masked Face Recognition with Identification Association," in *2020 IEEE 32nd International Conference on Tools with Artificial Intelligence (ICTAI)*, Baltimore, MD, USA, Nov. 2020, pp. 731–735. doi: 10.1109/ICTAI50040.2020.00116.

16. M. Geng, P. Peng, Y. Huang, and Y. Tian, "Masked Face Recognition with Generative Data Augmentation and Domain Constrained Ranking," in *Proceedings of the 28th ACM International Conference on Multimedia*, Seattle WA USA, Oct. 2020, pp. 2246–2254. doi: 10.1145/3394171.3413723.

17. N. Damer, F. Boutros, M. Süßmilch, M. Fang, F. Kirchbuchner, and A. Kuijper, "Masked Face Recognition: Human vs. Machine," *ArXiv210301924 Cs*, Jun. 2021, Accessed: Dec. 20, 2021. [Online]. Available: http://arxiv.org/abs/2103.01924

18. V. S. Patel, Z. Nie, T.-N. Le, and T. V. Nguyen, "Masked Face Analysis via Multi-Task Deep Learning," *J. Imaging*, vol. 7, no. 10, p. 204, Oct. 2021, doi: 10.3390/jimaging7100204.

19. [19] G. B. Huang, M. Mattar, T. Berg, and E. Learned-Miller, "Labeled Faces in the Wild: A Database forStudying Face Recognition in Unconstrained Environments," presented at the Workshop on Faces in "Real-Life" Images: Detection, Alignment, and Recognition, Oct. 2008. Accessed: Apr. 26, 2022. [Online]. Available: https://hal.inria.fr/inria-00321923

20. Z. Liu, P. Luo, X. Wang, and X. Tang, "Deep Learning Face Attributes in the Wild," 2014, doi: 10.48550/ARXIV.1411.7766.

21. T. Karras, S. Laine, and T. Aila, "A Style-Based Generator Architecture for Generative Adversarial Networks," 2018, doi: 10.48550/ARXIV.1812.04948.

22. I. Goodfellow *et al.*, "Generative Adversarial Nets," in *Advances in Neural Information*

Processing Systems, 2014,vol.27. [Online]. Available: https://proceedings.neurips.cc/paper/2014/file/5ca3e9b122f61f8f06494c97b1afccf3-Paper.pdf

23. A. Anwar, *MaskTheFace - Convert face dataset to masked dataset*. 2022. Accessed: Apr. 26, 2022. [Online]. Available: https://github.com/aqeelanwar/MaskTheFace

24. Z. Hui, J. Li, X. Wang, and X. Gao, "Image Fine-grained Inpainting," 2020, doi: 10.48550/ARXIV.2002.02609.

25. Z. Wang, A. C. Bovik, H. R. Sheikh, and E. P. Simoncelli, "Image quality assessment: from error visibility to structural similarity," *IEEE Trans. Image Process.*, vol. 13, no. 4, pp. 600–612, Apr. 2004, doi: 10.1109/TIP.2003.819861.

Note: All figures and tables were created/edited by the author using algorithm outputs.

Recent Trends in Computational Intelligence and its Application – Sugumaran D. et al. (eds)
© 2023 Taylor & Francis Group, London, ISBN 978-1-032-48410-5

5

Deep Learning-Based Heart Sounds Classification Using Discrete Wavelet Transform Features

Sabeena Yasmin Hera* and Mohammad Amjad[1]

Department of Computer Engineering,
Jamia Millia Islamia, Delhi, India

Abstract—Cardiovascular disease is one of the major causes of mortality round the globe. Early detection and diagnosis are the only ways to combat it. Modern automated technical solutions are now crucial as a result of this. Artificial intelligence-based automatic diagnosis systems have proved their significance in this cause. Different image and sound modalities are being used in the diagnosis of heart diseases. This paper focuses on the prediction of heart disease using heart sounds. In this study two different architecture of deep learning based on convolutional neural with discrete wavelet transform is presented to deal with multi-labelled heart sounds dataset. First, the data are acquired from the repository and the features are extracted from the heart signal using discrete wavelet transform. These features are then fed as an input to the deep learning model for the identification of heart health. The performance of the model is validated on a phonocardiogram dataset that consists of 4 classes of heart disorder and is measured in the terms of accuracy, precision, recall and specificity. The experimental results show that the model with second architecture achieved highest accuracy with 99.5%, 100% precision, 97.36% recall and 100% specificity.

Keywords—Cardiovascular disease, Heart sounds, Discrete wavelet transform, Convolutional neural networks, Phonocardiogram

I. Introduction

According to WHO, cardiovascular disease (CVD) is one of the leading causes of death around the globe [1]. Unfortunately, the rates of CVD mortality and morbidity are rising year after year, especially in emerging countries. Similarly, the figures in Asia are not much different. In India, according to the Global Burden of Disease, over 25% of all deaths were due to CVDs [2]. CVD can be broadly classified as coronary artery disease, stroke, angina and arrythmia. Some of these diseases are associated with electrical activity of heart, that is, arrhythmias. Such diseases are diagnosed with electrical activities based on

*Corresponding author: sabeenayasminhera@gmail.com
[1]mamjad@jmi.ac.in

DOI: 10.1201/9781003388913-5

physiological signals called electrocardiograms (ECG) [3]. Similarly, other very common heart diseases are associated with heart valves and are known as valvular diseases [4]. To diagnose such diseases, medical experts listen to the heart sounds using the stethoscope and diagnose different heart diseases. Machine learning techniques are popularly being used in the healthcare for diagnosis and prognosis purposes. This study focuses on heart sound classification using such techniques.

Celin and Vasanth [5] proposed a machine learning model to discover heart disease from ECG signals. The preprocessing of signals includes the removal of outliers using various types of filters and then extracting the useful feature through peak detection technique. These features were then used by support vector machine (SVM), Naïve Bayes, Adaboost and artificial neural network (ANN) to carry out classification. The experimental results show that the Naïve Bayes approach outperformed and achieved an accuracy of 99.7% followed by ANN with 94% accuracy. Rath et al. [6] presented the related work, using GAN-based LSTM model on MIT-BIH dataset of ECG signals. Authors employed GAN to get rid of the unbalanced dataset and hence improved the performance of the model. The experimental results show that the system achieved an accuracy of 99.2%, F1 score of 98.7% and an area under curve of 98.4%. Reference [7] suggested pre-trained deep learning models to classify heart sounds. They used CHSC dataset to validate the performance of the model using total, normalized and class-based precision as metrics for evaluation purposes.

Authors in Ref. [8] presented a system to detect cardiac disorders using SVM, random forest (RF), Naïve Bayes and k-nearest neighbor (KNN). They used a statistical test with Wilcoxon score to identify the prominent features and classified the signal as either normal or abnormal to predict patients with heart disease. In Ref. [9], Xiao et al. proposed a novel 1-D convolution neural network (CNN) architecture leveraging the channel and spatial features and achieved above 90% scores of accuracies and specificity. Authors in Ref. [10] carried out the multilabel classification of sound signals using a WaveNet model. They used a dataset of five classes to detect different categories of valvular disease and obtained an accuracy of 98.20% for one of the classes with 10-fold cross validation. Yuenyong et al. [11] proposed a principal component analysis (PCA) based classification of unsegmented heart sounds approach by using time-frequency features and applying PCA before the classification with the neural network algorithm with 92% of accuracy. Potes et al. [12], deployed Adaboost and CNN-based classification algorithm on time and frequency domain features and achieved a specificity of 77.81% and sensitivity of 94.24%.

Despite of all the advancements, no single method is believed to provide consistent result as it may vary due to size and type of data, algorithm used for pre-processing of the raw data, method used for feature selection, techniques used for learning and diversity of metric used for evaluation of end results. In this paper, we have evaluated two architectures of CNN to detect the type of heart disease using sound signals. Initially, features are extracted, and feature scaling is done. After this, the 1-D-CNN model is trained, and its performance is evaluated using different performance metrics.

The rest of the paper is organized as follows: The proposed methodology that includes feature extraction, feature scaling and classification, is detailed in Section II. In Section III, the obtained results are discussed and analyzed. In the last section IV, the paper is concluded, and the future direction is provided.

II. Proposed Methodology

In this section, the proposed methodology is detailed that includes the data description, feature extraction, feature scaling and

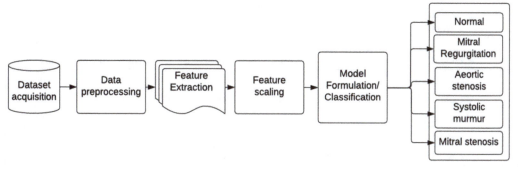

Fig. 5.1 Architecture of proposed system

classification. The architecture of our proposed model for the identification of heart abnormalities is shown in Fig. 5.1.

A. Dataset Description

This study uses the publicly available dataset by Yaseen et al. [13]. The detail of the dataset is illustrated in Table 5.1. The dataset contains five classes: Normal (N), Mitral Regurgitation (MR), Aortic Stenosis (AS), Murmur in Systole (MVP) and Mitral Stenosis (MS). A total of 1000 sounds that is 200 of each class are recorded and utilized in this study.

Table 5.1 Dataset description

Categories	N	MR	AS	MVP	MS
Number of samples	200	200	200	200	200

B. Features Extraction

Discrete wavelet transform (DWT) is a process in which wavelets are sampled discretely. It uses same technique as Fourier Transform. A DWT captures both frequency and location while FT captures frequency only. The signal is decomposed in several levels into two different subsections based upon high-pass and low-pass filters and the features associated with them are known as approximation and detailed features, respectively. Multisignal 1-D wavelet decomposition-based features are considered for this study. Two-level decomposition is

performed with the Daubechies 6 Wavelet (db6).

C. Features Scaling

Features scaling is one of the important parts of the implementation of machine learning algorithms. In this study, data standardization is performed to have zero mean and unit variance using the formula given below, where x represents the original features and y is the standardized one.

$$y(i) = \frac{x_i - \text{mean}(x)}{std} \qquad (1)$$

D. Classifications

In this paper, an efficient and robust 1-D CNN architecture is presented. 1-D Convolutional neural networks are like general convolutional neural networks that are mainly used on 1-D signals and texts. In this work, two different architectures are presented in Tables 5.2 and 5.3, respectively, and their performances are compared and analyzed. The purpose of different architectures is to analyze the impact of the results with the complexity of the architecture.

In the first architecture, that is, architecture 1, a total of 7 convolutional layers are deployed with sigmoid as activation function. Max-pooling layers are introduced after every two convolutional layers. Before the dense layers, global average pooling is performed. Tanh

Table 5.2 Architecture 1

Name	Layer Type	Input	Output	Activation Function
conv1d_19	Conv1D	(None,210,1)	(None, 206, 64)	Sigmoid
conv1d_20	Conv1D	(None, 206, 64)	(None, 204, 32)	Sigmoid
max_pooling1d_8	MaxPooling1D	(None, 204, 32)	(None, 68, 32)	-
conv1d_21	Conv1D	(None, 68, 32)	(None, 66, 32)	Sigmoid
conv1d_22	Conv1D	(None, 66, 32)	(None, 64, 32)	Sigmoid
max_pooling1d_9	MaxPooling1D	(None, 64, 32)	(None, 21, 32)	-
conv1d_23	Conv1D	(None, 21, 32)	(None, 19, 32)	Sigmoid
conv1d_24	Conv1D	(None, 19, 32)	(None, 17, 32)	Sigmoid
max_pooling1d_10	MaxPooling1D	(None, 17, 32)	(None, 5, 32)	-
conv1d_25	Conv1D	(None, 5, 32)	(None, 3, 32)	Sigmoid
global_average_pooling1d_3	GlobalAveragePooling1D	(None, 3, 32)	(None, 32)	-
dense_6	Dense	(None, 32)	(None, 100)	tanh
dense_7	Dense	(None, 100)	(None, 5)	Sigmoid

Source: Author

Table 5.3 Architecture 2

Name	Layer Type	Input	Output	Activation Function
conv1d_14	Conv1D	(None, 210,1)	(None, 206, 64)	Relu
conv1d_15	Conv1D	(None, 206, 64)	(None, 204, 32)	Relu
max_pooling1d_6	MaxPooling1D	(None, 204, 32)	(None, 68, 32)	-
conv1d_16	Conv1D	(None, 68, 32)	(None, 66, 16)	Relu
conv1d_17	Conv1D	(None, 66, 16)	(None, 64, 16)	Relu
max_pooling1d_7	MaxPooling1D	(None, 64, 16)	(None, 21, 16)	-
conv1d_18	Conv1D	(None, 21, 16)	(None, 19, 16)	Relu
global_average_pooling1d_2	GlobalAveragePooling1D	(None, 3, 16)	(None, 16)	–
dense_6	Dense	(None, 16)	(None, 100)	–
dense_7	Dense	(None, 100)	(None, 5)	Sigmoid

Source: Author

is used as an activation function with a dense layer prior to the output layer.

Similarly, in architecture 2, the same types of layers are used but with different parameters and network complexity. In both the cases, 64 neurons were selected in the input layer with the kernel size of 5 and unit step stride. In the rest of the layers, 32 neurons are used. The descriptions for both the architectures are illustrated in Tables 5.2 and 5.3. Moreover, the optimizer and loss functions were kept fixed for both the cases which were adam and categorical_crossentropy, respectively.

III. Results and Discussion

In this section, the setups for and the obtained results are discussed and an in-depth analysis is presented. In this study, DWT and 1-D-CNN-based approach is proposed. Two different architectures of CNN are selected to analyze the impact on the results due to their complexity.

Five different classes, that is, AS, MR, MVP, MS and Normal are considered in this study. Initially, Architecture 1, as illustrated in Table 5.2, is considered and the performance metrics are measured for all of the five classes. The overall achieved results for architecture 1 scores, in this case, are over 86% for each performance metric as illustrated in Table 5.4. Considering each class individually, the algorithm performed well in the case of AS detection and achieved 97.80% of accuracy, 97.61% of precision, 93.18% of sensitivity and 99.27% of specificity. In the case of normal, although, the architecture could achieve specificity of over 97% but at the cost of poor sensitivity of 67.74%, which is the lowest individual score of any performance metrics.

For architecture 2, the MS class identified with the highest individual scores of 99.45%, 100.0%, 97.36% and 100.0% for accuracy, precision, recall and specificity respectively. Overall, 7% to 9% of increase is observed

using architecture. In some cases, even more than that. For instance, the sensitivity for the normal is increased by over 30% from 67.7% to 98.6%. Similarly, in the case of MVP, over 20% increase is observed. It is also observed from the results of AS and MS that different architectures may work well for different classes.

The obtained results are also compared with the literature as indicated in Table 5.5, in which the same was utilized and classification algorithms were deployed. In this study, the results are compared in terms of accuracy and sensitivity for classification algorithm. It can be seen that the proposed model has outperformed the previous work by over 6% in term of accuracy and 2% in case of sensitivity.

IV. Conclusion

In this study DWT and 1-D CNN-based algorithm are proposed to identify abnormalities in phonocardiogram. Main focus of the study was to analyze the impact of different architectures on the results. Two different architectures are presented and their results are evaluated. It

Table 5.4 Performance of two architecture of CNN

Category	Architecture 1				Architecture 2			
	Accuracy (%)	Precision (%)	Recall/ Sensitivity (%)	Specificity (%)	Accuracy (%)	Precision (%)	Recall/ Sensitivity (%)	Specificity (%)
AS	97.80	97.61	93.18	99.27	98.35	95.0	97.43	98.60
MR	94.50	94.11	80.0	98.59	93.95	82.5	89.18	95.17
MVP	93.95	77.77	97.22	93.15	96.70	97.05	86.84	99.30
Normal	91.80	80.76	67.74	96.71	98.32	93.54	96.66	98.35
MS	95.60	82.85	93.54	96.02	99.45	100.0	97.36	100.0
Overall	86.81	87.54	86.81	-	93.41	93.69	93.41	-

Source: Author

Table 5.5 Comparison with literature

Reference	Dataset	Features	Classifier	Accuracy	Sensitivity	Specificity
Yaseen et al. [13]	[13]	DWT	DNN	87.8%	91.6%	97.4%
Proposed	[13]	DWT	CNN	**93.41%**	**93.41%**	-

Source: Author

is observed that suitability of architecture is very important not the complexity. As in this study case, architecture 1 was much larger and complicated compared to the architecture 2. But the results for architecture 2 were 7% to 9% improved and even up to 30% in some cases.

In the future, the authors would like to dive much deeper into the deep neural network analysis to find the relationship between different hyperparameters of neural networks with the performance in identification of different abnormalities.

References

1. H. Thomas et al., "Global Atlas of Cardiovascular Disease 2000-2016: The Path to Prevention and Control," *Global Heart*, vol. 13, no. 3, p. 143, Sep. 2018, doi: 10.1016/j.gheart.2018.09.511.
2. D. Zhao, "Epidemiological Features of Cardiovascular Disease in Asia," *JACC: Asia*, vol. 1, no. 1, pp. 1–13, Jun. 2021, doi: 10.1016/j.jacasi.2021.04.007.
3. J. L. Garvey, "ECG Techniques and Technologies," *Emergency Medicine Clinics of North America*, vol. 24, no. 1, pp. 209–225, Feb. 2006, doi: 10.1016/j.emc.2005.08.013.
4. A. K. Dwivedi, S. A. Imtiaz, and E. Rodriguez-Villegas, "Algorithms for Automatic Analysis and Classification of Heart Sounds–A Systematic Review," *IEEE Access*, vol. 7, pp. 8316–8345, 2019, doi: 10.1109/ACCESS.2018.2889437.
5. S. Celin and K. Vasanth, "ECG Signal Classification Using Various Machine Learning Techniques," *Journal of Medical Systems*, vol. 42, no. 12, p. 241, Dec. 2018, doi: 10.1007/s10916-018-1083-6.
6. A. Rath, D. Mishra, G. Panda, and S. C. Satapathy, "Heart disease detection using deep learning methods from imbalanced ECG samples," *Biomedical Signal Processing and Control*, vol. 68, p. 102820, Jul. 2021, doi: 10.1016/j.bspc.2021.102820.
7. F. Demir, A. Şengür, V. Bajaj, and K. Polat, "Towards the classification of heart sounds based on convolutional deep neural network," *Health Information Science and Systems*, vol. 7, no. 1, p. 16, Dec. 2019, doi: 10.1007/s13755-019-0078-0.
8. A. Yadav, A. Singh, M. K. Dutta, and C. M. Travieso, "Machine learning-based classification of cardiac diseases from PCG recorded heart sounds," *Neural Computing and Applications*, vol. 32, no. 24, pp. 17843–17856, Dec. 2020, doi: 10.1007/s00521-019-04547-5.
9. B. Xiao, Y. Xu, X. Bi, J. Zhang, and X. Ma, "Heart sounds classification using a novel 1-D convolutional neural network with extremely low parameter consumption," *Neurocomputing*, vol. 392, pp. 153–159, Jun. 2020, doi: 10.1016/j.neucom.2018.09.101.
10. S. L. Oh *et al.*, "Classification of heart sound signals using a novel deep WaveNet model," *Computer Methods and Programs in Biomedicine*, vol. 196, p. 105604, Nov. 2020, doi: 10.1016/j.cmpb.2020.105604.
11. S. Yuenyong, A. Nishihara, W. Kongprawechnon, and K. Tungpimolrut, "A framework for automatic heart sound analysis without segmentation," *BioMedical Engineering OnLine*, vol. 10, no. 1, p. 13, 2011, doi: 10.1186/1475-925X-10-13.
12. C. Potes, S. Parvaneh, A. Rahman, and B. Conroy, "Ensemble of Feature:based and Deep learning:based Classifiers for Detection of Abnormal Heart Sounds," Sep. 2016. doi: 10.22489/CinC.2016.182–399.
13. Yaseen, G.-Y. Son, and S. Kwon, "Classification of Heart Sound Signal Using Multiple Features," *Applied Sciences*, vol. 8, no. 12, p. 2344, Nov. 2018, doi: 10.3390/app8122344.

Recent Trends in Computational Intelligence and its Application – Sugumaran D. et al. (eds)
© 2023 Taylor & Francis Group, London, ISBN 978-1-032-48410-5

6

Optimal Allocation of Resources in Data Center using Artificial Intelligence

D. Kalpanadevi*, P. Babysudha[1]
Department of Computer Applications, Kalasalingam Academy of Research and Education, Krishnankoil, Tamil Nadu, India

K. Kartheeban[2]
Department of Computer Science and Engineering, Kalasalingam Academy of Research and Education, Krishnankoil, Tamil Nadu, India

M. Mayilvaganan[3]
Department of Computer Science, PSG College of Arts and Science, Coimbatore

Abstract—This research can focus on virtualization technology supporting green computing in virtualization to allocate data center resources during the application process. By developing heuristics to avoid system overload for attained system efficiency. Experiment results and Trace-driven simulation demonstrate good performance. This proposed research work has been enhanced by the name as hybridized of multi-objective optimization dynamic resources implementation. The simulation findings demonstrate that obtained how to deploy a virtual machine by scheduling has a longer life span than the scheduling approach for energy savings and multi-virtual machine redistribution overhead. Without keeping the server hardware or any adjustments in configuration, an infrastructure can deliver the model to provide industry users with remote access to server resources aimed by resource scheduling within side the contemporary cloud computing surroundings can be attained in resource allocation for energy-saving systems. In this research work, the space-shared policy and time-shared policy can be integrated with multi-objective optimization dynamic resources allocation and an experimental result show that scheduling can converge faster and produce comparable multi-objective optimization outcomes at the same calculation size.

Keywords—Allocate data center resources, Cloud the space-shared policy and time-shared policy, Dynamic resources allocation computing, Hybridized of multi-objective optimization, Multi-objective optimization, VM allocation optimization, Artificial Intelligence

*Corresponding author: dkalpanadevi@klu.ac.in
[1]p.sudha0100@gmail.com, [2]k.kartheeban@klu.ac.in, [3]mayil24_02@yahoo.co.in

DOI: 10.1201/9781003388913-6

I. Introduction

Cloud computing is a discipline of resource sharing, distributed and interconnected over the Internet. Although it has been adopted, its improvement is far from accurate manner. Technical specifications be required no longer but have been formed, and are nevertheless withinside the length of growth. It can able to hold an area withinside the destiny cloud computing discipline [2]. In order to commercial programs, many great properties of cloud computing have made it a crucial issue of army education, meteorology, astronomy, and different fields.

In cloud infrastructure, resource allocation can be processed for the task of sharing the data effectively. The intention of the modern era is to support customers with offers that fall under a pay-per-use charge approach [3]. Maintaining in thoughts the stop purpose to inspire an ultimate goal of stimulating the strategic progress and use of cloud computing.

Cloud computing resource allocation includes the timing schedule and useful resource provision at the same time as maintaining in view the to-be-had infrastructure, provider-level consensus, cost, and power factors [1] [4]. For example, cloud providers manage their sources in step with the on-call for pricing approach at the same time as making sure to satisfy the customer and Quality of Service. Similarly, without exceeding the limit of the cloud environment, the sources should be assigned so that each utility receives the sources.

II. Resource Allocation in Cloud Computing

Before dividing the Virtual Machine (VM) useful resource allocation, checking to satisfy the condition over the resources among the different cloud users depends on the cross over the overall definition and cause of useful resource allocation in industrial administration. Virtual allocation [6] is the act of provisioning

digital device processors onto bodily hosts or cores, inclusive of your CPU, to growth networking centers and opportunities.

Figure 6.1 represents the allocation of cloud computing by VM resource, it can allocate, Virtual gadget allocation trouble is one of the demanding situations in cloud computing environments, in particular for the non-public cloud design. The user is the Demand Request and end result is the cloud brokering, cloud broker can manage the use of resources, overall performance, and shipping of cloud offerings, and agree on the relationships between cloud vendors and the clients.

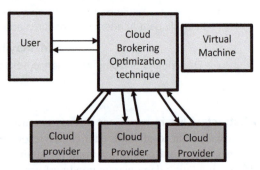

Fig. 6.1 Virtual machine allocation in cloud environment

Source: Author

By appearing of cloud computing platform virtualization generation carries new possibilities for recognizing power saving withinside of an environment. This generation permits a couple of customers to percentage the computing assets pool consistent with the real needs. It has significantly progressed as an asset for systematic approaches. Thereby growing the earnings of the cloud provider as shown in Fig. 6.1, useful resource allocation issues, in the long run, boil right all the way down to a couple of utility mapping family members among the digital system, via affordable allocation, adjusting the distribution of utility digital machines at the bodily nodes, making complete use of the provider circumstance of the server idle assets, in order to lessen the quantity of activation server, and thereby gain the motive of decreasing power consumption. Allocation of the sources is the system wherein the right sources are allotted to the responsibilities required with the aid of using the client in order that those responsibilities are completed proficiently [7] [8]. In cloud computing, designating a digital device enjoyable home is described with an aid of using the consumers. Users ought to put up their challenge which might also additionally have its very own time imperatives.

III. Related Work

With the advancement of cloud computing, humans discover that despite the fact that cloud computing intake of strength is likewise massive in comparison with the conventional carrier mode. In cloud computing, the finest Virtual Machine can follow a distribution scheme that can efficiently save strength through nice use in resource deploying VM assets on bodily nodes [15,16]. Therefore, which approach is followed to recognize the location of digital machines to reap the cause of strength saving has always been a warm study subject matter withinside the discipline of cloud computing [9, 10].

Network-aware allocation of virtual devices in a reflection photo of multiple useful resource requirements within an information center to limit the power consumption of the information center [14]. Most techniques are primarily based on the useful resource allocation awareness of network awareness regarding reassets and verbal exchanges at the fact center.

An all-new network-aware virtual gadget deployment method aimed at reducing access to the average network site through the most beneficial deployment of virtual gadgets [11] [12]. Allocate virtual machines alongside large verbal interactions to reduce costs and connect website visitors. Similarly, network-aware allocation of virtual devices to reflective photos of multiple useful resources within a fact center requires limiting the power consumption of the fact center [13].

IV. Problem Definition

Today, many energy-saving resource planning techniques for discovering cloud platforms are on the road to reducing an active server, saving energy, and leveraging inexperienced computers. These useful scheduling schemes described above ignore the dynamics of the payload as a result of consumer demand in a cloud environment and do not take into account the stability of payload distribution at each physical node. Distributing utility VMs across physical nodes in modern countries reduces the number of active physical nodes, but by distributing consumer needs, you can make adjustments within a load of each utility. In this case, the dynamics of the specified load will cause hotspots of body node resources that are useful in the fate of payload extrusion. The reputation of those bodily nodes may reappear withinside the close to destiny, to be able to at once cause a brand new spherical of dynamic useful resource configuration necessities. Based on the resource scheduling method, can solve an issue and place ahead of bodily nodes on the stable.

V. Proposed Methodology

Virtualization Machine is a commonly used technique for making dynamic and resource allocation in virtualized environments. Therefore, the technique of walking on a set of physical hardware or using a larger logical computer system.

In this research work, there are three physical resources Desktop1, Desktop2, and Desktop3 are taken for computing sources are pooled to sever more than one purchaser the use of a multi-tenant model, with one kind of bodily and digital sources dynamically assigned and reassigned in line with patron demand.

The infrastructure layer is 3 kinds IaaS, software program layer, Platform layer. IaaS is the first actual and simple layer of cloud computing Infrastructure as a carrier (IaaS). The software

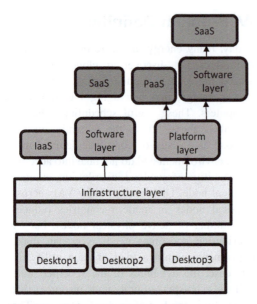

Fig. 6.2 Framework for proposed methodology

Source: Author

layer is the SaaS the second layer of the cloud is the platform – the PaaS (Platform as a carrier). This layer is an improvement and deployment surroundings withinside the cloud and offers the sources to in reality construct applications. The platform layer is the PaaS and the software program layer and tender layer are the SaaS.

A. Scheduling

Scheduling set of guidelines is the approach used to determine which of several processes, each of that world as it should have a beneficial useful resource allocated to it, will honestly be granted use of the sources.

B. Scheduling Policies

There are policies that are probably defined in this segment the ones are Space-Shared scheduling insurance and Time-Shared scheduling insurance. In Space-shared scheduling insurance its time desks one assignment on a virtual machine at a given instance of a time and after its final touch it time desks a few different assignments on a virtual machine. This identical insurance is used to time desk the virtual machines on the host. This insurance behaves identically due to the fact the primary comes first to serve a set of guidelines (FCFS) [6]

The requested property is allocated dedicatedly to the inquiring for workload for execution and can be released only on completion. Space-shared is also referred to as batch approach scheduling.

Time-sharing cloudlet scheduling allows running multiple cloudlets in parallel from a single VM. When a cloudlet is assigned to a VM for execution, space-shared cloudlet scheduling can complete a particular cloudlet with the resource that uses the virtual machine until the latest cloudlet will not be dispatched with the resource that uses the VM.

VI. Implementation: Hybridized Method of Multi-objective Optimization Dynamic Resources Allocation

The hassle turned into solved the use of a multi-goal optimization scheduling technique. The simulation findings display that the digital gadget distribution technique received via way of means of scheduling has an extended balance time than the scheduling technique for power financial savings and multi-digital gadget redistribution overhead. This study gives a multi-goal optimization dynamic useful resource allocation method primarily based totally on area shared and time shared for digital gadget distribution to deal with this hassle. The experimental simulation indicates that scheduling can converge quicker and bring similar multi goal optimization consequences on the identical calculation size.

Cloud platforms require a large number of physical services to be provided in order to generate resource services. These VMs typically run with excellent performance, and a set of policies guarantees the fastest and most convenient resource allocation for new requests. If the physical server pj can be allocated in VM vi, the physical server must satisfy $xij = 1$ and the mapping element xij between that virtual machine. Otherwise, xij is set to 0. The mapping matrix from VM request queue V to physical server organization P can be represented in equation 2. Therefore, the number of physical servers used can be calculated by summing all the elements of the matrix y into Equation 1 as follows: $\sum_{i=1}^{m} yij$. Therefore, in hybridized multi-objective optimization, dynamic resource allocation based on minimum resource physical server usage can be established as follows:

$$min\left\{\sum_{i=0}^{m}\sum n_{j=1}MDij\right\}$$

$$min\left\{\sum_{i=0}^{m}\sum n_{j=1}MPij\right\} \qquad (1)$$

The constraint must satisfy the relationship in Equation 2 can represent the resource requirements of all Virtual Machines located on the physical server pj cannot exceed their free resources as follows:

$$\sum_{i=0}^{m} vic.aij \le pidf \ (1, 2 ..., n)$$

$$\sum_{i=1}^{m} vim.aij \le pidf \ (1, 2 ..., n)$$

$$\sum_{i=1}^{m} vid.aij \le pidf \ (1, 2 ..., n) \qquad (2)$$

And if one loose useful resource capability of a bodily server can't fulfill the useful resource call for of any Virtual Machine, the ratio of a different kind of loose useful capability of an entire useful resource capability in multi-objective.

In this proposed research work, the space-shared policy and time-shared policy can be integrated with multi-objective optimization

dynamic resources allocation has been enhanced and named a hybridized multi-objective optimization dynamic resources implementation exactly known the task scheduling problem in cloud environments with positive implications.

A. Integration of Space Shared Policy

Step 1: Accepted obligations are organized in a queue.

Step 2: The first challenge is the timetable for the given digital gadget.

Step 3: It completes the first challenge after which takes the subsequent challenge from the queue.

Step 4: If the queue is empty, it's assessments for a brand new challenge.

Step 5: Then repeats step 1.

Step 6: End

B. Integration of Time-shared Policy

In Time-Shared scheduling coverage its timetable of all responsibilities on the digital devices at the identical time. It shared the time amongst all responsibilities and timetable concurrently on the digital device. This coverage is likewise used to timetable the digital device at the host. The idea of a round-robin (RR) scheduling algorithm [6] is used in this coverage.

Step 1: All general assignment is organized below the queue.

Step 2: Then time table the assignment concurrently at the digital device.

Step 3: When queue is empty it exams for brand new assignment.

Step 4: If new assignment arrives it time table further as within side the step 2.

Step 5: End.

Cloud Sim implements time-shared and space-shared scheduling policies. The distinction among those policies and their impact on the utility performance display an easy scheduling technique.

VII. Result and Discussion

Figure 6.3 shows the server access Disk capacity 100 GB Monthly transfer total 50 GB for Total account 25 Server named as Access1. Welcome server 1 disk space allocates 100 GB

Fig. 6.3   Server access page

Source: Compiled by Author

Fig. 6.4   Server access page

Source: Compiled by Author

and Monthly transfer 50 GB to Server Access1 and shared Total account 25. Welcome server 1 and Welcome server 2 disk capacity 100 GB monthly transfer 50 GB to server name access 1 and welcome server 1 and shared total account 25.

Figure 6.4 shows the server access Disk capacity of 10000 KB, Monthly transfer total of 384 KB for a Total account of 25 Servers named Access1. WelcomeServer1 and Welcome server 2, disk space allocates 10000 KB and Monthly transfer 384 KB to Server Access1 and shared Total account 25.

Figure 6.5 shows the cloud services offered to cloud customers and provided to cloud customers based on the relationship between cloud customers and cloud consumers. When the user needs a username and password to access something, you are signing in page.

Figure 6.6 and Fig. 6.7 shows, the User web page of the server velocity to Deploy documents to the cloud server document id, document name, and upload server velocity: 500kbps purchaser name, used space, area space, loose space.

The Server can edit the Username, password, server, server size, domain, domain size, date form, and planning on the server detail edit page. The provider can store and access the data in a location separate from the end user, also allows the server through the device connected with the network for accessing in online and acts as a centralized server resource.

A domain name, domain size, used space, free size, total-provider storage, availability provider storage, and used provider storage on the server can be maintained. The server document list for the welcome server. File ID, file name, file information, download, and delete on the server.

The following Table 6.1, represents the load process server at the time of resource allocation when assigning the hosted site to the server. The host site of jawbone can be serviced by the level of agreement for 180 days and the usage of resources is 3 KB from their allocation. Similarly, hosts of cool service and eBay be serviced by the level of agreement of 90 days and the usage of resources is 20 KB from their allocation.

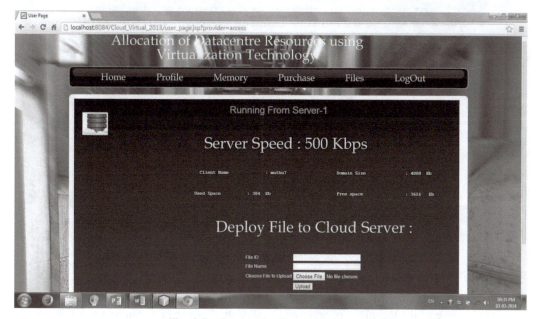

Fig. 6.5 Deploy the file to cloud server

Source: Compiled by Author

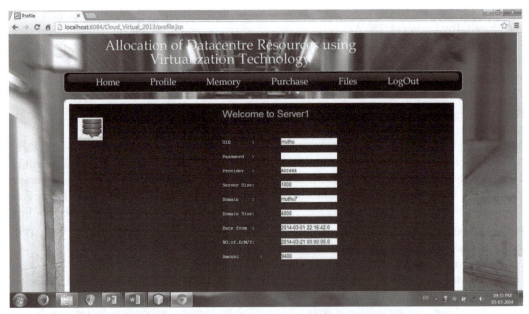

Fig. 6.6 Server details edit, view and document list

Source: Compiled by Author

Fig. 6.7 Server details edit, view and document list

Source: Compiled by Author

Figure 6.8 represents Cloud load balancing is described because of the technique of splitting workloads and computing homes in cloud computing. For data sharing, it can be taken 500 GB assigned server Usage, the resource allocation in server A can assign 8 GB, server B can assign 10 GB and server C can assign 15 GB for resource allocation.

Table 6.1 Resources allocation details

Host Site Names	Date	Service Level Agreement	Used	Allocated	Server Name
WWW.jabong.com	2022-06-25	180 days	3KB	20KB	Server B
WWW.cool.net	2021-08-26	90 days	2KB	20KB	Processing
WWW.ebay.com	2022-02-08	90 days	2KB	20KB	Server B

Source: Compiled by Author

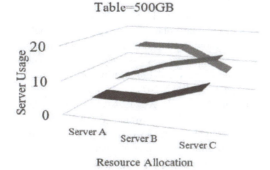

Fig. 6.8 Resource allocation for 500 GB to several servers

Source: Compiled by Author

VIII. Conclusion

In this research work, it can be concluded that the space-shared policy and time-shared policy can be integrated with multi-objective optimization dynamic resources allocation, and the set of rules achieves each overload avoidance and computation for multi-aid constraint structures. Aiming at the deficiency of some contemporary resources scheduling energy-saving resources, and scheduling withinside the contemporary cloud computing surroundings may be attained resource allocation, without keeping the server hardware or any adjustments in configuration, an infrastructure can deliver the model to provide industry users with remote access to server resources in order to power applications and store data.

In the future, the use of desktop, data, and network virtualization technologies is expected to grow by double digits within the next two years.

References

1. Zhen Xiao, Senior Member, IEEE, Weijia Song,and Qi Chen-"Dynamic Resource Allocation Using Virtual Machines for cloud computing environment" – IEEE Transactions on Parallel And Distributed Systems. Vol. 24, No. 6, June 2013.
2. I. Odun-Ayo, R. Goddy-Worlu, J. Yahaya, and V. Geteloma, "A systematic mapping study of cloud policy languages and programming models," Journal of King Saud University-Computer and Information Sciences, vol. 33, no. 7, pp. 761–768, 2021.
3. H. A. Alobaidy, M. J. Singh, M. Behjati, R. Nordin, and N. F. Abdullah, "Wireless Transmissions, Propagation and Channel Modelling for IoT Technologies: Applications and Challenges," *IEEE* Access, vol. 10, pp. 24095–24131, 2022.
4. H. Ma and J. Wang, "Application of artificial intelligence in intelligent decision-making of human resource allocation," in International Conference on Machine Learning and Big Data Analytics for IoT Security and Privacy, 2020: Springer, pp. 201–207
5. Y. Ning, X. Chen, Z. Wang, and X. Li, "An uncertain multi-objective programming model for machine scheduling problem," International Journal of Machine Learning and Cybernetics, vol. 8, no. 5, pp. 1493–1500, 2017.
6. Thein, T., Myo, M.M., Parvin, S. and Gawanmeh, A., 2020. Reinforcement learning based methodology for energy efficient resource allocation in cloud data centers. Journal of KingSaud University-Computer and Information Sciences, 32(10), pp. 1127–1139
7. T. Sandholm and K. Lai, "Mapreduce Optimization Using Regulated Dynamic Prioritization," Proc. Int'l Joint Conf.

Measurement and Modeling of Computer Systems (SIGMETRICS '09), 2009.

8. Singh, M. Korupolu, and D. Mohapatra, "Server-Storage Virtualization: Integration and Load Balancing in Data Centers," Proc. ACM/IEEE Conf. Supercomputing, 2008.

9. Y. Toyoda, "A Simplified Algorithm for Obtaining Approximate Solutions to Zero-One Programming Problems," Management Science, vol. 21, pp. 1417–1427, Aug. 1975.

10. D. Meisner, B.T. Gold, and T.F. Wenisch, "Powernap: Eliminating Server Idle Power," Proc. Int'l Conf. Architectural Support for Programming Languages and Operating Systems (ASPLOS '09), 2009.

11. T. Das, P. Padala, V.N. Padmanabhan, R. Ramjee, and K.G. Shin, "Litegreen: Saving Energy in Networked Desktops Using Virtualization," Proc. USENIX Ann. Technical Conf., 2010.

12. Y. Agarwal, S. Savage, and R. Gupta, "Sleepserver: A Software Only Approach for Reducing the Energy Consumption of PCS within Enterprise Environments," Proc. USENIX Ann. Technical Conf., 2010.

13. N. Bila, E. d. Lara, K. Joshi, H. A. Lagar-Cavilla, M. Hiltunen, and M. Satyanarayanan, "Jettison: Efficient Idle Desktop Consolidation with Partial VM Migration," Proc. ACM European Conf. Computer Systems (EuroSys '12), 2012.

14. K. Karthiban and J. S. Raj, "An efficient green computing fair resource allocation in cloud computing using modified deep reinforcement learning algorithm," Soft Computing, pp. 1–10, 2020.

15. S. Shamshirband et al., "Game theory and evolutionary optimization approaches applied to resource allocation problems in computing environments: A survey," Mathematical Biosciences and Engineering, vol. 18, no. 6, pp. 9190–9232, 2021.

16. Liu, N., Li, Z., Xu, J., Xu, Z., Lin, S., Qiu, Q., Tang, J. and Wang, Y., 2017, June. A hierarchical framework of cloud resource allocation and power management using deep reinforcement learning. In 2017 IEEE 37th international conference on distributed computing systems (ICDCS) (pp. 372–382) IEEE.

Recent Trends in Computational Intelligence and its Application – Sugumaran D. et al. (eds)
© 2023 Taylor & Francis Group, London, ISBN 978-1-032-48410-5

7

Flower Leaf Image Classification using Deep Learning Techniques

Bittu Kumar Aman[1] and Vipin Kumar[2]

Dept. Computer Science & Information Technology,
Mahatma Gandhi Central University, Bihar, India

Abstract—Statista reports that there are over fifty thousand flower species worldwide; the challenge here is to identify each variety so that we may know the flower plants' actual and absolute natural goodness. To recall, expertise or prior knowledge is required. Without it identifying the flower plants is challenging. Therefore, it is crucial to make the effect and automated systems to classify the different flowers using their leaf images. In this research, we collected twenty-five different categories of flower plant leaves images; each class contains more than 250 images, totaling 6619 RGB images. This research applies eight classical deep learning models like ResNet, AlexNet, VGG, SqueezeNet, GoogLeNet, ShuffleNet, ResNext50, and DenseNet to a given dataset. The comparative study of the model performances has been done based on model accuracy, train accuracy, train loss, validation accuracy, and validation loss. This research aims to find a practical deep-learning classification approach that can be utilized for automation. The analysis of the results shows that the ResNext50 has the highest classification accuracy, i.e., 99.6243%. The comparative performance analysis is based on the accuracy of deep learning models and similar Machine Learning.

Keywords—Classification, Flower plant leaf, Deep learning, Machine learning, RGB image, Computer vision, Augmentation

I. Introduction

More than 18,000 flower plants exist in India, where 6-7% of species are shared by India, i.e., more than 50,000 species of plants. Flowers do much more than look good in our homes or increase the beauty of our humble aboard. Medical practitioners have recognised the medicinal properties of flowers and plants for millennia Das and Islam (2019), serving the medical needs of Alrefaee et al. (2021), Kumar et al. (2022) Kumar and Kumar (2022b). For instance, it can assist in relieving anxiety and headaches and refreshing the mood Pang et al. (2021). The problem first arises with identifying each category of plants Dhingra et al. (2018) because, in several flower plants, flowers come in only in a specific season or for a minimal

[1]b2kraman@gmail.com, [2]rtvipink@gmail.com

DOI: 10.1201/9781003388913-7

period. The leaf of the flowering plant contains much information about the plant. Which is usually available in the plant, but people must depend on those who have prior knowledge or expertise about the flowers or must be aware of the features or natural goodness of the flowers and flower plants Sharif et al. (2021) Aman and Kumar (2022). This research work has prepared the dataset of 25 different flower plant leaves in the form of RGB. More than 250 images of each category have been collected for the dataset of this research work. The research aims to identify a Machine Learning and Deep Learning approach for flower plant leaf classification based on 25 categories and analyze the performances of each model based on accuracy. The different deep learning models Wang et al. (2021), and Sahana et al. (2022) have been deployed like ResNet, AlexNet, VGG, SqueezeNet, GoogLeNet, ShuffleNet, ResNext50, and DenseNet. The analysis of the results shows that the ResNext50 has the highest classification accuracy, i.e., 99.6243 %.

The Novelty of the proposed research work:

- *To the best of my knowledge, this is the first multiple flower plant leaves image dataset that has been created for 25 categories of flowers with a minimum of 250 of each category where a total of 6619 RGB images.*
- *A comparative study of the classification performance of the models has been done where the best performer has been identified successfully with model accuracy, train accuracy, train loss, validation accuracy, and validation loss.*
- *The reason for misclassification has been identified successfully*
- *Comparative analysis of deep learning model performance based on accuracy using augmentation techniques.*

This paper is organized as follows. Section 2 deals with the literature review. In section 3 Proposed methodology is introduced. Section 4 explains the dataset description and experimental designs in section 5, results, and analysis of the proposed work using various visualization methods. And the last section contains the conclusion and future research direction.

II. Literature Review

In the paper by Atabay (2016), the authors discuss the significance of planting and agriculture; the bulk of the population depends on agriculture since agricultural raw materials are utilized to feed many people. Plants are lost to various causes, including natural disasters, pests, and infections; rural or less educated farmers are disproportionately affected by agricultural production issues. The leaf is crucial in learning about the plant's development and production; the suggested task is pre-processing the dataset and disease categorization and detection using the Convolution neural network (CNN). Another research author Vilasini (2020) mentioned the development of a unique Convolution neural network architecture for leaf identification. The framework is built from grayscale photographs from the Flavia and Swedish leaf collections, totaling 160 by 160 pixels. The grayscale version of the bright leaf photographs was picked from the green channel. The parallel dataset augmentation effect was demonstrated when a better model would not need the augmentation procedure. In addition, the study by Aochi et al. (2013) examined CNN-based algorithms for recognizing Indian leaf species from white backgrounds using mobile phones. CNN model variations of over-familiar shape, texture, venation, and color, as well as extra microscopic elements of uniformity of edge patterns, margin, leaf tip, and many more statistical factors, are examined for successful leaf categorization. In another research using CNN, the author Yalcin and Razavi (2016) proposes a Convolutional Neural Network (CNN) architecture for classifying plant types from picture sequences acquired from smart agro-stations.

In the paper by Jasim and Al-Tuwaijari (2020), the author discussed the diseases of specific types of plants; tomatoes, paper, and potatoes; he proposed a system using a convolutional neural network (CNN) that uses a plant village dataset. There were 15 classes, including 12 courses for illnesses of various plants found, such as bacteria, fungus, and so on, and three classes for healthy leaves.

The authors give a thorough study on disease diagnosis, and classification of plant leaves using image processing techniques in Ni and Wang (2018), Iqbal et al. (2018), and Prakash et al. (2017). In the paper, Kumar and Kumar (2022a) author created 25 categories of vegetable leaf image datasets and apply several ML models to classify the vegetable leaf plant.

In the paper by Lee et al. (2017), the author talks about how Deep learning is used to extract and learn leaf attributes for plant categorization. Convolutional Neural Networks (CNN) are used to acquire meaningful leaf features from raw input data representations directly. In contrast, Deconvolutional Networks (DN) are used to develop an understanding of the chosen features. The author Liu et al. (2018) proposed a novel approach based on Convolutional Neural Networks (CNN) in the research paper. For plant leaf categorization, a ten-layer CNN was built, and to enhance the categorization, a sample augment for leaf was applied to the photos to expand the database.

III. Methodology

In this research work, the dataset has been prepared of 25 categories of flower plant leaves using a plain white background, each type containing more than 250 RGB images, then preprocess the leaf images and deployed deep learning model and the comparative study to find the best classification of the model. Therefore, a stepwise description of the methodology has been mentioned in subsections, and the Architecture diagram is shown in the Fig. 7.1.

- **Step-1 (Preparation of Dataset):** Let, $i = \{I_i, y_i\}_{i=1}^{m}$ is a set of captured RGB images, where I_i represents the ith image in the dataset with their corresponding label $y_i \in Y$ i.e., $Y = \{y_i\}_{(i=1)}^{m}$ for $i \in \{1,2,3,....,m\}$ and m is the number of samples in the dataset. Each image $I_{(i)} \in I$ can be represented as 2D matrix $M \in R^{(H \times W)}$, where $M_{(i,j)}$ is the value at i utilized w and j^{th} column of the matrix M. The tensor for deep learning is represented as $T \in R^{H \times W \times C \times b}$ where H, W, C, and b are the height, width, color channel, and batch size respectively. The required resized dataset $I^r = \{I_i^r, y_i\}_{i=1}^{m}$ may be obtained with resized function $Re()$ shown in eq.1:

$$I^r = Re\{I_i^r, y_i\}_{i=1}^{m}, p \times q) \qquad (1)$$

where $p \times q$ is the required resized dimension of each image $I_i \in I$.

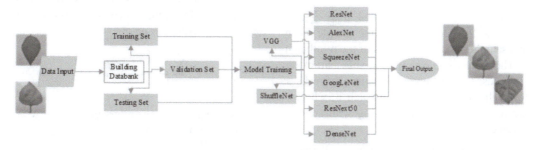

Fig. 7.1 Architecture diagram of flower leaf image classification

Source: Author

- **Step-02 (Dataset sampling and splitting):** Each experiment iteration must rearrange the whole dataset before partitioning it into the train, validation, and test datasets. let dataset I^r has split into training, validation, and test dataset i.e., represented as $T_r \subset I^r$, $V_a \subset I^r$, where $I^r = T_r \cup V_a \cup T_e$ and $\{T_r \cap V_a\} \cup \{T_r \cap T_e\} \cup \{V_a \cap T_e\} = \phi$. Now, the CNN input can be represented by a string of four i.e., tensor T has rank-4 with four axes. The length of the axes has a value at the index in the tensor which is a logical or real-world feature of the input data. The height (H), width (W), and color channel (C) of the RGB input image may be represented in tensor T corresponding to three arises i.e., (H,W,C). The batch size is the first axis in the tensor i.e., the number of images in a single batch. So, an ith image of the dataset $I^r_i \in I^r$ can be represented in tensor T s $I^r_i = [H, w, c, i]$ where $i \in \{1,2,3,.....,b\}$ and b is the batch size as shown in eq. 2

$$[X_T, Y_T] \qquad (2)$$

- **Step-03 (Model Training and selection):** Batch normalization is essential during deep network training in each iteration to achieve efficient convergence of complicated deeper networks without overfitting. The batch normalization (BN) of an input $x \in \text{ß}(\text{ß} = minibatch)$ is transformed as eq. 3

$$BN(x) = \gamma \odot \frac{x - \mu_\text{ß}}{\sigma_\text{ß}} + \text{ß} \qquad (3)$$

where $\mu_\text{ß} = \frac{1}{|\text{ß}|}\sum x \in \text{ß}^{(x)}$ and $\sigma^2_\text{ß} \frac{1}{|\text{ß}|}\sum x \in \text{ß}^{(x-\mu_\text{ß})^2} + \in$, where $\in > 0$ are the sample mean and sample standard deviation of the minibatch ß. The scale parameter (γ) and shift parameter (β) is used to maintain the same shape as x for each element in the minibatch ß, where γ are required to be learned along with another parameter while training.

Batch normalization differs between convolutional and fully linked layers. The batch normalizing function is used before the non-linear activation function, then the fully connected layer with batch normalization for input x can be written for output h as shown in eq. 4

$$h = \psi(BN(wx + b)) \qquad (4)$$

where ($wx + b$) is affine transformation with weight w and bias b parameters, and ψ is the activation function. Similarly, batch normalization is applied in between the convolution and non-linear activation functions. The deep network was made up of several levels. Let's, T^l tensor representation for the input at l^{th} layer of the network is $x^l \in R^{H^l \times w^l \times C^l \times b}$. The output of the ith layer will be the input of the next layer in the network which can be represented as $x^{l+1} \in R^{H^{l+1} \times W^{l+1} \times C^{l+1} \times b}$. The description of utilized layers is as given below:

- **Rectified Linear Unit (ReLu) Layer:** This layer has no learnable parameters and so enhances the nonlinearity of the convolution neural network Nair and Hinton (2010). It performs truncation for individual input as $x^{l+1}_{i,j,k,b} = \max\{0, x^l_{i,j,k,b}\}$ where $H^{l+1} = 0 \le i \le H^l, W^{l+1} = 0 \le i \le W^l$, $C^{l+1} = 0 \le i \le C^l$, and $B^{l+1} = 0 \le b \le B^l$

- **Convolutional Layer:** Jmour et al. (2018) The kernel was necessary to alter the input during the convolution procedure. At the I_{th} layer, the kernel size for four orders of the tensor is written as $H \times W \times C^l \times C$. So, the convolution size for the input size $H^l \times W^l \times C^l \times b$ will be $(H^l - H + 1) \times (W^l - W + 1) \times C$. Then, the convolutional procedure can be written as eq. 5

$$y_{(i^{l+1}, j^{l+1}, k, b)} = \sum_{i=0}^{H} \sum_{j=0}^{W} \sum_{k^l=0}^{C^l} w_{i,j,k^l,b} \times x_{(i^{l+1}, j^{l+1}+j, k^l, b)} \qquad (5)$$

where $x_{(i^l+1, j^l+1+j, k^l, b)}$ refers the element x^l at l^{th} layer. Equation 5 is iterated $\forall 0 \le c \le C = C^{(l+1)}$.

- **Backward propagation:** Two sets of gradients are necessary for error backpropagation, which is $\dfrac{\partial_z}{\partial_{vec(x_i^l)}}$ for input x_i^l at l^{th} layer and partial derivatives of z under the control of parameters w^l i.e., $\dfrac{\partial_z}{\partial_{vec(x^l)}}$.

As a result, the lth layer's parameter update rule may be expressed using convolution parameters, i.e., $\phi(X^l)^T$ and supervision signal transferred from the $(l+1)^{th}$ layer i.e., $\dfrac{\partial_z}{\partial(x^{l+1})}$ is shown in Eq. 6

$$\frac{\partial_z}{\partial_{vec}(w^l)} = \phi(x^l)^T \times \frac{\partial_z}{\partial_{x^{l+1}}} \qquad (6)$$

- **Pooling Layer:** The Pooling operation does not have the parameter i.e., $w^l = null$. The lth layer of pooling is the spatial extent of input $X^l \in H^l \times W^l \times C^l \times b$. The pooling output $X^{l+1} \in H^{l+1} \times W^{l+1} \times C^{l+1} \times b$ will be calculated as $H^{l+1} = \dfrac{H^l}{H}$, $W^{l+1} = \dfrac{W^l}{W}$, $C^{l+1} = \dfrac{C^l}{C}$. It strides in horizontal and vertical directions which are smaller regions than $H \times W$.

- **Fully connected layer:** In the convolutional layers $(l+1)^{th}$ layer output value $x(l+1)$ is computed based on the previous l^{th} layer input value x^l only, but in the fully connected layer, the output y i.e., $x^{l+1} \in R^D$ with tensor size $H^l \times W^l \times C^l \times b$ requires all input values (*i.e.,* x^l) of the previous l^{th} layer. The final output at the last L^{th} layer can be written like the convolutional layer output shown below eq. 7

$$y_L = \sum_{i=0}^{H} \sum_{j=0}^{W} \sum_{k^L=0}^{C^l} w_{i,j,k^L,b} \times x_{(i^{L+1},j^{L+1}+j,k^L,b)} \qquad (7)$$

Let F be a network architecture function with learning rate, hyperparameters, bias, and so on. There are sets of parameters $\forall f \in F$, weights, and biases that may be

obtained when training on appropriate training datasets. However, it is necessary to discover the best function $f* \in F$ (with optimal parameter) to get the best-fit network with the minimum overfitting/underfitting. The optimal function can be written corresponding to dataset X and labels y, as shown in eq. 8

$$f* = argmin_f(\mu(X,y,f), subject to f \in F) \qquad (8)$$

Therefore, $f*$ is the selected model for a given network. A better architecture F' can also be obtained as $f_F^* \notin F$.

IV. Experimental Design

A. Flower Leaf Image Dataset Description

This research work has prepared the dataset of 25 different flower plant leaves of 6619 total RGB images. More than 250 images are taken of each category using a white background to avoid anonymous or unusual data during experimentations. The image of leaves was taken with 64 megapixels (3456×4608) camera of the Realme Six. All images were captured with different intensities of natural daylight and distance ranging to 4 inches to 10 inches from the leaf's surface. The different intensities of light and distance will yield a more generalized performance of the model in the practical situation of image capturing. Each leaves image has been resized using python programming as 250×250 and stored in the categorized folders. Figure 7.3(a) shows flower leaf data set information.

B. Experimental Setup

The prepared dataset has been split into training, testing, and validation. The number of images in training is 60% of the total data, and then from the remaining 40%, 20% of the data is for testing, and 20% data is for validation Chang et al. (2022). During this research work, the image processing, resizing, cropping, normalization, and classification tasks were performed by an anaconda (spyder) using python programming. Pytorch library is used to implement deep learning models. The Configuration of the workstation is processor Intel Xeon Silver 4210 CPU-2.19GHz, ten crores, Ram - 128.0 GB, Windows 10 server edition 64 Bit, and GPU-Quadro RTX 4000.

V. Result and Analysis

A. Description of Results

Bar Plot depiction of Deep learning and Machine learning models with comparative analysis of before and after implementation of augmentation: Figure 7.2 shows the performance of models where the x-axis and y-axis are model and classification accuracies, where models are ResNet, AlexNet, VGG, SqueezeNet, GoogLeNet, ShuffleNet, ResNext50, and DenseNet.

Line Graph of Training & Validation Accuracy: Figure 7.3(b)shows the performance of models where the x-axis and y-axis are model

and training and validation accuracies and loss accuracies.

B. Analysis of Results

Bar Plot depiction of Deep learning and Machine learning models with comparative analysis of before and after implementation of augmentation: From Fig. 7.2, it can be observed that ResNext50 has performed the best among others, with 99.6243% classification accuracy before augmentation and after augmentation, it 99.7005%, and the GoogleNet and ShuffleNet are the second and third highest performers with accuracy of 99.023% and 98.797% respectively, and the least accuracy is noted of model AlexNet with 93.238%. The comparative study shows that ResNext50 performs best based on accuracy measures. In machine learning improvement of accuracy before and after augmentation technique can observe clearly, where the highest accuracy of the MLP model before augmentation is 89.616% and after augmentation, it is 96.031% respectively we can observe in other models.

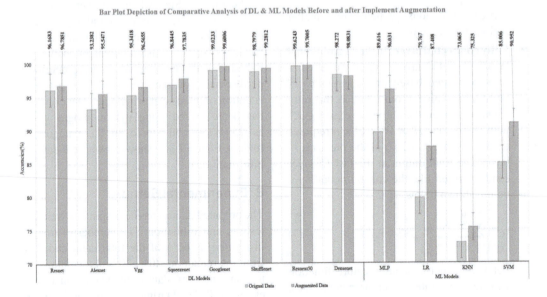

Fig. 7.2 Bar plot depiction of deep learning and machine learning models with comparative analysis of before and after implementation of augmentation

Source: Author

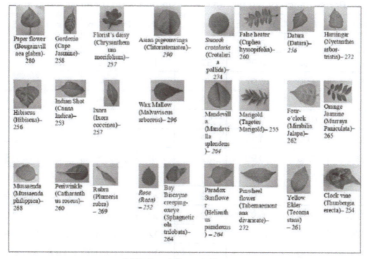

(a) Dataset Description

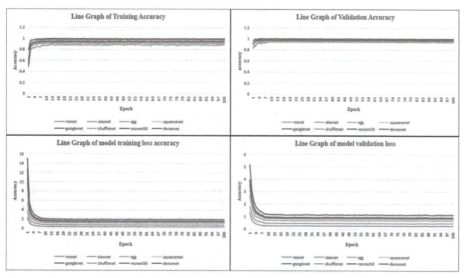

(b) Line Graph of Training & Validation Accuracy

Fig. 7.3 Dataset description and line graphs

Source: Author

Line Graph of Training & Validation Accuracy: Figure 7.5 shows the Training Loss and Validation Loss of the models, where we can see the training accuracy and validation accuracy start increasing. This is also fine as the model built is learning and working fine, similarly in Training Loss and Validation Loss of the models where we can see the training loss and validation loss start decreasing. This also

means the model built is learning and working fine Hedman et al. (2021).

Precision, recall, and F1-Score based comparison: Here, from the Table.1 classification report of models, we can analyze or observe the best or worst classification happenings in each category. It is observable that category 11, i.e., Malvaviscus arboreus has the highest value of 1.00, 1.00, and 1.00

Table 7.1 Performance of the model based on precision recall F1-score and support

Category	Precision	Recall	F1-Score	Support	S.N.	Precision	Recall	F1-Score	Support
0.	0.95	1.00	0.97	56	12.	1.00	1.00	1.00	52
1.	1.00	0.92	0.96	52	13.	0.96	1.00	0.98	55
2.	1.00	1.00	0.97	81	14.	0.94	1.00	0.97	51
3.	0.98	0.98	0.99	58	15.	0.95	1.00	0.97	54
4.	1.00	0.98	0.99	55	16.	0.96	0.98	0.97	52
5.	1.00	1.00	1.00	52	17.	0.98	0.98	0.98	55
6.	0.98	1.00	0.99	52	18.	0.98	1.00	0.99	53
7.	1.00	0.98	0.99	51	19.	1.00	0.98	0.99	51
8.	0.96	1.00	0.98	53	20.	0.98	0.90	0.94	51
9.	0.96	0.98	0.97	53	21.	1.00	1.00	1.00	51
	1.00	1.00	1.00	53	22.	0.98	0.96	0.97	56
	1.00	1.00	1.00	55	23.	1.00	0.92	0.96	59
					24.	1.00	1.00	1.00	51
Accuracy								0.98	1331
Macro Avg.		0.98					0.98	0.98	1331
Weighted Avg.		0.98					0.98	0.98	1331

Source: Author

in terms of precision, recall, and f1-score, respectively, so we can say this category is the best classification performer. At the same time, Mirabilis Jalapa's level of misclassification is very high. It has been observed that this is due to the visual similarity between other flower Plants, i.e., Tecoma stans.

(a) Fouro'clock
(Mirabilis Jalapa)

(b) Yellow Elder
(Tecoma Stans)

Fig. 7.4 Leaves Images from those categories in which misclassification is higher

Source: Author

VI. Conclusion

This research utilized 25 new categories of flower plant leaf RGB images for learning Deep Learning models like ResNet, AlexNet, VGG, SqueezeNet, GoogLeNet, ShuffleNet, ResNext50, and DenseNet. And the classification performance of the model has been observed with training accuracy. The performance analysis shows that ResNext50 has a better performance with 99.62%. Among other classifiers over the accuracy, precision, recall, and F1-score. The investigation has also identified the reason for misclassifying some shape and texture similarities categories where the learner got mistaken for determining the correct types.

Future work: The analysis indicates that visual similarities increased misclassification. Therefore, a novel method/approach can be developed to minimize misclassification. Moreover, advanced Deep learning models can also be used for comparative study for even more categories of the flower plant leaf.

References

1. Alrefaee, S. H., Rhee, K. Y., Verma, C., Quraishi, M., and Ebenso, E. E. (2021). Challenges and advantages of using plant extract as inhibitors in modern corrosion inhibition systems: Recent advancements. *Journal of Molecular Liquids*, 321: 114666.

2. Aman, B. k. and Kumar, V. (2022). Flower leaf classification using machine learning techniques. In *Third International Conference on Intelligent Computing, Instrumentation and Control Technologies (ICICICT-2022)*. IEEE Explore.

3. Aochi, H., Ulrich, T., Ducellier, A., Dupros, F., and Michea, D. (2013). Finite difference simulations of seismic wave propagation for understanding earthquake physics and predicting ground motions: Advances and challenges. In *Journal of Physics: Conference Series*, volume 454, page 012010. IOP Publishing.

4. Atabay, H. A. (2016). A convolutional neural network with a new architecture applied on leaf classification. *IIOAB J*, 7(5): 226–331.

5. Chang, Y.-Y., Li, P.-C., Chang, R.-F., Yao, C.-D., Chen, Y.-Y., Chang, W.-Y., and Yen, H.-H. (2022). Deep learningbased endoscopic anatomy classification: An accelerated approach for data preparation and model validation. *Surgical Endoscopy*, 36(6): 3811–3821.

6. Das, U. and Islam, M. S. (2019). A review study on different plants in malvaceae family and their medicinal uses. *American Journal of Biomedical Science and Research*, 3: 94–97.

7. Dhingra, G., Kumar, V., and Joshi, H. D. (2018). Study of digital image processing techniques for leaf disease detection and classification. *Multimedia Tools and Applications*, 77(15): 19951–20000.

8. Hedman, D., Rothe, T., Johansson, G., Sandin, F., Larsson, J. A., and Miyamoto, Y. (2021). Impact of training and validation data on the performance of neural network potentials: A case study on carbon using the ca-9 dataset. *Carbon Trends*, 3: 100027.

9. Iqbal, Z., Khan, M. A., Sharif, M., Shah, J. H., ur Rehman, M. H., and Javed, K. (2018). An automated detection and classification of citrus plant diseases using image processing techniques: A review. *Computers and electronics in agriculture*, 153: 12–32.

10. Jasim, M. A. and Al-Tuwaijari, J. M. (2020). Plant leaf diseases detection and classification using image processing and deep learning techniques. In *2020 International Conference on Computer Science and Software Engineering (CSASE)*, pages 259–265. IEEE.

11. Jmour, N., Zayen, S., and Abdelkrim, A. (2018). Convolutional neural networks for image classification. In *2018 international conference on advanced systems and electric technologies (IC ASET)*, pages 397–402. IEEE.

12. Kumar, C. and Kumar, V. (2022a). Vegetable plant leaf image classification using machine learning models. In *International Conference on Advances in Computer Engineering and Communication Systems (ICACECS-2022)*. Lecture Notes in Networks and Systems (LNNS), Springer Nature, August 2022. (Accepted).

13. Kumar, G. and Kumar, V. (2022b). Herbal plants leaf image classification using deep learning models based on augmentation approach. In *The 4th International Conference on Communication and Information Processing (ICCIP-2022)*. SSRN conference series Elsevier. (Accepted).

14. Kumar, G., Kumar, V., and Hrithik, A. K. (2022). Herbal plants leaf image classification using machine learning approach. In *International Conference on Intelligent Systems and Smart Infrastructure*. CRC Press, Tailor & Francis Group.

15. Lee, S. H., Chan, C. S., Mayo, S. J., and Remagnino, P. (2017). How deep learning extracts and learns leaf features for plant classification. *Pattern Recognition*, 71: 1–13.

16. Liu, J., Yang, S., Cheng, Y., and Song, Z. (2018). Plant leaf classification based on deep learning. In *2018 Chinese Automation Congress (CAC)*, pages 3165–3169. IEEE.

17. Nair, V. and Hinton, G. E. (2010). Rectified linear units improve restricted boltzmann machines. In *Icml*.

18. Ni, F. and Wang, B. (2018). Integral contour angle: an invariant shape descriptor for classification and retrieval of leaf images. In *2018 25th IEEE International Conference on Image Processing (ICIP)*, pages 1223–1227. IEEE.

19. Pang, C., Wang, W., Lan, R., Shi, Z., and Luo, X. (2021). Bilinear pyramid network for flower species categorization. *Multimedia Tools and Applications*, 80(1): 215–225.

20. Prakash, R. M., Saraswathy, G., Ramalakshmi, G., Mangaleswari, K., and Kaviya, T. (2017). Detection of leaf diseases and classification using digital image processing. In *2017 international conference on innovations in information, embedded and communication systems (ICIIECS)*, pages 1–4. IEEE.

21. Sahana, M., Reshma, H., Pavithra, R., and Kavya, B. (2022). Plant leaf disease detection using image processing. In *Emerging Research in Computing, Information, Communication and Applications*, pages 161–168. Springer.

22. Sharif, K. O. M., Tufekci, E. F., Ustaoglu, B., Altunoglu, Y. C., Zengin, G., Llorent-Mart´ınez, E., Guney, K., and Baloglu, M. C. (2021). Anticancer and biological properties of leaf and flower extracts of echinacea purpurea (l.) moench. *Food Bioscience*, 41: 101005.

23. Vilasini, M. (2020). The cnn approaches for classification of indian leaf species using smartphones. *Computers, Materials & Continua*, 62(3): 1445–1472.

24. Wang, P., Fan, E., and Wang, P. (2021). Comparative analysis of image classification algorithms based on traditional machine learning and deep learning. *Pattern Recognition Letters*, 141: 61–67.

25. Yalcin, H. and Razavi, S. (2016). Plant classification using convolutional neural networks. In *2016 Fifth International Conference on Agro-Geoinformatics (Agro-Geoinformatics)*, pages 1–5. IEEE.

Recent Trends in Computational Intelligence and its Application – Sugumaran D. et al. (eds)
© 2023 Taylor & Francis Group, London, ISBN 978-1-032-48410-5

8

Vehicle Monitoring System Detection Using Deep Learning Technique

K. Arthi*, S. Suchitra, A. Shobanadevi and K. Sharon Babu[1]

Department of Data Science and Business System, School of Computing,
SRM Institute of Science and Technology, Kattankulathur, Chennai

Abstract—Given the increase in population, vehicle tracking is no longer practical. Both time and resources are wasted in doing it. The enormous daily growth in the automotive industry has made tracking individual automobiles an extremely challenging undertaking. This research suggests a system for automatically tracking moving cars using roadside security cameras. License plate recognition systems are used in toll collection, parking fee, and residential entry control in contemporary smart cities. In addition to being helpful in people's daily lives, these electronic technologies also give management access to secure and effective services. The suggested technique now includes a useful method for identifying Indian license plates on cars. The suggested technique can work with number plates that are obtrusive, dimly lit, cross-angled, and have unusual fonts. The effective deep learning-based ALPR (Automatic License Plate Recognition) model presented in this study uses character segmentation and a CNN-based recognition model. The experimental finding yields a 94.94% accuracy percentage for the f1 score.

Keyword—Image binarization, Number plate detection, Edge detection, K-Nearest neighbors, CNN, Character extraction

I. Introduction

The increased Computer Vision (CV) technology known as Vehicle Number Plate Recognition (NPR) or License Plate Recognition (LPR) links vehicles without a direct human affiliation through their license plates. The number of vehicles on the road continues to rise every day[1-3]. As a result, news reports of a vehicle being stolen from a parking garage or another location inside the city or having an accident and escaping are very common now. Vehicle establishments should put variety plate detection and acknowledgment equipment on CCTV (Closed Circuit Television) at each corner in each region in order to recognize these. By using this technique, the police are better able to identify illegal actions that include driving. In all areas of safety, examination, and traffic supervision

---
*Corresponding author: arthik1@srmist.edu.in
[1]sk9977@srmist.edu.in

DOI: 10.1201/9781003388913-8

tenders, provincial institutions and production teams employ NPR systems efficiently.

Presently, simple deep learning techniques and traditional machine vision techniques are the two main categories for vision-based vehicle object recognition. Conventional machine vision techniques use the vehicle's movements to separate it from a background image that is immobile. The three approaches—backdrop removal, continuous video frame difference, and optical flow—can be grouped into one of three groups. Using the video frame difference approach, the variance is calculated using the pixel values of two or three subsequent video frames. The threshold additionally divides the moving foreground zone. By using this method and muzzling noise, the car stoppage can also be detected.

The background data is utilized to create the backdrop model after the background image in the movie has been repaired. After that, the backdrop model is used to compare each frame image, and the moving item can also be split. The motion zone in the video can be found using the optical flow technique. The speed and direction of each pixel's motion are represented by the optical flow field that is generated. Scale Invariant Feature Transform (SIFT) and Speeded Up Robust Attributes (SURF) approaches are two common vehicle detection techniques that use vehicle features. 3D models, for instance, have been employed for vehicle identification and classification tasks. The vehicles are separated into three groups based on the correlation curves of 3D ridges on their exterior: cars, SUVs and minibuses.

II. Scope

By keeping the number plate details in a database and cross-checking them, NPR can be used in housing societies or apartments to only allow the resident's vehicle inside. This could lessen the number of people around the gate [4]. NPR can be used in parking lots to find cars that break the regulations. Ticketless parking, automated parking, charging for parking and

ticket fraud are some other applications for NPR. NPR can also be applied on motorway tolling, where successful tolling lowers non-payment related frauds, makes fees effective, and reduces the need for labor to handle exclusionary events. NPR can also be used to measure travel duration, with the data it gathers being used to teach drivers about how to improve traffic safety, route traffic more efficiently, and cut travel time and costs[5].

III. Number Plate Detection

The following variables should be taken into account when detecting number plates:

1. *Plate size:* Different vehicles will have varying plate sizes.
2. *Background color:* The color of the number plate background will vary depending on the type of vehicle, for instance, a taxi will have a yellow background while a private vehicle will have a white background.
3. *Plate placement:* The car will have various plates at various places.

We use picture segmentation to extract license plates. Image binarization is the most popular technique. Images can also be converted to grayscale using Otsu's technique[6].

A. Image Binarization

Figure 8.1 illustrates how an image can be rendered in black and white using image

Fig. 8.1 Image converted to grayscale

Source: Author

binarization. To distinguish between pixels that are white and those that are black, we select a specific threshold value. Although a challenge, this can be handled via adaptive thresholding[7].

B. Edge Detection

Edge detection is used to choose when to remove features. Edge detection in complex images should be avoided at all costs because it could lead to object boundaries without associated curves, as demonstrated in Fig. 8.2.

Fig. 8.2 Edge detection

Source: Author

C. Character Segmentation

We check the characters after locating the number plate's position. Character segmentation can be done in many different ways. As seen in Fig. 8.3, we can use the previously stated picture binarization technique for character segmentation. Blob coloring is inappropriate for Indian license plates [8,12]. Image scissoring should be used, in which the plate is vertically scanned, and when a white pixel is not detected, it is scissored at that row, with the values being saved in a matrix.

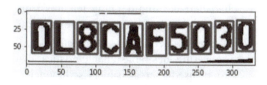

Fig. 8.3 Segmented characters

Source: Author

D. Character Extraction

The visual text is recognized and transformed into real text by character extraction. Character extraction is used by all number plate identification techniques, some of which are covered here [9].

Optical Character Recognition (OCR), Artificial Neural Network (ANN), and Template matching are the three most popular techniques utilized in character extraction.

OCR is a character extraction technique. Several character extraction programs are available, including Google's open-source OCR tesseract. Tesseract has been able to extract characters at a rate of 98.7%.

Artificial neurons are interconnected in ANN. There are numerous ANN-based algorithms. Character classification is done using the MLP ANN model. The input layer, hidden layer, and output layer are its three layers. The decision-making process is handled by the input layer while complex relationships are computed by the hidden layer, and the outcome is handled by the output layer. We use feed forward back propagation to train ANN. Character extraction using ANN was successful at 99% of the time [10]. Characters with fixed sizes can be used through template matching. It is mostly used in medical immaturity processes and facial detection. It consists of two parts: feature-based matching and template-based matching. Feature-based matching can be employed if the template image has prominent features. Using template matching, it was possible to extract characters at a rate of roughly 95.7%. Figure 8.4 displays a sample.

```
In [32]: print(show_results())
```

DL8CAF5030

Fig. 8.4 Character extraction

Source: Author

IV. Proposed Methodology

The issue of transportation has received attention due to the growing population and technological developments. One of the main reasons of traffic congestion and infractions is the rapid increase in the number of cars on the road. NPR was developed to automate traffic management while reducing traffic infractions. Different NPR strategies are employed in India; however, they are not very effective. The suggested model is focused on detecting and identifying vehicle license plates. Python is used to implement various algorithms using the Russian blurring approach, OpenCV and CNN[11].

Understanding the data must be done first. Our goal is to extract the character pictures when we input the plate image. Character segmentation, the most crucial phase in identifying the number plates, is what we do for that.

In a nutshell, after entering the image, we transform it to grayscale or black and white. Following image blurring to eliminate obtrusive noise, image binarization is performed. Then, using OCR, template matching, or ANN[13], we extract characters one by one utilizing character segmentation to remove every single character.

Learning the Libraries SKlearn, Tensorflow, OpenCV and Numpy

A Python library called NumPy is utilized in many different applications. It has a variety of purposes in addition to the obvious scientific ones. Common data can be kept in a multidimensional container using NumPy.

Following are the characteristics of NumPy:

NumPy Array is a rapid and compact multidimensional array with vectorized arithmetic. Advanced broadcasting features and procedures.

Standard mathematical processes can be utilized to quickly process large data sets.

Data may be gathered without the usage of loops.

Memory-mapped files and reading/writing data can be done from/to an array to a disc

Tools for adding capabilities like random number generation and linear algebra to C, C++, and Fortran *programs.*

- An open-source software library for computer vision and machine learning is called open-source computer vision. It has a variety of specializations, including the ability to recognize different objects, faces and actions in photographs and movies.

- With the aid of Keras, TensorFlow is another Python library used in deep learning models. It includes a number of tools, including MaxPool, Dense, Flatten and Image Data Generation.

- Scikit-learn (Sklearn) offers a variety of tools for modeling machine learning methods, including classification, regression, clustering, SVM, PCA, and many more. The Matplotlib, SciPy, and NumPy libraries are used to implement these.

Here are a few of Sklearn's most important characteristics:

- Clustering is a technique for categorizing unlabeled data.

- Data can be summarized, displayed, and given a set of features by using a technique called dimension reduction to cut down on the number of characteristics in the data.

- As suggested by their name, ensemble methods combine the results of various supervised models.

- A technique for extracting data characteristics and defining features from text and images is called feature extraction.

- Feature selection is a method for discovering desirable characteristics that may be used to create supervised models.

A. Flowchart

The image is entered, turned to grayscale, and then put through a number of filtering stages. Here, the term "filtering" refers to the blurring of an image to remove distracting edges and louds from the image. As indicated in Fig. 8.5, these actions are regarded as image pre-processing actions. After filtering, we apply cunning edge detection techniques to conduct edge detection. Following plate detection, we go on to character segmentation, which is regarded as a crucial phase since it is vital to identify each individual character picture. After that, character extraction is formed utilizing a variety of character extraction technologies, including OCR, ANN, template matching, etc..

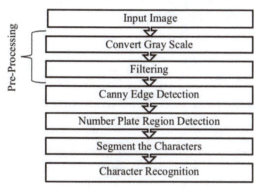

Fig. 8.5 Flowchart

Source: Author

B. ANN

The input layer, hidden layer and output layer are the three primary layers of an ANN.

As shown in Fig. 8.6, the decision-making process takes place at the input layer, while complex associations are computed at the hidden layer, and the outcome is displayed at the output layer. We use feed forward back propagation to train ANN. Character extraction rates of 99% were accomplished via ANN.

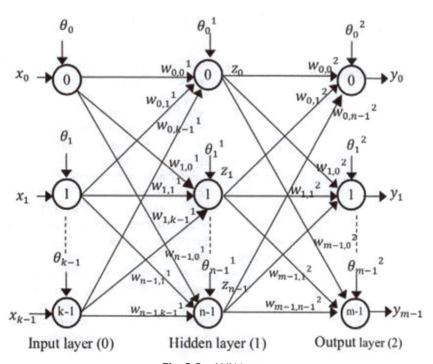

Fig. 8.6 ANN layers

Source: Author

Each input will have its own weight, and after entering the input layer, it goes through a number of preprocessing steps before being transferred to the hidden layer for convolution and pooling. Once the processed segmented image has reached a certain threshold value, it is then transferred to the output layer, where we will be getting the results.

C. CNN

Multiple artificial neuronal layers make up CNN. Diagonal, vertical, and horizontal edges are first detected by the first layers. The more intricate elements, such as corners and edges, are picked up by the second layer. The deeper layers find complex features, faces, objects, etc.

After input, the vehicle image is subjected to image pre-processing, which includes grayscale conversion and blurring to muffle sound. Then edge detection is applied; in this case, clever edge detection. After filtering, it goes through horizontal detection, vertical detection, character segmentation to identify the character pictures, and character extraction to extract the vehicle number plate.

D. KNN (K-Nearest Neighbors)

For programmed automotive authorization number plate extraction, a KNN machine learning structure was developed. To recover the registration plate from the image, the KNN classifier is used second-hand in accordance with the sides of the license. They required an associate's degree to complete, they had to reply to complaints that an authorization code was incorrectly labeled using the information they had provided, and they achieved an accuracy rating of more than 90%. Moving neural network image extraction techniques can recognize everything, including text, thoughts, radio wave-transmitted audio and video files, and videos. A mesh of neurons or perceptron-named growths makes up neural networks. Every bud uses a single recommendation, typically a face feature, and employs a clear estimation known as incitement performs that

yields results. Additionally, every somatic cell has a mathematical score that determines the attract influence. For the extraction of vehicle license plates, a high-routine system has reached maturity. In addition to using a three-layer feed forward fictitious interconnected system with a learning invention for back propagation, they have also employed edge-based countenance preparation techniques for identification of license plates. Figure 8.7 depicts the system architecture.

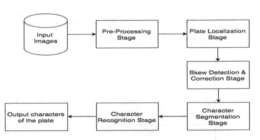

Fig. 8.7 System architecture

Source: Author

V. Conclusion

Applications for NPR-programmed number plates are rapidly expanding on Indian platforms because to the astonishing rise in the number of automobiles, motorcycles, and commercial vehicles. NPR can be used for a variety of tasks, including as toll collecting, managing parking places at malls, and managing vehicles there. It can also be integrated with traffic cameras, which can be used to monitor and control traffic infractions, among other things. This project has a lot of advantages. The number plates' numbers were extracted using character segmentation. Finally, the outcome was determined using the F1 score precision and recall for the input image. The model's lack of advanced machine learning or deep learning techniques is one of the project's constraints, but it still performs well for the majority of use cases, and it will be taken into account as future work in growing the project to a higher level. The goal of this project was to demonstrate how capable free and open-source technologies are for usage in machine-learning applications. The device functions brilliantly in a variety of lighting conditions and with different sorts of number plates that are frequently encountered in India. Although, it has acknowledged shortcomings, it is undoubtedly better than current proprietary alternatives. This was created in a notebook program called Jupiter utilizing the Python language and system-provided photos. It provides an f1 score accuracy of 94.94% after 80 epochs.

VI. Future Enhancements

Future military threats will be identified in large part by the automatic vehicle recognition system. Since they can quickly identify the number plate before using cabs or other services, it can also increase women's security. The system's durability can be improved by using a bright and clear camera. When employed effectively in a variety of circumstances, this technology is both economical and environmentally friendly, so the government should be interested in developing it.

The method can be improved by using sensors at the entrance. The image of a car's license plate is taken as it approaches the gate. The image is then sent to MATLAB, where it is categorized before being sent to a database. The gate will open if the license plate matches; otherwise, an alarm will sound.

The most important feature of this technology is undoubtedly the accuracy of the recognition. As a result, this application needs to be updated and enhanced in order to overcome the accuracy limitations. We must incorporate specific preprocessing techniques to remove interferences in order to increase the recognition's precision. In addition, we would carry on our investigation into license plate recognition in more challenging circumstances, such as in cars at night or in bad weather, and so forth. If we were to succeed in all of our objectives, the future of our application would be very promising.

References

1. Abhay Singh, Anand Kumar Gupta, Anmol Singh, Anuj Gupta, Sherish Johri, "VEHICLE NUMBER PLATE DETECTION USING IMAGE PROCESS- ING", Department of IT, Volume: 05 Issue: 03 | Mar- 2018
2. Ganesh R. Jadhav, Kailash J. Karande, "Automatic Vehicle Number Plate Recognition for Vehicle Park- ing Management System", IISTE, Vol.5, No.11, 2014.
3. Mutua Simon Mandi, Bernard Shibwabo, Kaibiru Mutua Raphael," An Automatic Number Plate Recognition System for Car Park Management", In- ternational Journal of Computer Applications, Vol. 175 – No.7, October 2017
4. M.T. Shahed, M.R.I. Udoy, B. Saha, A.I. Khan, S. Subrina, "Automatic Bengali number plate reader," IEEE Region 10 Annual International Conference, Proceedings/TENCON, 2017-Decem, 1364–1368, 2017, doi:10.1109/TENCON.2017.8228070.
5. Regtransfers.co.uk, "Number plates rules". [Online]. Available: https://www.regtransfers.co.uk/acrylic-number-plates/number-plates-rules
6. R. Islam, M.R. Islam, K.H. Talukder, "An efficient method for extraction and recognition of bangla characters from vehicle license plates," Multimedia Tools and Applications, 79(27–28), 20107–20132, 2020, doi:10.1007/s11042-020-08629-8.
7. S. Saha, "A Review on Automatic License Plate Recognition System," ArXiv, 2019.
8. M. Rajeev Kumar and K. Arthi "An effective Non- Cooperative Iris Recognition system using hierarchical collaborative representation-based classification", "Journal of Supercomputing, July 2019.
9. B Kanisha, S Lokesh, PM Kumar, P Parthasarathy, G Chandra Babu,"Speech recognition with improved support vector machine using dual classifiers and cross fitness validation", Personal and ubiquitous computing 22 (5), 1083–1091, 2018.
10. Surampalli Ashok, Gemini Kishore, Velpula Rajesh, S Suchitra, SG Gino Sophia, B Pavithra," Tomato leaf disease detection using deep learning techniques", 979–983, 2020.

11. R Thiagarajan, R Ganesan, V Anbarasu," Optimised with secure approach in detecting and isolation of malicious nodes in MANET", Wireless Personal Communications 119 (1), 21–35, 2021.

12. M. Rajeev Kumar and K. Arthi "An effective Non-Cooperative Iris Recognition system using hierarchical collaborative representation-based classification", "Journal of Supercomputing, July 2019. https://doi.org/10.1007/s11227-019-03007-0.

13. K.Arthi and Sudha Rani,"Secured Message Transmission between Vehicles for Reducing Delay and Collision in VANET", International Journal of Innovative Technology and Exploring Engineering (IJITEE), Volume-8 Issue-8, June, 2019.

Recent Trends in Computational Intelligence and its Application – Sugumaran D. et al. (eds)
© 2023 Taylor & Francis Group, London, ISBN 978-1-032-48410-5

9

Edge Detection System for Object Detection and Alert Generation

E. Kamalanaban[1], P. Selvarani[2], S. K. Manigandan[3],
V. Sabapathi*, Roja K.[4] and Roshini P.[5]
Department of CSE,
Vel Tech High Tech Dr. Rangarajan Dr. Sakunthala Engineering College, Chennai, India

Abstract—In today's world, the emerging crime rate is terrifying and an immediate precautionary method should be taken to control the crime rate with the help of surveillance stream. An idea of developing an edge detection system with features including object detection and pitch detection in surveillance stream would play a vital role in crime prevention. Tracking the activities in public place is a time consuming and difficult process. Hence, an intelligence surveillance system would monitor the activities of human in real time and categorize it to be normal or suspicious and generate an alert. Through our surveillance system, human activities can be monitored in areas such as railway station, bus station, banks, malls, institution, companies and parking areas. It prevents terrorism, accidents, chain snatching, robbery, etc. But there are cases where the suspicious activity can remain unnoticed and recognized after the occurrence of emergence or crime situation. In this paper, we propose two features that will be performed simultaneously along with greater accuracy. Object detection includes foreground extraction and activity analysis. Pitch detection algorithm includes time domain detection and frequency domain detection. This paper proposes a brief introduction and the characteristics of detection algorithm, precision and threshold value with alert generation process.

Keywords—Object detection, YOLO generation, Alert generation, PDA algorithm

1. Introduction

An intelligent surveillance system would highly contribute in crime prevention and resolution. In accordance with that this project includes suspicious human activity detection using object and pitch detection. The real-time input video from surveillance stream is fed in an edge detection platform to perform both the detections simultaneously for acquiring a high accuracy rate in a quick span of time. Violating traffic rules, shooting, chain snatching,

---
*Corresponding author: sabapathi@velhightech.com
[1]drekamalanaban@velhightech.com, [2]selvarani.meena@gmail.com, [3]skmanikandan@velhightech.com,
[4]rojakadugumalai@gmail.com, [5]roshinisri1234@gmail.com

DOI: 10.1201/9781003388913-9

punching, or fire detection and accidents [1] are considered to be abnormal activities and [2] alert message is sent to related rescuing places like police station, hospital and owners mobile phone.

2. Literature Review

A. Real-Time Input Based Edge Surveillance

It proves that in most cases, the security video surveillance [3] feeds are used as only proof. Convolutional neural network is used in the project (CNN) [4]. Hand movements like pushing etc., are the various action identified.

B. Security Analysis of Intelligent System Based on Edge Computing

This paper has developed an authentication system which identifies face on edge computing [5]. It uses nearest neighbor algorithm.

C. Human Suspicious Activity Detection using Deep Learning

Detecting suspicious activities in public places [6] has become an important task. This paper includes deep learning for detection of suspicious activities [7] which consists of CNN.

D. Real-time Pitch Detection using a Digital Signal Processor

This paper focuses on presenting speech detection and uses a conventional digital signal processor and Discrete Running Fourier Series (DRFS).

III. Existing System

The work proposed in all the referred literature surveys identifies the hand movement/face of the person through the surveillance stream. The rate of accuracy is reduced whenever the distance between surveillance and object is too

far, hence [7] this paper uses CNN to extract the feature of the input image or video, and in face recognition system user privacy is protected using nearest neighbor algorithm. A message or alert is generated on abnormal activity.

IV. Proposed System

The proposed system identifies all types of body movements of a person and it also identifies suspicious object using image processing technique in real-time [8]. Suspicious sound recognition feature has been included in this system. Impulsive sounds (IS) are detected using pitch detection algorithm. On basis of prediction result, alert is generated. This project uses edge computing which places the resources closer to the end users device. This minimizes data dependencies and speeds up prediction. Since both detections are done simultaneously, the rate of prediction is done at a faster interval of time compared with the existing system.

A. Overall Architecture

See Fig. 9.1 on next page.

V. Methodology

A. Feature Extraction For Various Detection

Theft Detection

Blob trajectory: Blobs are separated for the detection of static and movable objects extracted. Dual foreground concept technique includes long-term and short-term learning rate, foreground masks, FL and FS, are created. Object is static if the (FL; FS) = (1, 0).

Violence Detection

In order to track moving objects, texture feature and shape feature are extracted. The computational similarity values between images are obtained using curvelet. The geometric image shape is obtained in Fig. 9.1.

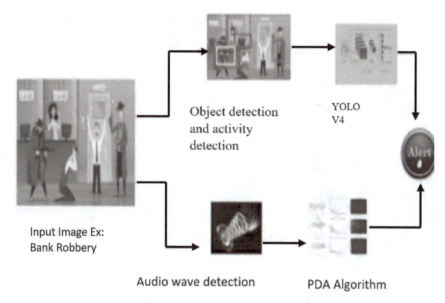

Fig. 9.1 Overall architecture [14]

B. Methods of PDA

Time- Domain: [9]

Repeated patterns are identified. Inputs are in fluctuating amplitudes.

1. *Autocorrelation:* Mathematically defined as in Eq.(1)

$$y(n) = \sum_{k-1}^{hd} u(k)u(k+n) \qquad (1)$$

2. *Adaptive Filter:* (1) Narrow band pass filter: The filtered, unfiltered input signals are given to the circuit. Output from the circuit is fed back again, as a result it converges the input signal frequency. (2) Adaptive enhancer: The below diagram is the result [10].

3. *Super Resolution Pitch Determination* [11]: The similarity degree between two adjacent intervals which are not overlapping and consists of infinite time resolution is examined

Frequency-Domain

1. *Harmonic Product Spectrum* [12]: The GCD compute frequency fundamentals. The peak frequency of frequency histogram is considered for performing GCD.

2. *Cepstrum:* The Fourier transform is now routed to Fourier spectrum of log magnitude. Frequency domain produces multiplied input signals, applying log

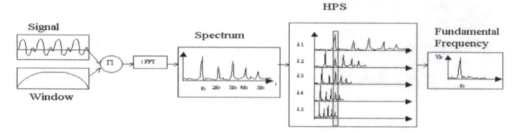

Fig. 9.2 Frequency domain [16]

to convert the multiples to adding the impulses chain. By applying FFT, we can de-convolute the original input signal.

VI. Example of Suspicious Activity Detection

VII. Process of Object Detection

A. Real Time Input

It refers to monitoring human activities at public places such as government buildings,

Fig. 9.3 Examples of suspicious activity detection [14]

flats, stations and complexes for preventing theft, terrorism and accidents.

B. Classification, Object Extraction, Feature Extraction

We use [13] YOLO (You Only Look Once) framework. It processes 45 frames per second. The input image is fed, each grid is classified and localized, bounding boxes, class probability are made by YOLO[11].

C. Impulsive Audio Classification, Feature-Extraction

We are choosing microphones in such a way that it has all specs like noise level, frequency response, SPL capacity, sensitivity and dynamic range.

VIII. Implementation

See Fig. 9.4 below.

IX. Results

The input is fed into YOLO V4 Framework. Images are spitted into frames. ResNet-34 recognizes the activity of detected people. If it is found to be suspicious then it is said as hostile behavior. Based on results from object detection and pitch detection, it will be confirmed with prediction score and alert will be produced. Compared to previous systems, our proposed system has high rate of accuracy.

X. Conclusion

The ultimate goal of this project is to design an efficient real-time suspicious activity detection software. Real-time input is given into YOLO v4 for activity detection and to pitch detection algorithm. The resultant prediction score and threshold value decides the suspicious activity prediction rate. This system produces comparatively more accurate prediction than

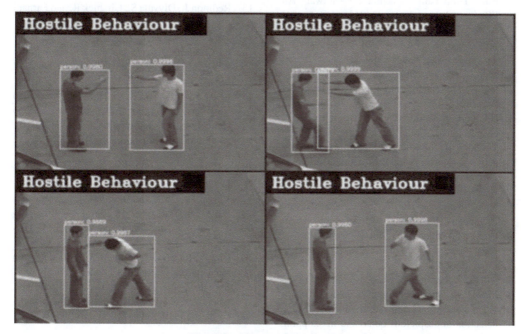

Fig. 9.4 Hostile behavior [15]

the existing systems and also in real time. Thus, as a conclusion, this system serves by predicting and alerting in case of crimes in areas under surveillance.

XI. Future Enhancement

In future, the software are data sets that are updated with periodical dataset. The proposed system is expected to be developed and used in nano robots. The system is also expected to detect more complex behaviors that occur in an environment.

──── References ────

1. Sri Krishna Chaitanya Varma, Poornesh, Tarun Varma, Harsha ,"Automatic Vehicle Accident Detection and Messaging system using GPS and GSM Modems", *International Journal of Scientific & Engineering Research,* Volume 4, Issue 8, ISSN 2229-5518, August-2013.

2. Tanaya Achalkar, Shrinath Panmand, Saurabh Naik, Dilip Patil, Rachna Sonkumwar "An Efficient Approach for Accident Detection System" *International Journal of Engineering Trends and Technology 67.4 4–7.* (2019)

3. Tsakanikas Vassilios, Dagiuklas Tasos "Video surveillance system-current status and future trends" Comput. Electr. Eng., 70 , pp. 736–753, (2018)

4. Nikouei Sayed Yahya, Chen Yu, Song Sejun, Xu Rongua, Choi BaekYoung, Faughnan Timothy R."Smart surveillance as an edge network service: from Harr-cascade, SVM to a Lightweight CNN" (2018)

5. Rajavel Rajkumar, Ravichandran SathishKumar., Harimoorthy Karthikeyan, Nagappan Partheeban "IoT-Based smart healthcare video surveillance system using edge computing" *J. Ambient Intel. Humaniz. Comput.* (2021).

6. Arroyo Roberto, Javier Yebes J., Bergasa Luis M., Dasa Ivan G., Almazan Javier "Expert video surveillance system for real-time detection of suspicious behaviors in shopping malls" Expert Syst. Appl., 42 , pp. 7991–8005, (2015)

7. Abdelmalek Bouguettaya, Ahmed Kechida, Amine Taberkit Mohammed "A survey on lightweight CNN-based object detection algorithms for platforms with limited computational resources" *Int. J. Inform. Appl. Math.,* 2, pp. 28–44, (2019)

8. Jha Sudan, Seo Changho, Yang Eunmok, Joshi Gyanendra Prasad "Real time object detection and tracking system for video surveillance system" *Multimedia Tools Appl.* (2020)

9. Yoan Medan, Eyal Yair and Dan Chazan " Super Resolution Pitch Determination of Speech Signals ", vol 39 No1 (1991)

10. Many attempts have been done in this line, see for instance: http://www.iua.upf.es/~xserra/articles/msm/pitch.html

11. "Details on the algorithm showed in Rabiner and Gold" *Theory and Applications of Digital Signal Processing. IEEE Transactions on Acoustics, Speech, and Signal Processing* AU-20; 322–337 (1975)

12. Moorer, J. A. " The optimum comb method of pitch period analysis of continuous digitized speech" *AIM-207. Stanford: Stanford Artificial Intelligence Laboratory.* (1973)

13. Redmon J., Divvala S., Girshick R. and Farhadi A., "You Only Look Once: Unified, Real-Time Object Detection," *IEEE Conference on Computer Vision and Pattern Recognition (CVPR), Las Vegas, NV,* 2016, pp. 779–788 (2016)

14. Zhou P., Ding Q., Luo H., Hou X. (2018) Violence detection in surveillance video using low-level features. PLoS ONE 13(10): e0203668. https://doi.org/10.1371/journal.pone.0203668

15. S. Adharsh et. al "Suspicious Activity Detection And Tracking In Surveillance Videos" © 2020 JETIR May 2020, Volume 7, Issue 5 JETIRDV06018 Journal of Emerging Technologies and Innovative Research (JETIR) www.jetir.org.

16. Chao He et. al "Salient object detection via images frequency domain analyzing" Signal image and video processing", 10, pages1295–1302 (2016) https://link.springer.com/article/10.1007/s11760-016-0954-x.

10

Pest Classification in Paddy by Using Deep ConvNets and VGG19

R. Elakya[1]

Research Scholar, SRM Institute of Science and Technology, Tamilnadu, India

T. Manoranjitham[2]

Associate Professor, SRM Institute of Science and Technology, Tamilnadu, India

Abstract—India is one of the largest producers of rice. About 3.5 billion people use rice as a food staple around the world. In India, rice is cultivated in all regions. About 60% of the food contribution is in the form of rice. The average production of rice during the year 2020–2021 was 121.46 million tons which is an increase by 9.01 million tons when compared to the previous year. Even though the production is high, farmers are still facing huge economic losses. Insects are causing more damage at all stages of rice cultivation and affect up to 30% of crop production loss. Insects that cause production losses are leafhoppers, borers, gall midge, plant hoppers, and grain-sucking bugs. Identification of these pests at an early stage is a major concern for all farmers. Early identification can reduce damage and production loss. Convolutional neural networks (CNN) are mostly employed for image classification. Throughout this framework, we categorized pests that affect rice production using the one, three, and four-layer convolutional neural networks as well as the VGG19 deep transfer learning model.

Keywords—Paddy crop, Pest attacks, Leaf disease, Deep learning model; CNN, VGG19

I. Introduction

Rice is a food crop that can be directly seeded or transplanted into soil. In the transplanting method, seeds are sown at one place named seedbeds then after it is grown they are planted in soil. In the seedling method, rice grains are directly sown in the soil field. A healthy crop will produce a large quantity of yield whereas it also depends on the soil fertility. On an average, around 37% of crop is damaged by pests and diseases. Early prediction of pests and diseases can reduce the loss. In this model, we are classifying different varieties of pests that cause damage and severe loss during the crop production in a nursery and main field stage.

All pests don't cause damage to crops. Many pests save the crop and we call them farmer-friendly pests. There are a few pests that cause more damage to crop and spread more infection

[1]er6022@srmist.edu.in, [2]manorant@srmist.edu.in

DOI: 10.1201/9781003388913-10

which results in low production of food grains. We have classified six pests that cause huge loss and damage in rice grains namely, thrips, green leafhopper, stem borer, leaf folder, rice hispa, and rice bug. We have collected all datasets of insect images from Kaggle and online. In this model, we have taken these following six pests that cause more damage to paddy during its growth stages namely:

Thrips: It affects the rice plant at the nursery stage. The major symptom is that a leaf color will be changed to yellow. Then the leaf will get dried from the top to bottom. The tip of the leaves is rolled.

Green leafhopper: At the early stage of this insect, green eggs can be located on the mid of the leaf and later they will look green with black color. A major symptom is the yellow color of the leaf tip which later gets dried.

Fig. 10.1 Pest that causes damage in rice namely (a) thrips, (b) green leafhopper, (c) stem borer, (d) rice hispa, (e) rice bug, and (f) leaf folder [21]

Source: https://www.nbair.res.in/index.php/node/1314

Stem borer: They affect the rice at any stage from seeding to maturity stage. The major symptoms are brown or yellow color in the leaf. Small tiny holes in the stem and plant can be easily pulled off from the soil due to the damage caused by the pest.

Leaf folder: Leaves will be rolled up and the insect is present inside the rolled leaf. The pest will eat the green tissues in the leaf and change the color of the plant.

Rice hispa: Hispa eats the upper surface of the paddy leaves. Symptoms include wheatish color leaves.

Rice Bug: Rice bug affects the vegetative stage, It sucks the milky–white substance present on the newly grown grain. This causes small, spotty, or empty grains.

II. Related Works

V. Malathi et al. [1] proposed work is based on transfer learning. Deep CNN is used to recognize 10 types of pests present in rice crops. The dataset consists of 3549 pest images that cause damage to rice crops. ResNet-50 classifier achieved the highest accuracy of 95.034%.

K. Thenmozhi et al. [2] suggested a Deep CNN-based model that incorporates pretrained learning algorithm. They gathered photos of insects from 40 different classes and divided them into three databases. Four deep learning classifiers, namely VGGNet, AlexNet, ResNet, and GoogLeNet were used to compare the results. The accuracy rate for identifying pests in food crops was 97.47%.

Chowdhury R. Rahman et al. [3] proposed the deep learning-based CNN model. They have done two stages of work, first, compared all the deep learning model accuracy, and second, proposed a two-stage CNN model with an accuracy of 93.3%. They used 1426 rice disease and pest images collected from the field.

Kusrini, Dr et al. [4] proposed data augmentation–based pest classification in mango leaf. They compared the accuracy of augmented dataset images with original dataset images. The proposed methodology for diagnosing illness in mangoes was accurate to within 76%.

Harikrishnan R et al. [5] proposed the model for identifying and classifying diseases in tomato leaves. Data is collected from Kaggle. They used two different augmentation techniques for classifying tomato disease. ResNet 101 model provided high testing and validation accuracy. The fully connected CNN model uses transfer learning as its classification method.

A deep learning approach for categorizing pests in paddy was proposed by Alfarisy et al. [6]. They created Alexnet and CaffeNet models using 4511 photos of pests they found in web resources. They grouped 4 pests and 9 classes of diseases in paddy and achieved an accuracy of 87%.

Dawei et al. [7] proposed the framework of a deep learning model for classifying 10 insects in paddy. The AlexNet model achieved a precision of 93%.

After reviewing different papers, deep learning models were widely used for classifying pest images in paddy. We can proceed with CNN with one convolutional, 2 Conv, 3 Conv, and 4 Conv layers and test the accuracy. Finally, VGG19 can be used along with fine-tuning the hyperparameters for better performance.

III. Proposed Method for Classifying Pest

A. Dataset Collection

Dataset consists of 6 major pests which cause more damage in paddy namely thrips, leafhopper, stem borer, rice hispa, rice bug, and leaf folder. The collection included 1520 images from Kaggle and from search engines. The dataset size is increased by using data augmentation. The remaining 20% of the data

was utilized to test the accuracy after training with the first 80% of the photos. Figure 10.1 displayed a few illustrations of the insects as examples.

B. Data Augmentation

Data augmentation is one of the methods to increase the size of the dataset. It will generate new data from sample data. We can propose a deep learning model effectively to train the model with augmented data. Transformations that are made in the original image are flipping, cropping, translation, rotation, scaling and zooming. Open-source albumentations Python packages are used for image augmentation packages.

C. Feature Extraction and Classification from ConvNets

To assess the performance, various classification models based on convolutional neural networks are developed. The keras framework is used to construct this model.

The dataset is divided into two parts in the first step: 20% for validation and 80% for training. About 15% of the training images should be kept for testing new data that the model has never seen. This makes it easier to determine whether a model is too well-fit for the trained data. The model is trained using the Adam optimizer for 100 epochs with a batch size of 32. The value for dropout is set to 0.1. This model uses the categorical cross-entropy loss function.

D. Proposed Deep Learning Model with CNN

After data augmentation is applied, the model generates new images. Now we need to train the model on an updated image dataset for another 50 epochs. The next step is preprocessing, where all the images were resized and the image dimension is 28×28, and scaling all values [0,1] interval.

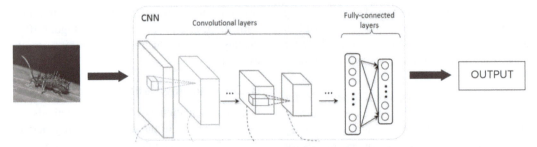

Fig. 10.2 VGG19 model for classifying Insects in Paddy [11]

E. Classification of Pests by Using VGG19

Classification of images is done by the ConvNet model. We proposed the VGG19 model which supports 19 layers which are grouped as 16 convolutional layers, 3 layers for fully connected, 5 for MaxPool, and 1 SoftMax layer. VGG19 takes input in the size of 224×224. The image center is cropped to 224×224 to keep the image consistent without scaling. It converts RGB images to BCR images. ReLu is the activation function used in the VGG19 model. Figure 10.2 shows the VGG19 model which is 19 layers deep. The input is given in the form of pests images. The model is trained with CNN layers and VGG model. After the training is over, we validated our model with some sample pests images.

IV. Results and Discussion

To calculate the training and validation performance metrics, a confusion matrix is used. If I have more than one classification of diseases in the dataset, the manual classification of myself will not be accurate. The confusion matrix will decide and predict the output with some parameters. The metrics which is set for performance calculation are Accuracy, Precision, Recall, and F1-Score.

ImageNet dataset consists of nearly 1520 images of majorly affecting pests in Paddy namely thrips, leafhopper, stem borer, rice

hispa, rice bug, and leaf folder. The accuracy of the proposed model was calculated by using the formula:

Accuracy = (TN + TP)/(TN + FP + TP + FN).

Where,

TN-True Negative, TP- True Positive, FP- False Positive, FN- False Negative.

Classification is done by ConvNet and VGG19 models for calculating whether the performance is,

TRUE POSITIVE: Pest is present in the image and the system classifies correctly that the pest which causes disease in the paddy is present in the given image (TP).

TRUE NEGATIVE: Pest is not present and the model classifies the pest as not present (TN).

FALSE POSITIVE: Pest is not present but the model classifies the pest as present (FP).

FALSE NEGATIVE: Pest is present in the image but the model classifies the pest as not present in the given sample image (FN).

Our proposed model calculates training accuracy, training loss, validation accuracy, and validation loss of 1,3,4 Convolutional layer, and the VGG19 Model is shown in Table 10.1. Normal VGG19 achieved less accuracy of 76.64 whereas after finetuning the VGG19 model by adding dropouts the overfitting problem can be reduced. The error rate is reduced by increasing the epoch value and by finetuning the value of the hyperparameter.

Table 10.1   Results of ConvNet with different layers and VGG19

Recognizing model accuracy	Training accuracy	Training loss	Validation accuracy	Validation loss
1 Layer ConvNet	0.9104	0.2484	0.9229	0.2210
3 Layer ConvNet	0.9079	0.2496	0.9117	0.2291
4 Layer ConvNet	0.9052	0.2656	0.9352	0.1776
VGG19	0.7431	0.7010	0.7664	0.6238

Source: Author

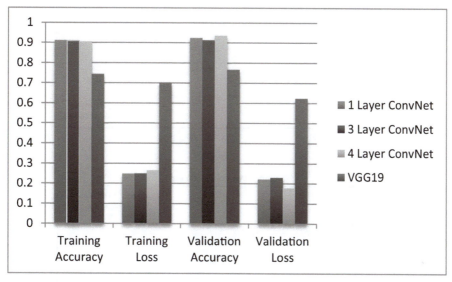

Fig. 10.3   Accuracy and loss of CNN model

Source: Author using Python

Figure 10.3 shows the overall performance metrics for the model. Figure 10.4, 10.5, and 10.6 show the graphical representation of training accuracy and loss and validation accuracy and loss of CNN with one, two, and three convolutional layer. Figure 10.7 shows training accuracy and loss and validation accuracy and loss of VGG19.

V. Conclusion and Future Work

We categorized six pests that harm rice productivity in our model. The training accuracy and loss of the 1-layer ConvNet, 3-layer ConvNet, 4-layer ConvNet, and standard Deep learning model VGG19 were compared in the creation of the model. Convent layers produced good accuracy with less loss whereas the standard VGG19 model produced less accuracy. So, we fine-tuned the VGG19 model by adding dropouts to avoid the overfitting problem. The modified VGG19 model produced a better accuracy rate of 91.33% by using 80% of training and 20% of validation data. Our model accurately classified the insects at the early stage which reduces the damage caused by pests. In the future, how diseases affecting rice crops can be incorporated, and can use other deep learning models to compare and check for better accuracy rates.

VI. Conflict of Interest

The authors are not connected to any companies that have a direct or indirect financial stake in the topics covered in the article.

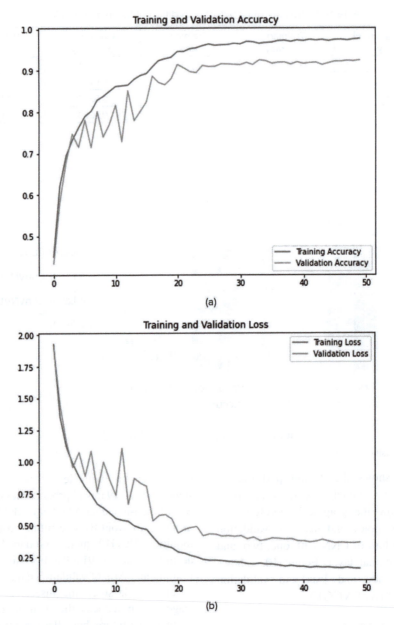

Fig. 10.4 Graph shows (a) training and validation accuracy and (b) training and validation loss of CNN with one convolutional layer

Source: Author using Python

References

1. V. Malathi & M. P. Gopinath., Classification of pest detection in paddy crop based on transfer learning approach, Acta Agriculturae Scandinavica, Section B—Soil & Plant Science (2021).

2. K. Thenmozhi, U. Srinivasulu Reddy, Crop pest classification based on deep convolutional neural network and transfer learning,

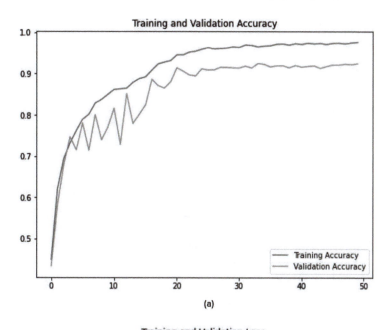

(a)

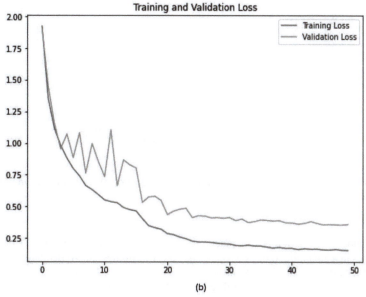

(b)

Fig. 10.5   Graph shows (a) training and validation accuracy and (b) training and validation loss for CNN with three convolutional layers

Source: Author using Python

Computers and Electronics in Agriculture, Elsevier B. V. (2019)

3. Chowdhury R. Rahman, Preetom S. Arko, Mohammed E. Ali, Mohammad A. Iqbal Khan, Sajid H. Apon, Farzana Nowrin, Abu Wasif, Identification and recognition of rice diseases and pests using convolutional neural network, Biosystems Engineering Volume 194, (2020)

4. Kusrini Kusrini, Dr., (Kusrini), Suputa Suputa, Dr. (Suputa), Data augmentation for automated

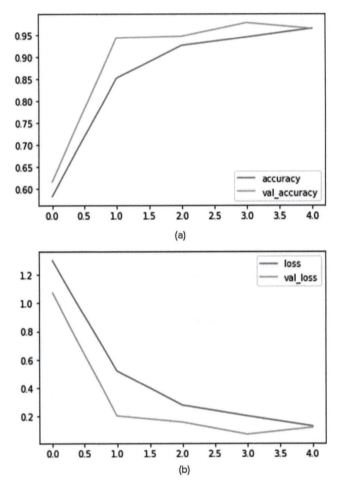

Fig. 10.6 Graph shows (a) training and validation accuracy and (b) training and validation loss for CNN with four convolutional layers

Source: Author using Python

pest classification in Mango farms, Computers and Electronics in Agriculture, (2020).

5. Madhavi G, Dr. A. Jhansi Rani, Dr. S. Srinivasa Rao, Pest Detection for Rice Using Artificial Intelligence. International Research Journal on Advanced Science Hub (IRJASH). May (2021).

6. Krishnamoorthy N. A., L.V. Narasimha Prasad, Rice leaf diseases prediction using deep neural networks with transfer learning, Environmental Research, 2017 Elsevier B. V(2021)

7. Alexander Johannes, Artzai Picon, Aitor Alvarez-Gila, Jone Echazarra, Sergio Rodriguez-Vaamonde, Ana Díez Navajas,

Amaia Ortiz-Barredo, Computers and Electronics in Agriculture, 0168-1699/2017 Elsevier B.V. (2017)

8. Md. Ashiqul Islam1, Md. Nymur Rahman Shuvo, Muhammad Shamsojjaman, An Automated Convolutional Neural Network Based Approach for Paddy Leaf Disease Detection, International Journal of Advanced Computer Science and Applications, Vol. 12, No. 1, 2021.

9. Elakya, R., Manoranjitham, T. (2022). A Novel Approach for Early Detection of Disease and Pest Attack in Food Crop: A Review. In: Gandhi, T. K., Konar, D., Sen, B., Sharma, K.

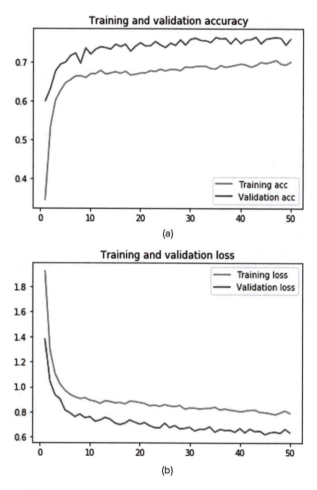

Fig. 10.7   Graph shows (a) training and validation accuracy and (b) training and validation loss for CNN with transfer learning model VGG19

Source: Author using Python

(eds) Advanced Computational Paradigms and Hybrid Intelligent Computing. Advances in Intelligent Systems and Computing, vol 1373. Springer, Singapore.

10. A. D. Nidhis, Chandrapati Naga Venkata Pardhu, K. Charishma Reddy and K. Deepa, Cluster Based Paddy Leaf Disease Detection, Classification and Diagnosis in Crop Health Monitoring Unit, © Springer Nature Switzerland AG (2019).

11. K. Thenmozhi, U. Srinivasulu Reddy, Crop pest classification based on deep convolutional neural network and transfer learning, Computers and Electronics in Agriculture, 0168-1699/ © 2019 Elsevier (2019).

12. Chowdhury R. Rahman, Preetom S. Arko, Mohammed E. Ali, Mohammad A. Iqbal Khan, Sajid H. Apon, Farzana Nowrin, Abu Wasif, Identification and recognition of rice diseases and pests using convolutional neural networks, Bio System Engineering, . Published by Elsevier (2020)

13. Yang Lua, Shujuan Yi, Nianyin Zeng, Yurong Liu, Yong Zhang, Identification of rice diseases using deep convolutional neural networks, Neurocomputing, Elsevier B. V. (2017)

14. T. Gayathri Devi1, P. Neelamegam, Image processing based rice plant leaves diseases in Thanjavur, Tamilnadu, Cluster Computing (2018)

15. Ziyi Liu, Junfeng Gao, Guoguo Yang, Huan Zhang & Yong He, Localization and Classification of Paddy Field Pests using a Saliency Map and Deep Convolutional Neural Network, Scientific Reports (2016)

16. S. Ramesh, D. Vydeki, Recognition and classification of paddy leaf diseases using Optimized Deep Neural network with Jaya algorithm, Information Processing in Agriculture, (2019)

17. Aydin Kayaa, Ali Seydi Kecelia, Cagatay Catalb, Hamdi Yalin Yalica, Huseyin Temucina, Bedir Tekinerdoganb, Analysis of transfer learning for deep neural network based plant classification models, Computers and Electronics in Agriculture, 0168-1699/ © 2019 Published by Elsevier B. V. (2019).

18. Yonghua Xionga, Longfei Liang,Lin Wang, Jinhua Shea, Min Wua, Identification of cash crop diseases using automatic image segmentation algorithm and deep learning with expanded dataset, Computers and Electronics in Agriculture, 0168-1699/ © 2020 Elsevier B. V. (2020)

19. Shriya V., Ishwarya R., Manoranjitham T., 'Probabilistic Neural Network for automatic detection of plant disease using DT-CWT and K-means feature extraction, International journal of Pharmaceutical research, Vol. 12, Issue 1, 2020, pp. 1327–1333

20. A. Subeesh, S. Bhole, K. Singh, N. S. Chandel, Y. A. Rajwade, K. V. R. Rao, S. P. Kumar, D. Jat, Deep convolutional neural network models for weed detection in polyhouse grown bell peppers, Artificial Intelligence in Agriculture, (2022) 47–54

21. M. Vaidhehi & C. Malathy (2022) An unique model for weed and paddy detection using regional convolutional neural networks, Acta Agriculturae Scandinavica, Section B— Soil & Plant Science, 72: 1, 463–475, DOI: 10.1080/09064710.2021.2011395

22. Elakya, R., Seth, J., Ashritha, P., & Namith, R. (2019). Smart parking system using IoT. International Journal of Engineering and Advanced Technology (IJEAT), 9(1), 6091–6095.

23. Rosle, Rhushalshafira & Che'Ya, Nik & Ang, Yuhao & Rahmat, Mohamad Fariq & Wayayok, Aimrun & Zulkarami, Berahim & Ilahi, Wan & Ismail, Mohd & Omar, Mohamad. (2021). Weed Detection in Rice Fields Using Remote Sensing Technique: A Review. Applied Sciences. 11. 10701. 10.3390/app112210701.

24. Arif, Sheeraz & Kumar, Rajesh & Abbasi, Shazia & Mohammadani, Khalid & Dev, Kapeel. (2021). Weeds Detection and Classification using Convolutional Long-Short-Term Memory. 10.21203/rs.3.rs-219227/v1.

Recent Trends in Computational Intelligence and its Application – Sugumaran D. et al. (eds)
© 2023 Taylor & Francis Group, London, ISBN 978-1-032-48410-5

11

Detection of Sink Hole Attack Using RIPEMD Algorithm in Wireless Sensor Networks

Stella K.[1], Nisha A. P.[2], Manjushree S.[3]
Veltech Hightech Dr.Rangarajan Dr.Sakunthala Engineering College, Chennai, India

Vethapackiam K.[4]
Government Polytechnic College, Kadathur, Dharmapuri District, India

Abstract—In recent years, the development of Wireless Sensor Network (WSN) technology has improved tremendously. WSN consists of a collection of sensor nodes that are used to monitor and record the different physical conditions such as sound, temperature, pollution levels, humidity and wind for transmitting the sensed information to a centralized station periodically. Typical applications of WSNs include tracking, healthcare monitoring, battlefield surveillance and many more. However, WSN is exposed to various attacks such as sinkhole attacks, wormhole attacks, blackhole attacks, etc. In this context, we mainly concentrate on sinkhole attack in which, the malicious nodes disseminate mendacious messages to the surrounding nodes. It may transmit false information to the sink node and avoids the node from obtaining the complete and correct sensing data. Hence, security is very important in WSNs. To overcome this security challenge, RIPEMD [RACE Integrity Primitives Evaluation Message Digest] algorithm is proposed to detect and neutralize sinkholes. The various parameters have been analysed for the performance evaluation of the proposed work.

Keywords—Wireless sensor network, Sink hole attack, Security, RIPEMD, Network layer

I. Introduction

Wireless sensor networks use a network of microsensors to track the actions of physical or environmental phenomena and transfer the data to a base station, which links the wireless sensor network and the outside world. Sensor nodes are essentially made up of sensing, data processing, and communication components. Small, light- weight, and portable sensor nodes are the norm. They detect and prepare a report based on the sensed and processed data, which they subsequently communicate to the base station through wireless networks. If numerous nodes in the vicinity identify the same occurrence, the report can be generated cooperatively.

The information sent by the sensor nodes is processed by the base station, which then sends it to the outside world via high-quality wireless

[1]stellaakk16@gmail.com, [2]purush.krb@gmail.com, [3]sudhamanju1972@gmail.com, [4]beulahece1981@gmail.com

DOI: 10.1201/9781003388913-11

or cable channels. In recent years, one of the most rapidly developing technologies is the wireless sensor network and it can be used in a variety of settings, including commercial, industrial, residential, and environmental fields.

Wireless sensor networks are frequently used in hostile and unattended situations in numerous applications. As a result, there is a critical requirement to secure sensing data and sensing readings against attacks such as sinkhole attacks among others. The proposed approach offers a way to deal with sinkholes. The algorithm's main goal is to identify sinkhole attack in WSNs, offer security against them, and increase network stability.

A. WSN Architecture

In WSN, there are two kinds of architectures such as layered and network architecture.

Architecture of Layered Networks

Hundreds of sensor nodes, as well as a base station, are used in this network. The network nodes are organized into concentric layers. The three gross layers are mostly utilized for network control and to improve overall network efficiency. When compared to other sensor network architectures, it consumes less power and is scalable.

Architecture of Clustered Network

One of the popular properties of this network is data fusion, every node collects the data and communicates with a base station via cluster head. As demonstrated in Fig. 11.1, the

construction of the cluster and cluster head is an independent autonomous distributed process. They can also lengthen the sleep durations of nodes by allowing cluster chiefs to coordinate and optimize activities by localizing data transmission inside established clusters.

Sensor nodes form clusters of their own volition in this type of design. It's a two-tier hierarchy clustering architecture in essence. Failover clusters, load-balancing clusters, and high-performance clusters are some of the most prevalent forms of clusters. It's commonly used for energy-conscious routing. Load balancing, lower energy consumption, resource reusability, high availability, and longer network life are all advantages of clustered network architecture.

B. WSN Characteristics

The following are the main characteristics of a WSN:

- *Application Specific Design*: WSNs are application specific, which means that their architecture is focused on the application.
- *Dynamic Network Topology*: In a few applications, nodes are free to travel at random rates and may occasionally fail to operate, add, or replace. As a result, different network topologies are possible.
- *Multi – Hop Communication*: In WSN, a large number of nodes can communicate with a base station via an intermediary node.
- *Scalability And Density*: In various applications, the number of nodes in WSNs is enormous and densely placed.
- *Communication Models*: Wireless sensor networks employ a variety of communication models, including flat, hierarchical, and distributed WSNs, as well as homogeneous and heterogeneous WSNs.
- *Small Physical Size*: Sensor nodes are often small in size and have a limited range, therefore their energy is constrained as well

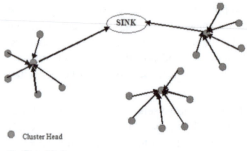

Cluster Head
Cluster Member

Fig. 11.1 Clustered network architecture

Source: Authors

- *Robust Operations*: Because the sensor nodes are used in hostile situations, they must be able to withstand faults and errors. As a result, sensor nodes must improve their ability to self-test, self-calibrate, and self-repair.

C. Network Layer Security Threats in WSN

Sensor networks use data-centric and attribute-based addressing at the network layer. The network layer is vulnerable to a variety of assaults such as,

- Spoofing and tampering with routing information
- Attack on a sinkhole
- Selective forwarding attack
- Wormhole attack
- Spoofing acknowledgments
- Sybil attack
- Hello Flood Attack

II. Related Works

Many attacks wreak havoc on mobile ad hoc networks. The sinkhole assault on the network layer is the most serious of all. In this, the vulnerable sinkhole node catches all mobile nodes by feigning routing and acquiring resources data from other neighbor nodes before sending modified data tampered in the subject of wireless sensor networks.

Neha Singh and Kamakshi Rautela [2016] made a modernistic addition. They've developed a novel application and prototype for sensing and broadcasting data from multiple settings, with the ability to deliver a wide range of desired outcomes at a cheap cost. A collection of small sensor nodes interacts through a radio interface in WSN. As a result, in order to establish balanced transmission and reception of data, low energy consumptions [1] are required at the base station.

Mohammad Wazid [2] et al. [2016] suggested a method for dividing hierarchical wireless sensor networking into separate clusters. Sinkhole node existence and identification techniques are used to detect anomalies and determine the type of sinkhole. Then CH adds those malicious nodes to a blacklist and alerts the rest of the cluster. The detection system utilized here involves fewer message exchanges, resulting in lower communication and computing costs, safeguarding the network against sinkhole attacks.

The RAEED protocol, introduced by Saghar et al. [2016], is used to find the penetration of basic and intelligent assaults. The DOS attack is reduced in this protocol by forwarding the data to the sink node. It can be improved in the future by using formal methods. [3]. Saghar et al. [2017] targeted wireless sensor network security concerns. During the routing process, it evaluates a Denial-of-Service attack. The attacker pulls traffic to the attacker's node and stops data from reaching the nearby node in this type of attack. It efficiently identifies both basic and clever tunnel attacks.

Jan, Mian [6], et al. [2017] suggested a cluster-based wireless sensor network with a lightweight payload-based mutual authentication technique. This is also known as the PAWN strategy. It is implanted in two stages during the implementation procedure. First, the authenticated cluster heads are chosen and then, they are allowed to communicate with nearby nodes in the network. Second, the chosen cluster head serves as a server, granting authentication to adjacent nodes. This scheme has been tested with a variety of schemes, and the results demonstrate that it performs admirably.

The formal models are combined to create an effective model. As the routes are selected over the dependable nodes, the adaptive technique is robust to network dynamism and has a low packet loss rate. In this approach rate of packet loss is less and routing is effective between dependable nodes. Furthermore, when the network converges, the number of false positive and false negative outcomes decreases.

Sumit Pundir [7] et al. [2020] proposed an edge-based IoT context. SAD - EIoT yields a detection rate of 95.83 percent, which is significantly higher than other equivalent existing schemes. It can be employed in very sensitive and important operations. In support vector machines (SVM) on the AODV protocol, Sihem [8] et al. [2021] suggested a unique algorithm underpins the two- way authentication system to prevent sinkhole attacks. Kumar [9] et al. [2017] devised a localization solution that protects wireless sensor networks from the Wormhole attack. When compared to current algorithms, the results reveal that this method outperformed them. For a cluster-based network, it's known as the payload-based mutual authentication technique [PAWN approach].

Zhaohui, Zhang [10], and colleagues [2019] proposed a model based on the RMHDS algorithm. The base station gathered information from all neighbor nodes which are based on their location in the whole network and created a database. The nodes are chosen at random and a smaller number of hop paths to the base station are created. If the count value of passed nodes and also the hop difference between neighbor nodes is greater than the threshold value then those nodes are considered devious nodes. In this method, the detection rate is improved while lowering false positive rates [11] [12].

Kenneth E. Nwankwo [13] and colleagues [2019] presented a two-stage framework. The first stage entails the conceptualization and planning of the problem, as well as the description and design of the dataset. The Ant Colony Detection method was implemented on the network simulator. In wireless sensor networks, our approach enhances the sinkhole detection rate and lowers false alarms [14].

Karthigadevi [15] et al. [2019] developed a disseminated sinkhole finding method based on density estimation and neighbor finding. Each node discovered neighbor nodes in the entire network and neighbor details is collected based on neighbor density. The sinkhole is discovered by observing the traffic patterns of all nearby residents. The overhead of gathering snapshots and routes is reduced using this strategy.

III. Sink Hole Attack Detection

The most dangerous assault in WSN is the sinkhole attack. It is carried out by either hacking a network node or introducing a manufactured network node. A sinkhole attack occurs when a malicious node (sinkhole) presents the best possible route to the sink node, deceiving its neighbor nodes, so that they can use that route more frequently. As a result, the rogue node intrudes on the data and disrupts the network operations. The attacker node then launches the attack using a compromised node. This guarantees that all network traffic will be routed via advertised route. Far from the base station the Sinkhole hub hopes to attract all system movement to itself by doing so. It then modifies the information packet or silently drops the parcel. It increases system overhead, reduces system lifetime by increasing energy use, and finally obliterates the system.

The sinkhole node announces to the sink node that the route via itself is a trust route. Then all the surrounding nodes within the communication range forward the collected data to the sinkhole node which may drop, modify or delete the packets received by them before sending them to the base station. So, we proposed the RIPEMD algorithm to detect the sinkhole node and redirect the packets via an alternate path which is shown in Fig. 11.2. When a new node claims a shorter route to the base station, it's important to figure out whether it's a trustworthy node or a sinkhole. The message was sent via the advertised route as well as the original route. Both routes would come together at a common point, which may be a regular WSN node, a destination, or the base station itself. The nature of the node, whether

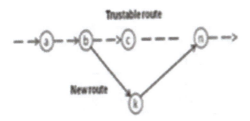

Fig. 11.2 Identification of sinkhole node

Source: Authors

trustworthy or malicious, can be determined by comparing the two messages at the common point of intersection.

IV. Ripemd Algorithm

A 160-bit cryptographic hash function is RIPEMD-160. It's designed to be a safe alternative to the 128-bit hash algorithms MD4, MD5 [16], and RIPEMD. RIPEMD-160, like all MD4 versions, 32-bit words are used. The basic operations are rotation of words by left, Boolean functions such as AND, OR, NOT and 2's complement with modulo 2^{32} functions are performed.

In the RIPEMD algorithm, the input bits are divided into blocks and each block consists of 512 bits. Each block is split into 16 strings and each string consists of 4 bytes. The little-endian convention is used to convert a 4-byte string into a 32-bit word. If the input bits are not multiples of 512 bits, padding bits such as 1 followed by 0's are added. The output of the RIPEMD algorithm contains five 32-bit words. Again, a little-endian convention is used to convert five 32-bit words into 16-bits strings.

In compression operation, five parallel rounds are performed and each round consists of 16 steps. Thus, the total number of steps involved is $5 \times 16 \times 2 = 160$. In each step, the new value is computed for each register which is based on the other four registers. After padding bits, the original message length of 64-bits is appended. The results of the intermediate and final hash values are stored in the 160-bit buffer. Five 32-bit registers such as A, B, C, D, E are initialized

with hexadecimal values. Based on little ending order, the values are stored. 10 rounds of processing are done for each module with 16 steps. The last-round output is added to the first-round input to get CV_{q+1}. The Lth stage output is the 160-bit hash value which is shown in Fig. 11.3.

Hash functions are a type of cryptographic method that addresses the needs for security, confidentiality, and validity in a variety of technological applications. Figure 6 depicts the general architecture of the RIPEMD- 160 core with its pipelined layout. It can be seen that RIPEMD-160 employs two five-round parallel processes, each with sixteen operations (5×16 operations for the process). As a result, we can safely assume that five pipeline stages will be used.

V. Results and Discussion

The performance of the sinkhole attack was analyzed by using NS2 simulation software. Node 1 is chosen as the source node and node 20 is chosen as the sink node. The sink node is represented by red color. The hash value was calculated by using the RIPEMD algorithm which is stored in each sensor node. When a new node announces a shorter route, we have to check whether that node is a trustable node or a malicious node. The sinkhole node is identified by comparing the hash value which is stored in the database and the present value. The sinkhole node is represented by blue color. The simulation output of the detection of sink hole attack is shown in Fig. 11.4.

The following parameters are used to evaluate the performance of network

A. Packet Interception Probability

The Packet interception probability is the probability of interception of shares. When the anomalous area becomes larger, the packet interception probability is more. In RIPEMD, the messages are transmitted via trusted paths, so the packet interception probability is very

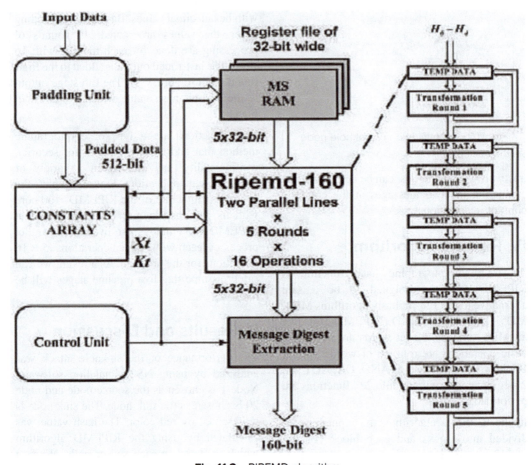

Fig. 11.3 RIPEMD algorithm

Source: https://ieeexplore.ieee.org/document/4676982

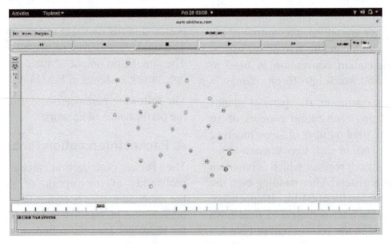

Fig. 11.4 Simulation output

Source: Authors

less. The proposed RIPEMD scheme acquires minimum packet interception probability which is observed and shown in Table 11.2.

Table 11.1 Packet interception probability

No of nodes	MD5	RIPEMD
50	0.008	0.002
100	0.011	0.004
150	0.014	0.006
200	0.020	0.010
250	0.025	0.013

Source: Authors

B. Network Lifetime

The network lifetime is the important metric in WSN. The residual energy and energy consumed in each node determine the lifetime of the network. The network lifetime is defined as the time interval until the first node or group of sensor nodes runs out of energy to send a packet. In RIPEMD, the probability of successful transmission is higher because the probability of intercepting packets by the adversary is less due to the transmission of data by trusted paths. It causes a reduction of number in the retransmissions of data and removes the burden of energy consumption due to retransmission. The network lifetime of the RIPEMD protocol is better compared to MD5.

Table 11.2 Network lifetime

No of nodes	MD5	RIPEMD
30	200	850
50	500	980
100	640	1100
150	730	1190
200	870	1330
250	950	1400

Source: Authors

C. Packet Delivery Ratio

The Packet delivery ratio is determined from the ratio of the number of successfully delivered

packets to the destination against the number of packets generated. The packet delivery ratio is better in RIPEMD compared to MD5. In RIPEMD, the probability of the message might be compromised is reduced due to trusted paths which are used to spread the information from the source node to the sink node. The selection of trusted node prevents node compromised attack and increases the chance of successful delivery of data to the sink node during its first transmission even increase of node density as well as compromised nodes.

Table 11.3 Packet delivery ratio

No of nodes	MD5	RIPEMD
30	92	95
50	93	96.2
100	91	94
150	86	92
200	84	91
250	83	90

Source: Authors

D. End-to-End Delay

The end-to-end delay of the data refers to the destination time minus origination time of the data. When the number of compromised node increases, stable and sustainable state is retained because RIPEMD uses a trustable path and find the shortest multipath for transmitting data to reach the destination. RIPEMD reduces the End-to-End delay by up to 10%. It shows that RIPEMD has sustainable performance for

Table 11.4 End-to-end delay

No of nodes	MD5	RIPEMD
30	64	56
50	76	62
100	98	76
150	155	98
200	178	114
250	187	123

Source: Authors

a large number of sensor nodes. Even though the network density is large, optimization in the delay is feasible because of trustable paths between sources and the sink nodes.

VI. Conclusions

Nowadays, WSN technology has developed tremendously. The malicious nodes disseminate mendacious messages to the surrounding nodes. It may transmit false information to the base station and avoids the sink node from obtaining the complete and correct sensing The sinkhole node announces to the node that the route via itself is a trust route. Then all the surrounding nodes within the communication range forward the collected data to the sinkhole node which may drop, modify or delete the packets received by them before sending them to the base station. The previous research work used the MD5 algorithm to detect sinkhole attacks which will detect with less detection rate. Using our proposed RIPEME – 160 algorithm hash value is calculated. By comparing the hash value of the new advertised node with the hash value stored in the database, the new advertised node is a trusted node or a malicious node can be found. The RIPEMD-160 algorithm is written in c++ as the backend. Using this algorithm hash value was calculated which was passed as arguments to the front end. We compared the hash value of a newly advertised node with the hash value received from the backend, If a mismatch occurs, then that particular node is identified as a malicious node else otherwise that node is a trusted node. In our proposed method, the packet delivery ratio and network lifetime are increased, and packet interception probability and end-to-end delay are reduced.

References

1. Neha Singh, Kamakshi Rautela [2016], International Journal of Engineering and Computer Science ISSN: 2319–7242, Volume 5, pp. no. 17544–17548.

2. Wazid, Mohammad & Das, Ashok Kumar & Kumari, Saru & Khan, Khurram [2016], Design of sinkhole node detection mechanism for hierarchical wireless sensor networks. Security and Communication Networks. 9.10.1002/sec.1652.

3. K. Saghar, M. Tariq, D. Kendall and A. Bouridane, "RAEED: A formally verified solution to resolve sinkhole attack in Wireless Sensor Network," *2016 13th International Bhurban Conference on Applied Sciences and Technology [IBCAST]*, 2016, pp. 334–345, doi: 10.1109/IBCAST.2016.7429899.

4. Mittal, Vikas, Sunil Gupta, and Tanupriya Choudhury. "Comparative Analysis of Authentication and Access Control Protocols Against Malicious Attacks in Wireless Sensor Networks." Smart Computing and Informatics. Springer, Singapore, 2017, pp. 255–262.

5. Yasin, N. Mohammaed, et al. "ADSMS: Anomaly Detection Scheme for Mitigating Sink Hole Attack in Wireless Sensor Network." Technical Advancements in Computers and Communications (ICTACC), 2017 International Conference on. IEEE, 2017

6. Jan, Mian, et al. "PAWN: a payload-based mutual authentication scheme for Wireless sensor networks." Concurrency and Computation: Practice and Experience 29.17 [2017].

7. Sumit Pundir et al. [2020] designed efficient sinkhole attack detection mechanism in egde based IoT deployment

8. Sihem Aissaoui & Sofiane Boukli, Sinkhole attack detection based on SVM in wireless sensor network, international journal of wireless networks and broadcast technologies, IGI Global, vol. 10(2), pages 16–31, July.

9. Kumar, Gulshan and Rahul Saha, "Securing range free localization against wormhole attack using distance estimation and maximum likelihood estimation in Wireless Sensor Networks." Journal of Network and Computer Applications 99 (2017): pp. 10-16.

10. Zhang, Z. et al. M optimal routes hops strategy: detecting sinkhole attacks in wireless sensor networks. Cluster Computer 22, pp. 7677–7685 [2019].

11. Dhivya M et al. vol.2, 2021, Detection and Prevention of Sinkhole Attack in Wireless

Sensor Network using Armstrong 16-digit Key Identity and GAN Network.

12. Abdulmalik Danmallam Bello, Dr. O. S. Lamba, 2020, How to Detect and Mitigate Sinkhole Attack in Wireless Sensor Network International Journal of Engineering Research & Technology (IJERT) Volume 09, Issue 05.

13. K. E. Nwankwo and S. M. Abdulhamid, "Sinkhole attack Detection in A Wireless Sensor Networks using Enhanced Ant Colony Optimization to improve Detection Rate," 2nd International Conference of the IEEE pp. 1–6.

14. S. Aryai and G. S. Binu, "Cross layer approach for detection and prevention of Sinkhole Attack using a mobile agent," *2017 2nd International Conference on Communication and Electronics Systems (ICCES),* 2017, pp. 359–365, doi: 10.1109/CESYS.2017.8321299.

15. K. Karthigadevi and M. Venkatesulu, "Based on Neighbor Density Estimation Technique to Improve the Quality of Service and to detect and Prevent the Sinkhole Attack in Wireless Sensor Network," 2019 IEEE International Conference on Intelligent Techniques in the control, Optimization and Signal Processing (INCOS), Tamilnadu, India, 2019, pp. 1–4.

16. Vidhya, S, "Sinkhole Attack Detection in WSN using Pure MD5 Algorithm." Indian Journal of Science and Technology 10.24 (2017).

Recent Trends in Computational Intelligence and its Application – Sugumaran D. et al. (eds)
© 2023 Taylor & Francis Group, London, ISBN 978-1-032-48410-5

12

DarkNet for Brain Tumor Detection and Classification

Deepa P. L.[1]

Research Scholar, Karunya Institute of Tech. and Sciences & Dept. of ECE,
Mar Baselios College of Engg. and Tech., Trivandrum, Kerala, India

D. Narain Ponraj[2]

Dept. of ECE, Karunya Institute of Tech. and Sciences, Tamilnadu

Sreena V. G.[3]

Dept. of ECE, Marian Engineering College, Trivandrum, Kerala

Abstract—The Central Nervous System (CNS) is the most important part of the human body. Brain tumor is one of the deadliest diseases that affect the CNS. The survival rate of this is very low when compared to other types of cancer. Thus early detection will help to increase the survival rate and diagnose it properly. In this paper, we have proposed an effective way for detecting and classifying brain tumors using different versions of DarkNet like DarkNet-19 and DarkNet-53. We have customized these networks via the method of transfer learning for implementing different networks. Two detectors are proposed for detecting brain tumors from MRI data and two classifiers are proposed for classifying the MRI data into 4 different classes. The detector using DarkNet19 is providing an accuracy of 90.8% and that using DarkNet53 is providing an accuracy of 94.6%. The classifier using DarkNet19 is providing an accuracy of 90.6% that using DarkNet53 is providing an accuracy of 94.1%. The performance parameters indicate that the networks are performing well when compared to the state-of-the-art methods and these can be used for developing computer aided diagnosis systems.

Keywords—Brain tumor detection, Magnetic resonance images, Brain tumor classification, DarkNet-19, DarkNet-53

I. Introduction

Central Nervous System (CNS) is the central processing unit of the human body which consists of the brain and spinal cord. Tumors affecting brain are known as a brain tumors and they are of different types according to the part which is affected by them. Tumors affecting glial cells are known as Glioma. Astrocytoma, Ependymoma, Glioblastoma and

[1]deepa.pl@mbcet.ac.in, [2]narainpons@karunya.edu, [3]sreenavg.ec@marian.ac.in

DOI: 10.1201/9781003388913-12

Oligodendroglioma are the main categories of glioma. Meningioma is the most common type of primary tumor inside the human brain which arises from the meninges. These are occurring inside the human brain and can be categorized mainly into 4 grades. Grade I tumors are the lower grade tumors with a very less growth rate. Grade II tumors and mid-grade tumors have a higher probability to come back after being removed. Grade III tumors are malignant and therefore growth rate is very high. Grade IV tumors are the fastest-growing type and have the highest probability of coming back after completing the treatment.

Different modalities are available for capturing the tumor area. The Magnetic Resonance Imaging (MRI) technique is most appropriate for capturing the tumor area inside the human brain as it is most suitable for soft tissues. Deep Learning is one of the best technology for designing efficient classifiers in terms of accuracy and other performance parameters. There are several pre-trained deep learning models like Alexnet [1], Squeezenet [2], VGGnet [3], Googlenet [4], Xception [5], ResNet [6], Inception [7], DarkNet [8], etc available in the literature. All these networks are Convolutional Neural Networks (CNN). The network performance fully depends on the design of the network, the amount of data used, network parameters, etc. These models can be developed from scratch by concatenating different layers according to the design or from an existing model after retraining. The second type of modeling is known as transfer learning [9]. Most of the existing models use convolutional neural layers as they can extract unique features from images [10].

Classification of brain tumors using ResNet-101 was proposed by Ghosal et. al in [11]. They have used transfer learning to make the network classify brain MRI data. The overall accuracy obtained was 93.83%. Mehrotra et. al. [12] also adopted transfer learning techniques for creating networks using existing models like

AlexNet, ResNet-101, SqueezeNet, etc for classifying brain MR image data. The accuracy was 99.05%, but the dataset consists only of 696 MR images. In [13], Saxena et al. compared VGG-16 and Inception-v3 with ResNet-50. The accuracy obtained was 95 % for the ResNet-50 model and that model outperformed others for a dataset consisting of 253 images. Ahmet et al. [14] compared the performance of ResNet50 with Alexnet, Googlenet, Densenet-201, and Inception-v3 and the accuracy was 97.2%. Another transfer learning-based approach used the pre-trained Densenet201 model for classification purposes [15]. They used a multiclass SVM cubic classifier and obtained an accuracy of 95%.

DarkNet versions are efficient and we can create new networks from them to detect and classify brain MR images. So we are proposing detectors and classifiers using the different versions of DarkNet like DarkNet19 and DarkNet53 for detecting and classifying brain tumors via the method of transfer learning. This paper uses the different versions of DarkNet for this purpose with more than 10000 images. Also, we can prove that data augmentation can avoid biasing problems and improve accuracy to an extent. Section 2 illustrates the proposed network models using DarkNet19 and DarkNet53 for detecting and classifying brain tumors. Section 3 describes the implementation details and section 4 with the details of the dataset we have used followed by the results obtained with detailed analysis in section 5 and conclusion in section 6.

II. Proposed Method

Automated systems need to be generated for identifying and classifying brain tumors. This will help the radiologist to diagnose the disease at an earlier stage and save the patient's life. In this paper, we are proposing deep learning networks for detecting and classifying brain tumors from MRI data. We have customized efficient classifiers like DarkNet19 and

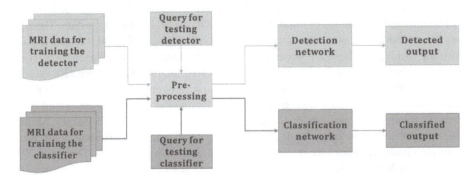

Fig. 12.1 Block schematic of the proposed method

Source: Author

DarkNet53, via the transfer learning method, to classify the MRI data. The block schematic representation of the proposed method is given in Fig. 12.1. The upper portion determines the data flow of the detector part and the lower portion determines that of the classifier part. Brain MRI data is taken for training and testing purposes. The input MRI data size may not be the same as the size or number of neurons of the input layer of the pre-trained networks. The input size of DarkNet19 and DarkNet53 is 256×256. So if the input image size is other than this, we need to resize it to 256×256. In order to resize the images we have incorporated a pre-processing stage. It does the same for the detection and classification stages. This step is also an augmentation of MRI data to equalize the number of samples in all the classes.

After pre-processing, the networks are trained with the training dataset. Two types of detection networks are there - one using DarkNet19 and the second using DarkNet53. Similarly, two types of classification networks are there. After completing the training process, the networks are tested with the pre-processed version of the testing query for getting the detected output in the case of detectors and classified output in the case of classifiers. Here, detection means the network is checking whether a tumor is present in the input MRI data or not and classification means classifying the input MRI data into any of the 4 classes Normal, Glioma, Meningioma and Pituitary tumor.

III. Network Implementation

In this paper, we have proposed 4 different architectures by customizing the existing DarkNet19 and DarkNet53 architectures using transfer learning. We have customized the network models for developing 4 different networks for the identification and classification of brain tumors from MRI data. The first 2 models are for detecting brain tumors to predict whether the tumor is present in the input MRI data or not and the next 2 models are for classifying the MRI data into 4 different classes as stated earlier.

DarkNet19 [7] is a multi-layer CNN which is the backbone of YOLOv2, one of the efficient object detectors. It is using global average pooling and 1×1 convolution filters for optimizing the feature space. This consists of 19 convolutional neural layers including 3×3 and 1×1 filters and 5 pooling layers. Its layered architecture is shown in Fig. 12.2. We have replaced the last 3 layers for detecting 2 classes - Normal and tumor. Then retrained the full network with the brain tumor detection dataset for finding the optimized network parameters. DarkNet19 is again customized into another network for classifying 4 classes. This is also done by replacing the last 3 layers for detecting one among the 4 classes.

DarkNet53 [16] is a multi-layer CNN that is the backbone of YOLOv3, one of the efficient object

Type	Filters	Size/Stride	Output
Convolutional	32	3 × 3	224 × 224
Maxpool		2 × 2/2	112 × 112
Convolutional	64	3 × 3	112 × 112
Maxpool		2 × 2/2	56 × 56
Convolutional	128	3 × 3	56 × 56
Convolutional	64	1 × 1	56 × 56
Convolutional	128	3 × 3	56 × 56
Maxpool		2 × 2/2	28 × 28
Convolutional	256	3 × 3	28 × 28
Convolutional	128	1 × 1	28 × 28
Convolutional	256	3 × 3	28 × 28
Maxpool		2 × 2/2	14 × 14
Convolutional	512	3 × 3	14 × 14
Convolutional	256	1 × 1	14 × 14
Convolutional	512	3 × 3	14 × 14
Convolutional	256	1 × 1	14 × 14
Convolutional	512	3 × 3	14 × 14
Maxpool		2 × 2/2	7 × 7
Convolutional	1024	3 × 3	7 × 7
Convolutional	512	1 × 1	7 × 7
Convolutional	1024	3 × 3	7 × 7
Convolutional	512	1 × 1	7 × 7
Convolutional	1024	3 × 3	7 × 7
Convolutional	1000	1 × 1	7 × 7
Avgpool		Global	1000
Softmax			

Fig. 12.2   The layered architecture of DarkNet19 [9]

Source: Author

detectors. This contains residual connections and more layers when compared to DarkNet19. Thus it is more efficient than DarkNet19 as well as famous classifiers like ResNet101. Its performance is almost similar to ResNet152. This can perform more floating point operations per second and thus it can utilize GPU most effectively. Thus it is faster than the above said networks. This consists of 53 layers as shown in figure 3. As discussed earlier we have replaced the last 3 layers for detecting 2 classes - Normal and tumor. Then retrained the full network with the brain tumor detection dataset for finding the optimized network parameters. DarkNet53 is again customized into another network for classifying 4 classes. This is also done by replacing the last 3 layers for detecting one among the 4 classes. The batch size is set to 200 and the initial learning rate to 0.0001. MATLAB 2021a is used for implementing all four networks. The resizing of images, training of the network and testing are also performed using the same.

IV. Dataset

The brain tumor detection dataset available in Kaggle is used for training and testing the developed networks. A total of 11722 MR images are there in this dataset with 3250 normal images and 8472 tumor images. As the number of normal images is less when compared to tumor images, we have applied affine transformation for balancing the normal class and thus it contains 6500 images. Data augmentation techniques like SMOTE, ADASYN, etc can also be adopted. Among these, 80% of images of each class are used for training and 20% are used for testing.

The figshare dataset is used for training and testing the developed classification networks. This dataset contains 1426 images belonging to the class glioma, 708 images belonging to the class meningioma and 930 images belonging to the class pituitary tumor. We have incorporated the normal images from the previous detection dataset into these classes to make it a 4 class classifier. As the number of images in each class is not comparable, we have applied affine transformation for balancing the data. Here also we can adopt SMOTE or ADASYN for data augmentation. Among these, 80% of images of each class are utilized for training and the remaining 20% for testing.

V. Results and Discussion

In this paper, we have proposed four different networks using DarkNet19 and DarkNet53 - two detection networks and two classification networks. The confusion matrix obtained for the DarkNet19 detector is shown in Fig. 12.4(a). Tumor class is taken as the positive class and normal class is taken as the negative class. Out of 1180 test images out of 1300 test images and 1538 are detected correctly and out of 1694 test images of the class tumor 1538 are detected correctly. The corresponding class accuracies are 90.77% and 90.99% and the overall accuracy of the detector is 90.8 %. The

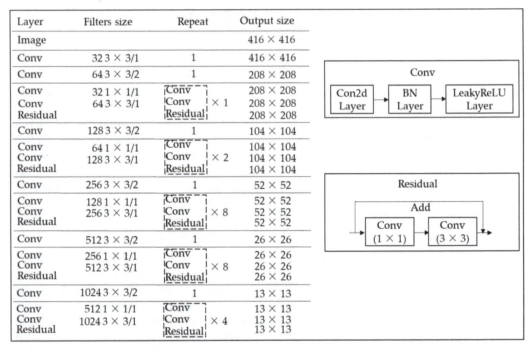

Layer	Filters size	Repeat	Output size
Image			416 × 416
Conv	32 3 × 3/1	1	416 × 416
Conv	64 3 × 3/2	1	208 × 208
Conv Conv Residual	32 1 × 1/1 64 3 × 3/1	Conv Conv × 1 Residual	208 × 208 208 × 208 208 × 208
Conv	128 3 × 3/2	1	104 × 104
Conv Conv Residual	64 1 × 1/1 128 3 × 3/1	Conv Conv × 2 Residual	104 × 104 104 × 104 104 × 104
Conv	256 3 × 3/2	1	52 × 52
Conv Conv Residual	128 1 × 1/1 256 3 × 3/1	Conv Conv × 8 Residual	52 × 52 52 × 52 52 × 52
Conv	512 3 × 3/2	1	26 × 26
Conv Conv Residual	256 1 × 1/1 512 3 × 3/1	Conv Conv × 8 Residual	26 × 26 26 × 26 26 × 26
Conv	1024 3 × 3/2	1	13 × 13
Conv Conv Residual	512 1 × 1/1 1024 3 × 3/1	Conv Conv × 4 Residual	13 × 13 13 × 13 13 × 13

Fig. 12.3 The layered architecture of DarkNet53 [16]

Source: Author

confusion matrix obtained for the DarkNet53 detector is shown in Fig. 12.4(b). Out of 1300 test images of class normal 1180 are detected correctly and out of 1694 test images of the class tumor 1538 are detected correctly. The corresponding class accuracies are 94.46% and 94.63% and the overall accuracy of the detector is 94.6 %. The performance matrices of the detectors are shown in the Table 12.1.

Table 12.1 Performance parameters of DarkNet used for brain tumor detection

Parameters	DarkNet19	DarkNet23
Sensitivity (Recall)	90.8%	94.5%
Specificity	90.8%	94.8%
Precision	88.3%	93.1%
F1 score	0.90	0.94
Accuracy of the network	**90.8%**	**94.6%**

Source: Author

Out of 650 test images of class normal, 570 test images of class glioma, 566 test images of class meningioma and 558 test images of class pituitary tumor 589, 519, 512, and 503 respectively are classified correctly. The corresponding class accuracies are 90.62%, 91.05%, 90.46% and 90.14% and the overall accuracy of the classifier is 90.6%. The confusion matrix obtained for the DarkNet53 4-class classifier is shown in Fig. 12.4(d). All are performing well in terms of various performance parameters. Out of 650 test images of class normal, 570 test images of class glioma, 566 test images of class meningioma, and 558 test images of class pituitary tumor 613, 540, 530, and 522 respectively are classified correctly. The corresponding class accuracies are 94.31%, 94.74%, 93.64%, and 93.55% and the overall accuracy of the classifier is 94.1%. The performance matrices of the classifiers are shown in Table 12.2.

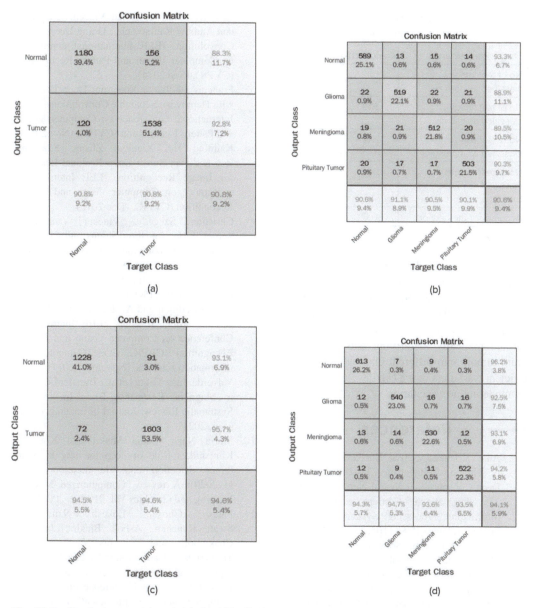

Fig. 12.4 Confusion matrix of (a) DarkNet19 2-class classifier, (b) DarkNet53 2-class classifier, (c) DarkNet 19 4-class classifier and (d) DarkNet53 4-class classifier

Source: Compiled by Author

VI. Conclusion

Automated identification and classification of brain tumors is a necessity in this digital era as disease diagnosis is a tedious process for the radiologist. In this paper, two networks are proposed for brain tumor detection using the efficient classifiers DarkNet19 and DarkNet53 and two networks for brain tumor classification using the above same models. All four networks are developed via transfer learning. The detectors using DarkNet19 and that using

Table 12.2 Performance parameters of DarkNet used for brain tumor classification

Parameters	DarkNet19	DarkNet23
No of test images classified as normal	589	613
No of test images classified as glioma	519	540
No of test images classified as meningioma	512	530
No of test images classified as pituitary tumor	503	522
Accuracy of the network	90.6%	94.1%

Source: Author

DarkNet53 are providing an accuracy of 90.8% and 94.6% respectively and the classifier using DarkNet19 and DarkNet53 are providing an accuracy of 90.6% and 94.1% respectively. As a future scope, we can develop networks for automatically identifying and classifying brain tumors into different classes using other efficient classifiers or we can develop networks by modeling new scenarios.

References

1. Alex Krizhevsky, Ilya Sutskever, Geoffrey E. Hinton, "ImageNet Classification with Deep Convolutional Neural Networks", ACM 25th International Conference on Neural Information Processing Systems (NIPS12), Volume 1, Pages 1097-1105, December 2012.
2. Forrest N. Iandola, Song Han, Matthew W. Moskewicz, Khalid Ashraf, William J. Dally, Kurt Keutzer, : Squeezenet: Alexnet-Level Accuracy with 50% fewer parameters and 0.5 mb model size. IEEE International conference on Computer Vision and Pattern Recognition (CVPR), (Feb 2016.
3. Simonyan, Karen, and Andrew Zisserman, : Very deep convolutional networks for largescale image recognition. IEEE International conference on Computer Vision and Pattern Recognition (CVPR), (Sep 2014).
4. Christian Szegedy, Wei Liu, Yangqing Jia, Pierre Sermanet, Scott Reed, Dragomir Anguelov, Dumitru Erhan, Vincent Vanhoucke and Andrew Rabinovich, : Going Deeper with Convolution. IEEE International conference on Computer Vision and Pattern Recognition (CVPR), (Sep 2014).
5. Francois Chollet, : Xception: Deep Learning with Depthwise Separable Convolutions. IEEE International conference on Computer Vision and Pattern Recognition (CVPR), (Nov 2017)
6. Kaiming He, Xiangyu Zhang, Shaoqing Ren and Jian Sun, : Deep Residual Learning for Image Recognition. IEEE International conference on Computer Vision and Pattern Recognition (CVPR), (Dec 2015).
7. Christian Szegedy, Vincent Vanhoucke, Sergey Ioffe, Jonathon Shlens and Zbigniew Wojna, : Rethinking the Inception Architecture for Computer Vision. IEEE International conference on Computer Vision and Pattern Recognition (CVPR), (Dec 2015).
8. Joseph Redmon and Ali Farhadi: YOLO9000: Better, Faster, Stronger. IEEE International Conference on Computer Vision and Pattern Recognition (CVPR), FOS: Computer and information sciences, (November 2017).
9. Valverde, Juan Miguel et al. : Transfer Learning in Magnetic Resonance Brain Imaging: A Systematic Review. Journal of imaging vol. 7, (Apr. 2021).
10. Maria Nazir, Sadia Shakil and Khurram Khurshid, : Role of deep learning in brain tumor detection and classification (2015 to 2020): A review. Computerized Medical Imaging and Graphics Vol. 91, (Jul. 2021).
11. Palash Ghosal, Lokesh Nandanwar, Swati Kanchan, Ashok Bhadra, Jayasree Chakraborty and Debashis Nandi, : Brain Tumor Classification Using ResNet-101 Based Squeeze and Excitation Deep Neural Network. IEEE International Conference on Advanced Computational and Communication Paradigms (ICACCP), (Oct 2019).
12. Rajat Mehrotra, M.A.Ansari, Rajeev Agrawal and R.S.Anand, : A Transfer Learning approach for AI-based classification of brain tumors. Elsevier Journal on Machine Learning with Applications, (Dec 2020).
13. Priyansh Saxena, Akshat Maheshwari, Shivani Tayal and Saumil Maheshwari, : Predictive modeling of brain tumor: A Deep learning

approach. Innovations in Computational Intelligence and Computer Vision, pp 275-285, (September 2020).

14. Ahmet Cinar and Muhammed Yildirin, : Detection of tumors on brain MRI images using the hybrid convolutional neural network architecture. Elsevier Journal on Medical Hypotheses, Vol. 139, (June 2020).

15. Sharif M.I., Khan M.A., Alhussein M. et al., : A decision support system for multimodal brain tumor classification using deep learning. Complex Intell. Syst. (Mar. 2021).

16. Joseph Redmon and Ali Farhadi: YOLOv3: An Incremental Improvement. IEEE International Conference on Computer Vision and Pattern Recognition (CVPR), FOS: Computer and information sciences, (April 2018).

Recent Trends in Computational Intelligence and its Application – Sugumaran D. et al. (eds)
© 2023 Taylor & Francis Group, London, ISBN 978-1-032-48410-5

13

IOT Based Smart Sprinkling System in Farming

Stella K[1], Ranjith Kumar S.[2], Sathish Kumar M.[3], Rajesh Kumar M.[4]
Veltech high-tech Dr. Rangarajan Dr. Sakunthala Engineering College, Chennai, India

Abstract—People worldwide have performed a vital role to improve irrigation. Mechanization and automation have increased the productivity of agriculture and greatly reduced the need for human labour. The main goal is to automate the cultivation of plants and trees without human intervention and also at the same time use low-priced components which could be used by everyone. The computerized irrigation gadget is totally based on different types of sensors, their design maximizes the broadcast quality and produces tremendous computerized structures. It assists human everyday activity, therefore this help to save time. The gadget makes use of a sensor, relay, DC motor, Microcontroller, and battery. This gadget switches ON/OFF when irrigation is needed and when not needed it will monitor the degree of dryness of the soil. The sensor reading is running on a monitor continuously. This will help to reduce water losses. The farmers are enjoying planting trees without fixing the irrigation and making plans. Computerized irrigation helps in the development of agriculture by reducing manpower by measuring numerous parameters and using it to increase the efficiency of irrigation.

Keywords—Irrigation, Sprinkling, IoT, Sensor, Automation

I. Introduction

Smart irrigation structures use sensors for real-time or historic information to tell watering workouts and adjust watering schedules to enhance efficiency. Soil is primarily based on totally clever irrigation structures that use neighborhood soil moisture information drawn from sensors inside the floor to guide knowledgeable selections approximately watering schedules. Any water re assets may be used for this irrigation gadget. It even senses the crop increase and might locate the water stage inside the close-by source. The fundamental benefits are it is far price effective, time-saving and automatic. This gadget is one of the finest benefits of a clever irrigation gadget.

In the past few years, the water techniques waste 50% of water by evaporation and overwatering and are ineffective for irrigating. It can offer excessive accuracy in water delivery and keep away from water from wastage. Due to routinely handling person calls for much less

[1]stellaakk16@gmail.com, [2]vh10990_ece20@velhightech.com, [3]vh10986_ece20@velhightech.com, [4]vh11008_ece20@velhightech.com

DOI: 10.1201/9781003388913-13

guy power. With the assistance of the sensors, it may correctly decide the soil moisture levels. It additionally lets in controlling the quality of water brought to the plant life while it's far needed. The environmental conditions are measured daily and they are updating it every moment which farmers can see. Using the CIMIS 8% of yield is increased and it reduces the use of water by 13%. They will receive an alert if the condition is a problem for farmers and they can able to control it.

2. Related Works

Agriculture is an essential aid for residing salary in Pakistan. A clever, sensible, and absolutely computerized agricultural machine become required and extraordinarily applicable in a few closings a long time whilst our populace grew exponentially in assessment to the herbal sources, we've got it in Pakistan and in our country. So IOT-primarily totally clever irrigating machine which will reduce the loss of wasted water. We are addressing our project to implement irrigation by less amount of water usage and reducing the usage time. There are specifics to this farming industry. Plastic tunnel farming is divided into three categories: low, excessive, and walk-in tunnels. The excessive tunnels due to their larger size make the process of sowing, irrigation, and harvesting much easier compared to the low and walk-in tunnels.

Compared to this traditional farming, however, would result in increased water waste due to its unforeseeable nature. One of the important which we would be addressing is that any software produced for the clever watering machine should be more environmentally friendly and precisely timed. Cloud computing alone is not simply sufficient for larger-scale IoT software. Quick software that uses the higher structure to deal with diverse records from sensors that is more environmentally friendly should be made available. The objective of this small and intelligent irrigation system is by taking the water sparingly which results in

meeting the needs of the plant more efficiently to store insufficient sweet water reservoirs.

Most of the sensor-related intelligent irrigation structures use many cellular applications that were built in extraordinary times to deal with this difficult problem, but their reliability may be questioned as records develop, increasing the latency rate of IoT devices. The basic factors of humidity, temperature, soil moisture, and mild depth were employed, as in earlier works, and the decision of whether to water plants or not was no longer based on fuzzy common sense [1]. The same hazy common sense has been applied to various healthcare facilities, where biosensors have been used to track temperature, oxygen, blood pressure, and wound infection status [2]. In the same way, in the hearth place worrisome packages, this era was quite helpful in 2018 [3] and 2019 [4]. We are now introducing you to a new age. This is the sum of the system's technique and interpretation for a few parameters which include the type of soil, kind of crop, and type of weather and also the output of the sensor, humidity, temperature, and soil moisture. Because the studies on irrigation structures so far haven't been very green, they can't be carried out on large-scale structures and have significantly lower performance for all sensing records. Due to overburdened sensors, a sophisticated irrigation machine with area computing software is required. As a result, a whole new intelligent and sensible machine must be created.

According to our research, there are a few reasons why changes to the current computer are required. Existing clever irrigation [5] structures either focus on less important readings like moisture, soil humidity, and air, or provide an uncleared function which is carried out in MATLAB to provide a choice in the output and use a plain system to study the set of rules used to predict water wanted for vegetation. However, a machine that does not discover the latency fee cannot provide a dependable answer. By ignoring critical criteria such as soil strata and crop type, a less-than-ideal watering mechanism for vegetation can be created. Unwanted records loading at the IoT server as a result of non-stop sensor records will cause the IoT server to perform poorly.

A sensible irrigation machine should never halt due to an overburden of records. As the most recent understanding has come to light as a result of development in every discipline, we should also e-trade our traditional irrigation technique to an advanced, clever, ideal, and simple expertise database. A GSM- based irrigation control system [6] [7] that had the ability to maintain consistent environmental conditions was discussed in Ayush Akhouri's paper. Similarly, for mobile devices, the Android operating system which gives you the tools and APIs you need to develop apps are utilized, which includes middleware, an operating system, and important applications. In an irrigation control system, the GPRS function is taken advantage of. However, this technique is not very cost-effective and could only cover a small portion of land [8]. The automatic control of closed circuits in Mansur impacts the growth and yield of yellow corn which is irrigated by the Rain gun system. The effect of the closed-circuit drip irrigation was studied by comparing it to a modified irrigation [9] [10] used in an I-Haze region of Saudi Arabia growing yellow corn.

The field experiment was carried out by using three lateral lines consisting of 80, 60, and 40 meters. The types of drip irrigation [11] on circuits were (a) closed circuits which had one manifold used for later lines (CM1DIS); b) CM2DIS, which is made of closed circuits having two manifolds are also used for lateral lines, it is done in order to compensate for salt leaching and ETC which results in more power wastage. A system that contributes to developing the production of greenhouses in Morocco was proposed. This system automates drip fertilizer irrigation in greenhouses.

The proposed approach entails the creation of an integrated system that automates drip fertilizer irrigation in greenhouses. A PC-controlled data acquisition card the PCC-812 PG played a role in the functioning of the system. A hydraulic circuit which is powered by an electric pump is used in irrigation. The soil status is measured using a soil humidity sensor to determine water requirements. Supplying water artificially to the crops growing in Purnima was proposed by S.R.N Reddy " The design of Automatic remote monitoring [12] [13] [14] and control system which employed the use of GSM-Bluetooth." Handpumps, canal water, and rainfall was the primary source of water used in irrigation. However, these methods posed serious problems like watering, over-irrigation, leaching and loss of nutrients in the soil.

Various changing circumstances resulting in environmental degradation and inadequate supply of water have created a need to develop a method for efficiently managing field irrigation. An automated irrigation system is a machine-based system that irrigates the land by automating and combining multiple hardware and software in field irrigation. This research examines a variety of GSM-based automated farm irrigation systems [15] in depth. GSM is a crucial component because it is in charge of managing irrigation systems and transmitting them to receivers via coded signals. Our research focuses on comparing different GSM techniques.

3. Proposed System Architecture

The connection for the smart irrigation system uses the power supply, relay module, microcontroller, and soil moisture sensor which indicates the moisture and humidity, it uses a water pump that can able to transfer water from one place to another and the pump is controlled by the relay module and the relay module controlled by the master control Arduino UNO consumes a power of 5v from the battery and the pir passive infrared sensor is used because of the accuracy and if the animals or other unwanted animals which are harmful to land can be controlled and intimated to the farmer by the message or it updated to IoT server the farmer can catch up the details the PIR sensor have three pins namely Vcc, ground and the digital output and PIR sensor is an output device [16] [17].

The system architecture of smart sprinkling system is shown in Fig. 13.1. The buzzer is used for to make sound which can make panic for the animals which have to get out of a land if the animals are get into the field which can be detected by the PIR sensor and Arduino can get the input form the output of the PIR sensor and make turn on the buzzer and also sending the information that is animals entering into the field the buzzer is connected to the terminal of pin 13 and ground to make use of a buzzer the moisture sensor is used for the purpose to maintain water in the field and the readings are noted by the Arduino by pin A0 the A0 pin is the analog pin for Arduino to have a input of analog devices the moisture sensor have a three pins namely vcc ground and analog output A0 the output of the soil moisture sensor is the input for the Arduino uno and by the collection of data if the water level below 10% the Arduino uno turn on the relay which is connected to the PWM pin2 in the Arduino uno and the relay is collected to the motor which can able to

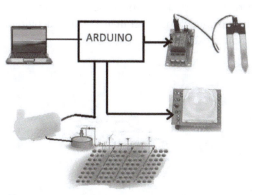

Fig. 13.1 System architecture of smart sprinkling system

Source: Authors

transmit water from one place to another place so the water is transmitted to ground and if the soil moisture senor have a value of greater than 80% the Arduino is came to know the water is filled to the field and Arduino turn off the pwm pin 2 which is connected to the relay the relay stop conducting power with a loss of power the motor cannot pump so the water not move to the field. The enterprise turns out to be extra aggressive and sustainable. Also in a dry region, in which there may be not enough rainfall, the right irrigation isn't always possible.

Hence via way of means of the usage of this irrigation device, a way of tracking the moisture content material in the soil saturation is dry agriculture is the spine of all advanced countries. It makes use of 85% of clean water sources internationally this percent remains dominant in water intake due to the population booming and elevated meals to triumph over those troubles and to lessen the person's energy plant communicator device has been used.

A. Proposed Algorithm

Step 1: Start.

Step2: Set initial values as soilMoistureValue = 0, percentage=0, led = 13, state = LOW, val = 0, soilMoistureValue = analogRead(A0); Serial.println(percentage);

percentage = map(soilMoistureValue, 490, 1023, 100, 0);

Step 2: Acquire data from sensor.

Step 3: Check the percentage values. if(percentage < 10)

{

  Serial.println(" pump on");

  digitalWrite(3,LOW);

}

if(percentage >80)

{

  Serial.println("pump off");

  digitalWrite(3,HIGH);

}

Step 4: Read sensor value

if (val == HIGH) { // check if the sensor is HIGH

   digitalWrite(led, HIGH); // turn LED ON

   delay(500); // delay 100 milliseconds

     if (state == LOW) {

        Serial.println("MOTION DETECTED NEAR THE GROUND");

   state = HIGH; // update variable state to HIGH

   }

 }

 else {

   digitalWrite(led, LOW); // turn LED OFF

   delay(500); // delay 200 milliseconds

   if (state == HIGH){

     Serial.println("MOTION DENIED");

     state = LOW; // update variable state to LOW

Step 5: If soil moisture level is less than threshold values then it start irrigation.

Step 6: Otherwise it stops irrigation.

Step 7: Stop.

IV. Current Scenario

The direction of making plant communicator connected to diverse trades which include fertilizers, natural fertilizers sprinkler irrigation machines, and extra plant communication with a one-of-a-kind characteristic which includes the measure of moisture of soil, temperature, and making irrigation of plant the moisture for an underneath level the thing in irrigation is sprinkler method as by comparing to that method the water source is used high as by using the drip irrigation the water reduces compare to sprinkler method though there remains to associate a validity for technology, this can be for the most part than to the cognitive content of the emergence of these technologies the project has to be adapting the place of the amendment and the cost of the drip irrigation is high and the project smart irrigation target of the agriculture sector and to target creating for less expensive with the international market

V. Results and Discussion

Set up a proposed method irrigation gadget. All of the additives started inside the block diagram are hooked up to the degree of the actual time values of the soil. The effects of our test inside the shape of s standard instruction of our computerized irrigation gadget examined the idea of a microcontroller and Arduino IDE integrated development environment and the moisture sensor started noting the dryness of the soil. When the dryness reaches a level lesser than 10% the Arduino started turning on the PWM pin which is the 2nd pin in Arduino UNO where the relay is connected the relay turns on the relay connected to the motor pump and makes it to off state to on the state when it reaches on state it starts pumping the water to the field.

If the moisture level reaches up to 80% of the value, then the Arduino turns off the PWM pin which is connected to the relay and the motor pump turns off and stops watering the field and the ph sensor is used to measure the ph value of water and check it is good or any mixture of unwanted content of minerals which will reduce the yield and the temperature is measured by using this ph sensor which is shown in Fig. 13.2. Once more and watering few seconds on the other hand stopped and given a regular cost. Irrigation inside the plant, and the moisture cost earlier than and after-acquired from an excel file.

Fig. 13.2 Output of smart sprinkling system

Source: Authors

VI. Conclusion and Suggestions

Irrigation is an important issue for monetary in any growing international location like Nepal, over the years, specialists worried about irrigation applied a guided approach to irrigation. The guide approach has masses of benefits and is pretty dependable in agriculture irrigation has a direct effect on the field. This gadget's goal is to get rid of the tractional guide approach of irrigation which wishes have to be progressed over time the challenge handed evolved after reading the marketplace requirements which makes it extremely appropriate with inside the content of the scenario. The gadget is too beneficial at the actual time state of affairs and quits customers are interested in the use of this gadget.

References

1. Sanku Kumar Roy, "IOT-Based Dynamic Irrigation scheduling system for water management of Irrigated crops", IEE Internet of things Journal, Vol. 8, No. 6, March 2021.
2. Rafael Gomes Alves, "Discrete-event simulation of an irrigation system using Internet of Things", IEEE Latin America Transactions, Vol.20, Issue. 6, June 2022.
3. Veem Divyaki, A Real time perpetration of a GSM grounded Automated irrigation Control System using drip Irrigation Methology, Vol. 4, No. 5 and May 2013.
4. Mansour H. A. Automatic control of unrestricted circuits main gun irrigation system on unheroic com growth and yield International Journal of Advanced Research (2013). Vol. 1, No. 10, pp. 33–42.
5. M. Guerbaoui Y., et al, Grounded automated drip irrigation system Vol. 5, No.1, January 2013.
6. Suresh, V. R. Balaji, K. Govindaraju. "GSM grounded Automated Irrigation Control using Rain gun Irrigation System", International Journal of Advanced Research in Computer and Communication Engineering, Vol. 3. Issue 2. February 2014.
7. Purnima S. R. N. Reddy. "Design of Remote Monitoring and Control System with Automatic Irrigation System using GSM Bluetooth", on IJCA 2012.
8. Chayan, C. H. (2014), Wireless Monitoring of Soil Moisture. Temperature & Moisture Using Zagbec in Agriculture. International Journal of Engineering Trends and Technology. 11 (10), 493–497
9. Dobbs, N. A. (2014). Assessing Irrigation applied and nitrogen percolated using different mart irrigation technologies on bahiagrass (Paspalum notatum). Irrigation Science, 32, 193–203
10. Nor Adni Mat Leh, et al, "Smart Irrigation system using Internet of Things", IEEE 9th international conference on System Engineering and Technology, October 2019.
11. Archana M. S., & Agrawal D. G. (2013). Bedded Controlled Drip Irrigation System International Journal of Emerging Trends and technology in computer.
12. Taslim Rea, Qider Newar. Jurnal Uddin, Touhidul Islam, and Jong-Myon Kim."Automated Irrigation System Using Solar Power" 2012 IEEE
13. Awasthi A., & R Romero S. R. N. (2013). Monitoring for Precision Agriculture using Wireless Sensor Network A review GICST-E Network. Web & Security. 13(7)
14. Pavithra D. S, "GSM grounded Automatic Irrigation Control System for Effective Use of Coffers and Crop Planning by Using an Android Mobile, I0OSR Journal of Mechanical and Civil Engineering (IOSR-JMCE), Vol 11, issue 1. Jul-Aug 2014, pp: 49–55
15. Laxmi Shabadi, NandiniPatil. Nikita, Smitha, P&S wati C, "Irrigation Control System Using Android and GSM for Effective Lise of Water and Power", International Journal of Advanced Research in Computer Science and Software Engineering. Vol. 4, Issue 7, July 2014
16. Shiraz Pasha B. R. Y. P. Patil, "Microcontroller Based Automated Irrigation System". The International Journal of Engineering and Science (IES), Vol. 3, Issue 7, pp. 06–09, June 2014.
17. S. R. Kumbhar, Ghanale, "Microcontroller grounded ControlledIrrigation System for Plantation". Proceedings of the International Multi Conference of Masterminds and Computer Scientists, March 2011.

Recent Trends in Computational Intelligence and its Application – Sugumaran D. et al. (eds)
© 2023 Taylor & Francis Group, London, ISBN 978-1-032-48410-5

14

Design and Development of LoRaWAN Based Module for Pulse Output Water Meter

Vivek B. M. E.[1], Shridhar M.[2], Snega S.[3], Sugamathi R.[4] and Kavin Prakash T.[5]
Electronics and Communication Engineering, Kongu Engineering College, Erode, India

Abstract—Water is an important resource for all the living things. Monitoring water resources become significant because of the multiple purpose for human beings. The rapid increase in population makes the water scarcity problem more serious. To control and monitor water use, a water monitoring process is needed. Water meters are used in the process of water monitoring. In the existing water meters, the user had to check and monitor the water usage manually by viewing the water meter at a close distance. In order to reduce the water scarcity problem and to monitor remotely, a LoRaWAN-based retrofit for the pulse output water meter is designed. By using the proposed design based on LoRaWAN, the manual checking of the water meter is not needed and the user can get the data for water usage from any place through the application server. LoRaWAN is defined as Long Range Wide Area Network protocol which is designed to establish a connection for battery-operated devices wirelessly through the internet. LoRaWAN focuses on the basic needs of Internet of the Things (IoT) such as two-directional communications, end-to-end encryption, and localized services. By using the LoRaWAN module-based water meter, the real-time monitoring of water meter readings can be monitored. The proposed water meter can be used in industries, educational institutions, agricultural purposes, apartments and individual residential houses to check the amount of water flowing from any distance. The components used in the proposed design are the retrofit Water meter, retrofit STM8 board and the DLOS8 LoRaWAN gateway. The output data can be transferred to the application server through the LoRaWAN gateway.

Keywords—Internet of Things (IoT), LoRaWAN, Pulse output water meter, STM8 board, LoRaWAN gateway

I. Introduction

Water metering is a process used to help the user for monitoring water use. Monitoring water use is essential for domestic, industrial agricultural, and other important purposes. An industry can understand and calculate the total volume of water used for production, cleaning machinery, and other purposes through the water monitoring process. A farmer can check the total volume of water used for growing crops, cleaning the land, and other agricultural

[1]vivekerode2005@gmail.com, [2]shridharmanivannan@gmail.com, [3]snegasm03@gmail.com, [4]sugamathisep17@gmail.com, [5]kavinprakasht@gmail.com

DOI: 10.1201/9781003388913-14

purposes through water monitoring. Educational institutions, apartments, and hospitals are some of the other important places that require a water metering process [1]. Water meters play a vital role in the water monitoring process. The existing water meters are the mechanical water meter, electrochemical water meter, and ultrasonic water meter. All the existing water meters require manual checking of reading by viewing the water meter at a close distance. The existing water meters will not send the day-to-day water usage readings and overall water usage for months and years. To check the water flow daily and to monitor the water usage from a long distance, the LoRaWAN-based retrofit pulse output water meter can be implemented [2-3].

II. Related Works

The following existing work mainly focussed on energy consumption and security configuration using LoRaWAN. To implement the work, the LoRaWAN module is designed with some important aspects such as transmission power, carrier frequency, bandwidth, and spreading factor. The simulation results of the work were compared with the results of other existing works. This work will be only used for the security and energy consumption of data and not for monitoring the water flow [4]. Other existing works used spliced convolution for recognizing the water meter. The work is ultimately based on image recognition and the Internet of Things (IoT). In the implemented work, the readings for water used can be collected. The design is not based on the LoRaWAN protocol [5]. The other proposed works were based on water consumption in different industries such as the pharmaceutical industry and cotton textile industry which mainly need water of purity. These works also focus on how an industry uses wastewater management and what are the possible ways to recycle the water for industrial use. These works give information on what are the needs for an industry to use water. By

analyzing the requirements of the industry, a new kind of water meter can be produced [6-7]. The other proposed work is based on geospatial technology for conserving soil water. The primary aim of the work is to help to preserve the soil water. The work is implemented in plot-wise priority zones by remote sensing. By using this work, we can conserve only water under the soil. The LoRaWAN technology is not used in this work [8]. The other existing work is mainly focused on fire monitoring using LoRaWAN in rural areas. The system also detects temperature, carbon dioxide level and wind speed in that particular place using LoRaWAN technology. Water monitoring is not possible by using this design [9]. Other existing works were based on monitoring the air quality using LoRaWAN [10-11] and water leakage detection using smart sensing [12].

III. Motivation & Research Gap

From the previous designs, monitoring water flow through water meters beyond several distances is not possible. Some of the existing hardware designs used sensors to monitor the water flow [13-14]. Most ordinary people find it difficult to calculate the usage of water flow using existing water meters. To overcome this, a LoRaWAN-based module for the pulse output water meter is designed. The user can calculate the water flow from any distance using the proposed design.

IV. Proposed Methodology

In the existing traditional water meters, the real-time monitoring of water flow cannot be achieved. It is not possible of monitoring the water flow through a water meter beyond several distances. The day-to-day usage of water cannot be calculated through the existing water meters. By using LoRaWAN based water meters on IoT, the real-time monitoring of water meter readings can be monitored. It can be used in industries, educational institutions, agricultural purposes, apartments, and individual residential houses to check the amount of water flowing daily from any distance. The components used for the proposed design are a retrofit water meter, a retrofit STM 8 board, and DLOS8 LoRaWAN Gateway. Connecting the retrofit water meter to the STM8 board through two wires. The STM8 board process the data and transfer it to the network server and application server through DLOS8 LoRaWAN Gateway. The amount of water flow can be monitored daily through computers, laptops and mobile phones. Figure 14.1 shows the block diagram of the proposed design.

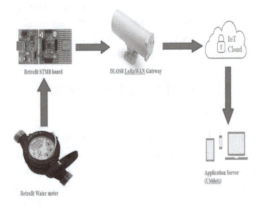

Fig. 14.1   Block diagram

A. Working Principle

The working principle explains the working of the proposed system The pulse output water meter consists switch having a magnet attached to a dial or drum which is inside the meter. The switch is normally in the open state. The switch gets closed when the magnet is in the nearest distance which pulls the reed to the closed position. When water flow occurs, the magnet attached to the drum or dial gets rotated. The position of the magnet gets altered and because of the change in the position of the magnet, the switch gets opened and closed. The closing and opening of the switch are the single pulses. Figure 14.2 explains the working principle of the proposed design.

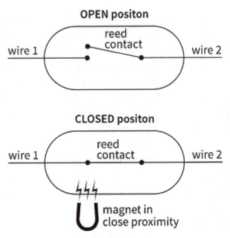

Fig. 14.2   Working principle

5. Components Description

The components description explains the hardware components and software used in the proposed design.

A. Retrofit Water Meter

The water meter used in the proposed design consists of needles inside. The needles get rotated due to pressure caused by the water flow. This results in producing an output pulse which is given as input to the STM8 board by connecting two wires from the water meter. The pulse output water meter measures the water flow using turbine rotation with a propeller. The volume of the water flow is directly proportional to the speed of the rotation of the blades.

The output pulse from the water meter is transferred as input to the STM8 board. The STM8 board will transfer the data to the application server through the DLOS8 LoRaWAN gateway.

B. Software Description

The software used in the proposed method is Serial port Utility, The Things Network and Ubidots stem. The Serial port Utility is used for serial monitoring and viewing any print statement. For getting the live data from the network server, The Things Network (TTN) software platform is used. Ubidots stem is used as an application server for the proposed experimental setup.

VI. Result and Discussion

The proposed design is implemented for the real-time application. The pulse output water meter is connected to the pipeline of a water tap along with the STM8 board. The setup is connected with the DLOS8 LoRaWAN gateway through Wi-Fi. The following Table 14.1 shows the water usage monitored by the proposed LoRaWAN-based retrofit pulse output water meter.

Table 14.1 The amount of water used

S. No.	Time (AM/PM)	Counts		Volume of Water Used in Litres
		Instant Count	Accumulated Count	
1	11:42	0	0	0
2	11:43	0	0	0
3	11:44	1	1	5
4	11:45	1	2	10
5	11:46	1	3	15
6	11:47	1	4	20
7	11:48	0	4	20
8	11:49	0	4	20
9	11:50	1	5	25
10	11:51	1	6	30
11	11:52	1	7	35
12	11:53	0	7	35
13	11:54	0	7	35
14	11:55	0	7	35
15	11:56	1	8	40
16	11:57	1	9	45
17	11:58	0	9	45
18	11:59	0	9	45
19	12:00	0	9	45
20	12:01	1	10	50

From the accumulated count and instant count of the live data, the volume of water can be calculated. The one complete rotation of the needle inside the water meter is considered as one count. The count which occurs immediately when the needle completes one full rotation is called an instant count. The count which gets increased, when there is a continuous water flow is known as the accumulated count. One complete count is equal to five liters as per the water meter used in the proposed design. From the instant counts and accumulated counts, the volume of water used can be monitored. The live data obtained from the network server (TTN) is transferred to the application server (Ubidots stem). The results obtained in the dashboard of the application server can be viewed in many types. From the dashboard, the amount of water used can be calculated. Figure 14.3 shows the water usage from the application server.

VII. Advantages

The advantages of the proposed design include good accuracy, real time monitoring and minimal cost.

VIII. Conclusion and Future Scope

To conclude the work, the LoRaWAN-based pulse output water meter is successfully completed. By using the proposed design in the real-time application, we can able to monitor the day-to-day usage of water through the application server from any place. The proposed water meter is more useful than the existing water meter since there is no need of monitoring the water meter manually. In the future, instead of using conversion from the instant count and accumulated count to volume in liters for measuring the water flow, the volume in liters can be sent directly from the water meter to the application server.

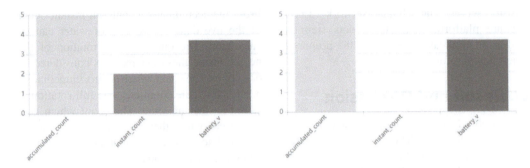

Fig. 14.3 Bar charts for water usage from the application server

Table 14.2 Advantages of the proposed design in comparison with existing design

S. No.	Existing Design	Proposed Design
1	Water flow cannot be monitored from several distances.	Water flow can be monitored from several distances because of implementing LoRaWAN.
2	The general people find it difficult to calculate water flow through the existing water meters. The people could not able to save daily data of water flow.	The general people can easily calculate water flow by viewing through the application server. The people can able to store day-to-day data on water flow.
3	Real-time monitoring of water flow cannot be achieved. Because of the long time usage of water motors for water purposes, the electricity bills got increased.	Real time monitoring of water flow can be achieved. Because of this, the usage of water motors will be efficient and the electricity bills are also got decreased.

References

1. Tomasz bergel, Tomasz Kotowski, Olga Woyciechowska. Daily Water Consumption for household purposes and its variability in a rural household. Journal of Ecological Engineering. 2016; 17(3).

2. Khutsoane O , Isong B and Abu-Mahfouz A. IoT Devices and Applications Based on Lora/Lorawan. The 43rd Annual Conference of the IEEE Industrial Electronics Society, Beijing. 2017; 6107–6112.

3. Zhou Q, Zheng K,. Xing J and Xu R. Design and Implementation of Open LoRa for IoT. IEEE Access. 2019; 7; 100649–100657.

4. Ala Khalifeh, Khaled Aldahdouh and Sahel Alouneh,: LoRaWAN Energy Optimization with Security Consideration. The International Arab Journal of Information Technology. 2021; 18(3A), Special Issue.

5. Chunsen Li, Yukun Su, Rui Yuvan, D. Chu, J. Zhu: Light weight spliced convolution network-based water meter. IEEE Access. 2019; 7.

6. Elina Strade, Daina Kalnina, Joanna Kukcycka: Water efficiency and safe re- use of different grades of water- Topical issues for the pharamaceutical industry. Water Resources and Industry. 2020; 2.

7. Sonaje. N.P., Chougule M.B.: Municipal Wastewater Recycling In Cotton Textile Wet Processing - A Review. International Journal on Recent and Innovation Trends in Computing and Communication. 2015; 3(3).

8. Jean Joy, Shruti Kanga, Sudhanshu, Suraj Kumar Singh: Cadastral Level Soil and Water Conservation Priority Zonation using Geospatial Technology. International Journal of Agriculture System. 2021; 9(1).

9. Sandra Sendra, Laura García, Jaime Lloret, Ignacio Bosch and Roberto Vega-Rodríguez: LoRaWAN Network for Fire Monitoring in Rural Environments. Electronics. 2020; 9(3).

10. Johnston S.J., Basford P.J., Bulot F.M.J., Apetroaie-Cristea M., Cox S.J., Loxham M.,

Foster G.L.: IoT Deployment for City Scale Air Quality Monitoring with Low Power Wide Area Networks. Global IoT Summit 2018; June; 4–7.

11. Steven J. Johnston, Philip J. Basford et al.,: City Scale Particulate Matter Monitoring Using LoRaWAN Based Air Quality IoT Devices. Sensors. 2018; 9(1).

12. Hsia S. C., Hsu S. W., Chang Y. J.: Remote monitoring and smart sensing for water meter system and leakage detection. IET Wireless Sensor Systems. 2012; 2(4).

13. Schantz C., Donnal J., Sennett B., Gillman M., Muller S., Leeb S.: Water Non-intrusive Load Monitoring. IEEE Sensors Journal. 2015; 15(4.23).

14. Shih Chang Hsia, Ming-Hwa Sheu, Yu-Jui Chang: Arrow-Pointer sensor design for lowcost water meter. IEEE Sensors Journal. 2013; 3(4).

Note: All figures and tables were created by the authors.

15

Face Emotion Recognition Using Histogram of Oriented Gradient (HOG) Feature Extraction and Neural Networks

Sheriff M.[1], Jalaja S.[2], Dinesh Kumar T. R.[3]

Assistant Professor/Department of ECE,
Vel Tech High Tech Dr Rangarajan Dr Sakunthala Engineering College, Chennai, India

Pavithra J.[4], Yerramachetty Puja[5], Sowmiya M.[6]

III yr Students Department of ECE,
Vel Tech High Tech Dr Rangarajan Dr Sakunthala Engineering College, Chennai, India

Abstract—Human facial expression recognition plays a mesmerizing role in each and every field of technological advancement. As our world, every time moves one step closer to technological advancement the dependency of humans on technology rapidly shoots ups. Generally, facial expression is a direct and phenomenal way of protecting human feelings for one another. Then what might be the outcome if the humans are understood by the machines? Here propose FER detection and extraction via a Histogram of Oriented Gradient (HOG). For starters, HOG upholds methods that can improve the accuracy and robustness of the algorithm used. Secondly, six universal emotions are detected by using lightweight Convolutional Neural Networks (CNN). This framework is used for the detection and transmission of the face coordinates to the facial expressions classification with additionally employing Multi-task Cascaded Convolutional Networks (MTCNN). This is mainly applicable in the medical research field, security, customized marketing, and augmented reality.

Keywords—FER, Multi-task convolutional neural network (MTCNN), Histogram of oriented gradient (HOG)

I. Introduction

In social communication, human facial expressions are crucial. In most cases, communication consists of both verbal and non-verbal elements. Facial expressions are used to communicate non-verbally. Face expressions [1] tacitly convey information. Non-verbal cues between both humans and animals include eye contact, gestures, nonverbal cues, mannerisms, and Para-language. Eye contact is a crucial component of communication

---

[1]msheriff@velhightech.com, [2]jalaja@velhightech.com, [3]trdineshkumar@velhightech.com,
[4]pavithrapaachu@gmail.com, [5]pujayerramachettypuja@gmail.com, [6]sowmiyamanickam248@gmail.com

DOI: 10.1201/9781003388913-15

since it enables the flow of thoughts. Directing contributions, dialogues, and establishing a connection with people begins with eye contact. Facial expressions include smiles, sadness, anger, disgust, surprise, and fear [2]. A curved form of the eye and a smile on a human face communicate contentment. A sense of unhappiness is conveyed by the sad face, which is characterised by rising slanted brows and a scowl. People that are frustrated often have uncomfortable or frustrating circumstances on their minds. [3]. Anger is expressed through tight eyebrows and thin, extended eyelids. Disgust is expressed through strained brows and a pleated nose. When something unexpected happens, astonishing or surprising expressions are used. This is an established expression defined by the broadening of the eyes and gaping of the mouth. [4]. Fear is associated with something like a surprised look that includes slanted brows. Automatic FER is an important interaction between human-computer systems such as robotics, machine vision, stress monitoring, and so on. It has emerged as an interesting and demanding region in computer vision, with developments ranging from mental state classification to safety, auto counselor structures, face expression formulation, lay detection, music playing, automated mentoring processes, and controller exhaustion recognition. In FER critical stages are classification stages and feature extraction. The feature extraction types

are geometric-based and appearance-based feature extraction [5]. Classification is a crucial procedure in which the expressions like "smile, surprise, anger, disgust, sadness, and fear" are categorised, whereas appearance-based feature extraction contains the exact region of the face. The eye, mouth, nose, brow, and other facial components are integrated into the geometric-based feature extraction. Pre-processing, face detection, facial component detection, feature detection, and classification are the five steps in FER. In the pre-processing stage, noise removal and enhancement are done by taking a picture or image sequence as input and giving it for further processing. Our extraction and detection methods are described in the following section.

II. Methodology

The motive behind this project is to develop a robust system that is capable of accommodating various human-artificial interrelated functions that can work on various kinds of real-time images.

The initial step in our project is to collect photographs of all six universal human emotions, which will be used to create our dataset. Our input images are these photographs from the datasets. Webcams and DSLR cameras can also be used to capture input photographs.

To begin, the input image is processed with detect faces and removal of noise. Preprocessing

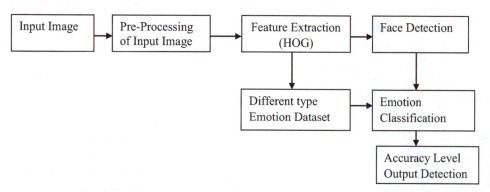

Fig. 15.1 Block diagram of face emotion detection and classification

Source: Proposed Block Diagram

has been accomplished by using various filters and data augmentations. Prior to preprocessing, the RGB (Red, Green, and Blue) image is converted into a greyscale image.

This conversion is highly crucial since facial features are hardly detected as well as the dimensionality is higher in RGB images. The multidimensional coloured image is reduced to a single particular photograph by applying the following formula:

$$Y = 0.229\text{Red} + 0.587\text{Green} + 0.114\text{Blue} \quad (1)$$

The categorised images with their corresponding grayscale and image size are shown in Table 1 for the sample test data set. The image then undergoes cropping and scaling, considering the nose of the face as the centre point of various facial features. The scaling is done on the input photographs using the Gaussian filter in order to give the images smoothness. After scaling, normalisation is done, altering the intensity of the pixels. It is also used to extract the eye

Table 15.1 Image extraction and pixel detection

Image	Grey Scale Image	Color	Pixel Dimensions
		R,B	365 x 325
		R,G,B	451 x 679
		R,B	600 x 330
		R,G,B	1280 x 720
		G,B	800 x 727

Source: Yale Face dataset, Kaggle

position on human faces in order to robustly differentiate personality differences and aid in histogram contrast stretching. Following normalisation, the Viola-Jones technique is used to perform localisation.

Face alignment is done by using a scale-invariant feature transform flow algorithm. Excessive pixels in an image are referred to as noise. Noise in images is of two types: salt and pepper noise and Gaussian noise. We moved all components with less than 5 pixels in our proposed method to simplify the removal of small unwanted pixel noise.

To overcome the lightning issues in the captured images, a pre-processing procedure called normalisation is employed. Normalization is the process that alters the intensity of the photograph according to the range of pixels in that data. It is also used to reduce the illumination and the variables in the captured dataset [6]. This procedure is also popularly known as histogram stretching.

After normalisation, localization is performed on the image using the Viola-Jones algorithm. The main function of localization is primarily used to determine the structure and the detection of the position of faces in an image.

The next step is a face alignment procedure, a pre-processing method that makes use of the Scale Invariant Feature Transform algorithm. (SIFT). The technique has three essential functions are such as controlling facial alignment; segmentation of the eye and other facial regions and lastly fragmentation of the face organs of the face image.

Excessive pixels in an image are referred to as noise. Here, the two kinds of noise that exist in any image are removed at this stage. This stage is of utmost importance as only after this stage can extraction procedures be implemented in any input image. For simply the removal of small unwanted pixel noise, we removed all components with less than 5 pixels in our proposed methodology.

The lighting challenges on the input database are reduced by the histogram equalisation method [7]. It is employed basically for improvising precise lighting and contrasting different intensities in an image. It is a vital preprocessing technique that increases differentiation.

These vast seven stages of preprocessing techniques are followed by Facial Expression Recognition (FER's) next stage which is Feature Extraction. Depicting and plotting positive as well as negative features of an image for further processing is all about feature extraction. Here we are intending on using the Histogram of Gradient (HOG) feature extraction technique.

It is a computer vision [8] feature extraction that helps in the transition from graphic to data graph depiction. There are primarily five different types of feature extraction techniques: patch-based methods employing the histogram of oriented gradient, edge-based techniques, global and local feature-based techniques, geometric feature-based techniques, and texture feature-based techniques.

A. HOG Feature Implementation Process

- The HOG feature implementation process is initiated by dividing the pre-processed image into smaller units called cells. The direction and edge orientation of every pixel in the cell is independently taken into account.
- After calculating the direction and orientation of every cell, every cell is further divided into angular bins depending upon their gradient orientation. Every isolated pixel contributes slope (gradient) to the corresponding angular bin.

- Cells are grouped together to form blocks. These blocks are the basic building blocks of histogram grouping and normalization.
- A group of histograms undergoes the normalisation procedure, and the resultant is called block histograms.

The HOG description operator can be obtained from an image by following the steps.

Step 1: Determine Gradient

Step 2: Histogram of Gradient in 8x8 cells.

Step 3: Determine amplitude and direction

Step 4: Calculation of HOG Descriptor Vector

Fig. 15.3 Output of histogram of oriented gradient feature

Source: https://iq.opengenus.org/object-detection-with-histogram-of-oriented-gradients-hog/

B. Feature Classification

Images have the capability of expressing and influencing people's expressions and feelings. From the captured image, we extract features using a fine-tuned histogram of oriented gradients in various places.

1. Convolutional Neural Networks (CNN)

In current years, Deep Convolution Neural Network (CNN) learning clearly shows

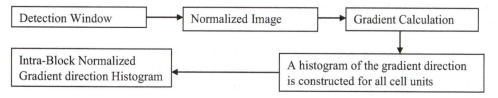

Fig. 15.2 Histogram of oriented gradient process

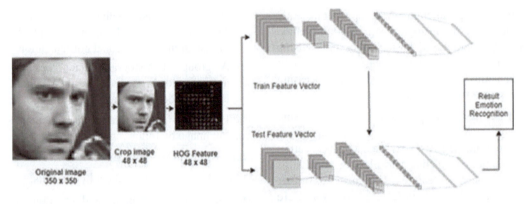

Fig. 15.4   Classification using convolutional neural network (CNN)

Source: https://www.researchgate.net/profile/Sahar-Zafar, Sahar Zafar, et al, 2019

exceptional results in the disciplines of original raw picture analysis, learning to machines, and pattern classification. CNN exceeds earlier classification algorithms when it comes to image classification. CNN employed a variety of approaches to extract the image from the feature.

In Fig. 15.4, a convolutional neural network (CNNs) is shown, a kind of neural network that has recently become popular. CNN's fundamental functionality builds complex systems for complex information representation by using multiple forms of information for data processing. The Face Emotion Detection CNN model structure is depicted. This model [9] takes as input 48x48 grayscale pictures using a Histogram of Gradient features, and the output provides the results of classifying into a single class of six categories.

Three Convolutional layers (From C1 to C3), three MaxPooling levels, and leakyReLU activation of four layers are present (from R1 to R4), there are three Convolutional layers, three Max-Pooling layers, and four leakyReLU activation layers. Furthermore, between the input and output levels, all of these layers are completely integrated. Convolutional C1 is a layer that filters a 48 by 48 input picture with 3×3 learnable kernels to create 32 metrics with a 62×62 dimension. When we will get non-zero in gradient and the unit will be not

active or deactivated, the previous layer's findings are sent to the leakyReLU activation layer, which merely converts the metrics into a narrow range. The outputs of the leakyReLU layer were sent into the P1 Max-Pooling layer [10], which contained 32 learnable kernels set up in a 2×2 configuration. This result was obtained by Max-Pooling P1 matrices of size 48×48. Its findings were passed to the second Convolutional layer C2, where 64, 3×3 learnable kernels are programmed to create 64, 24×24 matrices [12].

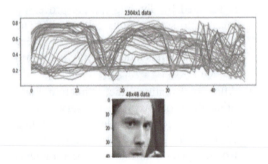

Fig. 15.5 48×48 pixel frame rate of the database image

Source: https://www.researchgate.net/profile/Sahar-Zafar, Sahar Zafar, et al, 2019

Then, in a continuous process, R2 results were transmitted to MaxPooling [12] layer P2 with 64 automated adjustable kernels of size 2×2, and 64 metrics of size 24×24 were built.

The C3 and P3 are then continued with 128 automated learnable kernels of sizes 3 × 3 and 2 × 2. Finally, they pass to the flattened layer, which generates 4608 values, three completely connected layers, the first of which has 2304 hidden units and the second of which has 1152, and seven classes to the output layers.

III. Experiment Output

The initial step in our system's processing is data collection, which is followed by some pre-processing, feature extraction, image storage in a data frame for quick access, and lastly the detection of various expressions. Using a histogram of directed gradients, we extract various facial elements from our acquired database (HOG) [14]. As a result, using MATLAB, a picture in extracted and HOG vector format is obtained.

A. HOG Extraction

HOG Extraction in surprised women.

In Fig. 15.6, the numerous face components such as eye curvature, nose, eyebrow curvature, and mouth openness are depicted in HOG vector format.

B. Expression Detection

The CNN classifier is used to detect expressions from the extracted HOG vector format. Various emotions are recognised using the CNN classifier in the Python computer language, as shown in Fig. 15.7.

The output ranges of different emotions of anger: 0.66, and disgust: 0.32 are shown in the image in Fig. 15.7. Fear is 0.01, happiness is 0.01, surprise is 0.01, the surprise is 0.01, and neutral is 0.0. According to the recommended picture analysis, the image most likely depicts the whole range of angry emotion.

The output ranges of several emotions are shown in the image in Fig. 15.8. angry: 0.0, disgust: 0.0, fear: 0.05, sad: 0.91, and Neutral: 0.04. According to the recommended picture

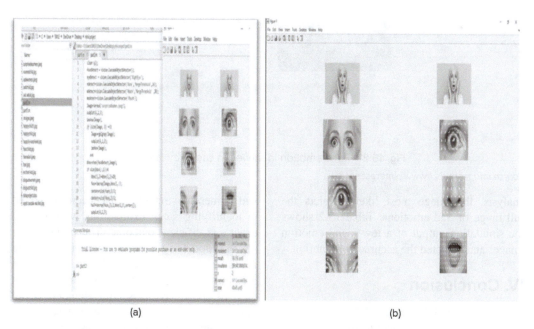

(a)　　　　　　　　　　　　　　(b)

Fig. 15.6 (a) and (b) HOG feature extraction for surprised woman image

Source: Sample simulation output of HOG operation for Yale face dataset images

```
test_image_one = plt.imread("disgustwomen.jpeg")
emo_detector = FER(mtcnn=True)
# Capture all the emotions on the image
captured_emotions = emo_detector.detect_emotions(test_image_one)
# Print all captured emotions with the image
print(captured_emotions)
plt.imshow(test_image_one)
```

```
ox': (37, 119, 355, 355), 'emotions': {'angry': 0.66, 'disgust': 0.32, 'fear': 0.01, 'happy': 0.0, 'sad': 0.01, 'surprise': 0.0, 'n
)lotlib.image.AxesImage at 0x7f66acdf3210>
```

Fig. 15.7 Face emotion identified in angry image

Source image: https://www.pinterest.es/pin/331155378854571466/

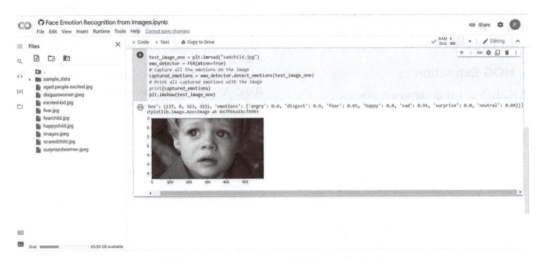

Fig. 15.8 Face emotion identified in sad baby image

Source image: https://www.pinterest.es/

analysis, the image most likely depicts the full range of Sad emotions. Table 15.2 shows the simulation output of a few facial emotion images and depicted the accuracy of emotion.

IV. Conclusion

This study uses HOG with CNN to introduce deep neural network architecture for recognising human-based facial expressions. The CNN model is the best network architecture for image processing in deep learning and FER system, and numerous strategies have been employed in identifying facial expressions with a more exact and efficient outcome. The initial phase of the model is feature extraction. There are 28,708 training samples available, as well as 3,589 testing and validation samples. HOG was used to characterise essential elements of an image in a greyscale range, and datasets were used to separate training, testing, and validation samples. FER-HOGCNN is used to train the model. This model uses the HOG Feature Extraction operator, thus the 48x48 picture

Table 15.2 Simulation output and emotion accuracy depicted

Image	Happy	Sad	Disgust	Anger	Neutral	Surprised	Fear	Accuracy
	0.0	0.05	0.01	0.17	0.0	0.03	0.74	74%
	0.0	0.01	0.66	0.32	0.0	0.0	0.01	66% (on mixed traits).
	0.0	0.91	0.0	0.0	0.04	0.0	0.04	91%
	0.0	0.0	0.0	0.01	0.01	0.20	0.78	78%
	0.0	0.01	0.0	0.02	0.03	0.01	0.93	93%

Source image: https://www.dreamstime.com/royalty-free-stock-photos-shocked-image2199858

runs through the feature extraction operator first, and then the operator's output maps the CNN network's input, which is then ordered to classify images of input facial expressions. Finally, they conducted a complete experiment with 97 percent training accuracy, 70 percent testing accuracy, and losses of 0.05 and 2.01 correspondingly, utilising a Facial Expression database. When the image size is small and the image quality is uncertain, HOG operators outperform other LPQ, LBQ, and machine learning models.

Reference

1. Lopes A. T., Aguiar E. D., Oliveira-Santos T., et al "A facial expression recognition system using convolutional networks" 28th SIBGRAPI Conference on Graphics, Patterns and Images; 2015.

2. Kołakowska, A. Landowska., et al., "Human-Computer Systems Interaction: Backgrounds and Applications", Emotion Recognition and Its Applications, pp. 51–62. Cham: Springer International Publishing, 2014.

3. Shankar, Sukritet al., et al., "DEEP-CARVING: Discovering visual attributes by carving deep neural nets" 2015 IEEE Conference on Computer Vision and Pattern Recognition (CVPR) (2015): 3403–3412.

4. Levi, Gil and Tal Hassner., et al., "Age and gender classification using convolutional neural networks"2015 IEEE Conference on Computer Vision and Pattern Recognition Workshops (CVPRW)(2015): 34–42.

5. Goodfellow, Ian J. et al., ETAL "Challenges in Representation Learning: A report on three machine learning contests" Neural networks: the official journal of the International Neural Network Society 64 (2013): 59–63.

6. Zeiler, M. D and Fergus R., et al "Visualizing and understanding convolutional networks" Published in Proc. ECCV, 2014.

7. Bashyal, S., Venayagamoorthy, G.K.V., et al., "Recognition of facial expressions using Gabor

wavelets and learning vector quantization" Artif. Intell 2021, 1056–1064.

8. Chollet, François.,et al., " Xception: Deep Learning with Depth Wise Separable Convolutions" 2017 IEEE Conference on Computer Vision and PatternRecognition (CVPR) (2017): 1800–1807.

9. Demir.Y., et al 2014. "A new facial expression recognition based on curvelet transform and online sequential extreme learning machine initialized with spherical clustering" Neural Comput. Appl. 27, 131–142.

10. Islam KTo, Raj RG., et al., "Performance of SVM, CNN, and ANN with BOW,HOG, and image pixels in face recognition" 2nd International Conference on Electrical and Electronic Engineering; 2017.

11. Mahoor M, Behzad H., et al., "Facial expression recognition using enhanced deep 3Dconvolutionalneuralnetworks" Proceedings of the IEEE Conference on Computer Vision and Pattern Recognition Workshops; 2017.

12. Suleiman A., et al., "Towards closing the energy gap between HOG and CNN features for embedded vision" IEEE International Symposium on Circuits and Systems; 2017.

13. Dachepally PR., et al., "Facial emotion detection using convolutional neural networks and representational autoencoder units" ArXiv preprint arXiv:1706.01509; 2017.

14. M. Anisetti, V. Bellandi, F.B., et al., 2005. "Accurate 3D Model based Face Tracking for Facial Expression Recognition" Proc. Int.

Recent Trends in Computational Intelligence and its Application – Sugumaran D. et al. (eds)
© 2023 Taylor & Francis Group, London, ISBN 978-1-032-48410-5

16

Historical Document Analysis Using Deep Learning

S. Devi[1], M. Rajalakshmi[2], R. Saranya[3], G. Swathi[4], K. Selvi[5]

Department of Information Technology, Coimbatore Institute of Technology, Coimbatore, India

Abstract—In recent years, the situation for dealing with managing and organising historical files have changed significantly in current times, which has caused a complicated call for additional scalable, accurate, and automated file category. Text processing for classification and small-scale databases has been the main topic of previous studies. For many years, automatic file textual recognition in documents has been a tough region of study for numerous years. The goal of automatic record processing is to digitise documents from the virtual world. Separating the text from the chronicle isn't easy as compared to normally printed documents. the most goal is to investigate information and recognize historical documents. The implementation shows within the following phases like a way for extracting text from a document as title, paragraph, header, footer, and caption; the Detected text is read within the sort of voice; Translating document language into a generalized global language. Finally, so as to accumulate more information about the document by giving details. The trained method presents a larger classification accuracy than those employing a single modality and a variety of other earlier text classification ways. The skilled method presents a large class accuracy than those employing one modality and a variety of other previous text classification ways. The skills will likely be adjusted to different global historical files with every textual content and image, that's beneficial for data and knowledge management in significant technology firms and organizations.

Keywords—Optical Character Recognition (OCR), Thresholding, Image enhancement, Denoising, TTS (Text to Speech)

I. Introduction

The field of Historical Document Analysis using Deep Learning is involved with the evaluation and reputation of files, with a focus on captured documents traditionally. Previously, the studies become basically targeted textual content reputation, first with revealed textual content after which with handwritten textual content. Digitizing historical materials ensures the preservation of a digital copy even if the original is lost or damaged. So, if someone needs to examine an ancient record, they don't have to visit the location or collect a noted

[1]devi.s@cit.edu.in, [2]rajalakshmi@cit.edu.in, [3]saranyaragu21@gmail.com, [4]swathi24.gt@gmail.com, [5]ivles002@gmail.com

DOI: 10.1201/9781003388913-16

transcript, photographic, or movie version if one is available [1].

The proposed method investigates an efficient pre-processing method to overcome trivial information in the raw text of documents, such as stop words and punctuation marks. Pre-processing modules such as image enhancement and document binarization were accomplished through the use of thresholding, which produces binary images that are useful for subsequent processing [2]. Following that, the OCR algorithm is used to obtain and recognise the textual content from the files. Then, to make the reading process easier, TTS is used. Text is detected and converted into audio with speed adjustment.

II. Related Work

Image Enhancement

Denoising historical documents is a tough step in photo processing and laptop vision. Adaptive thresholding methods are used in the proposed thresholding [3]. The primary goal of ancient record recuperation is to enhance image

exceptional to make future records evaluation easier. This project challenges repairing the damaged letters in ancient inscriptions, which are required before they can be read.

III. Literature Survey

In [4] Karthick, Ravindrakumar, Francis, and Ilankannan presented different processes engaged in textual content segmentation and studies in OCR, they have examined essential steps for OCR. Optical Character Recognition is an automatic identity strategy that satisfies the systematization needs in numerous forms. A device can analyze the facts contained in herbal sites or different substances in any shape with OCR. The scriptural and published individual reputation is clear-cut because of its well-described length and structure. The manuscript of people with inside the overhead feature. The several degrees in textual content separation, manuscript OCR shape kind in accordance to the text kind, take a look at on other languages textual content segmentation as nicely as paperwork oriented modern-day research in OCR has been cited here.

Table 16.1 Literature survey

Title	Year	Author Name	Journal Name	Methodology	Limitation
Classification of printed text and handwritten characters with neural networks	2020	K. Bramara Neelima, Dr.S. Arulselvi [5]	Journal of critical reviews	ANN, AVM, OCR, KNN	Graphics and fonts are not handled in this paper.
Automatic Deep Understanding of Tables in Technical Documents	2020	Michail S. Alexiou; Nikolaos Bourbakis[6]	International Journal of Engineering & Technology	Convolutional Neural Network(CNN) model, Optical character recognition (OCR)	This can not be applicable for mixed tables and no-line tables. Applicable only for simple, multidimensional and extended tables.
Text separation in document images through Otsu's method	2016	M.Siva Sindhuri; N. Anusha.[7]	Institute of electrical and electronics engineers	Otsu's method	Failed to retrieve the textual content with 50% of salt and pepper noise

Source: Authors

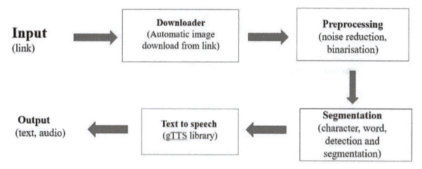

Fig. 16.1 Architecture diagram

Source: Author

4. Proposed Work

Figure 16.1 shows how the text is segmented and converted into audio from the historic image.

Dataset

The dataset contains various historical documents and images (as links) of different types. And also, content in old writings historical books, and other manuscripts. The link will automatically download the image.

Data pre-processing

The first step is to pre-process the input image. This task aims to improve document quality in order to improve either human examination of the work or automated recognition of future process steps. For pre-processing, Files picture improvement to eliminate noise, increase contrast, minimize blurring, and document binarization. This is possible with the thresholding technique.

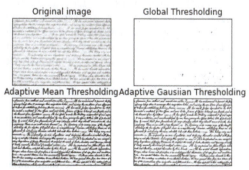

Fig. 16.2 Thresholding document image

Source: Author

Figure 16.2 tells about the Data pre-processing for English manuscript images.

Text Detection

To detect text, the OCR algorithm is used. OCR is divided into two parts. The first method is textual content identification, which identifies the textual part of a picture. This textual content distribution involved in the picture is critical for the next stage of OCR, text segmentation, in which the text part is derived from the picture. OCR Analysis takes a printed or handwritten digital image as input and converts it to machine-readable digital text format. OCR then divides the digital image into small components for the analysis of finding text, words, or character blocks. The character blocks are then broken down into elements and compared to a dictionary once more. The important thing is that the algorithm supports a wide range of languages, which could be very useful in analyzing historical data [8].

Text To Speech(TTS)

The machine computation of a person's voice is known as TTS. It converts natural language text to audio that sounds like human speech. To convert text files, gTTS offline library is used. Text-to-speech is very helpful for people who are struggling with reading.

Libraries

EasyOCR is a python package. EasyOCR supports 80+ languages detection purposes and increases accuracy. gTTS is a python library used for the conversion of voice into text.

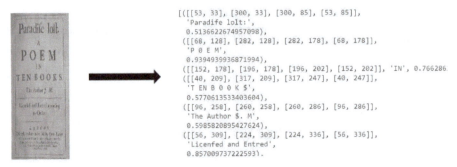

```
[([[53, 33], [300, 33], [300, 85], [53, 85]],
  'Paradife lolt:',
  0.5136622674957098),
 ([[68, 128], [282, 128], [282, 178], [68, 178]],
  'P 0 E M',
  0.9394939936871994),
 ([[152, 178], [196, 178], [196, 202], [152, 202]], 'IN', 0.766286
 ([[40, 209], [317, 209], [317, 247], [40, 247]],
  'T EN B O O K $',
  0.5770613533403604),
 ([[96, 258], [260, 258], [260, 286], [96, 286]],
  'The Author $. M',
  0.5985820895427624),
 ([[56, 309], [224, 309], [224, 336], [56, 336]],
  'Licenfed and Entred',
  0.857009737222593).
```

Fig. 16.3   Input and output for text detection

Source: Author

gTTS can able to detect more than a hundred languages. It works offline and works ideally and has multiple TTS-engine support.

V. Result and Analysis

Accuracy

The detected text achieved a slight increase in the previously detected accuracy which means that one out of one thousand characters is undetermined. Calculating OCR accuracy is performed by taking the result of an OCR moves for a picture and matching it to the real model of the identical text. It is measured depending on what number of characters that have been detected rightly and a number of phrases diagnosed rightly.

Table 16.2   Comparison of accuracy

Existing Techniques	Accuracy
Combined CR	85%
K-nearest neighbor classifier.	92.8%
Unicode mapping	91%
Proposed algorithm	93%

Source: Author

Table 16.2 shows the proposed algorithm gives good results compared to existing algorithms for text recognition.

According to Table 16.3, the downloaded image from various languages (e.g. Tamil, English, Telugu, Kannada, Korean) is transformed into a thresholding image and the text is extracted.

VI. Conclusion

The proposed system is useful, according to the detailed study and analysis. This is due to its close to 93% accuracy and the desired time. To improve the document, the thresholding method was used. Another aspect of our work is to digitise the script into audio output. The outcome demonstrates that the attempt to provide document analysis using deep learning was successful. There is the possibility of future implementation of converting the document language into any required language using a more thoroughly researched dataset.

References

1. James P. Philips, Nasseh Tabrizi, Historical Document Processing: A Survey of Techniques, tools and trends, 2020.
2. Francesco Lombardi, Simone Marinal, Deep Learning for Historical Document Analysis and Recognition—A Survey.mdpi-2020.
3. Neji, H.; Nogueras-Iso, J.; Lacasta, J.; Ben Halima, M.; Alimi, A. Adversarial Auto encoders for Denoising Digitized Historical Document, International Conference on Document Analysis and Recognition, ICDAR 2019.
4. Karthick, K., Ravindrakumar, K. B., Francis, R., & Ilankannan, S. "Steps involved in text recognition and recent research in OCR; a study" International Journal of Recent Technology and Engineering, 2019.
5. Neelima, K. B., & Arulselvi, S. "Classification of printed text and handwritten characters with

Table 16.3 Comparison of different languages

Languages	Historical documents	Pre-processing		Digitized text output
		Original image	Adaptive Gaussian Thresholding	
Case 1: English				[[[53, 33], [300, 33], [300, 85], [53, 85]], 'Paradife loit:', 0.513662267495709B), ([[68, 128], [282, 128], [282, 178], [68, 178]], 'P O E M', 0.939493993687199A), ([[152, 178], [196, 178], [196, 202], [152, 202]], 'IN', 0.766286: ([[40, 209], [317, 209], [317, 247], [40, 247]], 'T EN B O O K S', 0.5770613533403604), ([[96, 258], [260, 258], [260, 286], [96, 286]], 'The Author $. M', 0.5985820895427624), ([[56, 309], [224, 309], [224, 336], [56, 336]], 'Licenfed and Entred', 0.857069737222593).
Case 2: Telugu				[[[441, 115], [465, 115], [465, 129], [441, 129]], '30', 0.99519838978884449 ([[471, 115], [497, 115], [497, 129], [471, 129]], 'ఴ', 0.2312108004177595A [[503, 113], [541, 113], [541, 129], [503, 129]], '201', 0.38397060989200957S). ([[20, 144], [158, 144], [158, 174], [20, 174]], 'అమ్మ ఐమ్మ', 0.49957672089413917), ([[92, 176], [402, 176], [402, 206], [92, 206]], 'గాలి చామించు సమం చారి చేయు సం.', 0.74522211412911283), ([[399, 178], [585, 178], [585, 212], [399, 212]], 'ఎంఎంగ్గ్ (మాంఎంఎ)', 0.63802953189404479), ([[14, 208], [90, 208], [90, 234], [14, 234]], 'మంఎంఎ', 0.73865659956729524', ([[96, 209], [340, 209], [340, 238], [96, 238]], 'గంఎ కిఎం ఎంఎ ఎంఎంఎంఎంఎ',
Case 3: Kannada				[[[97, 0], [273, 0], [273, 36], [97, 36]], 'చంఎంఎంఎ, ఎంఎం ఎం', 0.3799912160090007), ([[0, 15], [65, 15], [65, 51], [0, 51]], 'ఴ ఴఴ.ఎ', 0.39987056604 ([[2, 37], [217, 37], [217, 74], [2, 74]], 'ఎంఎంఎంఎ ఎంఎంఎంఎంఎ', 0.16999099384578979), ([[1, 67], [231, 67], [231, 107], [1, 107]], 'ఎంఎ ఎంఎంఎంఎ, ఎంఎంఎం', 0.31803717663171827), ([[1, 103], [251, 103], [251, 141], [1, 141]], 'ఎంఎంఎం ఎంఎంఎ ఎం', 0.21837373860978831), ([[3, 132], [269, 132], [269, 171], [3, 171]], 'ఎంఎంఎంఎంఎ ఎంఎంఎ, ఎంఎ', 0.07906281545635987), ([[8, 162], [176, 162], [176, 194], [8, 194]], 'ఎ ఎంఎంఎంఎంఎ',

Source: Author

neural networks." Journal of Critical Reviews, 2020.

6. Alexiou, M. S., & Bourbakis, N. "Automatic Deep Understanding of Tables in Technical Documents.",International Conference on Tools with Artificial Intelligence (ICTAI), IEEE, 2020.

7. Sindhuri, M. S., & Anusha, N. "Text separation in document images through Otsu's method",

International Conference on Wireless Communications, Signal Processing and Networking (WiSPNET) IEEE, 2016.

8. Gilani, A., Qasim, S. R., Malik, I., & Shafait, F. "Table detection using deep learning", International conference on document analysis and recognition (ICDAR), IEEE, 2017.

Recent Trends in Computational Intelligence and its Application – Sugumaran D. et al. (eds)
© 2023 Taylor & Francis Group, London, ISBN 978-1-032-48410-5

17

IoT Based—Low Cost Smart Solar Panel Power Monitoring System Using Arduino and ESP8266

Prabaakaran K*

Department of Electrical and Electronics Engineering,
Easwari Engineering College, Chennai, India

Anto Bennet[1]

Department of Electronics and Communication Engineering,
Vel Tech Rangarajan Dr. Sagunthala R&D Institute of Science and Technology, Chennai, India

Senthil Kumar R.[2], Jayasurya E.[3], Kocherla Narendra[4], Saravanan R.[5]

Department of Electrical and Electronics Engineering,
Vel Tech High Tech Dr. Rangarajan Dr. Sakunthala Engineering College, Chennai, India

Abstract—The Internet of Things (IoT) is a new technology that supports the system to improve our daily lives in a cost-effective and efficient manner. The presently invading a new period of modernism, namely the Internet of Things in our daily activities. Solar energy may be considerably enhanced by using the Internet of Things to enhance monitoring of it. Performance and plant monitoring connect real things and gadgets to the internet system. As a result, it requires certain sources that can generate electricity organically and at no cost. Solar panels and solar plants are used to generate electricity in a natural way using the sun's energy. Therefore, a productive arrangement is required that naturally controls and screens the current, voltage, and different boundaries of nearby planet groups while likewise furnishing clients with continuous insights. This paper gives an IoT-based way to deal with sun-oriented power utilization and checking that permits clients to screen or control a sun-based plant from their cell phones.

Keywords—Power monitoring, Solar, Renewable energy, IoT, ESP8266

I. Introduction

Solar energy generation due to fluctuations in sun irradiation, temperature, and other factors, photovoltaic plants are inherently variable.

As a result, remote monitoring is critical. For the purpose of building a solar monitoring system that can be accessed remotely. This venture utilizes an IoT (Internet of Things) system to plan a sunlight-based power

*Corresponding author: prabaakaran031@gmail.com
[1]drmantobenet@veltech.edu.in, [2]rskumar.eee@gmail.com, [3]vh10412eee19@velhightech.com,
[4]vh10630eee19@velhightech.com, [5]vh10209_eee19@velhightech.com,

DOI: 10.1201/9781003388913-17

plant that predicts a not-so-distant future wherein normal spot items will be outfitted with sensors, computerized correspondence microcontrollers, and handsets. The Internet of Things (IoT) is a future innovation that permits a machine to be controlled or distinguished utilizing a cloud server. It influences advanced and actual innovations, machines, articles, people, and creatures, in addition to other things. Individuals can utilize the IoT framework to get to each of the elements of their machines, robots, and different devices. Smart urban areas, savvy autos, observing frameworks, brilliant street lighting, shopping frameworks, computerized homes, natural sensors, and different things are completely associated with the Internet of Things. Today, power is the backbone of the globe, and it can't work without it. Electricity is used in almost every aspect of our everyday lives, including house lighting, refrigerators, heaters, coolers, and transportation. As the demand for power rises, so should the generation of electricity. However, electricity is not produced in large enough quantities in our country or in other developing countries to meet demand. The expense of the power is moreover rising, and the vast majority can't manage the cost of it. Resolving this issue requires an exceptionally productive framework that creates power from sustainable assets. A few sun-oriented plants are situated in effectively open areas, while others are implicit places where individuals can't visit consistently to screen plant action. Therefore, this examination paper means to create a profoundly productive IoT-based sun-oriented power utilization and observing methodology that permits clients to screen the exercises of these sunlight-based plants from far off. In this paper, a model is made to test the proposed method and approve the discoveries that the framework produces while checking sunlight-based plants continuously.

II. Literature Review

A literature review is similar to a proof essay. It is a review of relevant literature in regard to a topic that has been assigned to us. We used Arduino Uno to construct a comprehensive IoT-based Solar Power Monitoring System. This chapter will help you go over all of the technical aspects of it. These internet-of-things-based photovoltaic frameworks are the cutting-edge arrangement that will further develop nearby planet group observing [1]. In the field of gadgets, the Internet of Things (IoT) is generally utilized. The voltage is measured by this system as well as the current of the PV system [2]. The webpage is used to access the system's data. Sunlight-based power boards are one of the non-customary power assets, and because of the huge fall in the expense of current innovations, these sun-oriented power frameworks are presently moderately reasonable and available [3]. Be that as it may, these frameworks should be minded a standard premise. Thus, individuals can really track and control.

The cloud-based nearby planet group checking procedure that sends nonstop records through the cloud after a specific measure of time [4]. Khobragade and his colleagues proposed yet another environmentally beneficial solar system. The amount of energy generated is constantly checked and updated on the server [5]. Zohora et al., discuss in great detail "how photovoltaic cells work" in order to show the solar system in order to keep up with the ever-increasing technical advances [6]. This research suggests a system that provides a solution and method for monitoring the accumulation of dust on solar panels. This gadget monitors the residue that structures on the sun-powered chargers and keeps the sun's beams from contacting them. Solar power plants should be checked on a regular basis to ensure maximum production. The settings are controlled by an Arduino UNO in this proposed study paper. This technology continuously checks solar plant performance and uploads data to the cloud. To monitor the voltage, they used an Arduino UNO controller with a QC0032 voltage sensor. The load current is measured with an ACS712.

III. Methodology

The fundamental objective of this undertaking is to get the best power yield from the sunlight-based chargers while there is a great deal of residue on them. Additionally, any failing solar panels will be highlighted, and it will be able to determine whether the solar or battery is connected to the loads. Whenever the framework perceives that it has fallen beneath the pre-characterized conditions, it sends an alarm to the user or administrator, which is displayed on the LCD. A solar panel is used to keep track of the amount of sunshine. Utilizing IoT innovation, a few qualities like the voltage, current, and temperature are introduced on the LCD. The sensors are utilized to detect the current status of the sunlight-powered charger, or at least, the current is estimated utilizing the ongoing sensor. The voltage sensor QC0032 is used for voltage sensing. The DHT11 Temperature and Humidity Sensor is used for temperature and humidity sensing.

Utilizing the WiFi module, a point of interaction is divided among the regulator and the cloud server, and the board boundaries like the voltage, current, and power created are then imparted to the server. The ongoing status of the board can be investigated somewhat thusly. Since the board's boundaries are saved money

on the server consistently and day, they might be analysed and contemplated. Information from different sunlight-powered chargers is consolidated through an Internet of Things stage, which utilizes examination to discuss the main information with applications intended to address explicit issues. The figure depicts a circuit schematic of the proposed strategy for integrating the system. Voltage, current, temperature, and humidity are the four sensors that are connected to both Arduino and a breadboard. There is a battery that is connected to an Arduino and a breadboard. The monitor is LCD. The board and Arduino are connected to the display and ESP8266.

A. Current Sensor (ACS712), Voltage Sensor (QC0032), Temperature and Humidity Sensor (DHT11) & Solar Panel

A gadget distinguishes electrical flow coursing through a substance and makes indistinguishable signs. This broadcast could be either simple or advanced signal range (0-5A). The voltage contrast between two spots can then be estimated utilizing this actual signal range (0-25V). The temperature sensor recognizes the temperature of the framework while it is in activity it permits us to screen the framework. It likewise screens the dampness in

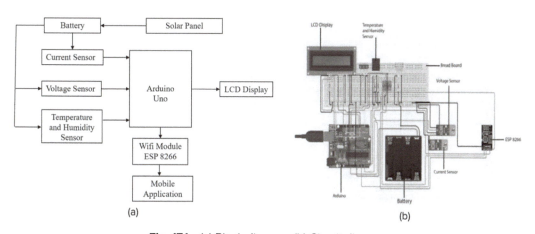

(a)

(b)

Fig. 17.1 (a) Block diagram, (b) Circuit diagram

the environment to decide how much radiation arrives on it. Range (0-100 %RH, - 40-80°C). PV panels with multiple cells change sunlight-based light or radiation into electrical energy when they come into contact with them. Photons from the sun's beams are utilized to produce power in these PV modules.

B. Battery and Blynk

A gadget has at least one electrochemical cell in it. It has two associations called cathode and anode, which are used to interface the battery to any gadget. The sun-powered chargers' electrical energy is put away in these batteries, which help in the arrangement of power to machines. Range (12 V, 1.3 Ah). Blynk used to build cell phone applications that let us speak with microcontrollers and, surprisingly, whole PCs like the Raspberry Pi. The Blynk stage's primary objective is to make creating cell phone applications as straightforward as could be expected. Each venture might incorporate graphical gadgets, for example, virtual LEDs, buttons, esteem shows, and, surprisingly, a text terminal, as well as the capacity to interface with at least one gadget. It is feasible to work Arduino or ESP32 sticks directly from your telephone utilizing the Blynk library, without composing any code.

IV. Result and Discussion

The findings of our system can be viewed directly on the LCD monitor that is connected to the entire system, as well as on a mobile device. A smartphone application has been created that pulls data from the cloud and displays real-time findings to the user. All parameters that were uploaded to the cloud are displayed on a mobile device, and the user can access them from anywhere. Figure 17.2 shows the output from the panel.

V. Conclusion

The demand for power is increasing every day, and existing energy sources are unable to supply this demand. The cost of electricity and human lives are both affected by this exponential demand. The internet of things is transforming people's lives in every aspect of their lives. Solar panels aren't your typical source of electricity, but they can help you meet your energy needs. An IoT-based method for checking sun-oriented power use is introduced in this exploration, alongside a model to mirror the outcomes. The key innovation utilizes sensors to record sunlight-powered charger boundaries like current, voltage, and temperature, which are then shipped off the cloud utilizing Arduino.

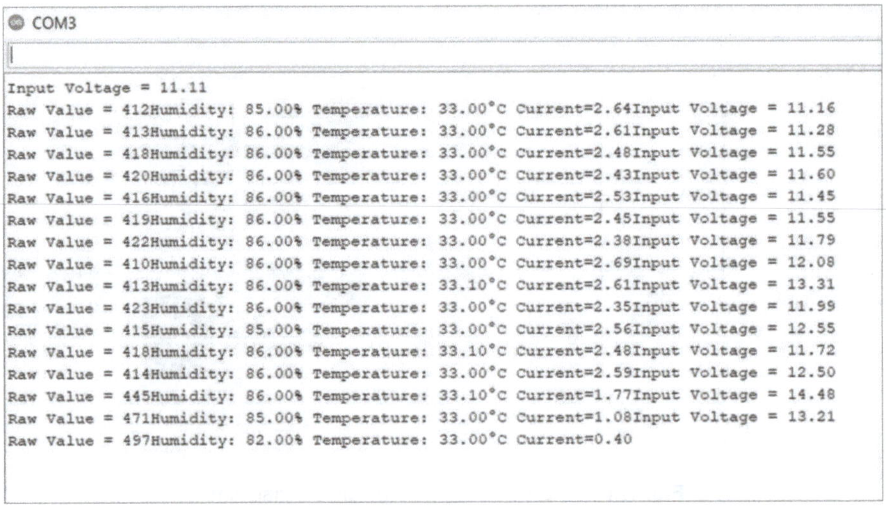

```
◉ COM3
|
Input Voltage = 11.11
Raw Value = 412Humidity: 85.00% Temperature: 33.00°C Current=2.64Input Voltage = 11.16
Raw Value = 413Humidity: 86.00% Temperature: 33.00°C Current=2.61Input Voltage = 11.28
Raw Value = 418Humidity: 86.00% Temperature: 33.00°C Current=2.48Input Voltage = 11.55
Raw Value = 420Humidity: 86.00% Temperature: 33.00°C Current=2.43Input Voltage = 11.60
Raw Value = 416Humidity: 86.00% Temperature: 33.00°C Current=2.53Input Voltage = 11.45
Raw Value = 419Humidity: 86.00% Temperature: 33.00°C Current=2.45Input Voltage = 11.55
Raw Value = 422Humidity: 86.00% Temperature: 33.00°C Current=2.38Input Voltage = 11.79
Raw Value = 410Humidity: 86.00% Temperature: 33.00°C Current=2.69Input Voltage = 12.08
Raw Value = 413Humidity: 86.00% Temperature: 33.10°C Current=2.61Input Voltage = 13.31
Raw Value = 423Humidity: 86.00% Temperature: 33.00°C Current=2.35Input Voltage = 11.99
Raw Value = 415Humidity: 85.00% Temperature: 33.00°C Current=2.56Input Voltage = 12.55
Raw Value = 418Humidity: 86.00% Temperature: 33.10°C Current=2.48Input Voltage = 11.72
Raw Value = 414Humidity: 86.00% Temperature: 33.00°C Current=2.59Input Voltage = 12.50
Raw Value = 445Humidity: 86.00% Temperature: 33.10°C Current=1.77Input Voltage = 14.48
Raw Value = 471Humidity: 85.00% Temperature: 33.00°C Current=1.08Input Voltage = 13.21
Raw Value = 497Humidity: 82.00% Temperature: 33.00°C Current=0.40
```

Fig. 17.2 Output of solar panel rating

The outcomes are introduced on the installed screen as well as on the versatile application. Clients will actually want to outwardly track, screen, and control their board to benefit from their power. Later on, we will utilize a support learning framework to conjecture future sunlight-powered charger use and power yield.

References

1. Prabaakaran et. al., Energy Management System for Small-Scale Hybrid Wind Solar Battery-Based Microgrid, International Conference on Computing, Communication, Electrical and Biomedical Systems (pp. 493–501), 2022.

2. E. Y. Song, et. al., "A Methodology for Modeling Interoperability of Smart Sensors in Smart Grids," in IEEE Trans. on Smart Grid, vol. 13, no. 1, pp. 555–563, 2022.

3. A. Rajaiyan and S. Sobati-Moghadam, "Optimized Power Consumption Formula for Designing IoT-Based Systems," 2022 2nd International Conference on Distributed Computing and High Performance Computing, 2022, pp. 74–77.

4. M. Rokonuzzaman, et. al., "IoT-based Distribution and Control System for Smart Home Applications," IEEE 12th Symposium on Computer Applications & Industrial Electronics, 2022, pp. 95–98.

5. P. Khobragade, et. al., "Advancement in Internet of Things (IoT) Based Solar Collector for Thermal Energy Storage System Devices: A Review," 2022 2nd International Conference on Power Electronics & IoT Applications in Renewable Energy and its Control, 2022, pp. 1–5.

6. F. T. Zohora Saima, M. N. Islam Rimon, T. I. Talukder, J. Ali, Z. Hossain and N. Sadia, "IoT and GSM Based Smart Grid Controlling and Monitoring System," 2022 International Conference for Advancement in Technology, 2022, pp. 1–7.

Recent Trends in Computational Intelligence and its Application – Sugumaran D. et al. (eds)
© 2023 Taylor & Francis Group, London, ISBN 978-1-032-48410-5

18

Segmentation of Corpus Callosum from MRI Images

A. Padmanabha Sarma[1], G. Saranya[2]

Department of Biomedical Engineering,
Vel Tech Rangarajan Dr. Sagunthala R&D Institute of Science and Technology, Chennai, India

Abstract—The largest white matter arrangement in the brain, the corpus callosum (CC), is crucial in disorders of the central nervous system. The severity and/or scope of neurodegenerative illnesses are inversely correlated with their size. Various strategies and procedures for CC fragmentation have been offered, despite the fact that the role of the Corpus Callosum has been progressively studied over the last centuries. CC is the vital biomarker for identifying various Autism Spectrum Disorders (ASD). Deep learning models are the fastest-growing algorithms in recent days due to their computational power and flexibility. We propose the segmentation of CC using the U-net model in this work. We trained our model for 14 epochs which took around 100 hours, as the accuracy of the model didn't improve much beyond 12 epochs, we stopped the training phase after 14 epochs. The trained model obtained an accuracy of 94% and when the model was given the new set of images the results were promising.

Keywords—Corpus callosum, Autism spectrum disorder, Deep learning, Convolutional neural networks, Image segmentation

I. Introduction

An essential organ for processing and integrating sensory, motor, and cognitive data is the corpus callosum (CC). The functions linked with the CC will be affected if it is missing or malformed. The impact can be mild to severe, depending on the individual's other medical issues. The genu (gCC), body (bCC), and splenium (sCC) are the three parts of the CC. The growth of each sub-area in CC varies depending on sex. The development of the CC in the first few years of life is very dynamic and non-linear in nature irrespective of sex [1]. The primary sub-region sensitive to the neuropathology of first-episode schizophrenia (FES), particularly in female patients, is the gCC. [2]. Significant regional fiber loss in the CC is reported in patients with Leukoaraiosis (LA) across the cognitive spectrum [3]. In [4] [5] [6] [7] the reduction in the size of CC with the autistic subjects is observed. T1

[1]padmanabhasarmaa@veltech.edu.in, [2]drgsaranya@veltech.edu.in

DOI: 10.1201/9781003388913-18

weighted MR images are used because the CC is the primary white matter structure in the human brain. All these studies require an accurate segmentation of the CC. Manual CC segmentation takes time and is prone to inter-user variations. The application of deep learning and machine learning techniques in the field of medical imaging made the segmentation much simpler and more accurate when compared with traditional methods. However, deep learning techniques will consume a large amount of time in the training phase.

II. Literature Survey

Image segmentation can be considered as the problem where the image is subdivided into several regions till the required portion of the image is highlighted. Various methodologies are available in the research but few are discussed here. Figure 18.1 shows the overview of segmentation methods.

[8] Proposed a method that uses the likelihood of the target image's sparse representation error and the probability map of CC derived via multi-atlas voting as prior information. On diffusion tensor data, a volumetric segmentation approach of the corpus callosum was presented utilizing a modified U-Net [9]. [10] Proposed an approach for evaluating segmentation that depends on the corpus callosum's form signature. These are the recent developments that happened in recent years. The U-net model is the most popular of the numerous models, so that's where we started to work.

III. Methodology

We considered OASIS (Open Access Series of Imaging Studies) and ABIDE (Autism Brain Imaging Data Exchange) two popularly known datasets in this study. The former contains T1-weighted MRI images of 416 people ranging in age from 18 to 96. The lateral contains functional and structural brain imaging data of people with Autism Spectrum Disorder (ASD) aged 5 to 64 years. By combining these two datasets we have a total of 2003 image files which are a combination of normal and ASD subjects. As the data is ready, the next step is to prepare the data for training a neural network. Figure 18.2 depicts a high-level summary of the process.

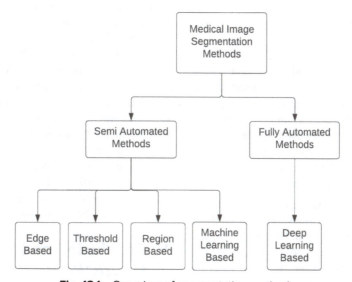

Fig. 18.1 Overview of segmentation methods

Source: Author

Fig. 18.2 High level summary of the process

Source: Author

To make the training simple in terms of computation the initial step in preparing data is to convert the given image to grayscale. The image is resized to 128 × 128 to reduce the resolution and speed up the training process. These images are then split into Training (80%) and Testing (D_{Te}- 20%) sets. To validate the training process, the training is further divided into training (D_{tr}- 70%) and Validation (D_{Va}- 10%) sets. To avoid the large gradient updates in the training process, the training dataset's mean (μ) and standard deviation (σ) are used to determine the z-score. The training, validation, and testing data are scaled as shown in the equation- 1, 2, and 3.

$$D_{tr} = \frac{(D_{tr} - \mu_{tr})}{\sigma_{tr}} \qquad (1)$$

$$D_{Va} = \frac{(D_{Va} - \mu_{tr})}{\sigma_{tr}} \qquad (2)$$

$$D_{Te} = \frac{(D_{Te} - \mu_{tr})}{\sigma_{tr}} \qquad (3)$$

Data Augmentation (DA) helps to avoid overfitting of the data and to build a robust model. Two affine transforms (translation and rotation) are chosen in our study to augment the data. DA will generate more images from the training data to ensure the model will not see the exact same image twice. This procedure will expose the model to large data which will help in obtaining a better generalization.

With the above-mentioned procedures, preprocessing of data comes to an end and the neural network must then be trained, which is a critical phase. We chose the U-Net [11] architecture by modifying the cost function as mentioned by [12] in this work. [12] has modified the loss function from weighted cross entropy to dice loss function, as the performance of the dice loss function is better than the weighted cross entropy function. The dice loss function calculates the dice coefficient which measures the overlap between the predicted and ground truth images. The return value of the dice loss function represents the overlap of the ground truth image on the predicted image.

The neural network is trained using Adam's Adam optimizer and mini-batch gradient descent with a batch size of 32 [13]. To avoid bias in presentation order affecting the final result, the training data is shuffled after each epoch in training. The final model is evaluated on the test set using the Structural SIMilarity (SSIM) index [14].

IV. Results and Discussions

We obtained an accuracy of 94% after successfully completing 14 epochs, beyond which the model did not improve in terms of accuracy, which made us stop the code forcefully. The training accuracy and the loss are shown in Fig. 18.3. The final result of the segmented image (sample) is shown in Fig. 18.4. The model spent approximately 100 hours in training, however when it is tested on the unseen images the results are quick and promising on the selected images. The F1 score, Accuracy, and precision were not that promising on the testing data, which can be due to the overfitting of data. When observed the reasons for the overfitting of data can be the complexity of data or the size of training data.

V. Conclusion and Future Work

As seen in the results section, the model has underperformed on the test data, however for

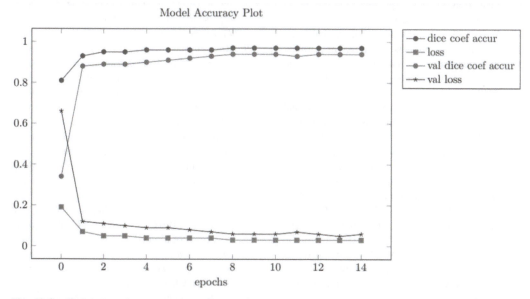

Fig. 18.3 Training and accuracy loss plot on the x-axis we have number of epochs and on the y-axis accuracy and loss plotted

Source: Author

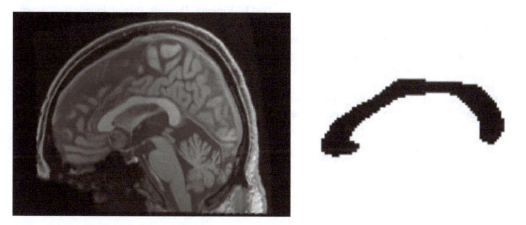

Fig. 18.4 Sample input and output images

Source: Author

the selected images results are promising, this cannot be considered a good sign which made us think developing the model, one way to fine-tune the design is by adding dropout and by trying to increase the training data set by including few more data augmentation methods. We would also like to combine CNN with Level Set Method (LSM) to observe the effectiveness of segmentation.

References

1. Tanaka-Arakawa, Megumi M., et al. "Developmental changes in the corpus callosum from infancy to early adulthood: a structural magnetic resonance imaging study." *PloS one* 10.3 (2015): e0118760.
2. Huang, Weikai, et al. "A Deformation-Based Shape Study of the Corpus Callosum in

First Episode Schizophrenia." *Frontiers in Psychiatry* (2021): 833.

3. Wang, Zhuonan, et al. "Corpus callosum integrity loss predicts cognitive impairment in Leukoaraiosis." *Annals of Clinical and Translational Neurology* 7.12 (2020): 2409–2420.

4. Chen, Bo. "A Preliminary Study of Abnormal Centrality of Cortical Regions and Subsystems in Whole Brain Functional Connectivity of Autism Spectrum Disorder Boys." *Clinical EEG and Neuroscience* 53.1 (2022): 3–11.

5. Guo, Yurui, et al. "Symptom-Related Differential Neuroimaging Biomarkers in Children with Corpus Callosum Abnormalities." *Cerebral Cortex* 31.11 (2021): 4916–4932.

6. Pristas, Noel, et al. "An observational report of swallowing outcomes following corpus callosotomy." *Epilepsy & Behavior* 123 (2021): 108271.

7. Giona, Federica, et al. "Another step toward understanding brain functional connectivity alterations in autism: An Editorial Highlight for "Neurobiological substrates underlying corpus callosum hypoconnectivity and brain metabolic patterns in the valproic acid rat model of autism spectrum disorder" on page 128." *Journal of Neurochemistry* 159.1 (2021): 12–14.

8. Park, Gilsoon, et al. "Automatic segmentation of corpus callosum in midsagittal based on Bayesian inference consisting of sparse representation error and multi-atlas voting." *Frontiers in neuroscience* 12 (2018): 629.

9. Rodrigues, Joany, et al. "Volumetric segmentation of the corpus callosum: training a deep learning model on diffusion MRI." *17th International Symposium on Medical Information Processing and Analysis*. Vol. 12088. SPIE, 2021.

10. Herrera, W. G., M. Bento, and L. Rittner. "Corpus Callosum Shape Signature for Segmentation Evaluation." *XXVI Brazilian Congress on Biomedical Engineering*. Springer, Singapore, 2019.

11. Ronneberger, Olaf, Philipp Fischer, and Thomas Brox. "U-net: Convolutional networks for biomedical image segmentation." *International Conference on Medical image computing and computer-assisted intervention*. Springer, Cham, 2015.

12. da Silva, Flávio Henrique Schuindt. *Deep learning for corpus callosum segmentation in brain magnetic resonance images*. Diss. Universidade Federal do Rio de Janeiro, 2018.

13. Kingma, Diederik P., and Jimmy Ba. "Adam: A method for stochastic optimization." *arXiv preprint arXiv:1412.6980* (2014).

14. Wang, Zhou, et al. "Image quality assessment: from error measurement to structural similarity." *IEEE transactions on image processing* 13.1 (2004).

Recent Trends in Computational Intelligence and its Application – Sugumaran D. et al. (eds)
© 2023 Taylor & Francis Group, London, ISBN 978-1-032-48410-5

19

Quality Indexing of Reclaimed Water using Internet of Things for Irrigation Purpose

S. Vishnu Kumar*

Centre for Autonomous System Research, Department of Electronics & Communication Engineering,
Vel Tech Rangarajan Dr. Sagunthala R&D Institute of Science and Technology, Chennai, India

E. Balasubramanian[1]

Centre for Autonomous System Research, Department of Mechanical Engineering,
Vel Tech Rangarajan Dr. Sagunthala R&D Institute of Science and Technology, Chennai, India

G. Aloy Anuja Mary[2]

Department of Electronics & Communication Engineering,
Vel Tech Rangarajan Dr. Sagunthala R&D Institute of Science and Technology, Chennai, India

M. Kumaresan[3]

Centre for Autonomous System Research,
Vel Tech Rangarajan Dr. Sagunthala R&D Institute of Science and Technology, Chennai, India

Abstract—Water recycling is gaining prominence in recent days due to the advancement in technologies. The primary goal of this work is to develop an Internet of Things-based Serverless Cloud Framework to measure the suitability of recycled water for irrigation purposes. Thus, it has developed a working prototype and applied it in a recycling plant to verify the working of the proposed model. In this research, a system for capturing both Physical and Chemical parameters like Temperature, Turbidity, Electrical Conductivity, pH, Dissolved Oxygen and Total Dissolved Solids of reclaimed water from domestic usage; Graywater and Stormwater is designed. Espressif -ESP8266 is used for telemetry purposes using 2.4 GHz Wi-Fi technology. Microsoft Azure is used as a cloud service to calculate the Water Quality Index (WQI) based on ISO 11624 - Bureau of Indian Standards. The system performance is tested on, an in-campus wastewater recycling plant and the results are compared between two types of sources over a period of six months. The analysis results will be useful for assessing the reclaimed-water quality and aid in decision-making for effective management of water.

Keywords—Reclaimed-water quality, Waste-water irrigation, Serverless IoT architecture, Microsoft Azure, Machine learning

*Corresponding Author: mail2jitvishnu@gmail.com
[1]esak.bala@gmail.com, [2]aloyanujamary@gmail.com, [3]kumaresh_m@outlook.com

DOI: 10.1201/9781003388913-19

I. Introduction

Water is the most vital natural resource and one of the main life-supporting commodities that accounts for 71% of the earth's surface, but only 3% of this is useable. Nearly 60% of the available freshwater is trapped in glaciers, rivers, streams, lakes, and dams (Yahans Amuah et al., 2022). According to (Payus et al., 2020) persistent periods of drought, changes in precipitation patterns and heatwaves, water security around the world is worsened. Thus, water conservation is an important aspect of comprehensive environmental management; emphasising the use of reclaimed-water for ancillary works like irrigation and urban-gardening. The most critical factor that affects crops is the quality of the water being used; the Salinity, pH, Alkalinity and the presence of chemical-ions influences the quality of the water.

(Khan et al., 2022) through their work have projected, an estimated 310 million hectares of land worldwide are currently irrigated, of which 20% are badly impacted by salinity and sodium hazards. It was also estimated, an average of 2000 hectares of irrigation land across 75 countries in the last 20 years have turned highly salt-contaminated. Microbial pathogens (Bauder et al., n.d.) are also affecting irrigation water.

The quality of irrigation water is a critical aspect of crop production and the scarcity of fresh-water for auxiliary works has resulted in adapting the usage of reclaimed-water for irrigation. The effluents not treated appropriately could carry more threats to soil and plants. The contaminated water can also affect the solubility of ions like magnesium, potassium and calcium in the soil, which could change the soil to more acidic.

Thus, an efficient system to monitor reclaimed-water before being used for irrigation is needed. The traditional laboratory setups are costly, time-consuming and non-real-time, which excerpts the modern-day concept of using the Internet of Things (IoT). The present work envisages incorporating an IoT platform to achieve a fast and real-time framework using edge devices to collect the required parameters and a cloud platform is used to classify the usability of the recycled water for irrigation.

This article is systematized in the following order: Section two discusses the literature review. In Section three, the methodology being employed is given. In Section four, the results of the proposed system are discussed. Section five, the conclusion and future scope of the work is provided.

II. Literature Survey

This work looks into the tactics that have been used to quantify the quality of water, specifically for irrigation. Water can be contaminated and deteriorated by various types of waste from homes, and companies (Ali et al., 2019). Food pulps, papers, chemicals, heavy metals, and pharmaceuticals wastes are found in industrial wastewater whereas mostly organic and inorganic content is usually found in domestic wastewater (Manasa & Mehta, 2020).

The authors (Widyarani et al., 2022) have outlined the sources of reclaimed water. They have defined 'Blackwater' as the wastewater collected from toilets and lavatories. 'Graywater', is released from the laundry, kitchen, and shower, and 'Stormwater', is the runoff water from the roads, pavements, and building roofs after precipitation.

In most work, traditional laboratory examination and arithmetical analysis are employed to determine WQI. Modern studies have explored the options of using Machine Learning algorithms to find an optimal solution for water quality management (Alqahtani et al., 2022; Lowe et al., 2022; Sheikh Khozani et al., 2022; Wang et al., 2022; Ziyad Sami et al., 2022).

(Al-Assaf et al., 2020) made a study to assess the value of the water being used to irrigate the trees and gardens. They collected twelve parameters from the wells to calculate WQI.

Thus, it is understood that the experiment was conducted only on fresh ground-water.

(Singh et al., 2020) conducted their work in Uttarakhand-India to find the suitability of groundwater for agricultural use. The study was conducted under a laboratory setup, using principles of chemical-reactions to analyse the results. This procedure is inferred to be time and cost-consuming and the results are non-real-time.

Earlier to the above-mentioned works, (Shafi et al., 2018) estimated WQI using three parameters: turbidity, temperature, and pH, and equating them against World Health Organisation's Standards. This study suffers from the limitation of using only three parameters.

The significant indexing parameters used on crop efficiency are the Salinity-Hazard (SH), pH, and soluble-salts (of Indian Standards, n.d.). SH is measured by Electrical Conductivity (EC_w), whose prime effect on vegetation is the psychological drought; the failure of a plant to absorb water due to a high concentration of ions in the water. As per the authors (Bakhtiar Jemily et al., 2019), temperature could affect the readings of EC_w of water, thus the data collection is to be performed at the ideal temperature; of 25°C. Total Dissolved Solids data can be also used as an indirect measurement to calculate EC_w.

According to (Mattson, n.d.), from 0 to 14, the pH scale assesses a liquid's acidity or alkalinity; 7 is neutral, anything below 7 is acidic, and anything above 7 is alkaline. To reduce the risk of dumping pollutants into surface or groundwater, recycled water must be evaluated for pH value else it may result in increased groundwater acidity. Most crops can withstand a pH of 6; but, when pH lowers, more delicate plants may develop sick and die.

III. Materials and Methods

It is understandable from the literature study that, more works are focused on potable/freshwater water and not much concentration has been given specifically to the usage of reclaimed-water aimed at irrigation purposes. The proposed framework enhances these notions which are discussed in the following sections.

A. Study Procedure

The study can be differentiated into two parts on a top-level abstraction; the Edge System and the Cloud Platform. The edge system concentrates on acquiring the physical and chemical values of the water and the cloud platform is used to calculate the WQI on real-time basis.

B. Parameters Analysed

The quality of the water can be assessed under three arenas; physical quality, chemical quality, and biological quality. Where in this study, both physical and chemical composition are taken into consideration. Temperature, Turbidity, and Electrical Conductivity are analysed under the physical field, and pH, Dissolved Oxygen, and Total Dissolved Solids are measured to assess the chemical properties of the reclaimed water as described in Fig. 19.1.

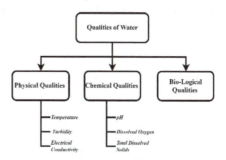

Fig. 19.1 Classification of water quality parameters

Source: Authors

C. Study Site

This study was conducted at a waste-water recycling station located at Vel Tech University, Chennai, Southern part of India as illustrated in Fig. 19.2. Where the recycled waste water is used for gardening purposes. The unit gathers "Graywater" water from domestic use and also collects "Stormwater" from pavements and building roofs after precipitation.

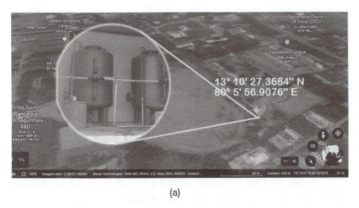 

(a)   (b)

Fig. 19.2 Field of study (a) Recycling plant-location (b) Prototype testing

Source: Authors

D. Sensor Module and Water sampling

In the current study, six different parameters have been measured using a custom-designed sensor module as shown in Fig. 19.3. The data collected is transferred to the cloud using the ESP8266 microcontroller. Message Queuing Telemetry Transport (MQTT) protocol is used as a telemetry standard. The working of the prototype model was tested under a laboratory setup to calibrate the sensors and verify the cloud communication as illustrated in Fig. 19.3.

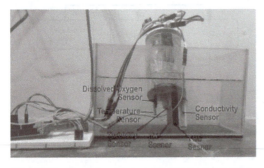

Fig. 19.3 Experimental verification under laboratory setup

Source: Authors

E. Cloud Server

Microsoft Azure platform was used for calculating the WQI. The data transmitted from edge devices are received by IoT

Hub and streamed into blob storage using Stream Analytics service, where the query to filter individual parameters is done. The mathematical calculations are also performed upon the received data using vscode and final values are presented through custom designed Azure Web-App service.

F. Mathematical Calculation

WQI is a single indicator used to quantify the water sample. It is calculated utilising different parameters based on the purpose like potable water or water for irrigation. We have utilised six parameters and the mathematical process used is based on the method proposed by (Mishra et al., 2009).

The data collected is subjected to Weighted Arithmetic WQI as illustrated in Fig. 19.4.

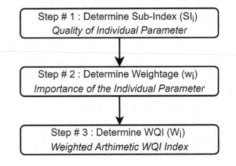

Fig. 19.4 Mathematical calculation

Source: Authors

WQI of each sample is calculated using six different parameters and their associated weights, as shown in Equation (1).

$$WQI = \sum_{i=1}^{n} W_i Q_i \qquad (1)$$

where Q_i is the sub-index or quality of the ith parameter and W_i is the unit weight for the ith parameter.

The steps to calculate WQI for irrigation purpose involves three stages; The first step is to identify the parameters required to calculate WQI. Followed by Q_i is calculated as stated in Equation (2), whose value ranges from 0 to 100,

$$Q_i = \frac{C_i}{S_i} * 100 \qquad (2)$$

where C_i is the actual present value of the ith parameter and S_i is the standard value of the ith parameter.

In step three, we calculate the unit weightage W_i, which is inversely proportional to the standard value of the particular variable as given in Equation (3).

$$W_i = \frac{k}{S_i} \qquad (3)$$

where $k = \dfrac{1}{\sum \dfrac{1}{S_i}}$

IV. Results

The physical and chemical analysis carried out over two different sources of reclaimed water for the period of six months is depicted in Figures 5 to 10. The study revealed that temperature varied from 20.8°C to 23.5°C in stormwaters and 22.53°C to 28.8°C for Graywater. The turbidity of both water sources was almost above 2.5 NTU. The value of DO from stormwater showed a higher concentration than Graywater. The EC_w and TDS both evident the treated water still had more pollutants that are not suitable for irrigation.

The pH values ranged from 7.4 to 8.2. Figure 19.11 illustrates the dashboard designed in the azure portal, which projects the recent average values of the data received in the cloud.

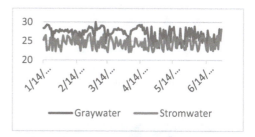

Fig. 19.5 Variations in temperature

Source: Authors

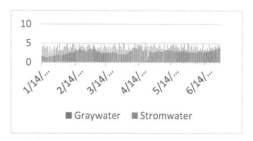

Fig. 19.6 Variations in turbidity

Source: Authors

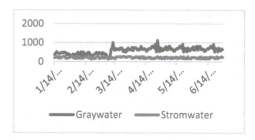

Fig. 19.7 Variations in electrical conductivity

Source: Authors

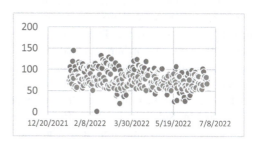

Fig. 19.8 Variations in dissolved oxygen

Source: Authors

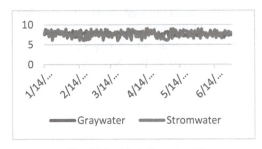

Fig. 19.9 Variations in pH

Source: Authors

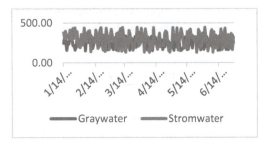

Fig. 19.10 Variations in TDS

Source: Authors

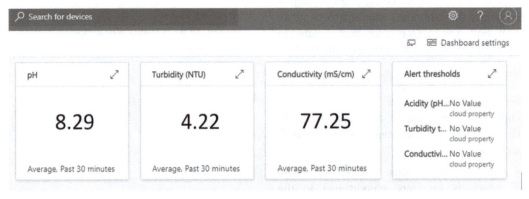

Fig. 19.11 Azure-dashboard

Source: Authors

V. Conclusion

This study was conducted to evaluate the suitability of reclaimed water for irrigation purposes. The variation in Salinity (EC_w and TDS), Turbidity, pH, and DO of reclaimed water were studied from the recycling plant of Vel Tech University. The present study concluded that the WQI of stormwater is better compared to Graywater for irrigation purposes. We suggest appropriate chemical treatment of domestic sewage water before being used for irrigation to maintain the TDS range at less than 1000 ppm. The authors conclude that the odour of water could also play a vital role in irrigating sensitive crops. Thus, future work could be concentrated on sensing the presence of Hydrogen Sulphide in the effluent, also the presence of Ammonium and Phosphate could be explored to deepen the suitability of reclaimed water for irrigation.

Reference

1. Al-Assaf, A. Y. R., Talat, R. A., & Al-Saffawi, A.-A. Y. T. (2020). Suitability of Water for Irrigation and Livestock Watering Purpose Using IWQI Model: The Case Study Groundwater Quality of Some Quarters of Mosul City, Iraq. *Plant Archives*, 20(1), 1797–1802.

2. Ali, I., Basheer, A. A., Mbianda, X. Y., Burakov, A., Galunin, E., Burakova, I., Mkrtchyan, E., Tkachev, A., & Grachev, V. (2019). Graphene based adsorbents for remediation of noxious pollutants from wastewater. *Environment International*, 127, 160–180. https://doi.org/10.1016/J.ENVINT.2019.03.029

3. Alqahtani, A., Shah, M. I., Aldrees, A., & Javed, M. F. (2022). Comparative Assessment of Individual and Ensemble Machine Learning Models for Efficient Analysis of River Water Quality. *Sustainability 2022, Vol. 14, Page 1183*, 14(3), 1183. https://doi.org/10.3390/SU14031183

4. Bakhtiar Jemily, N. H., Ahmad Sa'ad, F. N., Mat Amin, A. R., Othman, M. F., & Mohd Yusoff, M. Z. (2019). Relationship Between Electrical Conductivity and Total Dissolved Solids as Water Quality Parameter in Teluk Lipat by Using Regression Analysis. *Advanced Structured Materials*, *119*, 169–173. https://doi.org/10.1007/978-3-030-28505-0_13/COVER/

5. Bauder, T., Waskom, R., Sutherland, P., & Davis, J. G. (n.d.). *Irrigation Water Quality Criteria*.

6. Khan, M. M., Siddiqi, S. A., Farooque, A. A., Iqbal, Q., Shahid, S. A., Akram, M. T., Rahman, S., Al-Busaidi, W., & Khan, I. (2022). Towards Sustainable Application of Wastewater in Agriculture: A Review on Reusability and Risk Assessment. *Agronomy 2022, Vol. 12, Page 1397*, *12*(6), 1397. https://doi.org/10.3390/AGRONOMY12061397

7. Lowe, M., Qin, R., & Mao, X. (2022). A Review on Machine Learning, Artificial Intelligence, and Smart Technology in Water Treatment and Monitoring. *Water 2022, Vol. 14, Page 1384*, *14*(9), 1384. https://doi.org/10.3390/W14091384

8. Manasa, R. L., & Mehta, A. (2020). *Wastewater: Sources of Pollutants and Its Remediation*. 197–219. https://doi.org/10.1007/978-3-030-38196-7_9

9. Mattson, N. (n.d.). *Substrate pH: Getting it Right for Your Greenhouse Crops*.

10. Mishra, A. *, Mukherjee, A., & Tripathi, B. D. (2009). Seasonal and Temporal Variations in Physico-chemical and Bacteriological Characteristics of River Ganga in Varanasi. *Int. J. Environ. Res*, *3*(3), 395–402.

of Indian Standards, B. (n.d.). *IS 11624 (1986): Guidelines for the quality of irrigation water*.

11. Payus, C., Huey, L. A., Adnan, F., Rimba, A. B., Mohan, G., Chapagain, S. K., Roder, G., Gasparatos, A., & Fukushi, K. (2020). Impact of Extreme Drought Climate on Water Security in North Borneo: Case Study of Sabah. *Water 2020, Vol. 12, Page 1135*, *12*(4), 1135. https://doi.org/10.3390/W12041135

12. Shafi, U., Mumtaz, R., Anwar, H., Qamar, A. M., & Khurshid, H. (2018). Surface Water Pollution Detection using Internet of Things. *2018 15th International Conference on Smart Cities: Improving Quality of Life Using ICT and IoT, HONET-ICT 2018*, 92–96. https://doi.org/10.1109/HONET.2018.8551341

13. Sheikh Khozani, Z., Iranmehr, M., & Wan Mohtar, W. H. M. (2022). Improving Water Quality Index prediction for water resources management plans in Malaysia: application of machine learning techniques. *Https://Doi.Org/10.1080/10106049.2022.2032388*. https://doi.org/10.1080/10106049.2022.2032388

14. Singh, K. K., Tewari, G., & Kumar, S. (2020). Evaluation of Groundwater Quality for Suitability of Irrigation Purposes: A Case Study in the Udham Singh Nagar, Uttarakhand. *Journal of Chemistry*, *2020*. https://doi.org/10.1155/2020/6924026

15. Wang, S., Peng, H., & Liang, S. (2022). Prediction of estuarine water quality using interpretable machine learning approach. *Journal of Hydrology*, *605*, 127320. https://doi.org/10.1016/J.JHYDROL.2021.127320

16. Widyarani, Wulan, D. R., Hamidah, U., Komarulzaman, A., Rosmalina, R. T., & Sintawardani, N. (2022). Domestic wastewater in Indonesia: generation, characteristics and treatment. *Environmental Science and Pollution Research International*, *29*(22), 32397–32414. https://doi.org/10.1007/S11356-022-19057-6

17. Yahans Amuah, E. E., Boadu, J. A., & Nandomah, S. (2022). Emerging issues and approaches to protecting and sustaining surface and groundwater resources: Emphasis on Ghana. *Groundwater for Sustainable Development*, *16*, 100705. https://doi.org/10.1016/J.GSD.2021.100705

18. Ziyad Sami, B. F., Latif, S. D., Ahmed, A. N., Chow, M. F., Murti, M. A., Suhendi, A., Ziyad Sami, B. H., Wong, J. K., Birima, A. H., & El-Shafie, A. (2022). Machine learning algorithm as a sustainable tool for dissolved oxygen prediction: a case study of Feitsui Reservoir, Taiwan. *Scientific Reports 2022 12:1*, *12*(1), 1–12. https://doi.org/10.1038/s41598-022-06969-z

Recent Trends in Computational Intelligence and its Application – Sugumaran D. et al. (eds)
© 2023 Taylor & Francis Group, London, ISBN 978-1-032-48410-5

20

Energy Efficient Heuristic Best Search Approach for Scheduling Virtual Machines in the OpenNebula Cloud

Abisha D.[1], Aishwaryalakshmi R. K.[2]

Assistant Professor, Department of Computer Science and Engineering,
National Engineering College, Kovilpatti, India

Abstract—With the growing improvement of open-source cloud platforms, power consumption is an essential difficulty in cloud computing which offers a large wide variety of machines. Virtualization enables data centers to consolidate their services with smaller physical waiters and share tackle coffers with them. Multihued strategies have been developed in this field to conduct the scheduling of Virtual Machines (VMs) over Physical Machines(PMs). There's a necessity to enhance the energy effectiveness of statistics installations by using a waiter connection in order to limit the range of PMs. In our work, we present a novel approach termed Optimal Search-Open Nebula-based Cloud (OS-OC) scheduler within a realistic cloud infrastructure using the Open Nebula tool. The OS-OC scheduler aims at minimizing energy consumption and also satisfies clients' QoS (Quality of Service) by optimally assigning VMs. The consequences exhibit that the OS-OC strategy outperforms traditional heuristic primarily based methodologies and obtains exceptional scheduling with minimal strength consumption. The proposed OS-OC scheduler outperforms in terms of energy when compared to the existing schedulers in cloud computing. The existing real-time scheduler results and that of the OS-OC scheduler is compared and analysis is done for different numbers of hosts with varying number of VMs at different times. The experimental outcome of different hosts consumed less energy in VMs.

Keywords—Virtualization, Optimal Search-OpenNebula based cloud (OS-OC) scheduler, Energy efficiency, Scheduling, Server consolidation

I. Introduction

Cloud computing uses a significant share of power in today's world, particularly while managing distributed structure. Hence, power-aware processing is vital for worldwide systems that spend substantial amounts of energy [2]. Virtualization assists functionalities of the cloud which allocates various resources, and storage for devices to enable efficient computing. The accessible processing resources such as the processor, disk space, and memory

---
[1]abisha_cse@nec.edu.in, [2]aishwarya_cse@nec.edu.in

DOI: 10.1201/9781003388913-20

to applications can be assigned merely when needed and not allocated constantly.

A cloud supplier owns a cloud data center that contains a number of PMs. The operational cost in the data center is increased due to high energy consumption and it is seen that energy consumed by the required physical asset is about 45% of the total operational costs [3]. Cloud providers give resources as a service to their clients if they request them and charge them based on their usage Service providers must commit to reducing energy consumption to cut down functional costs [8]. Hence VM scheduling should be performed to minimize energy expenditure. Various energy conservation techniques such as workload prediction, VM placement, resource over-commitment, and workload consolidation have been proposed earlier. An inactive server can be prominent with about 50% of the server's highest power [4]. Switching the servers to lower power states when they are not in use can save power. In addition, the dynamic consolidation method that is used along with the reinforcement learning and Q-agent resolves the host power mode and includes about 19% power consumption of servers. Hence, these issues should be taken into consideration as a challenge to find the optimal scheduler for VM placement.

The proposed work aims to achieve optimal scheduling of VM in cloud computing using the real cloud tool. There is a trade-off between power efficiency and QoS in the process of allocating on-demand workload [8]. Hence the service provider should ensure QoS at any point in time to their customers. The growth of asset utilization along with the powerful effectiveness of cloud infrastructure is vital. Hence, optimal search-based mapping of virtual machines to their parallel physical machines is performed in the proposed work in order to provide QoS.

The paper is structured in the following manner. In Section II the affiliated work to our system is stated. Section III presents the summary of the problem linked to the being scheduling

mechanisms. Our new approach grounded on the original hunt heuristic is presented in Sections IV and V. The results of the zilches-OC scheduler are agitated in Section VII. The conclusion of the proposed work is given in Section VIII.

II. Related Works

A. Ant Colony Optimization

The Ant Colony Optimization Algorithm (ACO) helps to solve computational problems based on the probabilistic technique in order to search the ideal paths using graphs [11]. This heuristic method is used in distributed systems that have an extremely designed and controlled association. Consequently, this heuristic can achieve tasks greater than what an individual task can perform in terms of a dynamic environment. In the real situation, an ant steps ahead to examine food, during its search it secretes pheromone trails [12]. If an ant searches a path that is optimum then the concentration of pheromones will be high along that path. The remaining ants from the colony will follow the pheromone trail and will arrive at a single path ultimately.

B. Particle Swarm Optimization

Particle Swarm Optimization (PSO) is an efficient scheme that improves a crisis by using a candidate solution iteratively in terms of quality [13]. The problem is solved by taking a population of possible results with particles and transferring these particles along the search space using position and velocity [16]. Each patch movement is grounded upon its original optimum position but directed towards the finest position in the quest- space as better positions are estimated by remaining patches [15]. This makes the group shift towards the best solution.

C. Fuzzy Logic

Fuzzy Logic (FL) is considered to be an enhanced method for categorizing and managing data. FL provides an effortless way to turn up at a definite solution. If a VM is viewed as a three-dimensional object (CPU, memory, and storage), then the difficulty of scheduling VMs over hosts is considered related to the bin packing problem, however, it is not close the same [5]. In the bin packing problem, the objects are viewed in three-dimensional order to be scheduled inside containers. The definitive aim is to schedule as many as VMs in the same host if possible so that the required number of hosts is minimized [9]. This is due to the fact as soon as the aid is utilized by way of a VM then it can't be assigned to any different VMs.

D. Neural Networks

Neural networks (NN) are used to approximate features that count on a massive quantity of inputs that are commonly unknown. VMs are categorized into sets of CPU, memory, network, or disk based on the value of weights respectively. Thus, VMs from a similar set contest intensively for obtaining the consistently needed resource [7].

The VMs are given priority in placing at initial rather than resources to ensure the minimal requirement of resources is achieved [6]. So as to allocate the VMs with greater priority resources, improvement can be done effectively by choosing the physical hosts that have high capacity. A VM scheduling approach is presented that addresses issues to decrease resource wastage, energy consumption, and indulgence costs.

E. Server Consolidation

The number of physical machines that remain active in a data center should be optimal as per the requirement to satisfy the energy constraints [17]. VM placement can be achieved relying upon the last purpose into the power-based method and QoS-based approach. VM is consolidated very smartly in the server to reduce energy consumption while meeting the SLA requirements [18]. The algorithms are tested from small to large clusters whose

energy consumption is variably less. In [19], power consumption in a data center can be caused due to two reasons namely network communications and server operations.

The schedulers employ the column initiation method to deal with the integer quadratic encoding optimization problem. In [20], consolidation of VMs in a smaller number of servers is an intelligent concept to reduce the cost and energy.

III. Problem Description

Our problem statement can be briefly described as scheduling X applications on N hosts which are distributed in the network. The problem is formulated as a two-tier architecture in which the first level consists of cloud providers with N hosts. The second tier has clients who request for the VM to run the jobs accordingly. In general, we know that this problem of efficiently mapping jobs for VM in the hosts is NP-hard [10]. Therefore a meta-heuristic algorithm is designed to provide a suitably good solution to this scheduling problem.

The existing method followed in OpenNebula is that it operates by the client submitting VM requests along with certain constraints such as storage capacity, type of OS, the memory required, speed, etc. In the proposed methodology another constraint, time requirement is added to obtain the time period of the request. This time period is necessary to keep track of the reservation which helps the user to spend for extended reservation in order to achieve reliable completion of jobs. It also predicts whether PMs need to be ON or OFF since VM requirements appear in real-time, and should be allocated to hosts as they arrive.

IV. Architecture Diagram

Energy consumption of hosts can be exactly portrayed by means of a linear relationship of CPU utilization. Power consumption can therefore be described as a CPU exploitation characteristic and by way of the usage of the mathematical system complete power consumption is computed.

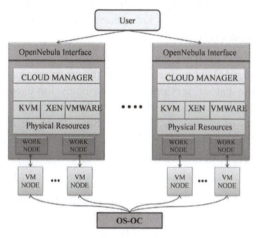

Fig. 20.1 OS OC scheduler

Source: Author

The architecture diagram of the anticipated system is as follows if a cloud supplier receives a VM request, the OS-OC scheduler first determines the cluster as shown in Fig 1. Once a cluster is selected, the next PM should be selected to put forward the demand. The virtualization concept allows placing multiple requests on the same PM. Workloads need to be consolidated within as fewer servers as possible, the rest servers are kept idle. The nation at which the selection is made to kind VM requests that must be assigned to hosts in a way that the wide variety of lively hosts is minimized is referred to as the VM consolidation disaster [10]. The difficulty in this approach is that the VM requests arrive in a dynamic environment and are assigned to hosts as they arrive. The basic idea is to schedule VMs with closer dispatch times together on the same PM, permitting PMs with small lifetimes to be turned OFF swiftly. The dispatch time should be stated directly by the client or expected by the sort of task the VM

will be carried out using the earlier activities of the client.

V. Algorithm and Proposed Methodology

A. Initial Search Algorithm

The initiation of the search algorithm is a significant phase in this proposed approach in which VM has to search its PM with its constraints. Each VM has to be scheduled to only one PM according to this approach. The selection of the initial solution is vital since it can influence the quality of future results. This search algorithm follows a random greedy method for each local search. The PMs are filtered from the total number of hosts based on the requirements and each VM obtains a set of PMs.

The optimal choice of PM is chosen from the set of hosts with certain constraints. If PM is not available with this approach then the greedy method is followed and allocates the VM to the following host in the PM list. Then the unassigned VMs are eliminated from the currently reserved pool and will be assigned in the next schedule.

B. Scheduling

The real-time cloud-based OpenNebula uses a novel scheduler in this proposed approach for the optimal placement of VM. The scheduler operates through many cycles in which it waits for a specified amount of time before each scheduling. The time interval is used to collect a pool of VMs to provide scheduling in the next cycle as shown in Algorithm 1.

This results in a list of hosts filtered with the specified constraints and the local search algorithm is implemented independently. Then the combination of the local search solutions is used to find the best scheduling based on the objective.

Algorithm 1: OS-OC Algorithm
INPUT: List of VMs and hosts along with the jobs to be scheduled
OUTPUT: Scheduling of VMs on hosts based on the constraints using optimal search heuristic algorithm
1. VM_list=readConfig()
2. **for** each vm i do
3.   int j=0
4.   PM_order=getOrder()
5.   **for** each scheduling cycle
6.   nextPM=PM_order[j++]
7.   **if** (constraints && energy) ==true
8.   Obtain the compatible list of hosts
9.   Allocate the jobs to VM
10.   **end if**
11.   **if** jobs allocated to VM then
12.   update the status of hosts
13.   else
14.   return null
15.   **end if**
16.   **end for**
17. return scheduling result

VI. Mathematical Analysis

A. Energy-Qos Model

Major elements which drain strength in a server are the entire consumption of a CPU, memory, disk, and community card. Above all, the CPU is the sole factor that consumes a lot of electricity in a server. Based on the above fact, the power mannequin of the server is given in equation (1).

$$P(E(t)) = k * P_{max} + (1 - k) * P_{max} * E(t) \qquad (1)$$

where $P(E(t))$ is the total energy consumed by the server at time t, P_{max} is the maximum energy consumed by a fully utilized server, k is the fraction of energy consumed when the server is idle.

Regarding QoS requirements, it carries many attributes, such as response time, reliability, throughput, delay, availability etc. There are two kinds of attributes namely: positive and negative QoS attributes.

Positive QoS attributes are the attributes that should have a high value to have a good impact on the performance of the server such as reliability, availability and so on. On the other hand, negative QoS attributes are the attributes that should have a low value such as response time, delay etc.

Consider there are r services to be provided $(S = \{s1, s2, ..., sr\})$ and S is the set containing the list of services. QoS attributes for each service s with v attributes are represented by a set $qs = \{qs1, qs2, ..., qsv\}$. Since there are u QoS attributes $(1 \leq m \leq u)$, the average QoS value of m-th attribute is given by,

$$Q_m^{avg} = Q_m^{max} + Q_m^{min} * 0.5 \qquad (2)$$

where Q_m^{avg} is the average value of the maximum Q_m^{max} and minimum Q_m^{min} values of the m-th attribute. The QoS utility features for the i-th provider running the VM of the j-th server is described as follows:

$$U(s_i) = \sum_{m=1}^{v} \frac{Q_{j,m}^{max} - q_k(s_i)}{Q_{j,m}^{avg}} \qquad (3)$$

VII. Experimental Results

We examined our work on the open nebula device which is a actual time cloud. In our experiment, we have labored with simply one data center. We tried with many hosts such as 6,80,420,1160 hosts on this data center. Each node contains one CPU core with 5 GB ram/ network bandwidth and a storage area of 500GB. The host includes 5000 MIPS of execution pace and a thousand IO MIPS. Disk I/O execution speed. For every virtual computer on host ram, measurement is 512 mb and bandwidth dimension is 2500 mb with 1000mips execution velocity and five hundred IO mips for disk I/O execution speed.

Table 20.1 shows the parameters taken for our work executed using the open nebula tool. We performed an initial search algorithm and compared the results with the global optimal solution using this heuristic approach. Our

Table 20.1 Estimated parameter

Parameter Value	Value
Maximum VM	6
Number of Cycles	50s
Number of VM	1,20,60,80
Number of Host	6,80,420,1160

Source: Author

approach OS-OC is compared with the existing scheduler for each scheduling cycle.

The results obtained by using our novel OS-OC algorithm are listed in Table.2 for the energy variable. The mentioned result outperforms the existing open nebula scheduler in terms of energy and efficient mapping is performed with respective jobs. The maximum virtual machine value is 6. The number of cycles needed is the 50s. The number of virtual machines is 12, 06,080. The number of hosts is 6,80,42,01,160.

The number of VM is 1 and the number of hosts varies from 6,80,420,1160, the OS-OC scheduler has 2 numbers of scheduled VMs. The OpenNebula scheduler also has 2 numbers of scheduled VMs. The consumed energy with 1 VM ranges from 453035 Joule to 171747.4 Joule for the OS-OC scheduler and respectively for 6 to 1160 hosts. The consumed energy with 1 VM ranges from 453035 Joule to 246517 Joule for the OpenNebula scheduler and respectively for 6 to 1160 hosts. The time taken for the OS-OC scheduler ranges from 10.0001 and 10.01 and for the OpenNebula scheduler 10.0001.

The number of VMs is 20 and the number of hosts varies from 6,80,420,1160, the OS-OC scheduler has 8 and 20 numbers of scheduled VMs. The OpenNebula scheduler has 7 and 20 numbers of scheduled VMs. The consumed energy with 20 VMs ranges from 3329294.6 Joule to 12145173.3 Joule for the OS-OC scheduler and respectively for 6 to 1160 hosts. The consumed energy with 20 VM ranges from 3624380 Joule to 13519200 Joule for the OpenNebula scheduler and respectively for 6 to 1160 hosts. The time taken for the OS-OC

scheduler ranges from 10.001 and 0.2 and for the OpenNebula scheduler 10.00001 and 10.001.

The number of VMs is 60 and the number of hosts varies from 6,80,420,1160, the OS-OC scheduler has 10.6 to 30 scheduled VMs. The OpenNebula scheduler has 12 to 30 numbers of scheduled VMs. The consumed energy with 60 VM ranges from 25837706.7 Joule to 59827060 Joule for OS-OC scheduler and respectively for 6 to 1160 hosts. The consumed energy with 60 VMs ranges from 18186400 Joule to 65313900 Joule for the OpenNebula scheduler and respectively for 6 to 1160 hosts. The time taken for the OS-OC scheduler ranges from 10.01 and 0.8 and for the OpenNebula scheduler 10.001.

The number of VMs is 80 and the number of hosts varies from 6,80,420,1160, the OS-OC scheduler has 14.9 to 80 scheduled VMs. The OpenNebula scheduler also has 24 to 80 numbers of scheduled VMs. The consumed energy with 80 VM ranges from 3110997.3 Joule to 135482200 Joule for OS-OC scheduler and respectively for 6 to 1160 hosts. The consumed energy with 80 VMs ranges from 41852300 Joule to 161012000 Joule for the OpenNebula scheduler and respectively for 24 to 80 hosts. The time taken for the OS-OC scheduler ranges from 10.01 and 2.4 and for the OpenNebula scheduler 10.001 and 10.01.

Table 2 truly explains the assessment of the effects acquired via the OS-OC algorithm and the open nebula scheduler.

Figure 20.2 shows that the comparison of energy and time with a single number of the host. It shows the number of hosts and the allocated number of scheduled virtual machines.

Figure 20.3 shows the comparison of energy and time with 20 number of the host.

Figure 20.4 shows the comparison of energy and time with 60 hosts. It shows the number of hosts and the allocated number of scheduled virtual machines.

Table 20.2 Comparison between the results obtained by the os-oc algorithm and the open nebula scheduler

Number of VMs	1					
Number of hosts	No. of Sched. VMs		Consumed Energy (Joule)		Time (s)	
	OS-OC	OpenNebula sched.	OS-OC	OpenNebula sched.	OS-OC	OpenNebula sched.
6	2	2	453045	453035	10e-04	10e-04
80	2	2	2506407.9	393660	10-e03	10e-04
420	2	2	2842140.6	3993660	10-e03	10e-04
1160	2	2	171747.4	246517	10-e02	10e-04

Number of VMs	20					
Number of hosts	No. of Sched. VMs		Consumed Energy (Joule)		Time (s)	
	OS-OC	OpenNebula sched.	OS-OC	OpenNebula sched.	OS-OC	OpenNebula sched.
6	8	7	3329294.6	3624380	10e-03	10e-05
80	20	20	22152200	19341900	10e-02	10e-04
420	20	20	22109773.3	34819800	10e-02	10e-01
1160	20	20	12145173.3	13519200	0.2	10e-03

Number of VMs	60					
Number of hosts	No. of Sched. VMs		Consumed Energy (Joule)		Time (s)	
	OS-OC	OpenNebula sched.	OS-OC	OpenNebula sched.	OS-OC	OpenNebula sched.
6	10.6	12	25837706.7	18186400	10e-02	10e-03
80	30	30	56091793.3	73657200	0.1	10e-03
420	30	30	68386373.3	84573000	0.2	10e-03
1160	30	30	59827060	65313900	0.8	10e-03

Number of VMs	80					
Number of hosts	No. of Sched. VMs		Consumed Energy (Joule)		Time (s)	
	OS-OC	OpenNebula sched.	OS-OC	OpenNebula sched.	OS-OC	OpenNebula sched.
6	14.9	24	3110997.3	41852300	10e-02	10e-03
80	93.7	64	175229600	181781000	0.7	10e-02
420	80	80	132939933	149688000	1	10e-02
1160	80	80	135482200	161012000	2.4	10e-02

Source: Author

Number of VMs =1

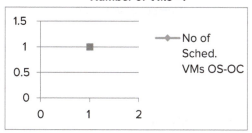

X axis : Number of VM's
Y axis: Time taken for each scheduler

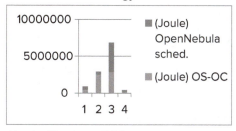

X axis : Number of VM's
Y axis: Time taken for each scheduler

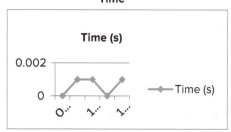

X axis : Number of VM's
Y axis: Time taken for each scheduler

Fig. 20.2 Energy comparisons for 1 host
Source: Compiled by Author

Figure 20.5 shows the comparison of energy and time with 80 hosts. It shows the number of hosts and the allocated number of scheduled virtual machines

The proposed OS-OC scheduler outperforms in terms of energy when compared to the existing schedulers in cloud computing. The existing real time scheduler results and that of the OS-OC scheduler is compared and analysis

Number of VMs = 20

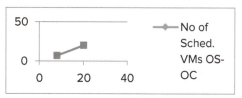

X axis : Number of VM's
Y axis: Time taken for each scheduler

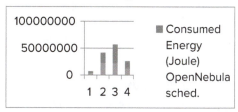

X axis : Number of VM's
Y axis: Time taken for each scheduler

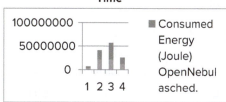

X axis : Number of VM's
Y axis: Time taken for each scheduler

Fig. 20.3 Energy comparisons for 20 hosts
Source: Compiled by Author

is done for different numbers of hosts with varying number of VMs at different times. The experimental outcome with the consumed energy is described in table 2 and the graphical representation of the comparison is shown in Fig. 20.6.

VIII. Conclusion and Future Work

In this paper, we developed a novel method termed Optimal Search-Open Nebula primarily

Number of VMs = 60

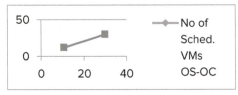

X axis : Number of VM's
Y axis: Time taken for each scheduler

Energy

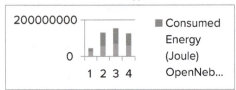

X axis : Number of VM's
Y axis: Time taken for each scheduler

Time

X axis : Number of VM's
Y axis: Time taken for each scheduler

Fig. 20.4 Energy comparisons for 60 hosts
Source: Compiled by Author

Number of VMs = 80

X axis : Number of VM's
Y axis: Time taken for each scheduler

Energy

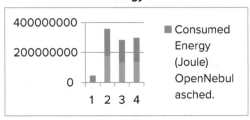

X axis : Number of VM's
Y axis: Time taken for each scheduler

Time

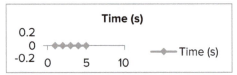

X axis : Number of VM's
Y axis: Time taken for each scheduler

Fig. 20.5 Energy comparisons for 80 hosts
Source: Compiled by Author

based Cloud (OS-OC) scheduler inside a sensible cloud infrastructure with the usage of the Open Nebula tool. The OS-OC scheduler targets minimizing strength consumption and additionally satisfies purchasers' QoS (Quality of Service) with the aid of optimally assigning VMs. The effects exhibit that the OS-OC strategy outperforms traditional heuristic primarily based methodologies and achieves pleasant scheduling with minimal electricity consumption. The proposed most reliable answer paves the way for much less cost, much less latency with environment friendly strength consumption with higher throughput as preferred via the cloud provider providers.

References

1. J. Zeng, D. Ding, K. Kang, H. Xie and Q. Yin, "Adaptive DRL-Based Virtual Machine Consolidation in Energy-Efficient Cloud Data Center," in IEEE Transactions on Parallel and Distributed Systems, vol. 33, no. 11, pp. 2991–3002, 1 Nov. 2022, doi: 10.1109/TPDS.2022.3147851.
2. MehiarDabbagh, BechirHamdaoui, Mohsen Guizaniy and AmmarRayes "Towards Energy-Efficient Cloud Computing: Prediction, Consolidation, and Overcommitment", IEEE

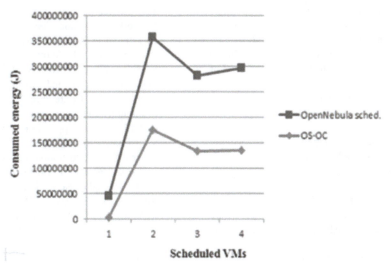

Fig. 20.6 Energy comparisons of OS-OC

Source: Compiled by Author

Communications Society, Volume: 29, Issue: 2, April 2015, pp. 56–61.

3. Xiong Fu, Chen Zhou "Virtual machine selection and placement for dynamic consolidation in Cloud computing environment", Springer, 2015, pp. 322–330.

4. Zheng Chang, Jie Gong, Zhenyu Zhou, Tapani Ristaniemi, Zhisheng Niu "Resource Allocation and Data Offloading for Energy Efficiency in Wireless Power Transfer Enabled Collaborative Mobile Clouds", IEEE Conference on Computer Communications Workshops, May 2015, pp. 336–341.

5. Mehiar Dabbagh, Bechir Hamdaoui, Mohsen Guizanit and Ammar Rayes "Efficient Datacenter Resource Utilization Through Cloud Resource Overcommitment", IEEE Conference on Computer Communications Workshops, May 2015, pp. 330–335.

6. Fahimeh Farahnakian, Pasi Liljeberg, and Juha Plosila "Energy-Efficient Virtual Machines Consolidation in Cloud Data Centers using Reinforcement Learning", IEEE 22nd Euro micro International Parallel, Distributed and Network-Based Processing (PDP), Feb 2014, pp. 500–507.

7. Mohammed Rashid Chowdhury, Mohammad Raihan Mahmud and Rashedur M. Rahman, "Implementation and performance analysis of various VM placement strategies in Cloud Sim", Springer, 2015, pp. 1–21.

8. Woo-Joong Kim, Dong-Ki Kang, Seong-Hwan Kim, Chan-Hyun Youn1, "Cost Adaptive VM Management for Scientific Workflow Application in Mobile Cloud", Springer, 17 april 2014, pp. 328–336.

9. Zohreh Sanaei, Saeid Abolfazli, Abdullah Gani, Rajkumar Buyya, "Heterogeneity in Mobile Cloud Computing: Taxonomy and Open Challenges", IEEE Communication Survey & Tutorial, Vol. 16, No. 1, First quater 2014, pp. 369–392.

10. Yating Wang, Ing-Ray Chen, Ding-Chau Wang, "A Survey of Mobile Cloud Computing Application: Perspectives and Challenges", Springer, 2014, pp. 1607–1623.

11. Maolin Tang, Shenchen Pan, "A Hybrid Genetic Algorithm for the Energy-Efficient virtual Machine Placement Problem in Data Centers", Springer, 18 Jan 2014, pp. 211–221.

12. Fahimeh Farahnakian, Adnan Ashraf, Tapio Pahikkala, Pasi Liljeberg, Juha Plosila, Ivan Porres, And Hannu Tenhunen, "Using Ant Colony System to Consolidate Vms for Green Cloud Computing", IEEE Transactions On Services Computing, Vol. 8, No. 2, March/April 2015, pp. 187–198.

13. LiliXu, KunWang, ZhiyouOuyang, XinQi, "An Improved Binary Pso-based Task Scheduling

Algorithm In Green Cloud Computing", IEEE Transactions On Communications and Networking, April 2014, pp. 341–352.

14. Zahid Sagu, Dinesh Kumar, "A PSO Based Vm Resource Scheduling Model For Cloud Computing", IEEE International Conference On Computational Intelligence & Communication Technology, 2015, pp .231–219.

15. Deepika Saxena, Silphi Saxena, "Highly Advanced Cloudlet Scheduling Algorithm Based On Particle Swarm Optimization ", IEEE, 2015, pp. 978–983.

16. Aruna, vasantha, "A Particle Swarm Optimization Algorithm For Power-aware Virtual Machine Allocation", IEEE, 18 Jan 2015, pp. 211–221.

17. A. Choudhary, M. C. Govil, G. Singh and L. K. Awasthi, "Energy-efficient resource allocation approaches with optimum virtual machine migrations in cloud environment," 2016 Fourth International Conference on Parallel, Distributed and Grid Computing (PDGC), 2016, pp. 182–187, doi: 10.1109/PDGC.2016.7913141.

18. W. Zhang, X. Chen and J. Jiang, "A multi-objective optimization method of initial virtual machine fault-tolerant placement for star topological data centers of cloud systems," in Tsinghua Science and Technology, vol. 26, no. 1, pp. 95–111, Feb. 2021, doi: 10.26599/TST.2019.9010044.

19. P. Ntambu and S. A. Adeshina, "Machine Learning-Based Anomalies Detection in Cloud Virtual Machine Resource Usage," 2021 1st International Conference on Multidisciplinary Engineering and Applied Science (ICMEAS), 2021, pp. 1–6, doi: 10.1109/ICMEAS52683.2021.9692308.

20. P. Gao, H. Li, X. Qu and Y. Cheng, "Research on Virtual Machine Performance Test Based on Cloud Platform," 2021 IEEE 4th Advanced Information Management, Communicates, Electronic and Automation Control Conference (IMCEC), 2021, pp. 392–395, doi: 10.1109/IMCEC51613.2021.9482364.

21. C. V. Marian, "DNS Records Secure Provisioning Mechanism for Virtual Machines automatic management in high density data centers," 2021 IEEE International Black Sea Conference on Communications and Networking (BlackSeaCom), 2021, pp. 1–5, doi: 10.1109/BlackSeaCom52164.2021.9527811.

22. P. Chinnasamy, B. Vinothini, V. Praveena, A. S. Subaira and B. Ben Sujitha, "Providing Resilience on Cloud Computing," 2021 International Conference on Computer Communication and Informatics (ICCCI), 2021, pp. 1–4, doi: 10.1109/ICCCI50826.2021.9402681.

23. Abisha D, "Perlustration on Authentication Protocols in 4G (LTE/LTE-A) Using Pro-Verif", Journal of Network Security Computer Networks, Vol.3, Issue 1, PP. 1–7, 2017.

24. R. N. Karthika, C. Valliyammai and D. Abisha, "Perlustration on techno level classification of deduplication techniques in cloud for big data storage," 2016 Eighth International Conference on Advanced Computing (ICoAC), 2017, pp. 206–211, doi: 10.1109/ICoAC.2017.7951771.

25. Karthika, R., Valliyammai, C., Abisha, D. (2018). Data Deduplication and Fine-Grained Auditing on Big Data in Cloud Storage. In: Reddy Edla, D., Lingras, P., Venkatanareshbabu K. (eds) Advances in Machine Learning and Data Science. Advances in Intelligent Systems and Computing, vol 705. Springer, Singapore. https://doi.org/10.1007/978-981-10-8569-7_38.

26. Lalitha, R. H., Weslin, D., Abisha, D., Prakash, V. R. (2022). A Hybrid Split and Merge (HSM) Technique for Rapid Video Compression in Cloud Environment. In: Pandian, A. P., Fernando, X., Haoxiang, W. (eds) Computer Networks, Big Data and IoT. Lecture Notes on Data Engineering and Communications Technologies, vol 117. Springer, Singapore. https://doi.org/10.1007/978-981-19-0898-9_72.

27. M. Xu, A. N. Toosi and R. Buyya, "A Self-Adaptive Approach for Managing Applications and Harnessing Renewable Energy for Sustainable Cloud Computing," in IEEE Transactions on Sustainable Computing, vol. 6, no. 4, pp. 544–558, 1 Oct.-Dec. 2021, doi: 10.1109/TSUSC.2020.3014943.

28. A. Marahatta, S. Pirbhulal, F. Zhang, R. M. Parizi, K. K. R. Choo and Z. Liu, "Classification-Based and Energy-Efficient

Dynamic Task Scheduling Scheme for Virtualized Cloud Data Center," in IEEE Transactions on Cloud Computing, vol. 9, no. 4, pp. 1376-1390, 1 Oct.–Dec. 2021, doi: 10.1109/TCC.2019.2918226.

29. A. A. Khan, M. Zakarya, R. Buyya, R. Khan, M. Khan and O. Rana, "An Energy and Performance Aware Consolidation Technique for Containerized Datacenters," in IEEE Transactions on Cloud Computing, vol. 9, no. 4, pp. 1305-1322, 1 Oct.-Dec. 2021, doi: 10.1109/TCC.2019.2920914.

30. A. Marahatta, Q. Xin, C. Chi, F. Zhang and Z. Liu, "PEFS: AI-Driven Prediction Based Energy-Aware Fault-Tolerant Scheduling Scheme for Cloud Data Center," in IEEE Transactions on Sustainable Computing, vol. 6, no. 4, pp. 655–666, 1 Oct.-Dec. 2021, doi: 10.1109/TSUSC.2020.3015559.

31. Z. Zhou, M. Shojafar, M. Alazab, J. Abawajy and F. Li, "AFED-EF: An Energy-Efficient VM Allocation Algorithm for IoT Applications in a Cloud Data Center," in IEEE Transactions on Green Communications and Networking, vol. 5, no. 2, pp. 658–669, June 2021, doi: 10.1109/TGCN.2021.3067309.

Recent Trends in Computational Intelligence and its Application – Sugumaran D. et al. (eds)
© 2023 Taylor & Francis Group, London, ISBN 978-1-032-48410-5

21

Fire Detection and Alert System Using Deep Learning

Sathishkumar B. R.[*], **Nivyadharsini N.**[1], **Niranjana T.**[2]
Department of Electronics and Communication Engineering,
Sri Ramakrishna Engineering College, Coimbatore, Tamil Nadu, India

Abstract—The principal purpose of a fire detection and alert system is to provide an early warning of fire so as to take immediate measures to evacuate people or human casualties. Physical sensors are being used by conventional fire detection systems to detect fire. The purpose of the fire detection and alarm system is to find the fire and send a voice or mail alert message with the location, temperature, and smoke density. It also finds the accuracy of the detected fire. The input of the fire detection process is taken in a video format, the input video is converted into several frames and a particular frame with fire is chosen to convert into a grayscale image, as a colored image consists of different intensity values and is inappropriate to detect the fire features. The converted gray image undergoes enhancement techniques, and image quality assessment, and the fire is detected in the given input video using edge detection. After the completion of all these processes, the accuracy of the detected fire is been calculated using five different algorithms like AlexNet, GoogLeNet, VGG16, ResNet50, and ResNet101. Finally, the alert message is sent through a mail and voice message with information like smoke density, temperature, location, date and time, and accuracy.

Keywords—Fire detection, Convolutional neural network, Deep learning, Voice and mail alerts, Accuracy

I. Introduction

While fire has both positive and harmful effects on the environment, it solely has negative side effects when it comes to cars, people, and animals. Fire in the home can cause an increased level of carbon dioxide and carbon monoxide this can cause damage to the respiratory system and several other health issues, and may also cause death. So, to overcome this issue, recent studies have brought early fire detection systems that could reduce the adverse effect. To date, there are many studies that bring a lot of advancement to the previous one to possibly avoid human casualties. In the early days, once when a fire is detected people inform the fire department either through phone calls or by visiting directly. But in recent days most

[*]Corresponding Author: sathishkumar.b@srec.ac.in
[1]nivyadharsini.1802162@srec.ac.in, [2]niranjana.1802131@srec.ac.in

DOI: 10.1201/9781003388913-21

developed countries have CCTV, Surveillance cameras installed on roads, streets, and buildings like schools, colleges, and shopping malls to take periodic images of the area. An essential part of the fire alarm system is played by this. In this paper, real-time implementation is being focused on. As to show the working, three different datasets have been chosen to find the accuracy and send an alert message in the form of mail and voice, to the required source like the fire department and the vehicle owner in the case of a car, the fire department and the forest officer in case of a forest, it purely depends on the information that is been fed for alert messages. It is also designed in such a way that the fire can be detected even if it is dark.

II. Literature Survey

Several researchers proposed diverse methods of fire detection and alert system strategies through MATLAB and various technologies. Some of the related ones are to be discussed below

To detect flames, Kewei Wang (2018) suggested a Deep CNN and SVM system [11] with three modules. The algorithm first creates IR flame datasets from input videos and expands them with data augmentation. Next, it creates a 9-layer IRCNN data set to extract key features from Infrared images. The algorithms used to detect fire were AlexNet, IRCNN+SGD, and IRCNN+Adam which undergo five experimental steps as IRCNN(A), AlexNet(B), IRCNN(C), IRCNN(D), IRCNN+SVM(E). The results obtained were in the form of data outputs with state-of-the-art performance levels such as Precision(P) of 98.82%, Recall(R) of 98.58%, and F1 score of 98.70%. Moreover, the classifying speed is 20 IR images per second.

Suhas G (2020) suggested a model that was split into two parts: gathering data and pre-processing it, and then using transfer learning to build a fire detection model[6]. The datasets in this instance are divided into two groups: fire-positive samples and non-fire-positive samples.

The datasets are segmented into train frames and test frames of large numbers as 1678 fire frames and 1368 non-fire frames. With the help of pre-trained models, the video frames are extracted. This model is trained with very large-scale video frames that have a huge number of images and give discriminative features. The feature vector size varies for different models. The main aspect of Transfer Learning is its use of complex pre-trained deep neural network models to find the solution for a simplified problem. The machine learning algorithms used to extract fire features are ResNet-50, Inception V3, and InceptionResNetV2. The best performance has been obtained from ResNet50.

Rabah N Farhan (2019) proposed a fire detection architecture with a structured process of dataset preparation, modified LeNet-based fire detection system, and ConvNet-based fire detection systems[10]. The dataset preparation process is done by collecting various image datasets from the internet which are resized uniformly to a dimension of 64x64. The labeling of fire and non-fire images is done manually. The modified LeNet-5-based fire detection system layers the images into 7 layers. The size of the proposed system was scaled back using ConvNet-based fire detection technology. This makes use of a CNN cascade classifier, which has 8 layers connected in a specific order based on the size of the input images. Finally, the average accuracy of LeNet and ConvNet is compared where LeNet has 81% accuracy with a high number of layers and is slower than ConvNet, the ConvNet has a 99% accuracy with its binary classification of the dataset as it has a very small layer number.

The survey shows that the available frameworks use are totally different algorithms for each dataset and each architecture. According to the fire detection system, the precision of 97% is acquired and the proposed systems were very efficient in predicting the fire accurately and the accuracy of 97%, 98%, and 99% is achieved successively. Comparing the detection error rates and accuracy rate, it is found that

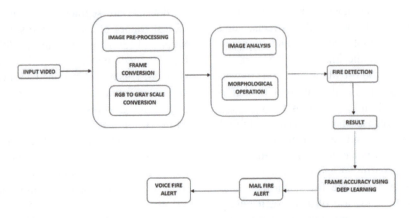

Fig. 21.1 Block diagram

the values are static and not dynamic, so it varies for each dataset with respect to different algorithms.

III. Implementation with Results

The project's flow is shown step-by-step in the block diagram in Fig. 21.1.

The given input video is taken from a surveillance camera implemented in different areas for regularly monitoring the activities. Figure 21.2 is segmented into several frames for effective analysis of fire features, this process is called Frame Acquisition.

A frame with a huge area of the fire is chosen to do the image pre-processing operations. Initially, the fire frames are of colored images i.e., RGB images. The selected frame is

Fig. 21.2 Frames acquisition image from input video

converted into a grayscale image for analyzing the features of fire as shown in Fig. 21.3.

The morphological operations like histogram equalization and adaptive histogram equalization are done with a parameter graph as shown in Figs 21.4 and 21.5.

This completes the process of fire detection with edge detection which clearly defines the

Fig. 21.3 Gray image

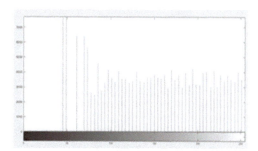

Fig. 21.4 Histogram equalization image and histogram equalization graph

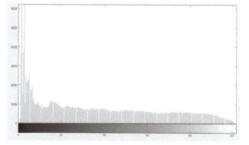

Fig. 21.5 Adaptive histogram equalization image and adaptive histogram equalization graph

boundary of fire in the selected frame. This frame also consists of information like the date and time of the fire explosion happened. Fig. 21.6 shows fire detected image with edge detection.

Fig. 21.6 Fire detected image with edge detection

After the fire is detected a voice message about the fire accident happened is immediately sent to the fire brigade. The next process of this project is to find the accuracy of the fire.

The accuracy process also takes input in the form of a video and converts it into frames. The accuracy process is shown in Fig. 21.7. The converted frames are further divided into two parts: training images and validation images. The training images are then pre-processed, and layer preparation is done to obtain features of fire which are used to train the desired architectures. The next step is where different types of architectures are used to comparatively find the one that gives the best accuracy. Then, the validation images are used to classify the data and calculated the values of true and false accuracy rates. The confusion matrix from all the architectures is used to find the accuracy values separately for training and testing processes. The algorithms used to find accuracy were, AlexNet, GoogLeNet, ResNet101, ResNet50, and VGG16. They are briefly explained below:

AlexNet: The AlexNet which is an 8-deep layered convolutional neural network that works with 3 fully connected layers and with 60 million parameters employed, has achieved a maximum accuracy of 100% which is 97.9% from the training images and 2.1% from the testing images which varies with every dataset.

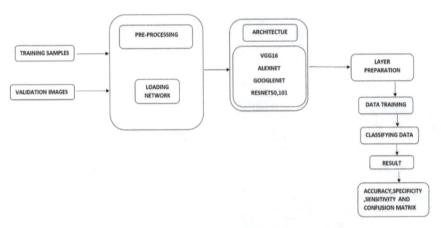

Fig. 21.7 Accuracy block diagram

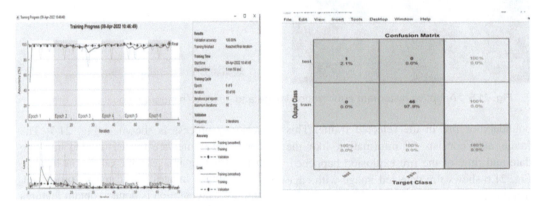

Fig. 21.8 Accuracy from AlexNet and confusion matrix of AlexNet

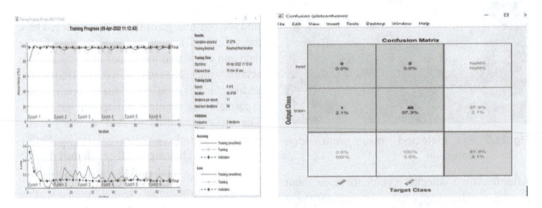

Fig. 21.9 Accuracy from GoogLeNet and confusion matrix of GoogLeNet

GoogLeNet: The GoogLeNet which is 22 layered convolutional neural network and is codenamed Inception with less than 12 million parameters employed, has achieved an accuracy of 97.9% and has an error rate of 2.1% from training images which varies for every dataset

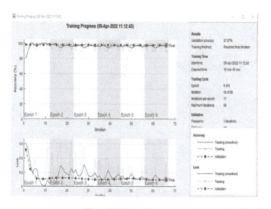

Fig. 21.10   Accuracy from ResNet101 and confusion matrix of ResNet101

and Accuracy from ResNet101 and Confusion Matrix of ResNet101.

ResNet101: The ResNet101 which is 101 deep layered convolutional neural networks with 11 million parameters employed, has achieved an accuracy of 97.9% and has an error rate of 2.1% from training images that are dynamic and change for every dataset, and Fig. 21.10 indicates the accuracy from ResNet101 and Confusion Matrix of ResNet101.

ResNet50: The ResNet50 which is 50 deep layered convolutional neural networks with 23 million parameters employed, has achieved an accuracy of 97.87% and has an error rate of 2.1% from training images that attain different

values for every dataset which is shown in Fig. 21.11.

VGG16: The VGG16 which is 16 deep layered convolutional neural networks with 138 million parameters employed, has achieved an accuracy of 89.8% and has an error rate of 6.1% from training images and 4.1% from testing images, which is shown in Fig. 21.12 and Fig. 21.13. In the case of VGG16, the accuracy values vary every time even for the same dataset.

Finally, after finding the accuracy of the fire from the frames, the output is sent as mail to the desired people with extended details such as the location of the fire, smoke density, and temperature. The mail alert output is shown in Fig. 21.14.

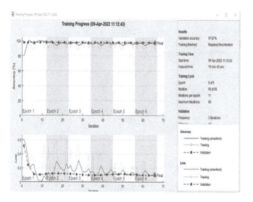

Fig. 21.11   Accuracy from ResNet50 and confusion matrix of ResNet50

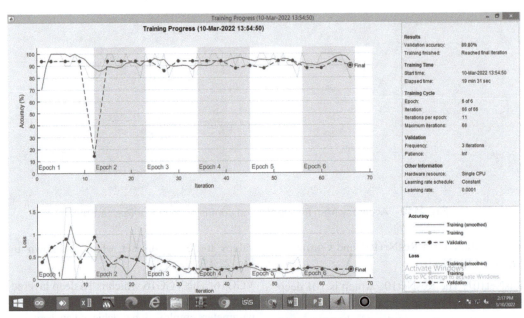

Fig. 21.12 Accuracy from VGG16

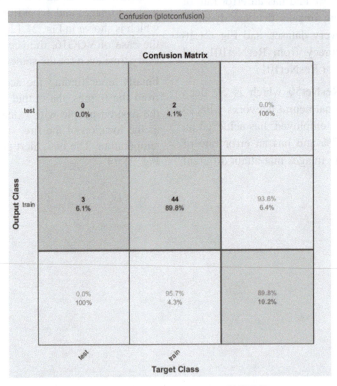

Fig. 21.13 Confusion matrix of VGG16

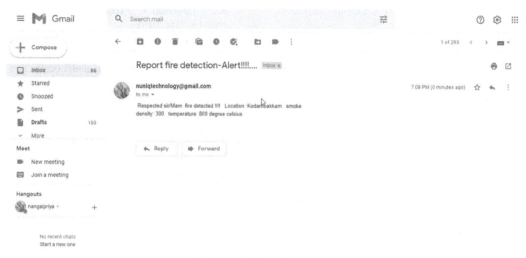

Fig. 21.14   Mail alert output

IV. Conclusions

The detection of fire for three different datasets (such as a house fire, a car fire, or a forest fire) is obtained and an alert about the detected fire is given through voice message and also a mail is sent with the details such as Location, Smoke Density, Temperature of Fire, etc. The accuracy of the fire area is detected and calculated with five different algorithms namely, AlexNet, GoogLeNet, ResNet50, ResNet101, and VGG16. The already existing methods of fire detection techniques use only images for detecting the fire and have not properly meant to detect the area of the fire. The proposed work is a step ahead by using real-time videos as input to convert frames and detect the area of fire accurately with the help of the Edge detection process. This is helpful in accurately detecting the density and area of the fire with different datasets. Advanced fire detection technologies could come in very handy to protect such infrastructure and people on a real-time basis. In future cases, a real-time implementation with cameras can be done to capture fire in advanced ways, the sensors, GPS, GSM, and Wi-Fi modules can also be included to get real-time features and details instantly and can also aim

in alerting the nearby fire brigades and police control room for the safety of the surroundings and people.

References

1. W. Li, S. Xiaobo, C. Jun and L. Ying, "Research on Forest Fire Image Recognition Algorithm Based on Color Feature Statistics," 6th International Conference on Intelligent Computing and Signal Processing (ICSP), pp. 346–349(2021).

2. Gopalakrishnan G., Arul Mozhi Varman S., Dinessh T. C., Divayarupa S., Benazir Begam R., "Automatic Forest Fire Detection and Suppression Smart Grid System", International Journal of Scientific Research in Science and Technology (IJSRST), Print ISSN: 2395-6011, Online ISSN: 2395-602X, Vol. 8, Issue 3, pp. 666–669, (2021).

3. C. Naimeng, Y. Wanjun and W. Xiaoyu, "Smoke detection for early forest fire in aerial photography based on GMM background and wavelet energy," 2021 IEEE International Conference on Power Electronics, Computer Applications (ICPECA), pp. 763–765(2021).

4. Y. Li, Y. Lei and M. Wang, "Design of Classroom Wireless Fire Monitoring and Alarm System Based on CC2530," 2020 IEEE 6th International Conference on Computer

and Communications (ICCC), pp. 2351–2355 (2020).

5. Sherstjuk, Vladimir & Zharikova, Maryna & Dorovskaja, Irina., "3D Fire Front Reconstruction in UAV-Based Forest-Fire Monitoring System". 243–248 (2020).

6. Suhas G., Chetan Kumar, Abhishek B. S., Digvijay Gowda K. A., Prajwal R., "Fire Detection Using Deep Learning", International Journal of Progressive Research in Science and Engineering Volume-1, Issue-5, pp. 1–5(2020).

7. L. Muduli, D. P. Mishra and P. K. Jana, "Optimized Fuzzy Logic-Based Fire Monitoring in Underground Coal Mines: Binary Particle Swarm Optimization Approach," in IEEE Systems Journal, vol. 14, no. 2, pp. 3039–3046, (2020).

8. H. Dang-Ngoc and H. Nguyen-Trung, "Evaluation of Forest Fire Detection Model using Video captured by UAVs," 19th International Symposium on Communications and Information Technologies (ISCIT), pp. 513–518(2019).

9. K. Muhammad, J. Ahmad, Z. Lv, P. Bellavista, P. Yang and S. W. Baik, "Efficient Deep CNN-Based Fire Detection and Localization in Video Surveillance Applications," in IEEE Transactions on Systems, Man, and Cybernetics: Systems, vol. 49, no. 7, pp. 1419–1434, (2019).

10. Nory, Rabah & Aljumaili, Mustafa & Ismat, Nezar, Fire Detection Using Convolutional Deep Learning Algorithms. AUS. 26. pp. 441–448, (2019).

11. Kewei Wang, Yongming Zhang, Jinjun Wang "Fire Detection in Infrared Video Surveillance Based on Convolutional Neural Network and SVM" IEEE 3rd International Conference on Signal and Image Processing (ICSIP), (2018).

Note: Author have produced all the figures and simulated image outputs.

Recent Trends in Computational Intelligence and its Application – Sugumaran D. et al. (eds)
© 2023 Taylor & Francis Group, London, ISBN 978-1-032-48410-5

22

EMO Player Using Deep Learning

Sathishkumar B. R.*, Shreya V.[1], Sujitha R.[2], Thilaga M.[3]
Department of Electronics and Communication Engineering,
Sri Ramakrishna Engineering College, Coimbatore, Tamil Nadu, India

Abstract—Many people like to listen to music and often they also have hobbies. As a result, music is very significant in one's life. People's emotions influence how they react to music. People listen to music according to their moods and interests. Listening to music is an important practice for reducing stress. However, if the music does not match the listener's current emotions, it can be counterproductive. There's no music playing within the display approach that can select melodies based on the user's feelings, such as sad, joyful, furious, or neutral. Agreeing later thinks about, people react and react to music, and music features a critical effect on a person's brain activity. Additionally, as there are more tracks, music listeners have a difficult time manually establishing and segregating playlists. Existing playlist-generating automation techniques are computationally inefficient, less precise, and not user-friendly to the users, and at times they even require the use of additional hard wares. The proposed method, which is based on the extraction of facial emotions, will automatically produce a playlist, saving time and effort. The goal of the project is to create an application that employs CNN and feature extraction from data sets for emotion recognition to recommend music to users based on their feelings by distinguishing between different types of emotions. By using the HTML CSS/JAVA code, a smart music player is created as the web page with the playlist set according to the emotions. It is based on the user's interests and feelings, and it has been designed as a web page for ease of usage. The system is framed as friendly to the user.

Keywords—Music, Emotions, CNN, learning, Data sets, Feature extraction

I. Introduction

Music is a crucial means of enjoyment for music lovers and listeners, and it plays a significant role in uplifting an individual's life. With the improvements in the field of multimedia and technology in today's society, different Music players have evolved to include capabilities like fast forward, reverse, and volume adjustment. Genre classification, Multics streaming playback, playback speed, and volume modulation. Emotions are triggered by a complicated mix of factors. The unconscious mind and hormones are difficult for a person to

---
*Corresponding Author: sathishkumar.b@srec.ac.in
[1]shreya.1802213@srec.ac.in, [2]sujitha.1802227@srec.ac.in, [3]thilaga.1802244@srec.ac.in

DOI: 10.1201/9781003388913-22

control. They cause mammals to change their behavior according to their current situation and mind. In such cases, they may resist their attempt to lead our lives in a logical way. The study of emotions has become one of Darwin's most famous books after human evolution. From the anatomy, the facial muscles and nerves were already known. The artistic nature of music means that these classifications are often respective to music and controversial, and some genres may overlap. The application would satisfy the basic requirement of the user, but the disadvantage is that he/she should manually browse the playlist of songs and select the songs that support the present mood of a person and their action. The data set used in the project includes four types of emotions such as happy, sad, angry, and neutral which are taken from Kaggle. There are many types of emotions that are classified as basic. i.e., Fear, anger, happiness sadness, disgust, boredom, indifference, surprise, other attacks, suspicion, fear, contempt, empathy, and depression. Music is divided into genres in various ways, such as art and popular music.

II. Related Works

A. Emotion-based Music Player

The literature review in this study presents an application that detects using smart band hardware and a camera. A smart band is connected to an Android smartphone or mobile phone in the associated search to retrieve the user's heart rate. The program measures the user's heart rate and determines what emotions have been felt. This study also suggests a straightforward technique for user emotion classification. If the user does not have a smart band, the application can assess the user's emotions from facial photos using the Microsoft Azure face detection API.

After separating the user's emotions, the proposed paper uses the identified emotion to select songs that are in keeping with the user's

mood. With the detection, an emotion-based playlist is created, which pulls music from a firebase data collection. There are now 200 songs available in the database. The admin system categorizes them into several emotional kinds.

The music is provided or recommended in accordance with the user's preferences: happy, upbeat or downbeat, and sorrowful. A piece of happy or upbeat music with a cheer-up song element, for instance, will be suggested by the application when a user's preference is set to positive and his present emotion is really depressing. The proposed study demonstrates that the proposed method may readily separate positive feelings from a person's emotional heart rate. [1]

B. Finding Patterns in Musical Databases to Detect Moods

The construction and development of a system for extracting moods from musical data make use of knowledge discovery in databases. Users can interact with a music player through the app, which utilizes a neural network to identify and categorize their emotional state. The findings demonstrate that real-time song suggestions can positively influence one's present emotional state, implying that greater use of the app will result in better results. The application provides users with an accuracy of more than 72.4%. The proposed solution also enables continuous data collecting, storage, real-time analysis, and song suggestions to positively affect the current emotional state, suggesting that increased use of the program will yield better outcomes. The implementation of Knowledge Extraction in Music Databases is a powerful tool whose potential has only been developed in an incipient way. The knowledge extraction methodology (Knowledge Discovery in Databases - KDD) emerged in the 1990s as an alternative to meet the challenge of efficiently processing the growing volume of data resulting from the digital revolution. The tool and model proposed in the survey constitute a valuable experimentation environment that, if proven effective, could become a means of mitigating the high degree of psychosocial risk experienced by the average citizen given exposure to various aggressor factors ironically enhanced by disruption. [2]

The systems suggested by many authors assist us in putting our notion into action. Many academics have developed systems that use a dynamic arrangement approach to meet a variety of needs. As a result, the proposed research concludes that it will be of great help to new researchers in finding new ways to develop emotion-based music players that employ deep learning and image processing.

III. Methodology

A. Existing System

Existing methods are designed such that the songs can be played manually based on the mood for customizing the playlist in the generation process. The existing methods are slow, not accurate and at times they even need the use of hardware like EEG or Sensors. These ideas have some limitations like not user friendly and manually operating.

B. Proposed System

The proposed system automatically generates playlists based on facial expression extraction, saving effort and time. By using Deep learning, a comparison model is created which is trained and the test image captured by the Open CV with the principle of Convolutional Neural Network is compared with the model with the help of Haar cascade vectors the emotions are detected. By using the HTML CSS/JAVA code, a smart music player is created as the web page with the playlist set according to the emotions. The playlist designated for that particular emotion will be played after the emotion has

been identified. The Emotion-based music player application is only based on interests and emotions depending on the user's mood which is created as a web page is the basic objective.

Working Principle

The block diagram is used to construct the full suggested model. The EMO music player is created using the CNN Haar cascade algorithm with various data sets loaded for all four emotions. It is created using a web page with python pip3 and HTML code.

General Block Diagram

Initially the face input is detected using the camera of the laptop or the external source. Once the image is detected, it is pre-processed to improve the quality of the image such that it analyses in a better way as shown in the block diagram in Fig. 22.1. It suppresses undesired distortions and enhances some features which are necessary for the particular application we are working for. After pre-processing, contrast stretching or histogram stretching happens through normalization for feature extraction. The Training Data sets are loaded from the Kaggle. The feature database is created and kept aside to extract the emotion of the user. They depict raw images in a simplified form to aid decision-making in areas such as pattern identification, classification, and image recognition. Emotions are then detected in Haar

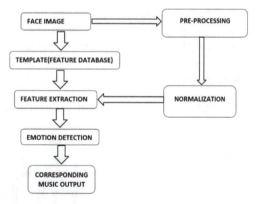

Fig. 22.1 Proposed work block diagram

Source: Author

Cascades. The Ada-Boost learning algorithm is used in Haar Cascades. To acquire efficient results from the classifier, this method takes a small number of significant features from a big set and employs cascading approaches to detect faces in the image. Later the corresponding music is played based on the emotion detected. Figure 22.1 displays a straightforward block diagram of the proposed work.

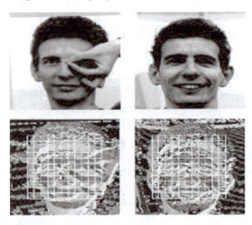

Fig. 22.2 Normalization of image

Source: Author

The normalization of the image is shown in Fig. 22.2. The Haar Wavelet technique from the classifier called Haar cascade is used to use the square function to evaluate pixels in an image. By using Tensor flow in deep learning this framework can be implemented, which recognizes and creates a comparison model. Python also plays a major role in executing the framework. Python eel is used as the user interface which connects the user, HTML CSS code with python. Over the past two decades,

the salience detection methods like CNN-based salience models were used which cannot detect blur objects. It primarily focuses on leveraging deep learning architectures to implement current salient object identification methods. The feature extraction flow is shown in Fig. 22.3.

By using Open Computer Vision, we create a test image captured by the camera of the device and the image is converted into grayscale. With the mat plot library, the image is been cut into cells of 480*581 pixels to compare it with the comparison model. Python pip3 is the technology used for the application.

Concept and Operation

In the proposed work, pre-processing has been done in order to improve the image quality such that it analyses in a better way. It suppresses undesired distortions and enhances some features which are necessary for the particular image. The Training Data sets are loaded from the Kaggle. Hence by using tensor flow in Deep learning, the data inputs are trained and the comparison model is created. The feature database is created and kept aside to extract the emotion of the user. Some of the data sets used in this project are neutral, angry, sad, and happy. Normalization is done for contrast stretching or histogram stretching. Stabilizing the distributions of layer inputs over the course of the training phase, enables deeper neural networks to be trained more quickly and with more stability. The feature extraction technique is useful when you have a huge data set and need to conserve resources without losing any important or pertinent data. The quantity of

Fig. 22.3 Feature extraction

Source: https://freecontent.manning.com/the-computer-vision-pipeline-part-4-feature-extraction/

data in a data collection that is unnecessary can be decreased with the aid of feature extraction. The main goal is to come up with a simple solution that takes less time to compute without sacrificing efficiency.

Development

A Smart music player is developed to play the songs for the emotions that have been detected. By using HTML CSS Java code which has been integrated using python eel software by creating a web page in Google Chrome. Once the program runs, the EMO music player opens, and the songs that have been set in the specified playlist for the emotion detected, and it is played accordingly.

IV. Results

In the proposed system, the EMO Player is introduced to play the songs for the emotions that have been detected. By using HTML CSS Java code which has been integrated using python eel software by creating a web page in Google Chrome. Once the program runs,

the music player is opened and the songs that have been set in the particular playlist for the emotion detected are played accordingly.

A. Implementation Output

See Fig. 22.5

Emotion Given by the user: Happy

Emotion Detected by the system: Happy

The Happy emoji is displayed in the music player and the corresponding playlist of songs is played.

See Fig. 22.6

Emotion Given by the user: Sad

Emotion Detected by the system: Sad

The Sad emoji is displayed in the music player and the corresponding playlist of songs is played.

See Fig. 22.7

Emotion Given by the user: Neutral

Emotion Detected by the system: Neutral

Fig. 22.4 Proposed music player

Source: Author

Fig. 22.5   Output for happy emotion

Source: Author

Fig. 22.6   Output for sad emotion

Source: Author

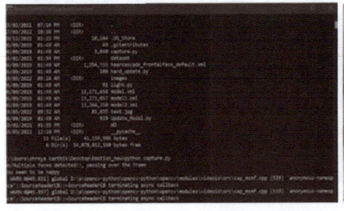

Fig. 22.7   Output for neutral emotion

Source: Author

Fig. 22.8   Output for angry emotion

Source: Author

The Neutral emoji is displayed in the music player and the corresponding playlist songs are played.

See Fig. 22.8

Emotion Given by the user: Angry

Emotion Detected by the system: Angry

The Angry emoji is displayed in the music player and the corresponding playlist of songs is played.

B. Accuracy Results

The accuracy results of the model are shown in Table 22.1.

Table 22.1   Accuracy table

Accuracy results of the Model	
Image used in the model	439
Overall accuracy of the model	86%
Highest individual emotion accuracy	97%
Lowest individual emotion accuracy	76%

V. Conclusion

The Emotion-Based Music Player is suggested as a way to reduce manual labor and improve the music-listening experience for users. It will be a good product for people who find it difficult to select songs manually. The accuracy results of the model are reduced the person's work by detecting the person's image with a camera, observing their expression, and providing a personal playlist with a more interactive and appealing method.

VI. Future Scope

The application can be improved by modifying and adding a few functions. Including more emotions that automatically create their own playlist according to the emotion, optimizing the EMO player with an algorithm that includes add-on or additional features which helps the system to categorize users with many other factors such as location and thus the user can travel to that location and listen to the songs accordingly. Making this application in-built for all mobile phones connected to artificial intelligence.

References

1. S. Chowdhury, V. Praher, and G. Widmer, "Tracing back music emotion predictions to sound Sources and Intuitive perceptual Qualities," vol. 14, no. 6, pp. 45–52(2021).

2. A. M. Proverbio, F. Benedetto, and M. Guazzone, "Shared neural mechanisms for processing emotions in music and vocalizations," European Journal of Neuroscience, vol. 51, no. 5, pp. 1987–2007, (2020).

3. I. Peretz, J. Ayotte, R. J. Zatorre et al., "Effects of vocal training in a musicophile with congenital amusia," Neuron, vol. 33, no. 2, pp. 185–191, (2020).

4. X. Tang, C. X. Zhang, and L. I. Jiang-Feng, "Music emotion recognition based on deep learning," Computer Knowledge and Technology, vol. 15, no. 11, pp. 232–237, (2019).

5. S. Swaminathan and E. G. Schellenberg, "Musical ability, music training, and language ability in childhood," Journal of Experimental Psychology Learning Memory and Cognition, vol. 46, no. 12, pp. 2340–2348, (2019).

6. Krittrin Chankuptarat, Raphatsak Sriwatanaworachai, and Supannada Chotipant, "Emotion-based music player", IEEE 5th International Conference on Engineering, Applied Sciences and Technology, (2019).

7. P. Sánchez, J. Cano, D. García, A. Pinzon, G. Rodriguez, J. García- González, and L. Perez, "Knowledge Discovery in Musical Databases for Moods Detection", IEEE Latin America Transactions, vol. 17, no. 12, (2019).

8. Ramya Ramanathan, Radha Kumaran, R Ram Rohan, Rajat Gupta, Vishalakshi Prabhu, "An intelligent music player based on emotion recognition", IEEE 2nd International Conference on Computational Systems and Information Technology for Sustainable Solution (CSITSS), 2017

9. Shlok Gilda, Husain Zafar, Chintan Soni and Kshitija, "Smart music player integrating facial emotion recognition and music mood recommendation", IEEE International Conference on Wireless Communications, Signal Processing and Networking, (2017).

10. S. Deebika, K. A. Indira, Jesline, "A machine learning based music player by detecting emotions", 5th IEEE International Conference on Science, Technology, Engineering and Mathematics, (2019).

Recent Trends in Computational Intelligence and its Application – Sugumaran D. et al. (eds)
© 2023 Taylor & Francis Group, London, ISBN 978-1-032-48410-5

23

Impact Assessment on the Factors Affecting the Crop Yield Using Data Analytics and Principal Component Analysis

Bharati Panigrahi[*]

Mukesh Patel School of Technology Management & Engineering (MPSTME),
NMIMS University, India

Krishna Chaitanya Rao Kathala[1]

University of Massachusetts, Amherst,
United States of America

Koteswara Rao Seelam[2]

Mother Teresa Institute of Science and Technology,
Khammam, India

Abstract—Everyone needs agriculture to survive. It's the only industry that benefits the country beyond agriculture. It provides food, raw materials, and jobs to a large section of the people. Archaic agricultural techniques have been supplanted by modern technology and equipment for thousands of years. Artificial Intelligence (AI) can provide farmers with real-time data from their crops to pinpoint regions that require irrigation, fertilizer, or insect control. Data analytics is critical for assisting decision-makers in making business decisions. The primary purpose of this research project is to use data analytics to provide useful insights regarding agricultural yields. Principal component analysis (PCA) is used with data analytics to discover the factor that most affects agricultural production. PCA found that total rainfall had the greatest impact on agricultural production since it captures the highest variability of 46.89%, followed by temperature which captures 25.09% of the variability. The three major factors influencing crop yield account for 90.16% of the total variance.

Keywords—Crop analytics, Agricultural analytics, Principal component analysis, Data analytics, Eigen values, Eigen vectors, Education, STEM, Policy

I. Introduction

Technological advances have made farm machinery larger, faster, and more productive, enabling more efficient cultivation of more land. Agriculture provides food, clothes, and shelter. Agriculture supports 58% of India's population. Due to its great value-adding

[*]Corresponding Author: bharatipanigrahi9901@gmail.com
[1]krishnakathala@gmail.com, [2]drkrseelam@gmail.com

DOI: 10.1201/9781003388913-23

potential, especially in the food processing business, the Indian food sector is growing rapidly and contributing more to global food commerce every year [1]. The new agricultural system must be efficient, resilient, diverse, long-term, and fast-growing. Artificial Intelligence (AI) and Machine Learning (ML) can tackle new paradigm difficulties with data. AI is transforming food production, delivery, and consumption [2]. AI encoding involves learning, reasoning, and self-correction [3]. Agriculture uses AI because of its flexibility, excellence, and cost-effectiveness. These technologies decrease water, pesticides, and herbicides, boost labour efficiency, and improve product quality [4]. Our research work focuses on leveraging Principal Component Analysis (PCA) to identify variables affecting agricultural productivity. PCA reduces the dimensionality of huge datasets while minimizing information loss. Generating uncorrelated variables successively optimizes variance [5]. PCA increases the variance of linearly combined variables [6]. PCA was performed to reduce data dimensionality [7]. It helps determine agricultural yield-affecting traits.

II. Literature Review

This paper by Abhijeet Pandhe, Praful Nikam, et al. [8] discusses the technique used to develop the 'Smart Farm' app, which estimates crop production based on 5 meteorological factors: precipitation, temperature, cloud cover, vapor pressure, and wet day frequency. The remaining agro-inputs are ignored because they vary by field. They combined columns by cropping season for each variable to reduce the dimensionality. 200 decision trees are used to train a Random Forest classifier, which provides poor splits and increases unpredictability. Data blending enhances forecasts. Accuracy was 87% after 10-fold cross-validation, indicating a high association between variables.

This research by Anna Fuzy et al. [9] attempts to select and assess plant growth, and structural, and biological elements at the ideal stage to optimize experiment performance and save research time. PC1 showed a distinct divergence between salt treatments in experiment I, accounting for 45.7% of the variation. PC1 separated control and drought-stressed plants without crossover in experiment II, accounting for 39.4% of the total variance.

This study by B.M. Nayana et al. [10] compares wheat production estimates for India and its major wheat-producing states. After defining the major features with Principal Component Analysis (PCA), they used Multivariate Adaptive Regression Splines (MARS). Comparative yield prediction results demonstrate Rajasthan's model to be superior to others. Multiple studies have shown that models with more attributes don't provide accurate predictions and their effectiveness fluctuates. Examining models with more and fewer features reveals the optimal model.

III. Methodology

In Fig. 23.1 below, the proposed method for determining the factor affecting agricultural yields of three crops, Bengal Gram, Groundnut, and Maize, in the Telangana districts of India is presented.

Fig. 23.1 Methodological approach for the PCA-based feature extraction

Source: Author

A. Research Question

The Data was acquired from the Open Data Portal of the State of Telangana, India [11].

Impact Assessment on the Factors Affecting the Crop Yield Using Data Analytics and Principal Component Analysis **177**

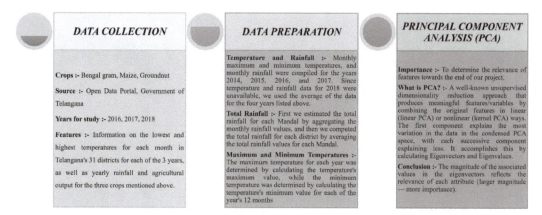

Fig. 23.2 Steps performed to identify variables that influence crop production

Source: Author

The research study is aligned with Zero Hunger as on the Sustainable Development Goals by constructing an optimum model to estimate agricultural output using the most recent information available for the state of Telangana to assist in the decision-making process. In this research work, the authors addressed the following research question: How can Data Analytics and Principal Component Analysis be employed to determine the significance of the factors influencing the crop production of Bengal gram, groundnut, and maize in the Telangana region of India from 2016 to 2018?

IV. Results

This research focuses on identifying which variables have the greatest influence on crop yield. The graphics below will help us visualize our data and acquire meaningful insights.

PowerBI was used to create these visualizations. Climate change makes agriculture particularly susceptible. The descriptive statistics and central tendency values for the crop yield dataset are represented in Fig. 23.3. It provides the Mean, Standard Deviation, Quartiles, and minimum and maximum values for each feature variable.

After visualizing temperature patterns in Fig. 23.4, the average maximum and minimum temperatures for the whole state of Telangana were 41.43°C and 14.84°C, respectively, and the line graphs depict temperature patterns for each of Telangana's districts during the last three years. Most crops have an "optimal" quantity of rainfall during any particular growing season. Fig. 23.5 depicts the rainfall distribution throughout all Telangana districts, with Bhadradri receiving the most and Nagarkurnool receiving the least.

	TOTAL RAINFALL	TOTAL YIELD	MINIMUM TEMPERATURE	MAXIMUM TEMPERATURE
MEAN	751.3821	2614.8602	14.8449	41.4250
STANDARD DEVIATION	207.9941	1850.3927	1.4678	1.0123
MINIMUM	422.9185	0	12.34	39.1
1ST QUARTILE	585.3026	1516.00	13.68	40.59
2ND QUARTILE	703.7388	2068.00	14.67	41.49
3RD QUARTILE	894.1818	3126.00	15.89	42.19
MAXIMUM	1290.1636	9282.00	17.75	43.34

Fig. 23.3 Descriptive analysis of the crop yield data

Source: Author

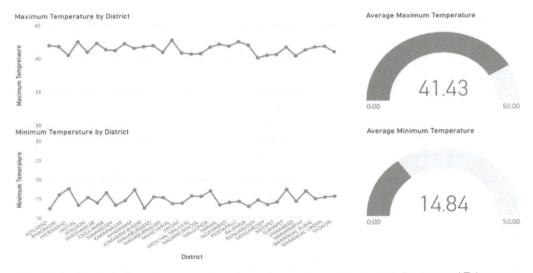

Fig. 23.4 Trends in the minimum and maximum temperatures across and districts and Telangana
Source: Author

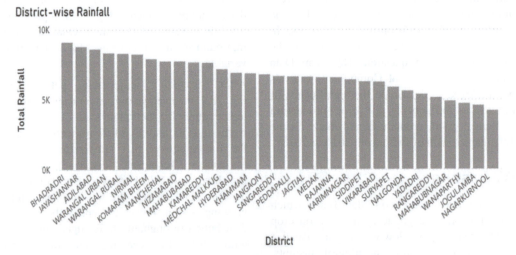

Fig. 23.5 District-wise rainfall distribution for the state of Telangana
Source: Author

Wanaparthy has the highest Bengal gram yield, followed by Adilabad and Nizamabad, according to Fig. 23.6(a), however, the rural region of Warangal district has the highest groundnut crop production, followed by Rangareddy and Wanaparathy, as per Fig. 23.6(b). Fig. 23.6(c) shows that the rural areas of the Warangal district have the maximum maize crop production, followed by Pedapalli and Nirmal.

Only the three principal components (PC) can represent 90.16% of the variability in the data, as seen in Fig. 23.7. Total Rainfall has the greatest eigenvalue in the first PC, making it the most critical factor impacting crop yield. Other important parameters that influence crop yield include minimum and maximum temperatures, as well as the crop cultivated.

Explained Variance refers to the degree of data set variability linked to each particular PC.

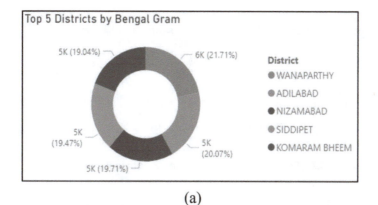

(a)

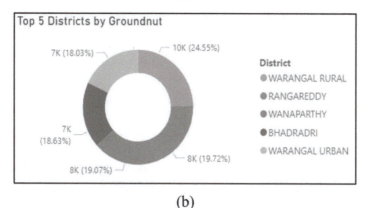

(b)

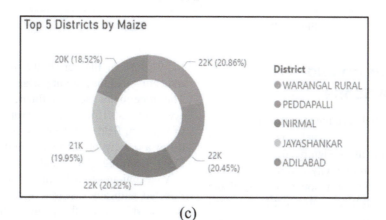

(c)

Fig. 23.6 (a) Top 5 districts by Bengal gram, (b) Top 5 districts by groundnut, (c) Top 5 districts by maize

Source: Author

In other words, it indicates how much of the overall variation each component explains. As shown in Fig. 23.8 taking into account the three PC can provide a cumulative explained variance ratio of 0.9016, when 1 is the maximum possible value. This reduces computations and preserves time.

PRINCIPAL COMPONENTS	FEATURE 1	FEATURE 2	FEATURE 3	FEATURE 4	VARIABILITY	CUMMULATIVE VARIABILITY
PC 1	Total Rainfall	Maximum Temperature	Minimum Temperature	Crop	0.4689	0.4689
PC 2	Minimum Temperature	Maximum Temperature	Total Rainfall	Crop	0.2509	0.7198
PC 3	Crop	Maximum Temperature	Minimum Temperature	Total Rainfall	0.1818	0.9016

Fig. 23.7 Feature importance for the 3 principal components

Source: Author

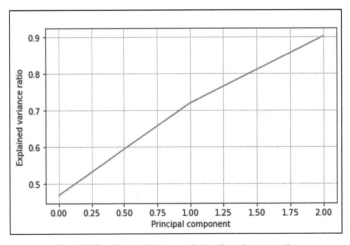

Fig. 23.8 Cumulative explained variance ratio

Source: Author

V. Conclusion and Future Scope

The research study is aligned with Zero Hunger as on the Sustainable Development Goals by identifying the most significant factors that influence crop productivity and making use of the most recent information that is available for the state of Telangana to assist in decision-making. Primarily, this research focus on implementing PCA to determine that, among the multiple factors influencing crop yield, total rainfall had the greatest impact, since it captures the highest variability of 0.4689 or 46.89%, followed by temperature which captures 0.2509 or 25.09% of the variability. Total Rainfall and Minimum Temperature together account for 71.98% of the variability in crop output, indicating that they are important determinants. The three major factors influencing crop yield account for 90.16% of the total variance. Images of the farm field and the crop can be analyzed using a variety of Artificial Intelligence (AI), Deep Learning (DL), and Computer Vision (CV) techniques to determine whether they are infected with any diseases or the presence of weeds, which affect the quality of the crop, and can be separated from the healthy crops as its technically viable.

References

1. "Agriculture in India: Industry Overview, Market Size, Role in Development", India Brand Equity Foundation (IBEF), 2022.

2. S. Y. Liu, "Artificial Intelligence (AI) in Agriculture," in IT Professional, vol. 22, no. 3, pp. 14–15, 1 May–June 2020.

3. JOUR Khan, Rijwan Ben Ayed, Rayda Hanana, Mohsen 2021 2021/04/22 Artificial Intelligence to Improve the Food and Agriculture Sector, Journal of Food Quality Hindawi.

4. Khan, R., Dhingra, N., Bhati, N. (2022). Role of Artificial Intelligence in Agriculture: A Comparative Study. In: Al-Turjman, F., Yadav, S.P., Kumar, M., Yadav, V., Stephan, T. (eds) Transforming Management with AI, Big-Data, and IoT. Springer, Cham.

5. Jolliffe Ian T. and Cadima Jorge 2016 Principal component analysis: a review and recent developments *Phil. Trans. R. Soc. A.*3742015020220150202

6. Vikas Singh, Alka Singh. Analysis of agriculture data using principal component analysis. International Journal of Multidisciplinary Research and Development, Volume 7, Issue 1, 2020, pp. 34–37

7. S. Mishra, D. Mishra, S. Das and A. K. Rath, "Feature reduction using principal component analysis for agricultural data set," 2011 3rd International Conference on Electronics Computer Technology, 2011, pp. 209–213.

8. Abhijeet Pandhe, Praful Nikam, Vijay Pagare, Pavan Palle, Prof. Dilip Dalgade (2019), Crop Yield Prediction based on Climatic Parameters, International Journal of Research in Engineering and Technology (IJRET), e-ISSN: 2395-0056, p-ISSN: 2395-0072, Vol. 06, Issue: 03

9. Füzy, A., Kovács, R., Cseresnyés, I, Selection of plant physiological parameters to detect stress effects in pot experiments using principal component analysis. *Acta Physiol Plant* 41, 56 (2019).

10. Nayana, B. M.; Kumar, K. R.; Chesneau, C. Wheat Yield Prediction in India Using Principal Component Analysis-Multivariate Adaptive Regression Splines (PCA-MARS). *AgriEngineering* 2022, *4*, 461–474.

11. Open Data Portal of the State of Telangana, Department of Agriculture and Cooperation, Government of Telangana, India.

Recent Trends in Computational Intelligence and its Application – Sugumaran D. et al. (eds)
© 2023 Taylor & Francis Group, London, ISBN 978-1-032-48410-5

24

Automatic Parking System with Dynamic Tariff and Allotment of Optimum Parking Space

Harshavardhan Naidu S.[1], Senthil Kumar R.*, Santhoshkumar H.[2], Kubera Rajan V. G.[3]
Vel Tech High Tech Dr. Rangarajan Dr. Sakunthala Engineering College, Chennai, India

Sugavanam K. R.[4]
Jaya College of Engineering and Technology, Chennai, India

Prabaakaran K.[5]
Easwari Engineering College, Chennai, India

Abstract—Notion behind the article is to develop a fully automated parking system that is smart in fixing tariffs based on peak time and off-peak time which would be fruitful in commercial places. The system offers another flexibility to the users by offering the freedom to select the slot based on the tariff with what there are comfortable. RIFD-based rechargeable cards provided in the system make the payment process simple and easy. The system offers a cost-effective solution to bottlenecks in conventional parking systems such as delays in locating parking slots, tariff calculations & payment processes. IR sensor in entry detects the vehicle it will display the tariff based on time (off-peak or peak tariff) if the user is comfortable with the tariff, then the available slot will be displayed. Users can select the slot with the keypad given. Time of entry is recorded with slot details for this future tariff calculation, the user's RFID card details, or a scan to get registered with the slot selected once this process over the servo motor operates to open the entry gate. When the IR sensor exists since the vehicle user has to scan the RFID card from which the system calculates the tariff based on the duration for which the slot is occupied at will be displaced.

Keywords—RFID, Ardiuno, Dynamic tariff, Parking system, Automatic system

I. Introduction

Population of the vehicle is shooting up along with the people population growth, especially in metropolitan cities it is a hectic task to find parking in a short time. It demands manual labour to organize and collect a payment, mostly parking issues rise in all commercial places. Hence it is required to develop a system that will be a cost-effective fully automated system, incorporating dynamic tariff fixing based on On-peak hours and Off-peak hours which will

---
*Corresponding Author: rskumar.eee@gmail.com
[1]sapineniharsha@gmail.com, [2]h.santhosh200102@gmail.com, [3]kuberaprt@gmail.com, [4]sugavanamkr@gmail.com, [5]prabaakaran031@gmail.com

DOI: 10.1201/9781003388913-24

be fruitful for parking providers and users. As it is required to eliminate manual labor, it becomes indispensable to develop a system that can manipulate fees to be collected based on the duration and period of hiring the slot and auto-detection of payment from the user.

II. Research Summary

In this article a system was been working on the IoT concept bringing all parking-related details to mobile in the form of a mobile application, though this system covers the objective helps users in finding a parking slot but it hasn't considered dynamic tariffs and collection of payment [1].

The outcome of the system was concentrating on providing secured allotment of parking slots by generating a unique OTP, though this worked on collecting payment, again in this system dynamic tariff was not taken into account [2].

This paper displays the type of sensor which is deployed to identify the empty parking slot, but they haven't worked on payment collection and dynamic tariff. Overall, many works are discussed to identify the available slot without a sensor, just by manipulating stored data [3]

most of the system developed considering only the end user perspective and wants to develop a system that is user-friendly, but the system has to be installed in aspect commercial point view, hence it is necessary to fix the tariff based on off-peak time and on peak time.

This article covers all the objectives discussed in the introduction that as locating the parking slot, dynamic tariff fixing, and payment collection, but the system works on the cloud and IoT, making the system very complex and expensive [4].

III. Proposed System

Figure 24.1 display the pictorial presentation of the proposed system, it hosts Ardiuno Uno, servo motor, keypad, LCD display, IR sensors, power supply battery and RFID reader, and rechargeable RFID card. IR sensors are used at the entrance and exit to detect the entering and leaving vehicle, the keypad is provided for the user to interact with the system. Ac servo motors are deployed to open and close the gate at exit and entrance, LCD display displays the required information to the users. RFID reader and card are provided to do initial registration

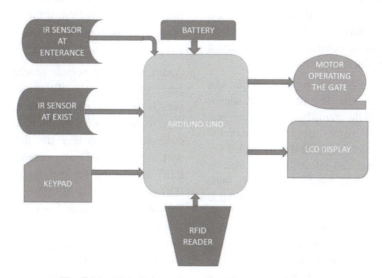

Fig. 24.1 Pictorial presentation of proposed system

Source: Author

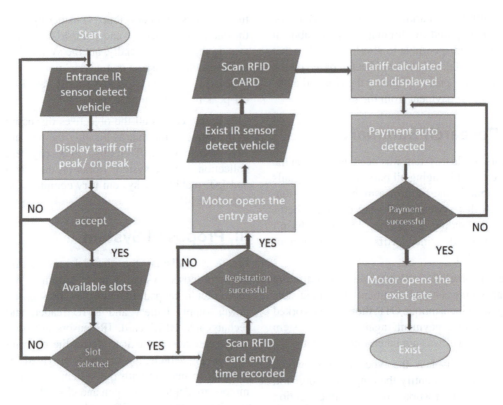

Fig. 24.2 Process flow diagram

Source: Author

with the slot allotted and for auto-detection of payment during the exit. Ardiuno Uno performs the role mastermind behind the system which do all tariff calculation considering all predefined conditions and slot registration.

Figure 24.2 discusses the process carried out by the system, at the time of entry IR sensor detects the vehicle immediately system gets activated and will display the tariff based on the location of the slot and period (Off peak / On peak), so that user can select the slot based on his convince and tariff what he can afford, once the user gives his acceptance to tariff via keypad and then available slots will be displayed. If the user feeds his selection through the keypad, his selection will be displayed for his confirmation then the user has to register the slot allotted with a prepaid RFID card, now the time of

allotment and tariff details will be recorded for future tariff calculation, if the registration was successful, the controller drives the servo motor to open the gate at the entrance.

Again, when the vehicle reaches the existing gate while leaving IR sensor detects the vehicle and initiates the system. Now users have to once again scan the RFID card, so that controller can retrieve the details registered with the registered card, with will calculate the tariff accounting for the duration and period (Off-peak/On-peak) of hiring the slot, once the calculation is completed duration and total amount will be displayed in the LCD screen. Since the RFID card prepaid card, the amount will be auto-detected. Figure 24.3 shows case the circuit interface of hardware components.

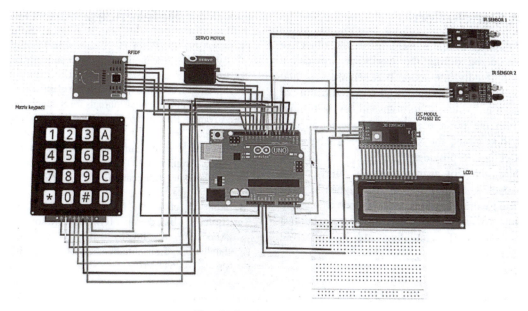

Fig. 24.3 Circuit interface

Source: Author

IV. Outcome of the System

Table 24.1 Inferences of the LCD display

Si. no.	LCD display	Inferance
1	TF-F1-PHT-1hr-50 F2-NHT-1hr-30	Once the vehicle is detected by the entrance IR sensor tariff will be displayed based on "PHT"- Peak hour tariff or "NHT"- Off peak hour tariff and based on slot location.
2	SL-FLOOR-1 S 6,4 FLOOR-2 S 1,2	After acceptance of the tariff, slot available in each floor will be displayed.
3	ENTER SLOT	Using the keypad given user can select the available slot based on his preference
4	ENTER SLOT FLOOR-2 SLOT-1	Once the slot is selected, for user confirmation it will be displayed, RFID card has to be scanned and the slot has to be registered with the RFID card
5	TIME PHT-2hr-100rs	Durning the exist once IR sensor detects a vehicle, the system gets activated, prepaid RFID card has to be scanned to retrieve the entry details, payment will be calculated based on duration and tariff period. The total amount will be displayed
6	PAYMENT Pay 100 succesfu	Amount on the prepaid RFID card will be auto detected and if the payment was successful it is displayed. Exist gate opens

Source: Author

V. Conclusion

Outcomes of the system fulfill the objective considered during the works; it provides a user-friendly interaction between the user and the system and provides all required details to the user, which offers transparency in billing. From the point of component selection and in operations, the system proves to be cost-effective, it provides freedom to the user to select the slot as per their convenience and budget what they can afford. Dynamic tariff fixing will please both users as well as parking slot providers. The system proves to be fast in both tariff calculation and in collecting payment.

References

1. Jayakshei D. B. et al. "Smart Car Parking Systems" Internationals Research Journals of Engg and Tech, 4(6), pp. 3036–3038, 2017.
2. Shruthi M. et al. "IoT Based Smart Car Parking System" International Journals for Science and Advances Research in Tech, 5(1), pp. 270–272, 2019.
3. Suvarna N. et al. "Smart Car Parking Systems using Arduino UNO" International Journals of Computer Application, 169(1), pp. 13–18, 2017.
4. Amir O. et al. "A New Smart CarParking Systems Based on Dynamic Resources Allocation & Pricing" IEEE Trans on Intelligent Transportations System, 17(9), pp. 2637–2647, 2019.

Recent Trends in Computational Intelligence and its Application – Sugumaran D. et al. (eds)
© 2023 Taylor & Francis Group, London, ISBN 978-1-032-48410-5

25

Parametric Study to Improve Forecasting Accuracy in Air Quality Data

Shobha K.*

Siddaganga Institute of Technology, Tumakuru, India

Abstract—The presence of harmful air pollutants has a direct impact on our health. It is important that we continuously monitor the air quality to improve our environment. Unfortunately, the cost of acquiring and maintaining air quality stations is very high. Therefore, it is important that we develop data-driven models to improve the accuracy of air monitoring. Air Quality Index (AQI) indicates how polluted the air is. AQI can increase due to traffic, industries, factories, or anything that can contribute to air pollution. The greater the air pollution index value, the greater the health concerns it can cause. Based on the types of pollutants that are most harmful to the environment, the index value is used to identify the most common sources of air pollution.

This work focuses on study towards parameters that affect the quality of air and an effective way to predict and forecast the AQI. Experimental work has been carried out using existing machine learning and deep learning techniques to predict and forecast the AQI. This will help the decision and policymakers to plan and reduce the harmful effects of the pollutants and control air pollution.

Keywords—Air pollution, AQI, Pollutants, Prediction, Forecast

I. Introduction

Without oxygen, it is impossible to comprehend how living beings would survive. Modern human culture has constant advancements that have a negative impact on the quality of the air. Daily transportation, industrial, and domestic operations churn up dangerous contaminants in our surroundings. In the modern day, air quality monitoring and forecasting have become crucial tasks, particularly in developing nations like India [4].

Air pollution is the contamination of air due to the presence of harmful substances which has a significant impact on social, economic, and other major sectors. Traffic and industries are major contributors to air pollution. Air is vital for all living beings on earth and clear links between pollution and health effects have been revealed. Hence, it is our responsibility to keep the air clean. In previous decades, the air we breathe in used to be less polluted but now due to the increase in population, vehicles, and industries the environment has been polluted to its peak.

*Corresponding Author: shobhak@sit.ac.in

DOI: 10.1201/9781003388913-25

Along with the advancement of smart cities, there is continuous improvement in air quality mapping. The prediction of pollutants for a given time and location in the atmosphere has become a hot research topic. With an accurate air quality forecast and knowledge, one can choose the right path for the best way to commute, establishing procedures to reduce pollution levels and the best time for outdoor and other daily activities. Awareness like this has the potential to create a cleaner environment and a healthier population.

The air changes rapidly with hourly data more uncertain compared with monthly and yearly trends. The effort has been made by researchers to create models to forecast the quality of air.

Half of the pollution in India is caused by industries, crop burning, and by other sources. Exposure to matter tinier than 2.5 microns (PM2.5) can cause deadly diseases such as breathing problems, lungs disorder, and heart disease, etc. Air Quality Index (AQI) is the metric used to designate how polluted the air is. The value changes based on various pollutants like PM10, PM2.5, NO_2, SO_2, CO_2, NH_3, and O_3.

Figure 25.1 shows the air quality index levels which tell how clean or unhealthy the air is.

AQI values ranging from 0 to 50 have minimal health impacts. Values ranging from 51 to 100 may cause minor breathing discomfort and AQI values above 100 are considered harmful and may cause respiratory illness.

Fig. 25.1 Air quality index levels

Source: https://www.business-standard.com/about/what-is-air-quality-index

Figure 25.2 indicates that air pollution has been controlled in recent times due to the recent hit in the pandemic and lockdown, but now the world population has started to reassume their activities, and as a result, air pollution has again started to rise. As per the world air quality report released by IQ Air in 2021, the top 5 cities in India that are top in air pollution are Bhiwadi (Rajasthan), Ghaziabad (Uttar Pradesh), Delhi, Jaunpur, Noida.

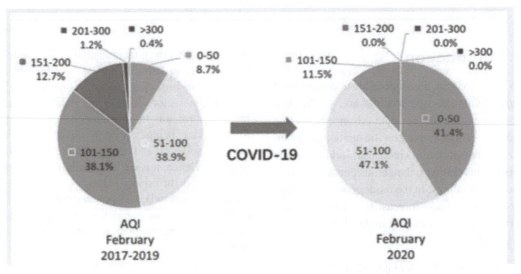

Fig. 25.2 AQI pre and post COVID

Source: https://doi.org/10.4209/aaqr.2020.04.0139

Considering the importance of measuring air quality this work aims at air quality prediction by exploring different machine learning and deep learning algorithms. Based on the predictions and pollution levels, the decision and policymakers can take appropriate decisions such as dust cleaning, plantation driving, carpooling, and avoiding outdoor activities, to reduce the dangerous consequences of air pollution.

The goal of air quality prediction is to evaluate and apply ideas and algorithms in the context of time series prediction to help decision-makers and individuals to take measures to reduce the health issues caused by air pollution.

This work aims to build models to predict AQI and different pollutants by looking at previously recorded data. To build the model various algorithms like Decision Trees, Random Forest (RF), Support Vector Regression (SVR), and Neural Networks are considered. The algorithm which gives the highest accuracy is considered for future prediction.

II. Literature Survey

There are enormous factors that influence the quality of air. Identifying the pollutants that cause the pollution and based on that forecasting the air quality index is a key issue. Traffic emissions are well known for contributing to higher levels of air pollutants in the ambient air. Many traditional methodologies have been proposed to capture the intricate link between various parameters and air pollution concentrations [2]. Approaches based on machine learning have been demonstrated to be promising.

PM2.5 forecasting is a critical topic for the development of smart cities, variations in the readings of meteorological variables, such as wind speed and direction, have an impact on PM2.5 propagation, demonstrating the complexity of forecast [9]. Researchers have created several PM2.5 prediction algorithms

based on statistical models and machine-learning approaches. Deep neural networks have recently become popular for predicting pollution concentrations. Deep learning's advantageous qualities make it ideal for modeling and predicting air pollution.

In the work proposed by Aditya et. al., logistic regression is used to determine if test data is contaminated or not, and autoregression is used to forecast future PM2.5 values based on previous PM2.5 readings. Predicting and forecasting the level of PM2.5 helps decision-makers to take precautionary measures to keep PM2.5 below the dangerous range. Centered on a data set of regular environmental situations in a certain city, this method is designed to predict PM2.5 levels and identify air quality [1].

Work proposed by Liu et. al., is based on two public datasets that use Support Vector Regression (SVR) and Random Forest Regression (RFR) to create a regression model that can predict the Air Quality Index (AQI) in Beijing and the Nitrogen Oxides (NOX) content in an Italian city. The Beijing AQI is used as the regression objective in this experiment [2]. The Radial Basis Function (RBF) is utilized as the kernel function for SVR-based model training. A grid search strategy is used to select the kernel value gamma and the penalty value. The regression model for the RF-based model was built using 100 regression trees. This research work shows that merging machine learning and prediction technique is a cost-effective and time-saving solution to handle several environmentally connected issues.

Fabiana et. al., shows in their work that predicting air quality is a challenging endeavor, due to volatile, vibrant, and highly variable kind of contaminants and particles with respect to both time and place. The noted significant impact of air pollution on people's health and the environment is making it increasingly important to model, predict, and monitor air quality, particularly in urban areas. In this analysis, SVR a well-liked machine learning technique is used to forecast pollutant and particle levels and to estimate the AQI. The radial basis function (RBF) was the kind of kernel among the investigated choices that gave SVR the most precise predictions. Their experimental outcome showed that SVR with RBF kernel gave promising results to precisely forecast hourly pollutant concentrations [3].

The work proposed by Thakur et. al., discusses how the ecology, climate, and human health are all negatively impacted by the presence of some harmful compounds in the air. The main reason for the addition of these harmful substances is human activity. Regular monitoring, determining the source of pollution, and implementing preventative measures are necessary for maintaining air quality. Bangalore, a city in India, has recently seen its population and size increase because of the expansion of the IT sector. The ecological services there have been impacted by its rapid growth and concomitant civic activity. This study's goal is to determine the trends in Bangalore's air pollution and investigate the causes of such trends. Without any changes, the analysis's data were taken directly from the website of the state pollution control board [5] [6].

In one of his work Dan Wei et. al., discusses how fine particulate matter (PM2.5) is an important component of the pollution index since its excessive levels in the air pose a serious threat to people's health. When levels are high, PM2.5, or particulate matter 2.5, reduces visibility and gives the air a hazy appearance. This project attempted to predict PM2.5 levels using machine learning techniques based on a dataset of daily weather and traffic parameters in Beijing, China. The value was chosen based on the Chinese Air Quality Level standard, which defines 115 ug/m^3 as mild pollution [7].

Q. Tao et. al., discusses how important it is to operate air pollution monitoring effectively and to make prevention plans, air pollution forecasting can offer trustworthy data about future pollution status. The complexity of interpreting changes in air pollutant concentration is increased by the dynamics of air pollution, which are typically mirrored by a number of variables, including temperature, humidity, wind direction, wind speed, snowfall, rainfall, and so on. In this paper [8], a deep learning-based short-term forecasting model for PM2.5 is put forth.

From the works referred to above, we can conclude that existing work aims at predicting the air quality index. But most of them concentrate on training one algorithm to predict and forecast. To overcome this in the proposed work, the accuracy of different algorithms is compared and the best one is chosen. The user can know the pollution level for the next 5 days as well. A larger dataset has been considered with more characteristics to support more precise statistical models for pollutants.

III. Proposed Model

To avoid the "garbage in garbage out" situations while using machine learning algorithms the collected raw data must be pre-processed and it must be analysed and explored. In this work, StandardScaler class of the sci-kit-learn library is used to standardize the data. The pre-processing techniques used in this work are as follows:

- **Encoding categorical data**: One important thing to do with categorical data is that it should be encoded into numerical data. If not done, the model cannot be trained for that attribute and there will be a greater chance of missing an important attribute. There are many types of encoding algorithms. In this work, Label Encoder has been used and categorical values are turned into numerical values.

- **Imputation**: It is an important aspect of the data pre-processing phase to take care of the missing values that are present in the dataset. There are many ways to fill in the missing values. In this work, missing values are replaced with zero.

A. Correlation Matrix

The relationship between two variables is called correlation. This matrix gives the idea of the interdependency of the attributes inside the dataset. Also, the categorical data attribute is neglected in this matrix. Based on the correlation matrix shown in Fig. 25.3, all the attributes of a dataset were considered in this work to train the model.

Once the pre-processing phase is complete the next step is to organize data into training and testing sets. In this work, 25% of the data is considered the test data, and the remaining 75% is used for training the model. Various algorithms of machine learning that is considered in this work are as follows:

- **Decision Tree Classification:** As the name says, a tree is constructed based on several calculations, in which leaf nodes correspond to the result and intermediate nodes represent attributes. An attribute is selected as the root of the tree based on the Information Gain (IG) that is earned by that attribute. An attribute that has high IG becomes the root node. Subsequently, subtrees are created based on the remaining attributes and information gained from sub-table. At last, the leaf node gives the target value of an instance.

- **Random Forest Classification:** This is the extension of the Decision Tree algorithm. The name of the algorithm itself is the key point of the concept. Several trees make a forest. When a new instance enters, according to majority voting the result of the new instance is classified. The advantage of Random Forest over Decision tree is that this algorithm reduces the overfitting of the model as many instances are created.

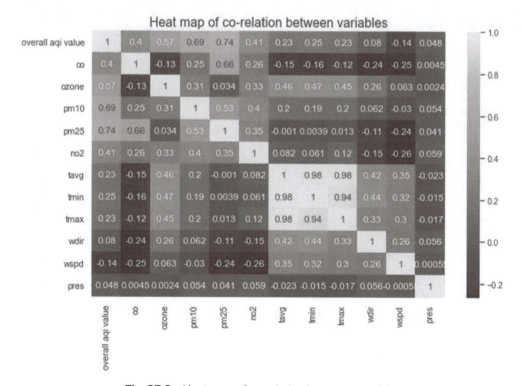

Fig. 25.3 Heat map of co-relation between variables

- **Neural Network:** It is an algorithm that mimics the way the human brain operates. It generates the simplest possible result with no need to revamp the output criteria as it can adapt to changing input.
- **Support Vector Regression (SVR):** The employment of Support Vector Machines (SVMs) in regression is termed SVR. The idea is to seek out a function that assesses mapping from an input domain to real numbers on the idea of a training model.

Deep learning is a clan of machine learning methods. They use numerous layers of nonlinear processing units for feature extraction and transformation. Deep learning algorithms use some style of gradient descent for training. It is seen as hierarchical learning with abstraction. This paper presents deep learning (DL) techniques to predict air pollution statistics. Air quality depends on statistic data captured at air monitoring stations because it can serve as the source of identifying population exposure to airborne pollutants.

With the increase in popularity of Artificial Intelligence, many deep learning algorithms have been proposed. This paper provides a forecasting model using LSTM.

IV. Results and Discussions

To show the importance of pre-processing and to make further predictions and forecasting the dataset chosen is the pollutant dataset and weather dataset which is taken from https://www.getambee.com/ and https://openweathermap.org/. Initially, datasets containing weather attributes and pollutant attributes of the years 2020 and 2021 are considered. The dataset contains the information for a year. The size of the dataset is 365 rows and 13 columns. Later we merge two datasets into a single dataset to train the accuracy. The weather data consists of attributes namely minimum temp, maximum temp, average temperature, wind speed, and pressure. The pollutant data has carbon monoxide, ozone, PM10, PM2.5, and nitrogen dioxide as its attributes.

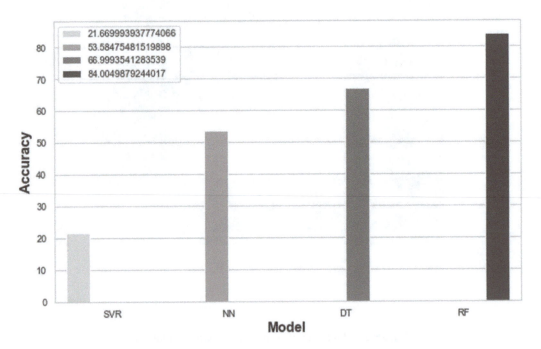

Fig. 25.4 Prediction accuracy of different models

Figure 25.4 shows the prediction accuracy of different models. The following observations are made from this figure. Firstly, the decision tree which has an accuracy of 66.99% is comparatively better than a neural network which has 53.58%, both methods can model the data that has nonlinear relationships between variables.

The other model which is the support vector regression attempts to fit the best line within a threshold value and doesn't perform well as the accuracy is low as 21.99%.

Based on the observations Random forests are chosen as the best model among all the foremost standard methods for the selection of features by contemplating the relative feature importance. This model performs the simplest at 84% which is the highest comparatively.

Figure 25.5 represents the future prediction done through the LSTM model in comparison with the sample values from the dataset. LSTM performs well for classifying, processing, and making predictions with a Mean Absolute Percentage Error of 29.6%. This model makes

forecasting more effective by addressing the computational efficiency needed for training large networks.

V. Conclusion and Future Scope

Due to the unpredictable nature and variety of pollutants in place and time, predicting air quality can be difficult. Continuous air quality monitoring and analysis are required, particularly in developing nations, due to the serious effects of air pollution on people, animals, plants, historical sites, the climate, and the environment. It has been observed that for the forecast of India's AQI, researchers have shown less interest. In the present work, air pollution data for different cities of Karnataka has been investigated. The dataset is initially cleaned and pre-processed by replacing the missing values with zero. Then, AQI-affecting contaminants are filtered for further investigation using a correlation-based feature selection technique. To uncover multiple unseen forms in the dataset, exploratory data analysis

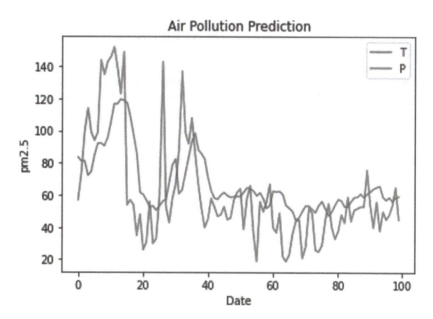

Fig. 25.5 Graph of predicted and tested values using LSTM model of year 2021 dataset

techniques are used. The dataset is divided between train and test subsets in a proportion of 75:25. A comparison study and ML-based AQI prediction are conducted and provided. Both the train-test subgroups' ML model outcomes are shown in terms of common measures like effectiveness. The SVR model has the lowest accuracy, while the Random Forest model has the best accuracy. Deep learning methods for AQI prediction can be used for further study. LSTM model performance is evaluated and contrasted using the traditional statistical error measurement, mean absolute percentage error whereas the LSTM model gave an error of 29.6%.

The following themes will be explored in future studies to enhance and better understand how algorithms are used to anticipate air quality: Choosing a big dataset with more measurements and characteristics to support more precise statistical models for airborne pollutants and particles, especially NO2 and PM2.5. Further, by utilising live API and continuing to make automatic predictions of future values, we may also enhance our project.

References

1. C. R., Aditya & Deshmukh, Chandana & K., Nayana & Gandhi, Praveen & astu, Vidyav. (2018). Detection and Prediction of Air Pollution using Machine Learning Models. International Journal of Engineering Trends and Technology.59.204-207. 10.14445/22315381/IJETT-V59P238.
2. Liu, Huixiang & Li, Qing & Yu, Dongbing & Gu, Yu. (2019). Air Quality Index and Air Pollutant Concentration Prediction Based on Machine Learning Algorithms. Applied Sciences. 9.4069.10.3390/app9194069.
3. Fabiana Martins Clemente, Aleš Popovič,Sara Silva, and Leonardo Vanneschi. "A Machine Learning Approach to Predict Air Quality in California.", Research Article, 04 Aug 2020.
4. K. Kumar B. P. Pande "Air pollution prediction with machine learning: a case study of Indian cities", International Journal of Environmental Science .and Technology (2022).
5. Thakur A. Study of Ambient Air Quality Trends and Analysis of Contributing Factors in Bengaluru, India. Orient J Chem 2017; 33(2).
6. H. S. Sudhira, T. V. Ramachandra, M. H. Bala Subrahmanya, Cities, Centre for Sustainable Technologies, Indian Institute of Science, Bangalore 560 012, India, October 2007.
7. Dan Wei "Predicting air pollution level in a specific city". Neural Comput 1997.
8. Q. Tao, F. Liu, Y. Li and D. Sidorov, "Air Pollution Forecasting Using a Deep Learning Model Based on 1D Convnets and Bidirectional GRU," in IEEE Access, vol. 7, pp. 76690–76698, 2019, doi: 10.1109/ACCESS.2019.2921578.
9. T. Banerjee and R.K. Srivastava, Evaluation of environmental impacts of integrated industrial estate-Pantnagar through appli-cation of air and water quality indices, Environmental Monitor-ing and Assessment 172(1–4) (2011), 547–560. doi:10.1007/s10661- 010-1353-3.

Recent Trends in Computational Intelligence and its Application – Sugumaran D. et al. (eds)
© 2023 Taylor & Francis Group, London, ISBN 978-1-032-48410-5

26

Smart Agriculture Monitoring System— Weed Detection and Predicting Soil Moisture and Weather

M. Senthamil Selvi*

Professor, Sri Ramakrishna Engineering College, Coimbatore-22

S. Jansi Rani[1]

Assistant Professor, Sri Ramakrishna Engineering College, Coimbatore-22

V. K. Nivetha[2], C. Sruthi[3], Rahman Shafeeq[4]

Student, Sri Ramakrishna Engineering College, Coimbatore-22

Abstract—In India Agriculture is one of the major sectors, contributing roughly 17% of the Manufacturing GDP and over 60% of the population. Indian Agricultural production and Allied Industries Industry Report (Size: 1.99 MB) (2021 November) The agriculture sector provides a livelihood for approximately 58 % of the country's population. Weeds compete with crop plants for sunlight, nutrients, water, space, and other growth requirements, decreasing crop yield and raising production costs, resulting in increased labor costs. Increases in agricultural production of 10% increase non-poor households' consumption growth by 2% on average and 0.8 percent for poor households. Agricultural output entails environmental aspects. Environmental calamities like Weather, Soil Moisture, are one of the major drawbacks for farmers in raising crops. Due to changes in weather conditions, Crops may not have healthy life support in case of prevention measures where not taken. So to avoid these situations we take preventive measures by removing the weeds and predicting the future weather report and take preventive measures to increase Agricultural Productivity.

Keywords—Auto regression model, Recurrence neural network, Recurrence neural network, World population growth, Gross domestic product

I. Introduction

The development of human civilization has proven to depend heavily on agriculture. Both art and science go into growing crops and raising livestock. Agriculture is the practice of raising plants for the purpose of supplying human necessities such as food, fuel, and fiber. Agriculture is a key source of employment and a significant contributor to the Indian economy.

*Corresponding Author: hod-it@srec.ac.in,

[1]jansi.sankar@srec.ac.in, [2]nivetha.1805103@srec.ac.in, [3]sruthi.1805144@srec.ac.in, [4]shafeeqrahman.1805165@srec.ac.in

DOI: 10.1201/9781003388913-26

In farming when the farmers grow something due to soil property and micro seeds available leads to additional growth of weeds, It spoils the actual outcome of farming as the farmers affect the growth of planted plants in the case of removing weeds. In this circumstance, Weed detection is an accurate solution, by identifying the area of weeds, and that specific area is targeted for spraying fertilizers by avoiding infecting other plants. IP approaches could be applied to solve this problem. IP is a method used to get the features of the plant and using RF the classification is done to detect the weed crops. As a result, agricultural scientists can supply services such as zoom modifications to find the 3D local map and converge it into the map. The most significant aspect of the system is the height gap between the weed and the crop. One of the system's drawbacks could be errors in weed recognition when crops and weeds are the same height..Identifying the area of weeds and that specific area are targeted for spraying fertilizers by avoiding infecting other plants. Image Processing approaches could be applied to solve this problem. Image Processing is a method used to get the features of the plant and using Random Forest Algorithm the classification is done to detect the weed crops. As a result, by employing images of crops that offer a clearer insight, agricultural scientists can offer better solutions. There are many weeds that are hard to spot early on and can cause social and financial harm. Therefore, to make it simpler and more affordable, we apply image processing, which helps farmers get over these obstacles. Due to the significant loss in agricultural productivity, we can quickly identify weeds by taking characteristics from the plant. Analysis of the data reveals that this approach is more advantageous and successful in identifying the condition at a minimal price. Farmers confront two problems: the first is the labor required for farming, and even after that, they must contend with natural calamities such as rain at precisely the time when the harvest is ready to be cut. For this, a mathematical model can be constructed that would assist agriculture professionals in providing an effective solution to the current problem in the crop., to increase the yield. This can be achieved by creating a method or technology to predict Soil moisture and weather report, water calamities are the major factor for natural calamities using the Machine Learning Algorithm.

II. Literature Review

1. **Sagarika Paul, Satwinder Singh, "Soil Moisture Prediction Using Machine Learning Techniques, November 2020, The 3rd International Conference on Computational intelligence and intelligent systems"**, This work accurately measures soil moisture, which is a crucial component of farm productivity and can kill plants if it is insufficient of it. The outcomes are root disease and water waste.

2. **Shikha Prakash, Animesh Sharma, Sitanshu Shekhar Sahu," SOIL MOISTURE PREDICTION USING MACHINE LEARNING, November 2020, International Journal of Advance Research, Ideas and Innovations in Technology"**,. Individual plant categorization has been demonstrated in this paper using either spectral or color imagery. The spatial resolutions of spectral approaches are typically insufficient for an accurate individual plant or leaf detection.

3. **Siddhesh Badhan, Kimaya Desai, Manish Dsilva, Reena Sonkusare, Sneha Weakey, "Real-Time Weed Detection using Machine Learning and Stereo-Vision, April 2021, 6th International Conference for Convergence in Technologies (I2CT)"**. This paper accurately uses the motion technique, as well

4. **Mansoor Alam1, Muhammad Shahab Alam2, Muhammad Roman1,2, Muhammad Tufail1,2, Muhammad Umer Khan3, Muhammad Tahir Khan, "Real-Time Machine-Learning Based Crop/Weed Detection and Classification for Variable-Rate Spraying in Precision Agriculture, August 2020, 7th International Conference on Electrical and Electronics Engineering"**

A real-time computer vision-based crop/weed detection system for variable-rate pesticide spraying are presented in this research.

5. Automated Weed Detection Systems: "A Review Saraswathi Shanmugam, Eduardo Assunção, Ricardo Mesquita, André Veiros, and Pedro D. Gaspar, December 2018, The Review of Advance Research," This paper examines and compares weed management approaches, with a focus on reporting current research in automated weed detection and control.

III. Methodology

A. Weed Detection

Weeds are one of the major nutrient restrictors to crops by consuming sunlight and nutrients provided to the crops due to these unwanted plants in the agricultural land. So to avoid this situation, In the traditional method, farmers use fertilizers to remove these weeds which may cause disease to the crops and loss in soil

properties, To prevent this situation we use a technique of Detecting the weeds using the images of the plants in the crop field and then spraying the fertilizers in the particular area where the weed is detected. This detection is done through Image Processing and Random Forest Algorithm.

This system follows the following steps to detect whether the plant is affected by weeds or not.

1. Image Acquisition (from database)
2. Image Pre-Processing to prepare the data
3. Feature Extraction (Important Feature)
4. Training & Testing
5. Classification /Predict

1. Image Acquisition

The first step is collecting photographs. It is a technique of retrieving an image from an external source and then using it in a subsequent process. In this situation, we must first capture the image, which will be stored as numerical data in the device that will be modified. This

Fig. 26.1 Block diagram

Source: Author

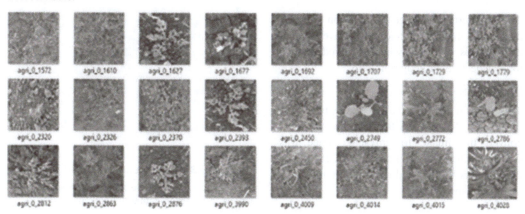

Fig. 26.2 Weed plants data set

Source: https://www.kaggle.com/ravirajsinh45/crop-and-weed-detection-data-with-bounding-boxes

data set contains 1300 images of sesame crops and various forms of weeds, with their own label. Each image is 512 × 512 pixels in size. These image labels are in the YOLO format.

2. Image PreProcessing

Image Preprocessing is the next step after the Image Acquisition process. In this process, we resize the images into a uniform size and remove the unwanted distortion of images to improve the quality of the image. This helps to classify the data set more accurately.

3. Feature Extraction

The color, edges, textures, and forms of weeds and crops differ slightly. As a result, all of these features are utilized to categorize weeds and crops. The following subsections cover feature extraction.

(a) Hu moments for shape matching.

(b) Feature of edge orientation

(c) Haralick textured.

(d) Histogram in color

(e) Training

(f) Testing

(a) Hu moments are used to match shapes.

Weeds and crops have leaves that have varied shapes and sizes. As a result, Hu moments and associated feature vectors are employed for shape analysis.

(b) Edge orientation

This property of edge orientation It was discovered that weeds and crops have distinct distributions of edges when observed from a top perspective, or geographically.

(c) Haralick Texture

This is used to quantify an image based on its texture. To perform this function first we convert the image to grayscale since this function expects a grayscale image. It utilizes the Mahotas Library and the function is image. mahotas.feature.()

(d) Histogram In Color

It's a technique for extracting color from an image that's usually utilized for color extraction. To extract Color Histogram features from the image, we utilise OpenCV's cv2.calcHist() function. Image, channels, mask, histSize, and ranges for each channel [typically 0-256] are the arguments supplied. The histogram is then normalised using.

(e) Training

The next step after extracting is Training the data using machine learning classification. The training procedure is a supervised machine learning method in which the model is fed labeled data and the system determines how to separate each image. The data is trained to make the model strong enough to differentiate crops and weeds, variations in sunlight, and blurring due to motion. It made use of a number of Scikit-learn learning models.

(f) Testing

Once the Random Forest model is trained, the image is predicted properly whether the image is Cropped or Weed correctly. We used 20% of trained data as test data for testing whether the image is Weed or Crop. The data has been predicted with an accuracy of 94% successfully through the Random forest Algorithm.

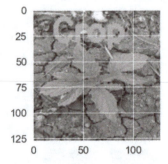

Fig. 26.3 Crop prediction tested image

Source: https://www.kaggle.com/ravirajsinh45/crop-and-weed-detection-data-with-bounding-boxes

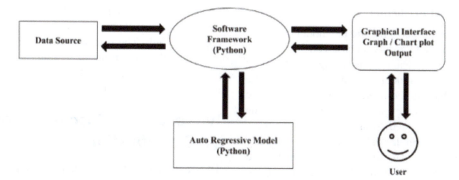

Fig. 26.4   Flow chart for prediction using autoregression model

Source: Author

B. Weather Prediction

Weather changes are one of the major natural factors which affect agriculture. The majority of field crops rely completely on the weather to give life-sustaining water and energy. The comfort and food supplies of livestock are likewise reliant on the weather. Bad weather conditions can occasionally result in lost productivity, especially if they occur during vital stages of growth. Individual weather elements have different effects on crops and cattle. However, the simultaneous occurrence of all meteorological factors might have additive impacts. Weather forecasting is an attempt to forecast future weather changes as well as expected weather conditions.

This system includes the following steps:

1. Data Set Collection
2. Data Set Training
3. Data Set Prediction

1. Data Set Collection

Data Set collection is the first step, The data set consists of the following parameters:

(a) Date

(b) Temperature

Based on the temperature of the past date makes the data set for the Weather Prediction. The size of the data set is 120 rows × 1 columns.

	Temperature
Month	
2010-01-01	21
2010-01-02	24
2010-01-03	30
2010-01-04	33
2010-01-05	36
...	...
2019-01-08	29
2019-01-09	29
2019-01-10	28
2019-01-11	26
2019-01-12	21

120 rows × 1 columns

Fig. 26.5   Data set of weather prediction

Source: Author

2. Data Set Training

Training a model refers to the process of iteratively refining your prediction equation by looping through the dataset numerous times and altering the weights and biases in the direction indicated by the cost function's slope (gradient). The training is complete when we reach an acceptable error level or when more training iterations fail to reduce our cost. The Auto Regression Algorithm was used to train us on the model.

- The auto regression algorithm forecasts future data using historical data.
- It's a straightforward approach that can yield accurate forecasts for a wide range of time series problems.
- It's a linear regression model using a lag as an input parameter.

3. Data Set Prediction

Once the data set is fit, and the model is trained we can predict the data set based on 3 categories Sunny, Hazy, Rainy, and using the function predict(). The model is trained and we get the rmse, which is the average squared difference between the value estimated and the true value is 2.02. The prediction system is accurate. All of the attribute values had been properly pre-processed. The method was constructed and trained using train data after all of the preprocessing was accomplished. Our precision was estimated to be about precise.

```
print (x, null )

25.299630887579934 Haze
20.814270256557563 Cloudy
19.062096091560925 Cloudy
20.963459987870078 Cloudy
26.575283047627117 Haze
30.985826022273823 Haze
32.43914344327642 Haze
32.17003266621991 Haze
30.362150047158188 Haze
30.034958383210956 Haze
30.541050238455078 Haze
29.040125187145325 Haze
25.69319287313301 Haze
21.23141300314746 Cloudy
```

Fig. 26.6 Predicted value and predicted category of weather for next 50 days

Source: Author

3. Data Set Prediction

Once the data set is fit, and the model is trained we can predict the data set based on 3 categories Sunny, Hazy, and Rainy, and using the function predict(). The model is trained and we get the rmse, which is the average squared difference between the value estimated and the true value

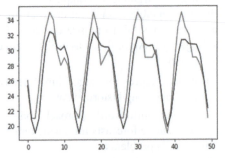

Fig. 26.7 Predicted value and predicted category of weather for next 50 days

Source: Author

is 2.02. The prediction system is accurate. All of the attribute values had been properly pre-processed. The method was constructed and trained using train data after all of the preprocessing was accomplished. Our precision was estimated to be about precise.

IV. Result and Discussion of Classification

The main goal of the system is to generate a model to classify the weeds among the plants. This can be achieved by Random Forest Classification which is one of the best algorithms used for classification.

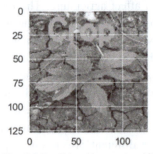

Fig. 26.8 Weed/crop classification

Source: https://www.kaggle.com/ravirajsinh45/crop-and-weed-detection-data-with-bounding-boxes

The crop is classified properly with an accuracy of 99.7% using Random Forest Classification successfully.

The Weather Detection is predicted using the Autoregression model with a root mean squared error of 0.2 and the weather is predicted

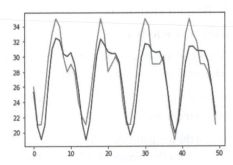

Fig. 26.9 Prediction of weather for next 7 days

Source: Author

for the next 7 days accurately. Auto regression is used for prediction because it helps to predict future data based on the past accurately.

```
rmse = sqrt(mean_squared_error(test, predictions))
print('Test RMSE:', rmse)

Test RMSE: 2.0296124387013044
```

Fig. 26.10   RMSE for weather prediction

Source: Author

V. Results

Table 26.1   Soil moisture prediction result

	Random Forest Regression	Linear Regression
Train	0.90	0.22
Test	0.78	0.24

Source: Author

The prediction of Soil moisture is done with a different Regression algorithm and has the best accuracy in RF Algorithm.

Table 26.2   Weed detection result

Random Forest Algorithm	Value
Accuracy	99.7%

Source: Author

Table 26.3   Weather prediction result

Auto Regression	Value
RMSE	2.02

Source: Author

VI. Conclusion

In conclusion, we have a combined agricultural system to detect weeds in agriculture, the way which helps the farmer to remove the weed without harming the other plants and also to predict the weather and soil moisture to increase the productivity of crop yield by taking necessary measures. The weather prediction was done using the auto-regressive model and are very essential for improving future performance. Weed detection also aids in the reduction or elimination of pesticide use, the reduction or elimination of agricultural environmental and health impacts, and the improvement of sustainability. so we have an accuracy of about 99.7% in Weed Detection using the Random Forest algorithm. Thus the agricultural sector is developed using this smart System and helps many farmers in an easy way, as a support to their lives and making our environment eco-friendly and creating an India a "NO HUNGRY INDIA" and "NO POVERTY" by making everyone have their meal and lives for farmers with profits.

References

1. Sagarika Paul, Satwinder Singh, Soil Moisture Prediction Using Machine Learning Techniques,November 2020,The 3rd International Conference on Computational intelligence and intelligent systems.
2. Shikha Prakash, Animesh Sharma, Sitanshu Shekhar Sahu, "Soil Moisture Prediction Using Machine Learning", International Journal of Advance Research, Ideas and Innovations in Technology-November 2020.
3. Siddhesh Badhan, Kimaya Desai, Manish Dsilva, Reena Sonkusare, Sneha Weakey, "Real-Time Weed Detection using Machine Learning and Stereo-Vision", 2021 6th International Conference for Convergence in Technology (I2CT) Pune, India. Apr 02–04, 2021.
4. Mansoor Alam1, Muhammad Shahab Alam2, Muhammad Roman1, 2, Muhammad Tufail1, 2, Muhammad Umer Khan3, Muhammad Tahir Khan, "Real-Time Machine-Learning Based Crop/Weed Detection and Classification for Variable-Rate Spraying in Precision Agriculture", 2020 7th International Conference on Electrical and Electronics Engineering
5. W. Baudoin, R. Nono-Womdim, N. Lutaladio et al., *Good Agricultural Practices for Greenhouse Vegetable Crops: Principles for Mediterranean Climate Areas (No. 217)*, Food and Agriculture Organization of The United Nations, Rome, 2013.
6. M. Raviv and L. Lieth, "Significance of soilless culture in agriculture," in *Soilless Culture: Theory and Practice*, M. Raviv and J. H. Lieth, Eds., pp. 117–156, Elsevier, Amsterdam, 2007. View at: Google Scholar

27

Extraction and Analysis of Snow Covered Area from High Resolution Satellite Imageries

Kodge B. G.*

Department of Computer Science, School of Science,
GITAM University, Hyderabad, TS, India

Abstract—Today the world is facing numerous environmental problems occurring due to improper climatic changes by global warming, natural imbalances, and other man-made things on earth. The global temperature is increasing day by day and therefore melting land snow or ice sheets are escalating global sea levels similarly. The increasing levels of melting of land snow or ice sheets become one of the biggest challenges in front of us and need to be monitored, controlled, or resolved by us as soon as possible. Therefore an attempt is made in this paper to study and develop a technique that can extract the snow-covered area using some digital image processing techniques and a few other geospatial techniques applied to high-resolution snow-covered satellite images. This paper further extended to study and understand the changes that occurred in snow-covered areas of Himalayan ranges during the years 1973, 1985 to 2020 using temporal satellite imageries.

Keywords—Snow covered area, Satellite images, Image segmentation, Image thresholding, Geo-spatial techniques, Comparative analysis

I. Introduction

There are several techniques developed by researchers to recognize and extract interesting patterns, colors, textures, morphologies, and regions of interest from images for their specific purposes. But, all these existing tools and techniques will not work on all kinds of problems related to imageries and patterns or regions of interest for expected results. Every tool and technique is basically developed to solve a specific problem and can have a desired and expected output as maximum as possible. There are several existing techniques and associated research papers are reviewed in this regard. Most of the researchers tried to develop and extract the snow cover area using some specific types of satellite imageries only, but those techniques are becoming unable to work with another type of input satellite images. [1] [3][4]

This paper deals with a kind of exercise that attempts to read high-resolution satellite images which are having snow-covered areas

---
*Corresponding Author: kodgebg@gmail.com

DOI: 10.1201/9781003388913-27

and will extract the only snow-covered area from those images as a result. This proposed technique is able to extract the snow-covered area using multiple types of high-resolution satellite images. However, the input image must be cloudless and non-noisy. [2]

II. Methodology

The primary data such as high-resolution satellite imageries or datasets are used from Google Earth Server, ESRI (Environmental Systems Research Institute), and Bhuvan-ISRO (Indian Space Research Organization). [5][6][7][8][9]

This proposed system uses several image processing, pattern recognition, and geospatial

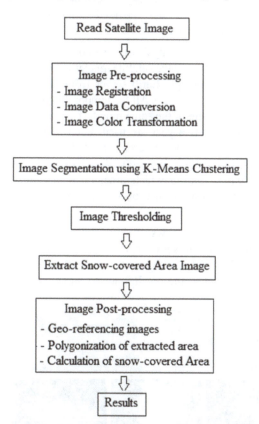

Fig. 27.1 Methodology for extracting snow covered area from satellite image

Source: Author

data processing and analysis techniques to extract the snow-covered areas and study the same using comparative analysis. The MATLAB tool is used to design and develop this proposed system and QGIS open-source software is used for geo-referencing and polygonization techniques. The step-by-step processes developed in this system is shown in Fig. 27.1. [10][11]

The system first read the input satellite image and starts the image pre-processing on it. The image pre-processing techniques such as

- Image registration for aligning the image at a specified space or location.
- The image data conversion method is used to convert the image data into the required data type. The 'double' data class image is used in this system.
- The image color transformation technique is used to convert the image color into LAB (**L**ightness, channel **A**, channel **B**) color space. Generally, the satellite images come with RGB (Red, Green, Blue) color spaces, therefore this image color conversion is applied.

After completing the image pre-processing processes, next, the system starts the image segmentation process to partition the image into multiple segments. The K-means clustering technique is used to partition the image segments in MATLAB. Before going to apply this K-means clustering technique for image classification, we have to manually recognize the number of major available colors in that image. In this technique further, we need to assign how many possible or required color segmentations can be done by it. Next, it results in the same number of classified images which we assigned to do partition/segment.

The thresholding technique is used to apply the interested segmented image to get or adjust the required level of extracted color intensity. Because, while doing the image segmentation process using K-means, some other patterns associated with the same colored intensity are also extract. To remove the same we need to adjust the required level of color intensity of our pattern or region of interest.

Next, the system displays the snow-covered area extracted image on the screen as output. But to calculate and measure the extracted snow-covered area, we need to apply some post-processing steps like geo-referencing, polygonization, and location-based area calculation techniques.

- The geo-referencing technique is used to assign/apply the geo-spatial coordinates (longitude/latitude) with a specific projection on the extracted image. For geo-referencing the extracted images EPSG:4326 WGS 84 coordinate reference system is used as shown in Figure 3.
- The polygonization technique is used to convert the raster image into its vector form. The converted vector image will be contained all the extracted each individual areas into a separate polygon. Each individual extracted polygon area will be calculated for further calculations and analysis.

III. Results and Discussions

A. Snow-covered Area Extraction

Figure 27.2 shows the extracted results of snow-covered areas from different types of images of different locations, different colors, patterns, and morphologies. The sample image inputs are collected from Google Earth Engine, Bhuvan (ISRO), and the ESRI server. [7][8][9]

B. Comparative Study on Extracted Snow Covered Areas

The second phase of this paper is to deal with the comparative analysis part of this study. The Himalayan ranges are taken into consideration as a study site for the analytical study of these extracted snow-covered areas. The geo-referenced and extracted snow-covered area map of the Himalayan ranges is compiled and shown in Fig. 27.3.

Figure 27.3 shows the geo-referenced snow-covered area (dark grey color) extracted map of Himalayan ranges. The India map is placed in Fig. 27.3 for an easy understanding of the location-based snow-covered area of the Himalayan ranges.

The temporal analysis of the said study site is processed using the satellite imageries collected from the Google Earth server from the years 1973, 1985 to 2020. All the temporal images of the same site of the same month (December) of that respective year are used. Each individual year's snow-covered area extracted images are used to analyze and see what sorts of changes occurred during which time. Each individual year's extracted list of images is shown in Fig. 27.4.

Figure 27.5 shows the year-wise results of the extracted snow-covered area from 1973, 1985 to 2020. There are major fluctuations are found in the result, and the snow-covered

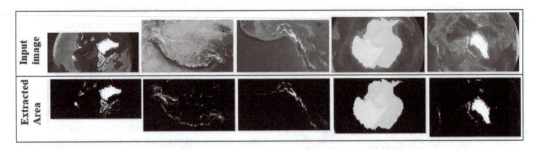

Fig. 27.2 Sample image inputs and extracted snow covered area

Source: Author

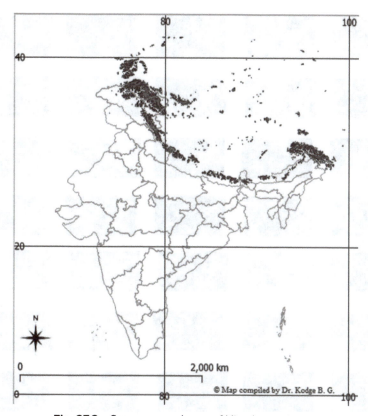

Fig. 27.3   Snow covered area of Himalayan ranges

Source: Author

area is decreasing year by year. In the year 1973, the snow-covered area was 136247.10 Sq. Km. and was in the range between 130000 to 120000 Sq. Km. till the year 1988. Next in the years 1990 and 1992, it increased to its highest level i.e. 151830.98 and 165969.96 Sq. Km. Further from the year 1993, it was found to be decreasing year by year to its lowest in years 2015 and 2020 i.e. 74525.01 Sq. Km. and 78512.92.

IV. Acknowledgement

I am very much grateful to Google, ESRI, and Bhuvan (ISRO) for providing primary data (high-resolution satellite images) for this study. Similarly, I am also thankful to the QGIS team for providing such a great tool for geospatial research and development.

V. Conclusion

The system developed and explained in this paper for extracting snow-covered areas and its analysis is very important to understand the current and past situations of the global environment. It also shows the impact of global warming and other artificial and natural imbalances occurring on earth which increase the melting of land snow or ice sheets. In response, the global sea level and temperature are rising day by day. It's a dangerous warning sign for the current as well as future living objects on earth.

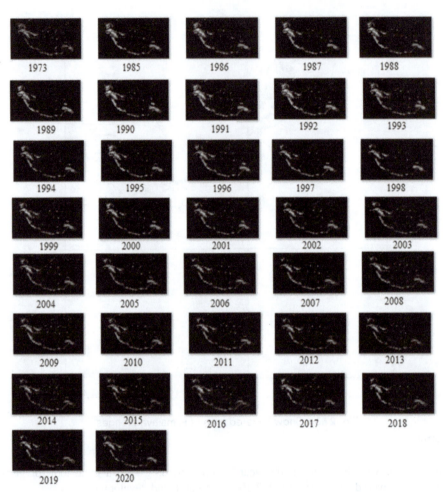

Fig. 27.4 Snow covered area extracted images from 1973, 1985-2020

Source: Author

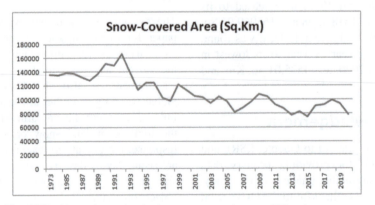

Fig. 27.5 Year wise extracted snow covered area of Himalayan ranges

Source: Author

References

1. Ma, Y. and Zhang, Y.: Improved on Snow Cover Extraction in Mountainous Area Based on Multi-factor NDSI Dynamic Threshold, Int. Arch. Photogramm. Remote Sens. Spatial Inf. Sci., XLIII-B3-2022, 771–778

2. Liu, Y., Chen, X., Hao, JS. et al. Snow cover estimation from MODIS and Sentinel-1 SAR data using machine learning algorithms in the western part of the Tianshan Mountains. J. Mt. Sci. 17, 2020, 884–897

3. Guangjun He, et al., Extracting snow cover in mountain areas based on SAR and optical data, IEEE Geoscience and Remote Sensing Letters, Vol. 12, No. 5, 2015, pp: 1136–1140

4. Yan Huang, et al., Snow cover detection in mid-latitude mountainous and polar regions using nightmare light data, Remote Sensing of Environment, Vol. 268, 2021.

5. Zhiming Liu, et al., The extraction of snow cover information based on MODIS data and spatial modeler tool, 2008 International Workshop on Education Technology and Training & 2008 International Workshop on Geoscience and Remote Sensing, 2008, pp. 836-839, doi: 10.1109/ETTandGRS.2008.252.

6. Hiremath P. S., Kodge B. G., "Automatic open space area extraction and change detection from high resolution urban satellite images", International Journal of Computer Application (IJCA), Special Issue on RTIPPR (2) 2010. Published By Foundation of Computer Science, USA, ISSN: 0975-8887 pp. 76–82.

7. https://bhuvan.nrsc.gov.in

8. https://www.esri.com/en-us/arcgis/products/imagery-remote-sensing/

9. https://earth.google.com/web/data

10. https://docs.qgis.org/3.16/en/docs/user_manual/index.html

11. https://www.iirs.gov.in/iirs/sites/default/files/upload/document/Long-Term_SCA_Mapping.pdf

Recent Trends in Computational Intelligence and its Application – Sugumaran D. et al. (eds)
© 2023 Taylor & Francis Group, London, ISBN 978-1-032-48410-5

28

A Customized Deep Convolutional Neural Network for Covid-19 Detection

M. Dhilsath Fathima*, R. Hariharan[1], Sachi Shome[2], Tamilarasan P.[3]
Department of Information Technology,
Vel Tech Dr. Rangarajan Dr. Sagunthala R&D Institute of Science and Technology,
Chennai, Tamil Nādu, India

V. Vijeya Kaveri[4]
Department of Computer Science and Engineering,
Sri Krishna College of Engineering and Technology, Coimbatore, Tamil Nādu, India

R. Natchadalingam[5]
School of Computing and Information Technology,
REVA University, Bengaluru, Karnataka, India

Abstract—The world is still experiencing the devastation caused by the covid-19 virus. It is vital not only to promote the vaccine and have people vaccinated as soon as possible, but it is also important to test more individuals and isolate those who are sick from the general population and stop the disease's spread. While the nasal swab test model is now used over the world to identify covid patients, radiography evaluation provides an alternative and more efficient method. In this suggested system, a covid-19 detection model is used to detect the virus more quickly and inexpensively using the proposed Customized Deep Convolutional Neural Network (CDCNN) approach and chest x-rays. The CDCNN method utilizes both positive and negative covid affected patient chest x-rays to train the model and help with early prediction of covid-19. The output accuracy of the proposed model is 97.93 percent when using openly accessible chest X-ray images.

Keyword—Customized Deep Convolutional Neural Network (CDCNN), Covid-19 detection, Chest X-Rays, Deep learning, Corona virus

I. Introduction

Covid-19 has shattered over millions of lives and declared so as one of the deadliest pandemic outbreaks ever since the Spanish flu, the virus belongs to the SARS COV-1 variant which was earlier detected in the year 2012. The virus mainly causes respiratory problems causing

*Corresponding Author: dhilsathfathima@veltech.edu.in
[1]hariharanr@veltech.edu.in, [2]sachishome26@gmail.com, [3]tamilarasanperumal2001@gmail.com,
[4]vijeyakaveriv@skcet.ac.in, [5]yraja1970@gmail.com

DOI: 10.1201/9781003388913-28

acute shortage of breathing leading to fever, cold and severe cough. Although vaccination camps have started worldwide with billions of doses already carried forward, still there's an abnormal spike in the cases in certain areas.

At present, the only form of detecting this virus is through the nasal swab testing method called Reverse Transcription Polymerase Chain Reaction (RT-PCR). It is considered to be accurate and would quickly identify a virus microparticle [1]. However, this test takes an ample amount of time to get the result and prepare a report, which keeps a patient under confusion and doubt, at times gets mistakenly detected, and is also an expensive process. These are also exposed to manual human errors. Apart from that, there exist some image detection technique that helps in detecting the virus through x-ray images but they lack precision because of the improper way of training the model due to which patients with a medical history of flu and lung disease and also people with smoking habits were being detected as a covid patient which leads to failure of existing model [2, 3].

A customized deep convolutional neural network (CDCNN) is an ensemble of many convolutions and deep neural network layers developed to build an accurate Covid-19 detection model with great precision. CDCNN is based on the concept that an ensemble of model predictions performs better than individual models. This ensemble model boosts the efficiency of the deep learning classifier, which combines several layers of convolution and neural network layers into a single predictive model for minimizing the variance, and bias, for enhancing model accuracy.

A. Motivation of CDCNN Covid-19 Detection Model

This proposed model intends to develop a fast and reliable Covid-19 identification system using the CDCNN model. The present nasal swab-based testing method is more prone to human error. As a result, our suggested method

overcomes the disadvantage of nasal swab-based testing. This suggested method detects Covid-19 with excellent accuracy using the CDCNN, saving time and effort.

B. Problem Formulation

The key contributions to the proposed research include: There have already been numerous studies on the diagnosis of coronaviruses using X-ray images, but almost all of them make use of pre-trained models, including ResNet-50, VGG 16, and Inception v3. The majority of these studies were carried out in 2020 when covid was still relatively new and datasets were not widely available, and the available data had to be supplemented to obtain a large dataset.

As a result, the proposed model used a larger dataset and aimed to achieve higher accuracy than the existing models. This proposed model reached 97.93 percent accuracy, which is higher than that of other models.

C. Organization of this Article

Section1 introduction discussed the idea in a little more detail and the problem with the current covid testing model and how our proposed method aims to solve it. The following are the remaining sections of the paper: Section 2 highlights recent research on Covid-19 detection models; Section 3 details the suggested CDCNN classifier; Section 4 includes experimental analysis of the research results, and Section 5 outlines the suggested work conclusion.

II. State-of-the-Art Research on Covid-19 Detection Model

Many authors have published their research on Covid disease detection using deep neural network (DNN) models. This section highlights the research on Covid-19 detection deep learning methods. Emrah Irmak et al. [4] developed Convolutional Neural Network, which has twelve weighted layers, two convolution layers, and one fully connected layer with the ReLU activation function. The fully connected layer makes a 2-dimensional vector, which is then fed into the softmax classifier, which generates the final prediction. The prediction has two possible outcomes, and the output layer has two neurons. The first convolutional layer is made up of 96 7×7 kernels with stride 4 and padding of zero. The second convolutional layer is made up of two sets of 128 kernels with stride 1 and padded 2. The input photos are 227×227×3 in size.

Jingxin Liu et al. [5] developed a DNN algorithm using a Computed Tomography (CT) image of 721 covid-19 patients from various hospitals, with 600 training images, for covid-19 detection. Asu Kumar Sing et al. [6] developed a Hybrid Optimized Support Vector model with Chest X-Ray input Images. The input features are retrieved using a modified social group optimization technique, which is classified using a Support vector algorithm to predict Covid-19. Kehran et al. [7] developed CNN model with X-ray images through ResNet-50 pre-trained with image augmentation techniques at a time when covid was relatively new and large datasets were not available.

Rakibul Islam et al. [8] developed a Novel LeNet-5 CNN model for CT images and detecting covid 19 cases, with 80% input utilized for model training and 20% used for model testing. Data Augmentation has been used to enlarge the pre-existing dataset. Taresh et al. [9] developed and optimised a transfer learning model with input X-Ray, with VGG16 and MobileNet getting the maximum accuracy, to improve model prediction accuracy and select the best performing model. Reshi et al. [10] utilized a CNN model, with data preprocessing techniques such as dataset normalization, image analysis, and data augmentation. Despite the lack of a dataset with enough high-quality chest X-ray images and appropriate size, the CNN classifier successfully performed well.

Boran Sekeroglu et al. [11] developed Convolutional Neural Networks (CNN) using Chest X-Rays. CNN classifier training and CNN classifier testing are the main stages of this approach. Eightfold cross-validation utilized in this model to measure model performance. Chen Li et al. [12] utilized a deep transfer learning model from chest X-rays, where 70% of images from each class are used for model training, 15% for model validation, and the remaining for model testing. This achieves the greatest classification accuracy of 89.3 percent, with average precision, recall, and F1 scores of 0.90, 0.89, and 0.90, respectively. Julia et al. [13] used lung ultrasound images with DNN architectures such as Xception, InceptionV3, VGG19, and ResNet50. For model training and optimization, a publicly accessible POCUS dataset of pneumonia patients and healthy COVID-19 patients was used.

III. Proposed Method

The proposed model uses the CDCNN classifier for Covid identification and distinguishes between Covid positive and Covid negative X-Ray samples. X-ray samples from patients who tested positive or negative for COVID are obtained and divided 80:20 for model training and testing. The training dataset is pre-processed and then fed into the CDCNN classifier for model training. The time taken for training and how well the model has been trained depends upon the number of epochs. The more the number of epochs, the better a model is trained. After the training part, the model is then tested against the test dataset. The image to be tested is preprocessed and passed as input and the output is generated as positive or negative.

A. CDCNN Classifier

The customized deep Convolutional Neural Network (CDCNN) extracts features from image input, assigns learnable weights and biases to input features, and is able to differentiate between normal samples and covid-19 samples. The main stages in developing a proposed CDCNN are feature extraction and feature classification. In feature Extraction, convolutional layers perform

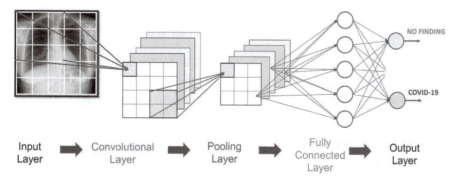

Fig. 28.1 Workflow of proposed Covid-19 detection model

Source: Authors

convolutional processes and identify a unique features of the input image. The input of this step is an input vector that is formed when all hidden layers are flattened. The flattened vector is coupled to fully - connected layers, where the feature classification process determines whether or not the patient has covid-19.

CDCNN requires significantly less pre-processing than other classification techniques. The x-ray images are processed by removing the background image and filtering the object in the foreground using an adaptive threshold to perform morphological operations in the grouped pixels and tracking it using the Kalman filter. The various layers of CDCNN are Convolutional, padding, Pooling, and Fully connected Layer, which are depicted in Fig. 28.1.

In CDCNN, the convolution layer reduces images into a format that is easier to handle while retaining essential features making the best prediction model. The convolution Operator retrieves the high-level attributes from the input image, such as edges. CDCNN has various layers. In CNN, the first layer captures low-level properties like gradient direction, colour, and edges. The model adapts to the high-level features as more layers are added, resulting in a network that understands the image feature in the dataset in the same manner that humans do.

The padding layer basically helps in increasing the area of the image which is processed by

CNN by adding rows and columns to the outer dimension of the images. It helps save the information which is present at the corner of the images from getting chopped off by the convolution process. This helps in maintaining the output size of the data. It helps with a better and more accurate analysis of the image. We have used 'same' padding in our proposed method which basically adds zero values in the outer frame of the image. The extracted features are passed through the Max pooling layer which reduces input dimensionality by picking the maximum value from a group of adjacent pixels.

Through the use of pre-processing filters, a CDCNN efficiently captures the spatial and temporal dependencies of input features. The suggested CDCNN model is suitable and efficient for the input dataset due to the reduced parameters and the reusable weights. Algorithm 28.1 depicts the steps needed in developing a CDCNN model.

Algorithm 28.1	Covid-19 detection using CDCNN algorithm
Input	Input dataset consists of N training samples.
Output	Covid-19 prediction model
Step 1	Fetching and creation of a dataset.
Step 2	Feeding the dataset into the program. The image vectors have been normalized from 0 to 255. It has been divided by 255 so that the input image pixel intensity ranges between 0 to 1.

Step 3	Dividing the dataset into a training dataset and testing dataset of 80:20.
Step 4	Create and train a CDCNN model with best fitting parameters and weights.
Step 5	Calculate each classifier prediction accuracy using testing samples.

IV. Experimental Results and Discussions

A. Experimental Setup of CDCNN Model

Dataset 1 [14] and dataset 2 [15] are combined to create the proposed model dataset. Initially, we started off with just dataset 1 and we got an accuracy of 96.02 percent. After combining both datasets, a total of 487 images were divided into training and testing models in a 4:1 ratio. We have tried various methods starting from increasing the hidden layers and the number of epochs to attain high accuracy of 97.93 percent. The output accuracy of the model is decided by a lot of factors like the epochs, the dataset size, and the hidden layers.

B. Performance Evaluation and Output of CDCNN Model

The Covid-19 detection model uses the customized Deep convolutional neural network (CDCNN) where the data to be trained is pre-processed and then fed into the model to train. The model accuracy is determined by the epochs. The confusion matrix evaluates the proposed model performance. True Positive (True Pos), False Positive (False Pos), False Negative (False Neg), and True Negative (True Neg) are determined as four components of the confusion matrix, and the model accuracy is calculated in following Equation:

$$\text{Model Accuarcy} = \frac{(\text{True Pos} + \text{True Neg})}{\begin{array}{c}(\text{True Pos} + \text{False Pos} \\ + \text{True Neg} + \text{False Neg})\end{array}}$$

The Model Accuarcy is the ability of the classier to differentiate between the classes correctly.

The results for various epoch(s) by the CDDNN model are given in Table 28.1.

Table 28.1 The output of the proposed CDCNN model

Epoch(s)	Training Accuracy	Training Loss	Test Accuracy	Test Loss
10	0.9700	0.0607	0.9251	0.1799
20	0.9500	0.0554	0.9535	0.1252
30	0.9800	0.0420	0.9328	0.1558
40	0.9600	0.0506	0.9690	0.0940
50	**0.9700**	**0.0504**	**0.9793**	**0.0598**

Source: Authors

The CDCNN classifier has a 97.93% accuracy at 50 epochs. Each epoch was obtained at an equal interval of 15s (1s/step). Figure 28.2 and 28.3 shows the accuracy and loss graph of the CDCNN model.

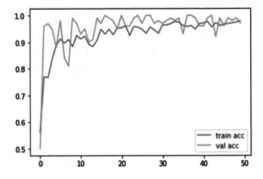

Fig. 28.2 Accuracy graph of CDCNN
Source: Authors

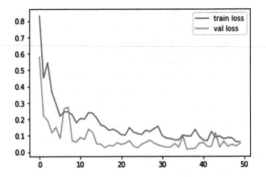

Fig. 28.3 Loss graph of CDCNN
Source: Authors

V. Conclusion

Covid-19 is mutating itself at a pretty rapid manner. There have been new mutants found every two to three months since its inception in December 2019. Even in December 2021, it shows no signs of slowing down. The public should respect and follow all the protocols imposed by the government and the government should set up more covid detection centers and speed up the vaccination campaign in order to efficiently battle the disease. The current system of Covid testing is an accurate method to a very good extent but has its own set of drawbacks. It takes a lot of time in the swab testing and then the sample is sent to the laboratory for RT-PCR testing. This makes it inconvenient for people in this fast-paced lifestyle. The project aims to introduce a novel method for the testing of coronavirus in India which is simpler, time efficient and more convenient than the existing method. Chest X-Rays are utilized for training the CDCNN model and differentiate between a covid-19 positive and a covid-19 negative sample image and after the model is trained, we can input test data to detect covid patients. The full process includes comprehending the problem, selecting the model, creation a dataset, creating and then training a model, implementing the model and then finally getting the output with an accuracy of 97.39 %.

Reference

1. Minaee, Shervin, Rahele Kafieh, Milan Sonka, Shakib Yazdani, and Ghazaleh Jamalipour Soufi. "Deep-COVID: Predicting COVID-19 from chest X-ray images using deep transfer learning." Medical image analysis 65 (2020): 101794.
2. Lee, Ki-Sun, Jae Young Kim, Eun-tae Jeon, Won Suk Choi, Nan Hee Kim, and Ki Yeol Lee. "Evaluation of scalability and degree of fine-tuning of deep convolutional neural networks for COVID-19 screening on chest X-ray images using explainable deep-learning algorithm." Journal of Personalized Medicine 10, no. 4 (2020): 213.
3. Basu, Sanhita, Sushmita Mitra, and Nilanjan Saha. "Deep learning for screening covid-19 using chest x-ray images." In 2020 IEEE Symposium Series on Computational Intelligence (SSCI), pp. 2521–2527. IEEE, 2020.
4. Irmak, Emrah. "A novel deep convolutional neural network model for COVID-19 disease detection." In 2020 Medical Technologies Congress (TIPTEKNO), pp. 1–4. IEEE, 2020.
5. Liu, Jingxin, Zhong Zhang, Lihui Zu, Hairihan Wang, and Yutong Zhong. "Intelligent detection for CT image of COVID-19 using deep learning." In 2020 13th International Congress on Image and Signal Processing, BioMedical Engineering and Informatics (CISP-BMEI), pp. 76–81. IEEE, 2020.
6. Singh, Asu Kumar, Anupam Kumar, Mufti Mahmud, M. Shamim Kaiser, and Akshat Kishore. "COVID-19 infection detection from chest X-ray images using hybrid social group optimization and support vector classifier." Cognitive Computation (2021): 1–13.
7. KARHAN, Zehra, and A. K. A. L. Fuat. "Covid-19 classification using deep learning in chest X-ray images." In 2020 Medical Technologies Congress (TIPTEKNO), pp. 1–4. IEEE, 2020.
8. Islam, Md Rakibul, and Abdul Matin. "Detection of COVID 19 from CT Image by The Novel LeNet-5 CNN Architecture." In 2020 23rd International Conference on Computer and Information Technology (ICCIT), pp. 1–5. IEEE, 2020.
9. Taresh, Mundher Mohammed, Ningbo Zhu, Talal Ahmed Ali Ali, Asaad Shakir Hameed, and Modhi Lafta Mutar. "Transfer learning to detect covid-19 automatically from x-ray images using convolutional neural networks." International Journal of Biomedical Imaging 2021 (2021).
10. Reshi, Aijaz Ahmad, Furqan Rustam, Arif Mehmood, Abdulaziz Alhossan, Ziyad Alrabiah, Ajaz Ahmad, Hessa Alsuwailem, and Gyu Sang Choi. "An efficient CNN model for COVID-19 disease detection based on X-ray image classification." Complexity 2021 (2021).

11. Sekeroglu, Boran, and Ilker Ozsahin. "<? covid19?> Detection of COVID-19 from Chest X-Ray Images Using Convolutional Neural Networks." SLAS TECHNOLOGY: Translating Life Sciences Innovation 25, no. 6 (2020): 553–565.

12. Chen, Joy Iong-Zong. "Design of accurate classification of COVID-19 disease in X-ray images using Deep Learning Approach." Journal of ISMAC 3, no. 02 (2021): 132–148.

13. Diaz-Escobar, Julia, Nelson E. Ordóñez-Guillén, Salvador Villarreal-Reyes, Alejandro Galaviz-Mosqueda, Vitaly Kober, Raúl Rivera-Rodriguez, and Jose E. Lozano Rizk. "Deep-learning based detection of COVID-19 using lung ultrasound imagery." Plos one 16, no. 8 (2021): e0255886.

14. Chest Xray for covid-19 detection [Internet] - [Cited 2021 Jan 23]. Available from: Chest Xray for covid-19 detection | Kaggle

15. COVID-19 Xray Dataset (Train & Test Sets) [Internet]. Available from: COVID-19 Xray Dataset (Train & Test Sets) | Kaggle

Recent Trends in Computational Intelligence and its Application – Sugumaran D. et al. (eds)
© 2023 Taylor & Francis Group, London, ISBN 978-1-032-48410-5

29

Real Time Face Mask Detection Using Mobilenetv2 Algorithm

M. Dhilsath Fathima*, R Hariharan[1], Prashant Kumar Singh[2], Praveen Kumar[3], S. Ramya[4]

Department of Information Technology,
Vel Tech Dr. Rangarajan Dr. Sagunthala R&D Institute of Science and Technology,
Chennai, TamilNadu, India

M. Seeni Syed Raviyathu Ammal[5]

Department of Information Technology,
Mohamed Sathak Engineering College, Kilakarai, TamilNadu, India

Abstract—The coronavirus outbreak is a life-threatening event that has resulted in a huge number of deaths and causing health concerns. Governments all across the world were forced to impose lockdowns due to the coronavirus's quick spread stop infection from spreading further. Wearing a face mask publicly and at work considerably reduces the chance of transmission during this epidemic, according to several health reports. When they leave the house, however, the majority of individuals do not use masks. According to a study, 80 percent of people do not use masks appropriately. The proposed MobileNetV2 algorithm, which builds a fast and accurate face mask identification model, is used in this study to provide a solution to the face mask detection problem. MobileNetV2 is the cutting-edge object identification algorithm to find the face mask in real-time. The proposed method efficiently handles object detection and contributes to high object detection accuracy of 99.6 percent and decreased object detection time.

Keyword—MobileNetV2, TensorFlow, Coronavirus, Deep learning, Mask detection

I. Introduction

The World Health Organization (WHO) reported, Coronavirus 2019 (COVID-19) virus killed over 6.44 million people and infected over 0.59 billion people globally [1]. Patients with COVID-19 have described a wide range of respiratory symptoms, including breathing difficulty, varying from mild warning signs to severe illness. Infection with Adults with lung illness is at higher risk for developing serious consequences from COVID-19 infection [2].

*Corresponding Author: dhilsathfathima@veltech.edu.in
[1]hariharanr@veltech.edu.in, [2]ps232259@gmail.com, [3]1121praveen@gmail.com, [4]rsramya32@gmail.com, [5]seenigood@gmail.com

DOI: 10.1201/9781003388913-29

The most prevalent coronaviruses that infect people worldwide include human coronaviruses 229E, HKU1, OC43, and NL63 [3]. People with respiratory conditions are more likely to spread infectious beads to anyone who comes into contact with them. Because virus-carrying droplets could settle on an infected person's surroundings, this could lead to contact transmission.

A clinical mask must be used to prevent the transmission of Coronavirus (SARS-CoV-2). Masks have the potential to reduce vulnerability to noxious people during the "pre-symptomatic" stage of infection, as well as stigmatize those who use masks to prevent viral transmission. Medical masks and respirators are given top priority by the WHO for healthcare assistants [4]. Detecting face masks has therefore become a key difficulty in today's global world. Face mask detection is the process of determining where a person's face is located and then determining whether or not they are wearing a mask. The issue is with generic object detection, which is used to recognize many sorts of things. The technique of identifying and differentiating a certain category of things, namely faces, is known as face identification. It may be used for a variety of things, such as autonomous driving, teaching, and surveillance [5]. In this suggested model, the Machine Learning (ML) libraries Scikit-Learn, OpenCV, TensorFlow, and Keras are utilized to achieve the proposed objective.

A. Motivation and Justification of the Real Time Face Mask Detection Model

This proposed model intends to develop a fast and reliable face mask identification model using the proposed MobileNetV2 algorithm. Existing systems lack an alarm sound system to alert the monitor; however, the proposed mobilenetv2 system has an alarm system and can recognize many faces faster and more accurately than existing versions. The mobilenetv2 model is more efficient and uses fewer model parameters, resulting in a lighter

deep neural network. MobileNetV2 models for face mask identification are faster and more accurate throughout the whole latency spectrum than prior systems. For example, the new models use 2x fewer processes and require 30% fewer parameters. This suggested method detects a person's face mask with excellent 99.6 percent accuracy using the proposed MobileNetV2, saving time and effort.

B. Organization of this Article

The concept and problem with the present face mask detection model are explained in Section 1 of the introduction, and how our suggested solution intends to solve them. The following are the remaining sections of the paper: Section 2 includes related work on object identification models which uses deep neural network (DNN) models; The proposed MobilenetV2 classifier is described in Section 3; Section 4 explains the experimental results of the proposed research finding; and Section 5 concludes the proposed face mask detection model.

II. Related Works

Pattern and object recognition are two processes that a computer vision (CV) must utilize. Object recognition includes both image segmentation and object detection [12]. DNN classifier from surveillance devices is used and identify the mask covering the face in the computer vision application, which is done with the help of an efficient object identification system. The many DNN classifiers used for object detection are highlighted in this section.

Ravindran et al. [6] suggested Deep Neural Networks (DNN), for multi-object identification and multi-object tracking for handling the key difficulties in driving situations for autonomous vehicles. In this system, sensing modalities such as cameras, RADAR, LiDAR, ultrasonic sensors, and GPS are combined with DNN. This study concludes that camera and RADAR fusion benefit from a more optimal approach with DNN.

Jiao et al. [7] suggested various single-stage and two-stage object identification models. In this object identification model, two components are used: Region of Interest pooling, which collects features from the input component, and classification and bounding-box regression. Two-stage object identification techniques are Region-based convolutional neural network (RCNN) and HOG-like features on PASCAL VOC datasets. The RCNN model has four components. The first component generates an independent category class for regions. From each indicated region, the second component retrieves a fixed-length feature vector. The third component uses a class-specific linear support vector algorithm (SVA) for the classification of the object in a single image. The final component, precise bounding-box predictions are made using a bounding-box regression technique.

Fast region-based CNN (FRCNN) [8] processes the region of interest (ROI) pooling layer after features extraction and uses bounding box regression fully connected layers for object classification. Faster RCNN [9] significantly enhances the region-based CNN. Slow and requiring the same amount of processing time as the detection network, the FRCNN uses selective search. Faster RCNN is a fully convolutional network, which accurately predicts ROI with a wide range of scales and aspect ratios. Faster RCNN is improved by Mask R-CNN [10], which is utilized for segmentation jobs. Mask R-CNN is considered a more accurate object detector despite the addition of a parallel mask branch.

The one-stage object detector is suitable for real-time devices and directly detects an object from input images without requiring a Region of Interest pooling stage. You Only Look Once (YOLO) [11] is a type of one-stage object detection model. YOLO performs object detection and predicts the object at every point on the actual image by splitting it into grids and predicting the bounding boxes and probability

of each region. YOLO performs object detection on the target image by dividing the image into grids and predicting the bounding boxes and probability of each grid. YOLO delivers a substantially faster object detection speed than two-stage detectors without a Region of Interest pooling stage. YOLO has been incrementally improved with the one-stage object detector algorithms YOLOv2 [12] and YOLOv3 [13].

Retina Mask, developed by Jiang et al. [14], combined a Feature Pyramid Network with a content attention technique and used ResNet for classification tasks to perform on both high- and low-computation hardware. Loey et al. [15] developed a hybrid transfer learning model for the Real-Time Face Dataset with an accuracy of 99 percent, while it is 99.49 percent for the Masked Face Dataset. Both earlier studies focused on the accuracy of mask recognition, and the speed of detection was never addressed. Furthermore, these systems could only identify masks and not determine whether they were applied correctly. Cabani et al. [16] developed a mobile application and demonstrated how to wear masks properly by determining whether or not a person's mask covered their mouth and nose using the MaskedFace-Net algorithm. This method increases training accuracy by utilizing a sizable dataset of 137,016 mask-based facial images.

III. Proposed Face Mask Detection Method

The MobileNetV2 [17] algorithm is used in this suggested face mask detection model, which aimed to distinguish between masked and unmasked faces. The proposed MobileNetV2 creates a face mask detection model using input features extracted from a freely available dataset collected using video and web cameras. This model detects whether a person is wearing a mask from the region of interest from the facial image (test image), and then classifies them as having a "Mask with id/name" and "No mask with id/name". The alarm will beep if no mask is worn. The steps taken to build this face mask detection model using the MobilenetV2 algorithm are discussed in Algorithm 1. Figure 29.1 displays the suggested face mask detection model workflow along with MobilenetV2 components.

Algorithm 29.1: Face mask detection using Mobilenetv2 algorithm	
Input	Input dataset consists of N training samples.
Output	Face mask detection model
Step 1	Initialize the video/web cam for detecting face mask that person wearing mask or not.
Step 2	Extract image from the video frame
Step 3	Load face detection model and identify the faces
Step 4	Apply image pre-processing and extract features (ROI) of facial image
Step 5	Build a Mobilenetv2 model to detect whether a person is wearing a mask or not.

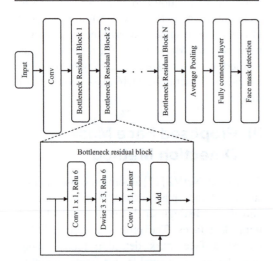

Fig. 29.1 Workflow of the proposed model and MobileNetV2 components

Source: Authors

A. Mobilenetv2

MobileNetV2 (MNV2) is a type of CNN with an inverted residual connection between bottleneck layers. The intermediate expansion layers have depth-wise convolutions, which enhances model accuracy. The MNV2 has 19 residual bottleneck layers after the fully convolutional layer with 32 filters.

Components of Mobilenetv2 model

Conv 1 x 1: In the convolution process, a small weighted matrix moves over the input data and multiplies the elements with the input, and adds the results to the output. A 1 x 1 convolution layer reduces dimensionality and provides efficient low-dimensional embeddings, and applies non-linearity convolutions. create output pixels with the desired output depth, and input pixels with all of their channels are converted.

Bottleneck Residual Block: MNV2 utilizes this Inverted Residual block for enhancing the model efficacy through a 1x1 convolution which reduces the number of input channels by combining input and output. An Inverted Residual Block performs a narrow -> broad -> narrow technique, hence the inversion. A 3x3 depth-wise convolution layer is placed after the 1x1 layer to further minimize the number of parameters.

Average Pooling: Average Pooling is a feature down-sampling method that computes the average value for regions and utilizes it to generate a down-sampled feature map. It is used after a convolutional layer and extracts input features more smoothly. Face mask detection is carried out using a fully connected layer, which multiplies the input by a weight matrix before adding a bias vector.

B. Procedure to Setup Face Mask Detection Model

A digital webcam camera, a computer, and a speaker make up the entire system. MobileNetV2 is used as the detection algorithm for identifying face masks. The model starts to work in real-time with high-resolution computation with the help of GPU when the camera observes the user removing the face mask. The speaker will be instructed to remind the user to wear their face mask in front of the camera if the system notices that they are not doing so. This will be done repeatedly until the user correctly wears their face mask.

IV. Experimental Results and Discussions

The face mask detection model uses the MobileNetV2 algorithm where the data to be trained is pre-processed and then fed into the model to train. The performance of this model is observed using the measures such as the epochs, the size of the dataset, and the number of hidden layers. The confusion matrix evaluates the model performance. True Positive (True Pos), False Positive (False Pos), False Negative (False Neg), and True Negative (True Neg) are determined as four components of the confusion matrix, and the model accuracy is calculated in following Equation:

$$\text{Model Accuarcy} = \frac{(\text{True Pos} + \text{True Neg})}{\begin{array}{c}(\text{True Pos} + \text{False Pos} \\ + \text{True Neg} + \text{False Neg})\end{array}}$$

The Model Accuarcy is the ability of the classier to differentiate between the classes correctly and this model attained high accuracy of 99.6%. The results for various epoch(s) by the MobileNetV2 model are shown in Table 29.1 and Fig. 29.2.

Table 29.1 Output of the proposed MobileNetV2 model

Epoch(s)	Model accuracy	Model loss	Test Accuracy	Test Loss
1	0.9628	0.4017	0.9922	0.0818
2	0.9736	0.1524	0.9935	0.0528
3	0.9789	0.0983	0.9922	0.0408
7	0.9809	0.0777	0.9935	0.0374
12	0.9845	0.0684	0.9935	0.0353
15	0.9845	0.0576	0.9922	0.0327
16	0.9819	0.0536	0.9922	0.0311
18	0.9855	0.0551	0.9935	0.0357
19	0.9871	0.0471	0.9948	0.0327
20	0.9885	0.0446	0.996	0.0311

Source: Authors

Fig. 29.2 Sample output of the suggested MNV2 face mask detection model

Source: Authors

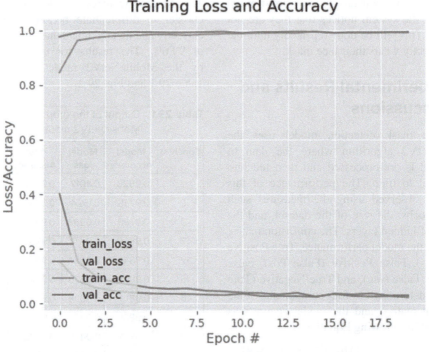

Fig. 29.3 Accuracy and loss graph of the proposed face mask detection model

Source: Authors

Figure 29.3 illustrates the model accuracy and loss plots of the suggested MNV2 model.

V. Conclusion

This proposed MNV2 architecture incorporates deep learning concepts to prevent the spread of the coronavirus in the community by recognizing masks covering faces in public places. The proposed method efficiently handles object detection and contributes to high object detection accuracy of 99.6 percent and decreased object detection time by using an ensemble of single and two-stage detection techniques.

Reference

1. Garg, Dweepna, Parth Goel, Sharnil Pandya, Amit Ganatra, and Ketan Kotecha. "A deep learning approach for face detection using YOLO." In 2018 IEEE Punecon, pp. 1–4. IEEE, 2018.

2. Yang, Wang, and Zheng Jiachun. "Real-time face detection based on YOLO." In 2018 1st IEEE international conference on knowledge innovation and invention (ICKII), pp. 221–224. IEEE, 2018.

3. Kim, Chan, Hyun Mi Kim, Chun-Gi Lyuh, Yong Cheol Peter Cho, Jeongmin Yang, Jaehoon Chung, Kyoung Seon Shin et al. "Implementation of Yolo-v2 image recognition and other testbenches for a CNN accelerator." In 2019 IEEE 9th International Conference on Consumer Electronics (ICCE-Berlin), pp. 242–247. IEEE, 2019.

4. Deore, Gayatri, Ramakrishna Bodhula, Vishwas Udpikar, and Vidya More. "Study of masked face detection approach in video analytics." In 2016 Conference on Advances in Signal Processing (CASP), pp. 196–200. IEEE, 2016.

5. Heydarzadeh, Yasaman, Abolfazl Toroghi Haghighat, and Nazila Fazeli. "Utilizing skin mask and face organs detection for improving the Viola face detection method." In 2010 Fourth UKSim European Symposium on

Computer Modeling and Simulation, pp. 174–178. IEEE, 2010.

6. Ravindran, Ratheesh, Michael J. Santora, and Mohsin M. Jamali. "Multi-object detection and tracking, based on DNN, for autonomous vehicles: A review." IEEE Sensors Journal 21, no. 5 (2020): 5668–5677.

7. Jiao, Licheng, Fan Zhang, Fang Liu, Shuyuan Yang, Lingling Li, Zhixi Feng, and Rong Qu. "A survey of deep learning-based object detection." IEEE access 7 (2019): 128837–128868.

8. Nsaif, Ahmed Khudhur, Sawal Hamid Md Ali, Khider Nassif Jassim, Asama Kuder Nseaf, Riza Sulaiman, Ammar Al-Qaraghuli, Omar Wahdan, and Nazrul Anuar Nayan. "FRCNN-GNB: Cascade faster R-CNN with gabor filters and naïve Bayes for enhanced eye detection." IEEE Access 9 (2021): 15708–15719.

9. Sun, Xudong, Pengcheng Wu, and Steven CH Hoi. "Face detection using deep learning: An improved faster RCNN approach." Neurocomputing 299 (2018): 42–50.

10. Bharati, Puja, and Ankita Pramanik. "Deep learning techniques—R-CNN to mask R-CNN: a survey." Computational Intelligence in Pattern Recognition (2020): 657–668.

11. Wu, Dihua, Shuaichao Lv, Mei Jiang, and Huaibo Song. "Using channel pruning-based YOLO v4 deep learning algorithm for the real-time and accurate detection of apple flowers in natural environments." Computers and Electronics in Agriculture 178 (2020): 105742.

12. Loey, Mohamed, Gunasekaran Manogaran, Mohamed Hamed N. Taha, and Nour Eldeen M. Khalifa. "Fighting against COVID-19: A novel deep learning model based on YOLO-v2 with ResNet-50 for medical face mask detection." Sustainable cities and society 65 (2021): 102600.

13. Hurtik, Petr, Vojtech Molek, Jan Hula, Marek Vajgl, Pavel Vlasanek, and Tomas Nejezchleba. "Poly-YOLO: higher speed, more precise detection and instance segmentation for YOLOv3." Neural Computing and Applications 34, no. 10 (2022): 8275–8290.

14. Jiang, Mingjie, Xinqi Fan, and Hong Yan. "Retinamask: A face mask detector." arXiv preprint arXiv:2005.03950 (2020).

15. Loey, Mohamed, Gunasekaran Manogaran, Mohamed Hamed N. Taha, and Nour Eldeen M. Khalifa. "A hybrid deep transfer learning model with machine learning methods for face mask detection in the era of the COVID-19 pandemic." Measurement 167 (2021): 108288.

16. Cabani, Adnane, Karim Hammoudi, Halim Benhabiles, and Mahmoud Melkemi. "MaskedFace-Net–A dataset of correctly/incorrectly masked face images in the context of COVID-19." Smart Health 19 (2021): 100144.

17. Sandler, Mark, Andrew Howard, Menglong Zhu, Andrey Zhmoginov, and Liang-Chieh Chen. "Mobilenetv2: Inverted residuals and linear bottlenecks." In Proceedings of the IEEE conference on computer vision and pattern recognition, pp. 4510–4520. 2018.

30

Artificial Intelligence in the Field of Textile Industry: A Systematic Study on Machine Learning and Neural Network Approaches

Dennise Mathew[1]

Research Scholar, Department of Computer Science and Engineering,
Kalasalingam Academy of Research and Education

N. C. Brintha[2]

Associate Professor, Department of Computer Science and Engineering,
Kalasalingam Academy of Research and Education

Abstract—Artificial Intelligence with Machine Learning and Neural Network approaches has explored major application areas from marketing to research. AI is being used in a variety of business areas (medical care, Information Technology, power, agribusiness, clothing, engineering, intelligent cities, touristy, and transportation), as well as administration and marketing operations (Human Resources, client services, reverse engineering, sound body and security, project administration, recommendations, organizational monitoring, and automation implementation)[2]. Nowadays, the Textile Industry is highly automated with different AI algorithms. In textile fabric inspection, artificial intelligence techniques such as neural networks and artificial vision are utilised to improve the performance of productive systems[3]. This study presents in detail how various Intelligent Learning Techniques such as Support Vector Machine, Bayesian Classification, Decision Tree, and K-Nearest Neighbour are implemented in the textile industry to overcome the problems with the traditional methods they are implementing. It also illustrates clearly the various Neural Network approaches such as Artificial Neural Networks, and Convolutional Neural Networks to provide better performance in textile applications. We conclude this work with comparisons of different Machine Learning and Neural Network Approaches, their advantages and how they can be used to overcome the current challenges in the field of the Textile Industry. The detailed survey represents the optimal strategy that can be implemented in the textile industry as neural network approaches.

Keywords—Support vector machine, Bayesian classification, Decision tree, K-Nearest neighbour, Artificial neural network

---

[1]dennisemathew@gmail.com, [2]n.c.brintha@klu.ac.in

DOI: 10.1201/9781003388913-30

I. Introduction

To optimize the design, production and quality processes in the textile industry, different Machine Learning and Neural Network approaches can be implemented. The simulations can be developed with suitable algorithms with existing textile-based data to erase the problems in the traditional methods of design, production and quality processes of the end product [1]. AI can be integrated with the textile industry applications for fault detection, pattern checking and colour matching for production processes. Most of the traditional computational and analytical methods have been utilised to process textile data in numerous textile research [3]. In recent years, Artificial Intelligence (AI) is being discussed as they possess a significant effect on manufacturing processes. However, the majority of research has concentrated on different AI algorithms, rather than the required features for AI implementation at the corporate level [1]. There are researches that give a general overview of intelligent machine algorithms implemented in many disciplines, like machine learning in education and deep learning in healthcare; but less review of machine learning and deep learning uses in the textile industry has been found [10].

II. Literature Survey

The different algorithms reviewed for the need of the textile industry in the areas such as design, production, and testing are SVM, Bayesian Classification, Decision Trees, K-Nearest Neighbour and ANN.

A. SVM in the Fabric Industry

Zhang et al. proposed a yarn-dyed woven fabric defect model using an SVM classifier. 180 defective images of yarn-dyed fabrics are used in the model. The test results are more than 91% accurate in detecting the defects in yarn-dyed fabrics [13]. Sun J et al. proposed a model

related to quality management for predicting the wrinkling appearance of the fabric. The model is implemented using 300 images with various weave architectures, fibre contents, colours, and laundry cycles. The validation results demonstrate that this model accurately classifies 78% of occurrences.[14]. Zhang et al. used SVM techniques for forecasting colour variations in the examined dyed clothes [13]. Nurwaha et al. proposed an AI-based agent for calculating grade check-in yarn by comparing six different techniques with SVM. When the performance is estimated, the lowest error value was provided with SVM [12].

B. Bayesian Classification

It is a well-known classification algorithm, that implements Bayes theorem to estimate conditional probabilities to find the unknown class values of samples [13]. It can be used to solve various problems in the textile industry such as estimation of fabric roughness, colour variations in dyed clothes etc[13]. Kim et al. suggested a model for evaluating fabric pilling that combined multiple classification techniques such as Naive Bayes, minimum distance, k-nearest neighbours, and Neural Networks. Using these classification methods, it is possible to predict the grades of various fabrics with an accuracy rate of >95% [14]. Hu et al. created a model that classifies fabric grades using morphological fractals and Bayesian Classification[15].

C. Decision Trees

Hsu et al. proposed a model with a decision tree called a new pants sizing system for manufacturing pants for army soldiers in Taiwan by determining different pant sizes [16]. The dataset used consists of features collected from 610 soldiers which results in 160000 pieces of data. In this case, the Classification and Regression algorithm is used in the Decision Tree to discover the different body patterns. Zakaria also proposed a model with a decision tree to design a uniform measurement system for the students [17]. The dataset consists of features collected from 1001 school girls from 29 different schools. The different challenges undergone when decision trees are used in the textile industry are overfitting, tree pruning, noisy data etc. The Random Forest algorithm can be used in future to overcome these problems in the textile industry.

D. K-Nearest Neighbour

Mariolis et. al., developed a model named scam quality control system to deal with surface roughness estimation using the KNN algorithm [18]. The dataset includes a total of 211 seam varieties collected from two classifications of cloths. When production is evaluated, the accuracy rate is 81.04%. Alternative performance metrics such as Euclidean, City-Block, and others can be utilised to discover the classification performance accuracy for the dataset when choosing the appropriate distance measure. Before implementing the KNN algorithm, PCA or filter-based feature selection methods can be employed to decrease the features. By weighting features in textile research, it is feasible to improve classification accuracy by identifying the weights of features.

III. Neural Network in the in the Fabric Industry

Neural Networks play a major role in the field of the textile industry for fault diagnosis [30]. The application of neural networks can be broadly categorised as follows:

1. *Prediction:* Neural networks are used to predict an output from inputs. For example, picking the best stocks in the market.

2. *Classification:* used to find an unknown pattern in data

3. *Data Association:* used mainly for Feature Extraction.

4. *Data Conceptualization:* used to infer relationships from the input data.

5. *Data Filtering:* used to remove noisy data or errors from the input data.

A. ANN

An Artificial Neural Network can be implemented in the textile industry in various fields such as production, maintenance, cost effectiveness, statistical data analysis, quality control etc. It comprises the manipulation of provided data or measurements, which are frequently represented as a signal, an image, or other visual representations.

B. Applications of Artificial Neural Network (ANN) in Fiber Engineering

Application of artificial neural network in the fibre sector

One of the most common issues is the classification of animal fibres. ANN is more precisely utilised in cotton grading. If we consider synthetic fibres, ANNs have helped to discover production control parameters and predict the qualities of melt spun fibres. For the recognition of fabric fibres, ANNs were utilised in conjunction with NIR spectroscopy.

ANN Application in yarn sector

Controlling yarn quality requires the application of artificial intelligence (AI) in the improvising of top roller diameter and the analysis of the rotating balloon during the lead rotating process. Spinning staple fibres for yarn manufacture is a multistage process with many variables that direct the qualities of the finished yarn, namely the spun yarn. The hairiness of worsted wool and cotton yarns has been predicted using artificial neural networks (ANNs). ANNs are used to anticipate the uniformity of ring-spun worsted and cotton yarns, as well as the uniformity of blended rotor yarns. ANN is implemented for more precise hooking of two yarn ends. To extract faults in yarn packages, image processing technology is combined with neural networks, which are then utilised to differentiate the quality rates of the yarn bundles [29].

ANN Application in the fabric sector

The identification and perception of designs on a fabric belong to the common complicated range of issues that can be tackled by using ANNs. The ANN algorithm is used to forecast the cloth drape. Fabric handling is increased as a result of the use of ANNs. When an ANN is used in the textile sector to forecast and analyse the drying process, the system is trained using gradient descent and optimised using Adam with default parameters[19] [32]. The worsted fabrics' shear stiffness and compression characteristics have been satisfactorily simulated.

ANN Application in apparel

ANN systems have been used to predict the production of materials in the fabric industry and fit fabric design.

ANN Application in nonwoven

Wavelet texture analysis and Bayesian ANNs were used to create a visual inspection system. Nonwoven fabric structure properties relationships, the development of an improvised prediction system with high performance and accuracy, the compression properties of needle-punched nonwoven fabrics, the simulation of the spinning of spun bonding nonwoven process, and the objective examination of pilling on nonwoven fabrics are all areas where ANN methods are used.

B. CNN

The trained neural network based on CNN to identify cloth faults obtains the location of defects samples with high accuracy. For this model, Yang. et. al., divided the network into two divisions, segmentation network and decision network. The results from the segmentation network are given to the decision network for further processing such as training and learning to detect the spots of faults in fabric with higher precision [20]. To increase the precision of garment detection categorization, an ensemble learning-based CNN is utilised to categorise five types of fabric defects. [21]. Based on the

Table 30.1 Comparison of different application areas in the textile industry based on ANN [15] [16] [17]

Application Area	Advantages
Application of artificial neural network in fibre sector	• Forecast copolymer composition. • Classification of animal fibres. • Cotton grading. • Discover production control parameters and predict the qualities of melt spun fibres.
ANN Application in yarn sector	• Improving yarn quality control throughout the carding process. • The rate of warp breaking during weaving is reduced. • Cotton fibre cost minimization.
ANN Application in fabric sector	• Inspecting clothing for flaws. • Forecast the cloth drape. • Forecast of tensile strength and the initial stress-strain curve of textiles.
ANN Application in apparel	• Forecasts about the performance of the fabric-making process, the use of sewing thread, the comfort of the fashion sense, the length of the cutting process, and clothing sales. • Online fabric classification, classification of fabric flaws, classification of fabric uses, and classification of seam features. • Used to assess seam puckering and optimise the sewing thread.
ANN Application in nonwoven	• Inspected to guarantee that the given material is of high quality. • The recognition and classification of nonwoven online photos.

traditional deep learning methods, the YOLOv4 algorithm is proposed to be used in many textile industries. In the neural network model, a series of convolutional layers are added for reducing the channel numbers to improve the image quality for detection[22]. The image enhancement technique is employed with the convolution neural network (CNN) algorithm to extract effective fabric features, decrease superfluous feature extraction, and finally reduce the fault rate caused by the overfitting of the neural network model [23]. The Cascade R-CNN can be used to design defect detection model design and network structures. It uses the detection characteristics of FPN to detect textile defects with increased defect detection of 4.09% to 95.43% [24]. Mask R-CNN based on instance segmentation can be used to recognise silhouette attributes [25][31]. Using IoT and Alexnet deep learning algorithms, Sun-Kuk Noh created an intelligent categorization system for used garments. [26]. The CNN along with the pyramid algorithm can be used in the textile industry [27] for object recognition. Cascade R-CNN is used to detect fabric defects with improved precision [28].

IV. Conclusion

As a result, over the last decade, the textile sector has created AI-based information technology solutions. This paper provides an overview of ML and DL approaches created expressly for textile applications, as well as a description of some experimental work. It illustrates how clustering and classification techniques in the textile business might be utilised to address an issue. In the realm of textile engineering, the use of Artificial Neural Networks (ANN) and Convolutional Neural Networks (CNN) opens up new possibilities. In the field of fabric engineering, artificial neural networks are rapidly being employed to solve a variety of difficulties. Textile businesses in wealthy countries have begun to take advantage of these approaches. Because most textile processes and related quality assessments are nonlinear, artificial neural networks are used in textile technology.

References

1. Steffen Kinkel, Marco Baumgartner, Enrica Cherubini: Prerequisites for the adoption of AI technologies in manufacturing – Evidence from a worldwide sample of manufacturing companies, Technovation, 2021
2. Marija Cubric, Drivers: Barriers and social considerations for AI adoption in business and management: A tertiary study, Technology in Society, Volume 62, 2020
3. Pereira F., Carvalho V., Vasconcelos R., Soares F. (2022) A Review in the Use of Artificial Intelligence in Textile Industry. In: Machado J., Soares F., Trojanowska J., Yildirim S. (eds) Innovations in Mechatronics Engineering. icieng 2021. Lecture Notes in Mechanical Engineering. Springer, Cham. https://doi.org/10.1007/978-3-030-79168-1_34
4. Pelin Yildirim,Derya Birant2, Tuba Alpyildiz. Data mining and machine learning in textile industry.WIREs Data Mining Knowl Discov 2017.
5. Han, J., Kamber, M., Pei, J.: Data Mining: Concepts and Techniques. ElsevierbInc., Amsterdam (2012). https://doi.org/10.1016/C2009-0-61819-5.
6. Domingos, P.: A few useful things to know about machine learning. Commun.bACM 55(10), 78–87 (2012).
7. Gibert, K., Izquierdo, J., S`anchez-Marr`e, M., Hamilton, S.H., Rodr´ıguez-Roda, I., Holmes, G.: Which method to use? An assessment of data mining methods in environmental data science. Environ. Model. Softw. 110, 3–27 (2018).
8. Neha Chauhan, Nirmal Yadav and Nisha Arya. 2018. Applications of Artificial Neural Network in Textiles. Int.J.Curr.Microbiol. App.Sci. 7(04): 3134-3143. doi: https://doi.org/10.20546/ijcmas.2018.704.356
9. Kumar C. and Dhinakaran M. 2015. Scope for Artificial Neural Network in Textiles. IOSR Journal of Polymer and Textile Engineering. 2(1). 34–39.
10. Zhang J, Yang C. Evaluation model of color difference for dyed fabrics based on the support vector machine. Text Res J 2014, 84: 2184–2197. https://doi.org/10.1177/0040517514537372.

11. Sun J, Yao M, Xu B, Bel P. Fabric wrinkle characterization and classification using modified wavelet coefficients and support-vector-machine classifiers. Text Res J 2011, 81: 902–913. https://doi.org/10.1177/0040517510391702.

12. Nurwaha D, Wang XH. Using intelligent control systems to predict textile yarn quality. Fibres Text East Eur 2012, 20: 23–27.

13. Cichosz P. Naïve Bayes classifier. In: Data Mining Algorithms: Explained Using R. Chichester: John Wiley & Sons, Ltd; 2015. https://doi.org/10.1002/ 9781118950951.ch4.

14. Kim S. C., Kang T. J. Fabric surface roughness evaluation using wavelet-fractal method. Part II: fabric pilling evaluation. Text Res J 2005, 75: 761–770. https://doi.org/10.1177/0040517505059209.

15. Hu J, Xin B, Yan H. Classifying fleece fabric appearance by extended morphological fractal analysis. Text Res J 2002, 72: 879–884. https://doi.org/10.1177/004051750207201005.

16. K. Vijay, R. Vijayakumar, P. Sivaranjani, and R. Logeshwari, "Scratch detection in cars using mask region convolution neural networks," Adv. Parallel Comput., vol. 37, pp. 575–581, 2020.

17. R. Vijayakumar, K. Vijay, P. Sivaranjani, and V. Priya, "Detection of network attacks based on multiprocessing and trace back methods," Adv. Parallel Comput., vol. 38, pp. 608–613, 2021.

18. Mariolis IG, Dermatas ES. Automated assessment of textile seam quality based on surface roughness estimation. J Text Inst 2010, 101:653–659. https://doi.org/10.1080/00405000902732883.

19. K. Taur, X. Deng, M. Chou, J. Chen, Y. Lee and W. Wang, "A study on Machine Learning Approaches for Predicting and Analyzing the Drying Process in the Textile Industry," 2019 International Automatic Control Conference (CACS), 2019, pp. 1–5, doi: 10.1109/CACS47674.2019.9024364.

20. Y. Huang, J. Jing and Z. Wang, "Fabric Defect Segmentation Method Based on Deep Learning," in IEEE Transactions on Instrumentation and Measurement, vol. 70, pp. 1–15, 2021, Art no. 5005715, doi: 10.1109/TIM.2020.3047190.

21. X. Zhao, M. Zhang, and J. Zhang, "Ensemble learning-based CNN for textile fabric defects classification", International Journal of Clothing Science and Technology, vol. 33, no. 4, pp. 664–678, Jan. 2021.

22. N. Natheeswari, P. Sivaranjani, K. Vijay, and R. Vijayakumar, "Efficient data migration method in distributed systems environment," Adv. Parallel Comput., vol. 37, pp. 533–537, 2020.

23. W. Yu, D. Lai, H. Liu and Z. Li, "Research on CNN Algorithm for Monochromatic Fabric Defect Detection," 2021 6th International Conference on Image, Vision and Computing (ICIVC), 2021, pp. 20–25, doi: 10.1109/ICIVC52351.2021.9526981.

24. Y. Li and Z. Wang, "Research on Textile Defect Detection Based on Improved Cascade R-CNN," 2021 International Conference on Artificial Intelligence and Electromechanical Automation (AIEA), 2021, pp. 43–46, doi: 10.1109/AIEA53260.2021.00017.

25. S. Sreesubha, R. Babu, R. Vijayakumar, and K. Vijay, "An efficient data hiding approach on digital color image," Adv. Parallel Comput., vol. 37, pp. 552–557, 2020.

26. Sun-Kuk Noh, "Recycled Clothing Classification System Using Intelligent IoT and Deep Learning with AlexNet", Computational Intelligence and Neuroscience, vol. 2021, Article ID 5544784, 8 pages, 2021. https://doi.org/10.1155/2021/5544784

27. V. K, K. Jayashree, V. R and B. Rajendiran, "Forecasting Methods and Computational Complexity for the Sport Result Prediction," 2022 International Conference on Electronic Systems and Intelligent Computing (ICESIC), 2022, pp. 364–369, doi: 10.1109/ICESIC53714.2022.9783514.

28. Li Feng, Li Feng. Bag of tricks for fabric defect detection based on Cascade R-CNN. Textile Research Journal. 2021; 91(5-6):599-612. doi:10.1177/0040517520955229

29. V. R, R. S, V. K and S. P, "Integrated Communal Attentive & Warning System via Cellular Systems," 2022 International Conference on Electronic Systems and Intelligent Computing (ICESIC), 2022, pp. 370-375, doi: 10.1109/ICESIC53714.2022.9783481.

30. Funa Zhou, Shuai Yang, Hamido Fujita, Danmin Chen, Chenglin Wen, Deep learning fault diagnosis method based on global optimization GAN for unbalanced data,Knowledge-Based Systems,Volume 187,2020,104837,ISSN 0950-7051,https://doi.org/10.1016/j.knosys.2019.07.008.

31. K. R. Sowmia, S. Poonkuzhali," Artificial Intelligence in the field of Education:A Systematic Study of Artificial Intelligence Impact on Safe Teaching Learning Process with Digital Technology", Journal of Green Engineering, Vol 10, Issue 4, April 2020, pp no: 1566–1583.

32. Sowmia, K. R., Poonkuzhali, S., Jeyalakshmi, J. (2023). Sentiment Classification of Higher Education Reviews to Analyze Students' Engagement and Psychology Interventions Using Deep Learning Techniques. In: Zhang, YD., Senjyu, T., So-In, C., Joshi, A. (eds) Smart Trends in Computing and Communications. Lecture Notes in Networks and Systems, vol 396. Springer, Singapore. https://doi.org/10.1007/978-981-16-9967-2_25

Recent Trends in Computational Intelligence and its Application – Sugumaran D. et al. (eds)
© 2023 Taylor & Francis Group, London, ISBN 978-1-032-48410-5

31

An Adaptive Feature Weight Using Hybrid Fire Fly with Harmony Optimization Algorithm on Feature Selection and Evaluating the Performance on Multi-Layer Perceptron with Hepatocellular Carcinoma Data

C. Saranya Jothi[1], Carmel Mary Belinda[2]
Vel Tech Dr. Rangarajan Dr. Sagunthala R&D Institute of Science and Technology

Eskandhaa K. Vel[3]
SRM Institute of Science and Technology

Karthiha K. Vel[4]
Anna University, Department of Computer Science and Engineering

Abstract—Hepato Cellular Carcinoma (HCC) is one of the primary liver cancer. It typically affects persons who have chronic liver diseases like cirrhosis, which is brought on by hepatitis B and C. It is the most common type of liver cancer also it is of long-term disease. The motivation of the work is to in early stages wants to identify the diseases because the annual mortality rate is increasing. By recognizing early on, we can lower the death rate. The objective of this work is to reduce the features with identify the diseases in their early stages. In the first phase, we find out the attribute weight for each feature by using Information Gain (IG). In the second phase, we applied the hybrid of Fire Fly with Harmony Optimization Algorithm (FFHOA) to reduce the high-level feature space to low dimensional space to improve the performance. In the third phase, Multi-Layer Perceptron (MLP) classifier is used for evaluating the performance with 12 metrics. It is justified that the above-mentioned FFOHA-MLP outperform well on reduced feature with good performance. Datasets are taken from the UCI machine learning repository. The dataset contains 165 instances and 50 attributes.

Keywords—Fire fly algorithm, Harmony algorithm, Feature selection, Multi-layer perceptron, HCC data

[1]saranyajothi22@gmail.com, [2]carmelbelinda@gmail.com, [3]kvdkishorekumarrc@gmail.com, [4]carmeltitus1971@gmail.com

DOI: 10.1201/9781003388913-31

I. Introduction

Hepato Cellular Carcinoma is sometimes called HCC. It is the third most prevalent type of cancer in the world and one of the main reasons why people die from cancer. The body's largest organ, the liver, is in charge of numerous crucial metabolic processes. It cleans toxins from the blood. Detoxifies chemicals and metabolizes drugs, which leads to the secretion of bile. And it maintains the right level of blood sugar in your body. HCC develops from underlying chronic liver diseases and cirrhosis. Alcohol consumers have a high risk of hepatocellular carcinoma than persons who have a stock of fat in the liver. HCC can be diagnosed by various methods like blood tests, imaging tests, liver biopsy, etc. It can be cured if it is discovered early stage to be cured. Various technologies are used to cure HCC. The primary treatment is a surgical method of removal of the cancerous part of the liver. The next treatment for HCC is liver transplant surgery. The remaining HCC treatments include targeted medication therapy, direct delivery of chemotherapy or radiation to cancer cells, radiation therapy, and the use of heat or cold to kill cancer cells [15]. Immunotherapy. Clinical trials are another technique that enables testing potential liver cancer treatments. In India, approximately 0.7 males and 0.2 percent of every 100,000 people are impacted annually, according to the WHO agency. And according to certain statistics, HCC affects 1.6 percent of the population annually. HCC is responsible for about 7 lakh deaths worldwide. Alejandro et al. [1] reviewed the key developments in therapeutic strategies and the current approaches which are used to sort patients with intermediate-stage HCC and assign the most appropriate treatment. Sahil et al. [2] discussed the latest trend of HCC all around the world and in the US. Also collected evidence about the commentary on the risk factor and listed some of the preventive methods to avoid the incident of HCC. David Anwanwan et al. [3] discussed the problems and current treatment for the bad outcomes for patients with liver cancer. Julien Calderaro et al. [4] developed to translate the scientific knowledge of cancerous biology into clinical practice, which improves medicine for highly aggressive malignant patients. Also, they reviewed the recent data on the current situation, future directions, and challenges. Wan-ShuYang et al. [5] collected and evaluated the observations of diet readings like nutrients, food, food groups, and dietary patterns associated with liver cancer. Ahmed A et al. [6] developed an intelligent analysis tool for feature selection to enhance performance. In this paper, the author used the slime mould algorithm with the firefly algorithm to enhance the convergence for increasing the quality output. Timea Bezdan et al. [7] discussed the flaws and deficiencies in the feature selection method, for selecting the features firefly algorithm is used to increase the accuracy. In this paper 21 datasets are used for implementation. Deepika et al. [8] describe the differential evolution with the wolf firefly algorithm to select the features. And to adjust the weight by the neural network to produce the optimal solution. Jin Hee Bae et al. [9] introduced the harmony search algorithm for feature selection to improve accuracy. For preprocessing the data z-normalization technique was used for cleaning the data. Xiaohua Li et al. [10] discussed the artificial intelligence tool for early recognition. A patient with a life-threatening condition could be diagnosed using artificial intelligence approaches.

A. Contribution of the Work

- In this paper, to decrease the computational time and enhance performance, we merged the Fire-Fly optimization technique with the Harmony algorithm in this study to determine the attribute weight using information gain. Using the HCC dataset, we applied the FFHOA approach to decrease 50 features to 10 features.

- We compared with original features with reduced features. The performance evaluation is done by the MLP algorithm with 12 metrics.

The rest of the paper is organized as follows: In section 2, we described the proposed framework with the hybrid of the harmony algorithm and firefly algorithm. Section 3 illustrated the experimental results with performance analysis. Section 4, shows the conclusion and future work.

II. Proposed Framework

The HCC dataset is provided as the input for the suggested framework. Preprocessing data is a critical operation that improves accuracy. This is because most data is erratic. It occasionally contains both false and missing values. Therefore, feature selection plays

a significant role in preprocessing, directly affecting the model's accuracy. Feature selection involves choosing important features and ignoring unimportant ones. In this work, we have given 50 features as input, applying the firefly optimization algorithm with harmony algorithm combination we get finally 10 features as output. The output of feature selection is given input as a classification model. Machine learning models that predict a class-type outcome are called classification models. An entity's class can be predicted using the attributes of any form of entity, including people, in a classification model. For this classification model, we employed the MLP method. Once the categorization is complete, we use 12 metrics to analyze the performance. Which is displayed in Fig. 31.1.

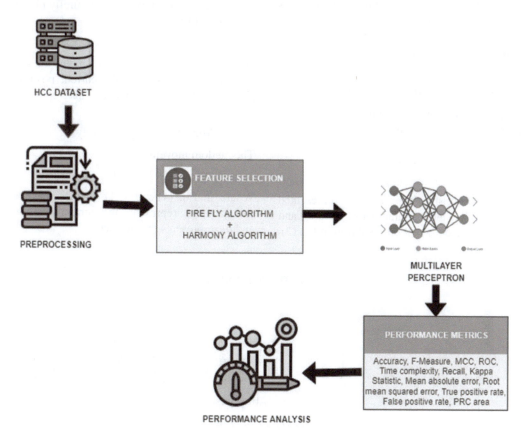

Fig. 31.1 Overview of the proposed architecture

Source: Author

A. Hybrid of FFOH Feature Selection Algorithm

A hybrid algorithm that combines the Fire Fly optimization method and the Harmony algorithm is called the proposed Fire-Fly with Harmony Optimization Algorithm (FFHOA)

Step 1: Initially, find the weight of each attribute by using the information gain formula in the given Eq. (1). Initialize the parameter and take the maximum weight as the first position in the n number of the hormonal search, space is illustrated in Fig. 31.2. This process is repeated until iterative to reach the optimal solution. Harmonies (n) value taken as 50 ($i = 1, 2, 3, 4...n$), Dimension (D) = 50×50 matrix ($j = 1, ... D$), Lower Bound (l) = −10, Upper Bound (u) = 10, Maximum Iteration = 20, PAR = 0.25, HMCR = 0.95. In Eq. (1) IG denotes the Information Gain, H(–class) is the entropy of the dataset, and H(Class/Attributes) is the conditional entropy of the dataset.

IG(Class, Attribute)

$$= H(\text{class}) - H(\text{Class/Attributes}) \quad (1)$$

Step 2: Calculate the fitness value (light intensity) for each firefly.

$$\text{fitness} = \Delta_{R(D)} + \frac{|X|}{|T|} \quad (2)$$

Where the classifier's error rate is represented by $\Delta_{R(D)}$. $|T|$ is the total number of features, and $|X|$ is the size of the subset displayed in Eq. (2).

Step 3: Check while (t<= maximum)

If condition true

Next step.

Here, 1<=50 (true)

Step 4: update the position for each firefly

for i=1 to n-1;

for j=i+1 to n;

if ($I_j > I_i$)

Update position (moving firefly i in the direction of firefly j); go to step 5

end if

end for

end for

If there are no brighter fireflies, move the firefly at random; proceed to step 6.

Step 5: Update position (moving firefly i in the direction of firefly j);

$$x_i(t+1) = x_i(t) + \beta e^{-\gamma r^2}(x_j - x_i) + \alpha \varepsilon_i \quad (3)$$

Where in the Eq. (3) $x_i(t+1)$ denotes the new solution, $x_i(t)$ represents the current position, $\beta e^{-\gamma r^2}$ denotes the attractiveness and represents the step size.

Step 6: Move the firefly randomly

The random movement is represented by:

$$x_i(t+1) = x_i(t) + \alpha \varepsilon_i \quad (4)$$

Where in Eq. (4) $x_i(t+1)$ denotes the new solution, $x_i(t)$ represents the current position, and $\alpha \varepsilon_i$ represents the step size.

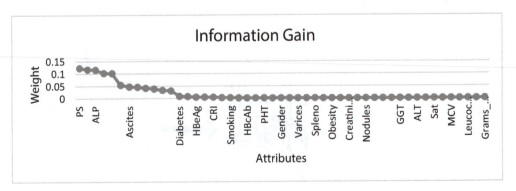

Fig. 31.2 Attribute weight by information gain

Source: Author

Step 7: To find the distance between two fireflies at positions x_i and x_j can be computed in Eq. (5)

$$r_{ij} = \sqrt{\sum(x_{j\_}x_i)^2} \qquad (5)$$

Step 8: Based on position attractiveness function is evaluated by using Eq. (6)

$$\beta = (\beta_0 \times \exp(-\text{gamma} \times r^2)) \qquad (6)$$

In the HCC dataset, we have taken 50 features by applying the FFHO algorithm the original features are reduced to 10 features. The symptoms, endemic, AHT, varices, ps, ascites hemoglobin, total_bil, sat, and class, are reduced from the dataset.

B. Classification Model

We employed an MLP classifier in this study. A feed-forward Artificial Neural Network (ANN) class includes MLP. In the MLP, more than one linear layer containing a mix of neurons is possible. If we use the straightforward example of three layers, the input layer comes first, the output layer comes last, and the middle layer is referred to as the hidden layer. As shown in Fig. 31.3, we feed input data like symptoms, endemic, AHT, varices, PS, ascites, hemoglobin, total_bil, sat, and class into the input layer and extract the output from the output layer. Depending on the complexity of our assignment, we can raise the number of concealed layers as much as we like.

$$h_n = a + \sum_{k=1}^{n}(i_k \times w_{k,n} + b_n) \qquad (7)$$

$$h_1 = a + \sum_{k=1}^{n}(i_1 \times w_{1,1} + i_2 \times w_{2,1} \ldots\ldots + b_1)$$

Here in Eq. 7, a denotes the activation function, w represents the weight in the layer, i denotes the input features, and b represents the bias vector.

III. Experimental Analysis

In the experimental analysis section A, explained the performance analysis of the feature selection and classification model, and

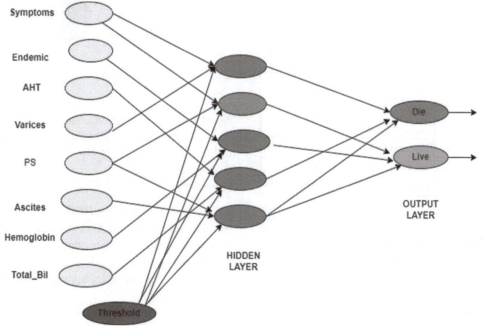

Fig. 31.3 Multilayer perceptron classifier

Source: Author

section B details performance evaluation with different parameter metrics.

A. Performance Analysis of Feature Selection and Classification Method

In the feature selection method, the Firefly-Harmony algorithm provides the best accuracy result of 85% compared to other feature reduction methods. The number of features is reduced from 50 attributes to 10 attributes, and the minimum number of features reduced compared to other search algorithms is shown in Table 31.1. And classification performance is compared with existing techniques as shown in Table 31.2.

Table 31.1 HCC feature performance

Feature Selection Method	MLP accuracy %	Reduced Features
Genetic Search	64.8	14
PSO search	79.2	12
Fire Fly	72	21
Firefly-Harmony	**85**	**10**

Source: Author

Table 31.2 HCC classification performance

Reference	Classification Model	Accuracy
[11]	Modified bat	69.07
[12]	Clinical Decision Support System	82.3
[13]	Decision Tree	69.40
[14]	ANN	68.49
Proposed	**FFOH-MLP**	**85.00**

Source: Author

B. Performance Evaluation With Different Parameter Metrics

In the performance evaluation, different parameter metrics are used in this work. We compared the original feature set with the reduced feature set. In Fig. 31.4(a). Clearly shows that reduced features are given good performance in accuracy, mean absolute error, root means squared, kappa statistic, true positive rate, false positive rate, precision, recall, F-measure, MCC, ROC area, and PRC area. In Fig. 31.4(b). The time to execute the original features has taken 4.63 sec and reduced features have taken 0.02 sec. while reducing the features we can save time for a prediction.

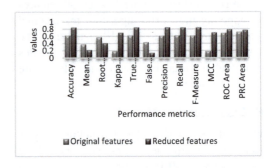

(a)

(b)

Fig. 31.4 Performance evaluation with original features and reduced features (a) Different performance metrics (b) Time complexity

Source: Author

In Fig. 31.5 represents the ROC curve for live and die, classes. Fig. 31.5(a) Denotes the live class, in the x-axis which contains the false positive rate with the y-axis showing the true positive rate. The curve lies between 0 to 1 the value which contains 0.79. Fig. 31.5(b) Denotes the die class. The curve lies between 0 to 1 on the y-axis.

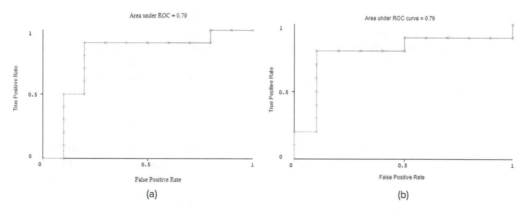

Fig. 31.5 ROC curve for alive and die class (a) ROC curve for live class (b) Roc curve for die class
Source: Author

IV. Conclusion

In this paper, a new FFH algorithm was introduced for reducing the features for performing good accuracy. The goal of the paper is to early-stage want to identify the diseases by decreasing the attributes we can reduce the time complexity. For evaluating the performance we applied different metrics to predicate the disease. Our hybrid algorithm was given good results compared to the existing algorithm. Initially, we have to find out the attribute weight by using information gain, from the position we implement the firefly optimization algorithm for memory storage we used the harmony algorithm. After implementation, the original attribute 50 is reduced to 10 attributes. The MLP algorithm is used for classifying the model, it achieved the best results. In future work, different optimization algorithms can implement to solve the problem.

References

1. Alejandro Forner, Marine Gilabert, Jordi Bruix and Jean-Luc Raoul. "Treatment of intermediate-stage hepatocellular carcinoma", nature reviews clinical oncology, volume 11, September 2014.
2. Sahil Mittal and Hashem B. El-Serag "Epidemiology of Hepatocellular Carcinoma Consider the Population" Clin Gastroenterol Volume 47, Supp. 1, July 2013.
3. David Anwanwan, Santosh Kumar Singh, Shriti Singh, Varma Saikam, Rajesh Singh, "challenges in liver cancer and possible treatment approaches" BBA - Reviews on Cancer (2019).
4. Julien Calderaro, Marianne Ziol, Valerie Parais, Jessic Zucman – Rossi, "Molecular and histological correlations in liver cancer" journal of hepatology volume 71, issue 3, September 2019.
5. Wan-Shui Yang, Xu-Fen Zeng, Zhi-Ning Liu, Qi-Hong Zhao, Yu-Ting Tan, Jing Gao, Hong-Lan Li, and Young-Bing Xiang, " Diet and liver cancer risk: a narrative review of epidemiological evidence", British Journal of Nutrition, volume 124 issue 3.
6. Ahmed A. Ewees, Laith Abualigah, Dalia Yousri, Zakariya Yahya Algamal, Mohammed A. A. Al-qaness, Rehab Ali Ibrahim & Mohamed Abd Elaziz, "Improved Slime Mould Algorithm based on Firefly Algorithm for feature selection: A case study on QSAR model", Engineering with Computers, 31 march (2021).
7. Timea Bezdan, Dusan Cvetnic, Luka Gajic, Miodrag Zivkovic, Ivana Strumberger, Nebojsa Bacanin, "Feature Selection by Firefly Algorithm with Improved Initialization of Strategy", ECBS 2021, May 26–27, 2021, Novi Sad, Serbia, ACM.

8. D. Deepika and Dr. N. Balaji," Effective heart disease prediction with Grey-wolf with Firefly algorithm-differential evolution (GF-DE) for feature selection and weighted ANN classification", computer methods in biomechanics and biomedical engineering, Published in online: 02 Jun 2022.

9. Jin Hee Bae, Minwoo Kim, J.S. Lim, and Zong Woo Geem, "Feature Selection for Colon Cancer Detection Using K-Means Clustering and Modified Harmony Search Algorithm", Mathematics, 9(5), 570, 7 March 2021.

10. Xiaohua Li, Jusheng Zhang and Fatemeh Safara, "Improving the Accuracy of Diabetes Diagnosis Applications through a Hybrid Feature Selection Algorithm", Neural Processing Letters, 27 March 2021, springer.

11. V. Yasaswini, "An Optimization of Feature Selection for Classification using Modified Bat Algorithm", I.J. Information Technology and Computer Science, 2021, 4, 38–46. DOI: 10.5815/ijitcs.2021.04.04

12. V. R. Elgin Christo, H. Khanna Nehemiah, J. Brighty and Arputharaj Kannan, "Feature Selection and Instance Selection from Clinical Datasets Using Co-operative Co-evolution and Classification Using Random Forest", 2020 IETE Journal of Research, https://doi:10.1080/03772063.2020.1713917.

13. Jin H, Kim S, Kim J, "Decision factors on effective liver patient data prediction", int J Biosci Biotechnol 2014, 6(4): 167–78.

14. Weng CH, Huang TCK, Han RP, "Disease prediction with different types of neural network classifiers", Telematics Inf 2016, 33(2): 277–92.

15. Ruba Abu khurma, Ibrahim Aljarah, Ahmad shariah, Mohamed Abd Elaziz, Robertas damasevicius and Tomas krilavicius,"A Review of the Modification Strategies of the Nature-Inspired Algorithms for Feature Selection Problem", Mathematics 2022, 10(3), 464.

Recent Trends in Computational Intelligence and its Application – Sugumaran D. et al. (eds)
© 2023 Taylor & Francis Group, London, ISBN 978-1-032-48410-5

32

Optimal and Dynamic Allocation of Efficient Data Center in Cloud Computing for Big Data Applications

D. Sugumaran[1]

Department of Information Technology, Vel Tech Rangarajan Dr. Sagunthala
R&D Institute of Science and Technology, Chennai, Tamilnadu, India

Rajesh G.[2], Syed Fiaz A. S.[3]

Department of Computer Science and Engineering
Vel Tech Rangarajan Dr. Sagunthala R&D Institute of Science and Technology
Chennai, Tamilnadu, India

Jayanthi K.[4]

Department of Information Technology, Vel Tech Rangarajan Dr. Sagunthala
R&D Institute of Science and Technology, Chennai, Tamilnadu, India

Abstract—Cloud computing plays an important role in sharing of resources concurrently with all big data applications. Handling multiple resources and regulating the path for information transfer to all the big data applications is a tedious task. In cloud computing, user request scheduling is take place by allotting the task to the best data center for execution. Locating the appropriate data center by determining efficiency value is one of the primary difficulties in cloud computing. Despite the fact there are many different types of scheduling algorithms still have certain limitations and a few such limitations have been quantified in existing approaches. In the proposed work an Optimal and Dynamic Allocation Policy (ODAP) has been introduced for allocating the task to the efficient data center (EDC). The effectiveness of the proposed approach was assessed utilising assessment metrics.

Keywords—Cloud computing, Task scheduling, Efficient data center, Big data applications, ODAP

I. Introduction

Cloud computing offers expeditious innovation, flexible resources, and economies of scale by delivering computer services through the Internet ("the cloud"). It aids in the reduction of running costs, the more efficient operation of infrastructure, and the scaling of the organization as the needs change. The cloud eliminates the need for local hardware. Maintaining a server can be time-consuming and challenging. Hardware maintenance may

[1]sugumaran_dhanda@rediffmail.com, [2]rajesh702me@gmail.com, [3]a.s.syedfiaz@gmail.com, [4]jayanthi2contact@gmail.com

DOI: 10.1201/9781003388913-32

increase the cost quite rapidly. The user only pays for the usage of cloud computing. The service provider is responsible for the physical server's load, security, and maintenance among other things. The current service broker policies were used to select the data center. Even if there are many service broker plans available, finding an efficient data center might be tough. Zhang et al. [1] introduced a Performance Optimized Routing strategy to determine the availability of a data center based on response time. Sharma et al. [2] suggested a round-robin routing scheme. Numerous data centers are chosen in the same region in the network area to spread the workload in parallel among all data centers. Manasrah et al. [3] introduced a Variable Service Broker Routing Policy (VSBRP) that is mainly focused on bandwidth for scheduling large-sized data sets tasks. Service brokers construct and organise data centers in various places, and based on the usage, service providers may choose to set up their data centers with various types of equipment. Existing service brokers have several restrictions in picking the most efficient data center while establishing the data center. To overcome this issue in the proposed work, an Optimal and Dynamic Allocation Policy (ODAP) has been implemented to identify the efficient data center.

II. Optimal and Dynamic Allocation Policy

Efficient Data center identification for task allocation is the objective of the proposed work. In the proposed method the efficiency value of each data center has been calculated and maintained in a list. This list will be updated dynamically based on the changes in the data center efficiency value. The data center consisting of maximum efficiency value will be considered as an efficient data center for directing the user request. Calculating the efficiency value helps in identifying a suitable efficient data center. The efficiency value is calculated for all the data centers. The steps to identify the efficient data center.

Step 1: The parameters to calculate data center efficient values are

(a) Utilization of CPU in terms of percentage is calculated as

The utilization of resources is given by dividing the total value of over the completion time of an application in a specific period of time.

$$U_k = \frac{\sum_{i=1}^{DC} B_i}{pc_{\text{time}}} * 100 \qquad (1)$$

where U_k CPU utilization percentage, B_i is the busy time of the processor and pc_{time} is the process complete time of the process.

(b) Throughput (Request/Sec) (TH): the amount of information delivered in a specific period of time.

$$\text{TH}(P_{(u,\,v)}) = \text{TH}(e),\, e\, \mathcal{E}\, P_{(u,\,v)} \qquad (2)$$

where $P(u,\ v)$ = path consists of throughput.

(c) Execution Time

Task execution time $\text{TE}_{\text{exe}}(k,\, l)$ represents the execution time of task k on virtual machine l in a specific period of time, according to the following equation:

$$ET_{\text{exe}}(k,\, l) = \frac{T_{\text{length}(k,\, l)}}{VM_{\text{cpu}(k,\, l)}} \qquad (3)$$

where $T_{\text{length}(k,l)}$ indicates the task required to execute the instruction length per second, $VM_{\text{cpu}(k,l)}$ represents the calculation ability l virtual machine

Step 2: Average of CPU utilization value

The average of CPU utilization is calculated by dividing the summation of CPU percentage of all data center with the total number of the data center.

$$U_{\text{Avg}} = \sum_{1}^{n} U_k / Tot_{DC} \qquad (4)$$

where U_{Avg} is the average value of the data cenetrs.

Step 3: Average of Throughput (TH) Rate

The average rate of throughput is calculated by dividing the summation of the TH rate of all data centers by the total number of data center's.

$$TH_{Avg} = \sum_1^n TH(P(u, v))/Tot_{DC} \qquad (5)$$

where TH_{Avg} is the average of throughput rate used by all the data center's.

Step 4: Calculating Data Center Efficient Value (η).

Efficiency (η) *per* unit of time

$$Eff(\eta)_{ij} = \sum_{i=1}^{pm} \sum_{j=1}^{vm} \left(\frac{CPU\ Util_{ij}}{CPU_{ij}} + \frac{TH\ util_{ij}}{th_{ij}} + \frac{ET}{ET_{ij}} \right)$$

$$(6)$$

where $CPU\ Util_{ij}$ is utilization of CPU of jth VM in ith PM,

$TH\ util_{ij}$ is utilization of throughput of jth VM in ith PM,

ET is execution time of applications in the jth VM of ith PM

Step 5: Threshold value Calculation

Threshold Value = Avg (η)

Pseudo code 1: To calculate the Data Center Efficiency Value

Input: ET_{exe}, TH, U_k, DC list

Output Efficiency value (η)

Start

For all DC_i

 For all PM_j

  For all VM_k

  $Eff(\eta)_i = DC_i$

  $$+ \sum_{j=1}^{pm} \sum_{k=1}^{vm} \left(\frac{CPU\ Util_{jk}}{CPU_{jk}} + \frac{TH\ util_{jk}}{th_{jk}} + \frac{ET}{ET_{jk}} \right)$$

  Update efficiency value in the data base for DC_i

  Loop End

 Loop End

Loop End

Calculate Threshold Value $(T_{val}) = \dfrac{\sum_{i=1}^n DC}{DC_{tot}}$

Update Threshold Value in the data base

End

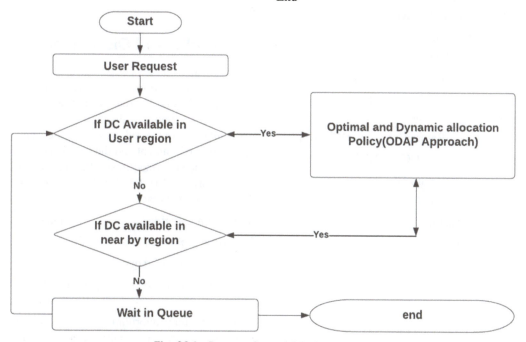

Fig. 32.1 Process flow of ODAP approach

Source: Author

Pseudo code 2: To generate the efficiency list

Input: DC list, U_{Avg}, TH_{Avg}

Output Efficiency list (Eff$_{list}$)

Start

For all DC_i

    If DC Util$_{cpu}$ <= U$_{Avg}$ then

        If DC util$_{TH}$ >= TH$_{Avg}$ then

            Eff(j) = DC_i

        End if

    End if

Loop end

Pseudo code 3: To find the optimized data center

Input: Region as RG, Efficiency list (Eff), T_{val}, Loc

Output: Best DC number

For any new request, ODAP do,

For all RG

If RG =UB (RG) then

    For all Loc in the RG

        If Loc (min$_{dist}$) then

            For all DC in the Loc

                If (DC_eff (i) > T_{val}) then

                    If Available (DC ← DC_lstval (i)) then

                    Return DC_ID ← DC_lstval (i)

                    Else DC ← DC_lstval (i-1)

                    Return DC_ID

                  End If

                End If

                Next for

        End If

    Next for

Else

    For all Loc in other RG

        If Loc (min$_{dist}$) then

For all DC in the Loc

    If (DC_eff (i) > T_{val}) then

        If Available (DC ← DC_lstval (i)) then

            Return DC_ID ← DC_lstval (i)

            Else DC ← DC_lstval (i-1)

            Return DC_ID

        End If

    End If

    Next for

  End If

Next for

User_Req ← wait in queue

End If

End If

Next for

Next for

Note: UB → User Base, DC_id → Data center identification number, Loc-Location in the region

III. Results and Discussion

The simulation-based performance evaluation approach is used to prove the effectiveness of the proposed work. In order to analyze different service broker policies, the cloud analyst tool was suitably configured to reflect the actual operating environment and the location of user bases has been defined in six different regions of the world as R0, R1, R2, R3, R4, R5 and their operating characteristics

Simulation Duration	: 30 Minutes
User Base grouping factor	: 200
Grouping Factor of request	: 200
Executable instruction length per request (bytes)	: 1000-5000
Task Scheduling Algorithms :	Existing methods-Service proximity-based routing, Performance optimization service, Weight-

Table 32.1 Comparison of existing and proposed methods

Broker Policies	Data center	Total No of process/ Sec	Throughput Rate (Mbps)	CPU Utilization (%)/Sec	Execution Time/Sec	Delay time/ms	Efficiency of Data center/Sec
Service Proximity Based Routing	DC5	100	750	99	40.34	25	889.34
Performance Optimized Routing	DC1	100	800	92	22.54	100	914.54
Weight-Based Data Center Selection	DC4	100	1000	78	33.21	100	1111.21
ODAP Approach	DC3	100	1200	72	16.54	22	1288.54

Source: Author

Based data center selection and proposed method Optimal and dynamic allocation policy (ODAP)

IV. Comparison Analysis

The comparison analysis between the existing and proposed methodology has been shown in Table 32.1. The proposed method has produced a satisfactory result when compared to the existing methods. The Throughput rate of the data center, CPU utilization percentage of the data center, execution time taken by each data center for the given user tasks, the delay time of the process and efficiency value for each data center are compared and displayed in Table 32.1.

V. Conclusion

Simultaneous requests from clients will arrive in cloud computing while processing large data applications. So efficient data center selection is critical for every user job to be completed in the desired time and at the lowest possible cost. This restriction is not met by the current server broker policy. An Optimal and Dynamic Allocation Policy (ODAP) has been developed to address this. Based on a comparison of the existing and proposed methods using common parameters, it was determined that the proposed method outperformed the existing method in processing the request as soon as possible, and

the user request was directed to that data center to complete their task with cost effectiveness and minimal response time.

References

1. Zhang, Q.; Cheng, L.; Boutaba, R.: Cloud computing: state-of the- art and research challenges. J. Internet Serv. Appl. 1(1), 7–18, 2010.
2. Sharma,V.; Rathi, R.;Bola, S.K.: Round-robin data center selection in single region for service proximity service broker in CloudAnalyst. Int. J. Comput. Technol. 4(2a1), 254–260, 2013.
3. Manasrah, A.M.; Smadi, T.; ALmomani, A.: A variable servicebroker routing policy for data center selection in cloud analyst. J. King Saud Univ. Comput. Inf. Sci. 29(3), 365–377, 2016.
4. Sahni, J.; Vidyarthi, D.P.: A cost-effective deadline-constrained dynamic scheduling algorithm for scientific workflows in a cloud environment. IEEE Trans. Cloud Comput. 6(1), 2–18, 2018.
5. A. Jyoti and M. Shrimali, "Dynamic provisioning of resources based on load balancing and service broker policy in cloud computing," Cluster Comput., vol. 23, no. 1, pp. 377–395, 2020.
6. Calheiros, R. N., Ranjan, R., Beloglazov, A., De Rose, C. A. F., & Buyya, R. CloudSim: a toolkit for modeling and simulation of cloud computing environments and evaluation of resource provisioning algorithms. Software: Practice and Experience, 41(1), 23–50, 2011.
7. L. Qi, W. Dou, and J. Chen. Weighted principal component analysis based service selection

method for multimedia services in cloud. Computing, 98(1-2): 195–214, 2016.

8. Roman Barták, Marec VLK, "A Scheduling-Based Approach to Multi-Agent Path Finding with Weighted and Capacitated Arcs", Scheduling and Planning, AAMAS July 10–15 2018.

9. Yuan Zhang, Nanjing, "Resource Scheduling and Delay Analysis for Workflow in Wireless Small Cloud", IEEE Transaction on Mobile Computing, volume: 17, Issue: 3, March 1, 2018.

10. Sugumaran, D. and Bharathi, C.R.,. AMBA: Adaptive Monarch Butterfly Algorithm based Information of Transfer Scheduling in Cloud for Big Information Application. International Journal of Advanced Computer Science and Applications, 12(2). 2021

Recent Trends in Computational Intelligence and its Application – Sugumaran D. et al. (eds)
© 2023 Taylor & Francis Group, London, ISBN 978-1-032-48410-5

33

Innovative IoT-based Predictive Control and Monitoring Algorithm for Industrial Safety

Gokul M.[1]
Assistant Professor, Department of Computer Science and Engineering,
Vel Tech Rangarajan Dr. Sagunthala R&D Institute of Science and Technology

Rajathi K.[2]
Associate professor, Department of Computer Science and Engineering,
Vel Tech Rangarajan Dr. Sagunthala R&D Institute of Science and Technology

Abstract—Everyone in India is alarmed when an accident occurs in one of the country's many industries. As important as providing high-quality products are sectors that address environmental protection, world-related safety, and security while in the workplace. All of these challenges necessitate the presence of a top-level executive, administrator, director, or officer to manage them. Environmental, Health, and Safety (EHS) divisions are referred to as EHS or HSE departments. Indian industry's senior management is often unable to implement proper safety measures in the workplace. A wireless sensor network based on the Augmented Data Recognizing (ADR) algorithm is proposed, in which each node is a computational platform that incorporates many industrial factors. A central BEAGLEBK controller server receives the measured data through WIFI and compares it with standard threshold readings to set crucial warnings in the event of a worker safety violation. The sensor node and beagle bone communicate via WIFI as a backbone. Using an augmented data recognising method, this study describes the system's hardware and software components in detail, allowing for real-time monitoring and reporting of the industry safety management system. This paper depicts a novel method for analysing safety requirements in industries. For researchers and industry professionals, this technique will serve as a guide for doing precise analyses and evaluating safety factors based on priority.

Keywords—IoT, ADR, Safety, Automation, BEAGLEBK

I. Introduction

Most of the accidents happening in Indian companies are a source of stress for everyone. Factories have to think about Health and Safety-related to the world in critical terms and company environmental reassurance as providing quality products. As no production delay will be accepted by directors, engineers, and supervisors in charge, in the same way,

[1]gokulm@veltech.edu.in, [2]rajathi@veltech.edu.in

DOI: 10.1201/9781003388913-33

implementation of safety practices in health and safety-related issues need to be addressed without delay[1]. The best management of Indian companies is going to forget to set up practices as well as customers, contract workers, and other interested parties who may have some interest in protecting delegates and management at their workplace, at least as satisfying as necessary[2]. Enough safety measures will create a positive impact on associations and many workers. This part is incorporated to compensate for medical costs, hardware damage, material damage, product delays, and interruptions. This research work is primarily considered to account for the reasons for this foundation [3]. In this research work, the proof of EHS (Environment, Health & Safety) practices are displayed in the establishment of different Indian companies and the goal is to distinguish any zones in which both the qualities and the regions can be moved. The more effective results have been achieved because of the conditions identified in the areas necessary for the development of those qualities. The recognizable proof of discernment and evidence, Health, Safety& Environment practices have been taken into consideration, and regions are not intended. The EHS overview diagram is shown in Fig. 33.1.

Fig. 33.1 EHS overview diagram

Source: Author

II. Literature Survey

This section deals with the general information on web-based automation, classification, and EHS safety solutions and problems related to industry-based systems. The present approaches for designing safety-related systems use qualitative analysis. Nowadays, it is broadly acknowledged that the dominant part of accidents in the industry is somehow inferable from humans and also specific factors as in activities by people started. From the review of different EHS safety measures for the industrial condition, it is anything but confusing to distinguish a few issues.

There were 319 notable industrial accidents around the world that [4] documented in their report. [5] created a database with data on 167 biogas plant mishaps. The authors found that safety standards, safety culture, and risk awareness need to be updated and implemented. In the years 1998-2014, a database of fuel ethanol industry accidents was compiled by [6]. According to the authors, mechanical failure is the most common cause of ethanol plant fires. For the alignment of 238 industrial incidents, [7] constructed a dataset dubbed the Multi-Attribute Technological Accidents Dataset (MATAD). 182,794 people were killed and $265.1 billion in property was damaged as a result of these incidents, according to the authors. Numerous studies have been conducted across a wide range of industries to better understand the causes of workplace accidents. The following are some of the types of study that can be done on the investigation of industrial accidents: The inspection equipment, according to [8], which examined 106 Accident Reports involving moving machinery parts, was found to pose various hazards for injuries and dangerous effects. Analyzing 567 accidents from 2009 to 2012, [9] found that these mishaps were caused by a lack of attention to minor flaws and noncompliance with safety requirements in a variety of industries, such as construction, manufacturing, agriculture, or service.

[10] looked at fertiliser plant explosions with ammonium nitrate. In his studies on industrial accidents, [11] identified the high-risk groups that led to accidents. Various researchers' efforts are lauded in the aforementioned literature. On the other hand, their findings appear nebulous and unreliable. According to [12], the dangers of working in dangerous industries are only going to get worse. There is no system or recommendations for evaluating these hazards. The risk assessment methodology is needed to fill in the gaps between the various analysis methods.

A. Problem Statement

After the review of various EHS measures in the industrial environment, several issues are identified which are mentioned below:

- Management of human factors is generally considered to be the most important task of carrying out harmful effects control.

- An absence of safety practices and proficient information for employees is the most well-known obstruction to small-scale and medium industries. Workers in a large portion of the Indian industries are either blind to advanced technology or getting along with the least compulsory necessities.

- EHS examination demonstrated that 80% of Multi-National Companies (MNC) and Large Scale classification industries conform to Minimum Mandatory Requirements and Compliance commitments.

- It is found from the EHS analysis that the small & medium-scale industries do not have a safety officer and a safety committee is constituted with the task of dealing with all environment, health and safety issues for the industry.

- It's possible that sensitive data collected wirelessly could be lost or tampered with because cloud computers are intended to employ all of their computing capacity

simultaneously. In order to properly implement this novel concept, trusted local computational resources are looking to the cloud as a viable alternative.

- However, it requires the invention of a new controller model and monitoring cloud framework that are probably safe and, at the same time, remain highly efficient. It is thus essential that expertise in IoT and efficient algorithms are to be combined to achieve EHS measures.

B. Objectives

The following are the encouraging factors for determining the need for environment, health, and safety practices (EHS) of various Indian companies.

- To develop new control algorithms for confirming EHS, integrity, and authenticity for cloud data storage.

- To identify the regions in which the industry can take corrective and preventive actions to strengthen its EHS practices.

- To analyze various available EHS solutions under the IoT framework.

- To analyze the self-assessment of the EHS practices by getting evidence, high risk, and areas for improvement in the implementation of methods.

- To analyze EHS issues and violations of employees involved in the industries based on the new control model of cloud computing.

- To categorize the industries based on their implementation of EHS practices.

- Our research targets to investigate these possibilities to develop next-generation compliance management systems

III. Methodology

Systems architecture, specifications, and implementations for the proposed IoT safety measures include extensive descriptions of the system's primary structural and flow diagrams in this part. For the Augmented Data Recognizing (ADR) algorithm, this paper presents a wireless sensor network system design in which each node of the computing platform combines several industrial factors. There is a central BEAGLEBK controller server that receives the measured data and compares it to established threshold values before determining whether or not an employee's safety is compromised. In order to connect the sensor node to the beagle bone wirelessly, WIFI is used. All the hardware and software components are explained in detail in this study, which is designed to enable real-time monitoring and reporting of the industry safety management system. The following are the main components:

The sensor module is the first step.

- Module of communication.
- Module for the centralized computer
- Module for the web application
- Module for sending out alert messages

The proposed research work reduces human intervention using the IoT, and it contains the beagle bone controller and course sensors in the Wi-Fi handset module. All owned belongs to the relation of the beagle bone controller and the concentration of the data using the USB connection. All owned by the beagle

bone controller continues to be updated with the newest data from the cloud center server sensors continuously through the WiFi handset module.

A. Interactive Safety Roles

Due to overwork or exhaustion, workers' reaction times can be slowed, increasing the risk of an injury. During the event, these systems should pay attention to their activities and work in harmony with those who are detrimental to human health care. The Augmented Data Recognition for Working Environment is depicted in Fig. 33.2. We can make important judgments from anywhere in the globe via the internet thanks to this system. The WiFi shield serves as a point of contact between the local network and the external network to which it connects. By keeping an eye on the system and connecting with the IoT network to shut down machines, the present work offers a preventative measure against these kinds of mishaps. In addition, an alert is triggered and a notification is sent to the authorities.

B. Security Systems

The panel made in these systems is made up of practical applications, which supervise our goods and property security. In practice, the

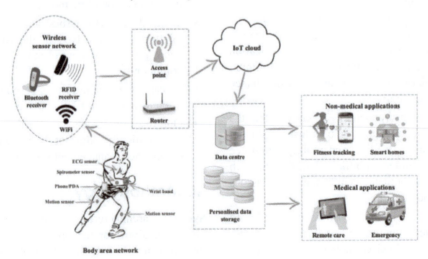

Fig. 33.2 Augmented data recognition for working environment (Example)

Source: Author

full range of engines use mechanical operating temperature, and because of the unwelcome development in separate zones, the following will provide fire controls for screening to come around.

C. ADR Control Procedure

Monitoring modern appliances is possible thanks to the industrial internet of things. Sensor data is used to train ADR's machine learning and big data technology. Devices in the workplace can be controlled using this technology:

In order to utilise the system for the first time, new users must register.

1. After credentials have been verified, the user must enter the necessary components. The system enables the user to log in.

2. Our cloud server database is where we keep all of the user data.

3. Listed below are the components of a cloud server:

4. As an administrator, you can monitor the entire system and send an alert to a cloud server.

5. A cloud server specialist can fix any issues that arise.

6. Using sensors, remote monitoring allows for hands-free operation.

7. Decision Maker: The system decides whether or not to send the warning through SMS/Mail.

IV. Results and Discussion

In this work, the ADR algorithm is used for EHS measures. Three different strategies are used for machine safety, workers' safety and environmental protection. The cost of trading can be averaged from various centers which is supposed to be reduced and Data can be ensured that the overall industrial safety and security level collected by different game plans are tested. As shown in Fig. 33.3, industrial safety and security software design is an important consideration. Here, we're going to talk about the population of the proposed industrial security and safety system. For more information on performance analysis, see the programs listed below. Figure 33.4 depicts the simulation design for the Industrial Safety and Security Measures. Proteus software

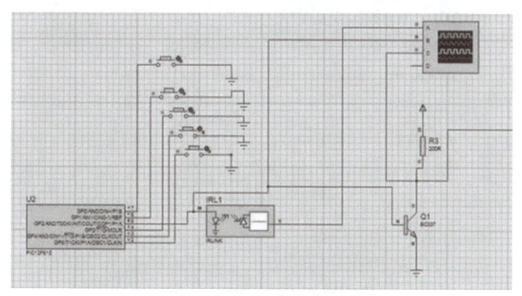

Fig. 33.3 Simulation design for industrial safety and security measures

Source: Author

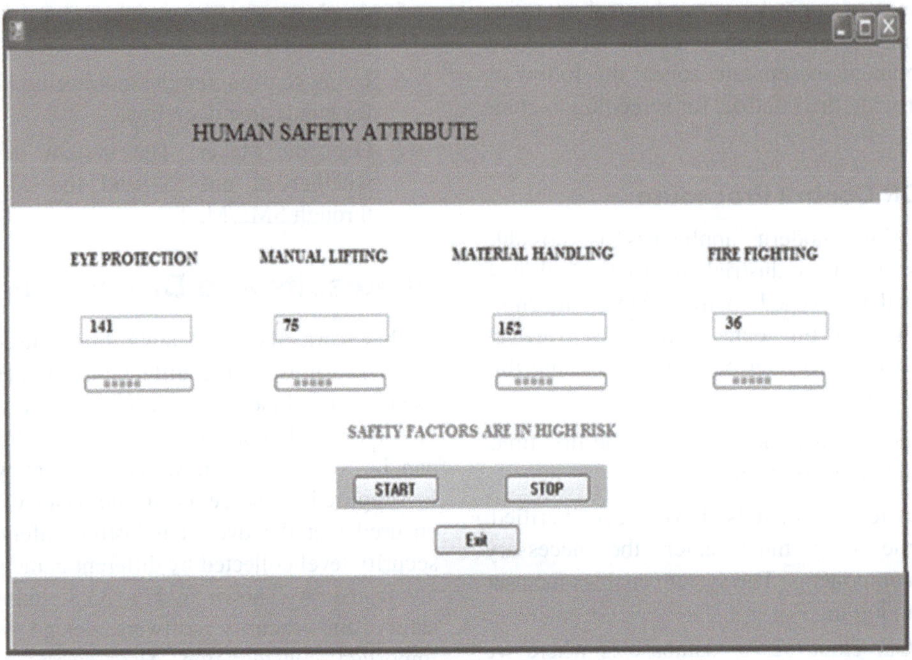

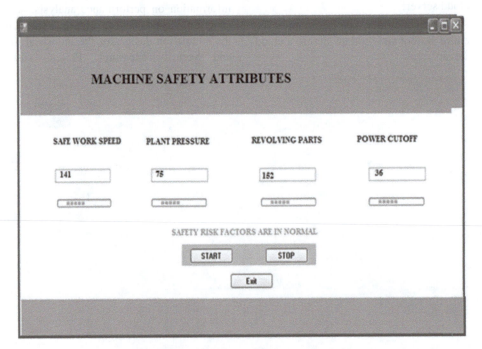

Fig. 33.4 Human safety measurement

Source: Author

Fig. 33.5 Machine safety measurement

Source: Author

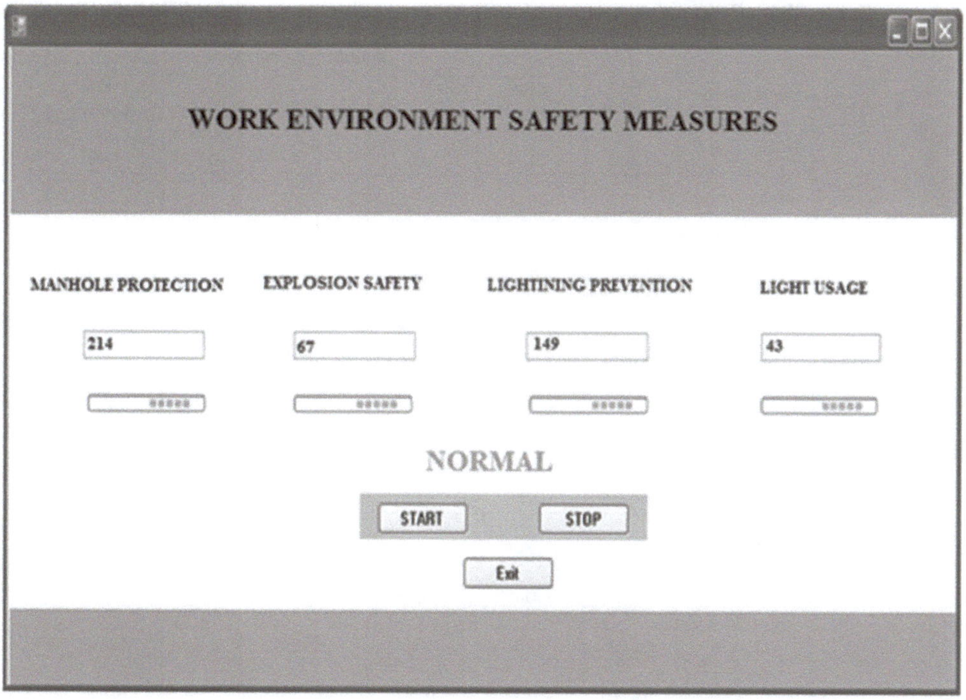

Fig. 33.6 Work environment safety measurement

Source: Author

was utilised in the creation of the simulation. Figure 33.5 shows the ADR controller-based factory security measures-using the Controller Security Attributes-demonstrates the population performance test. Different EHS measurements are analyzed based on different controller models. Advanced parameters are expected to evaluate a contact that combines measures of the EHS Monitoring Framework. As compared with Existing Fuzzy AHP, and proposed ARM controllers, the proposed ADR controller is more productive.

Figure 33.6 demonstrates the examination of proposed ADR execution measurements. In this way, an ADR-based data security approach is used as a piece of our proposed population-care. In Table 33.1, the result of this study will be more important for professional health care, security and environmental empire, which has started investigating further human behavioral angles.

V. Conclusion

Industrial management relies heavily on the automation of environmental health and safety (EHS) criteria. Various sectors rely on the measurement and control of certain safety features depending on parameters including human, machine, and workplace safety. Arduino and Processing Packages were used to construct an integrated hardware and software solution for EHS safety, while Visual Studio 2013a was used to increase communication through the Internet of Things (IoT). Both projects have been successful in lowering the number of people killed on the job, as well as the safety of the machines and the surrounding environment, by utilising wireless sensor networks to assess accuracy and safety factors. The results of this work demonstrate the IoT and sampling technique as a tool for assessing the environment, health, and safety practices of

Table 33.1 Respective metrics of main criteria, sub-criteria and industries

Main Criteria	Sub Criteria	Metric (per unit)	Industries (per unit)				
			Heavy Engineering	Automobile	Manufacturing	Foundry	Textile Industry
Human safety Attributes(db)	Eye Protection	0.1825	0.42	0.27	0.20	0.12	0.09
	Manual Lifting	0.4415	0.47	0.1457	0.2247	0.0784	0.0457
	Material Processing Practices	0.0573	0.0617	0.2303	0.191	0.3423	0.5448
	Firefighting Drills	0.0411	0.6322	0.3277	0.1238	0.1596	0.2517
	Training	0.0504	0.5315	0.126	0.2047	0.088	0.0498
	Safety Officer	0.291	0.4545	0.2434	0.1929	0.0707	0.0306
Machine Safety Attributes(db)	Fencing	0.1331	0.2695	0.1156	0.3873	0.1653	0.0623
	Revolving Parts Protection	0.1658	0.444	0.27	0.1885	0.0676	0.0298
	Safe Work Speed	0.3163	0.2843	0.4084	0.1745	0.0823	0.0506
	Pressure Plant Protection	0.5127	0.4365	0.1631	0.2853	0.0767	0.0384
	Power cut-off Devices	0.2438	0.418	0.1784	0.2557	0.096	0.0519
Work Environment Safety Attributes(db)	Manhole Protection	0.2143	0.4149	0.1771	0.2539	0.1086	0.0455
	Explosion Safety	0.2406	0.4408	0.1534	0.2507	0.1079	0.0473
	Lightening Protection	0.1723	0.3873	0.1653	0.2695	0.1156	0.0623
	Flammable Dust Prevention	0.2592	0.418	0.096	0.2557	0.1784	0.0519
	Pits, Sumps Protection	0.0508	0.4592	0.1509	0.2471	0.0928	0.0499
	Portable Light Usage	0.1028	0.4609	0.141	0.282	0.0759	0.0402

Source: Author

various Indian industries. With the use of the methodology, several objectives are achieved which are as follows.

- Getting the staff of multiple sectors involved in the assessment.
- The self-assessment of the safety practices by getting evidence, strong points, and areas for improvement in the implementation of methods.
- Categorization of various industries based on their implementation of safety practices.
- The identification of the areas in which the industry can take corrective and preventive actions to strengthen its safety practices.

VI. Scope for Future Work

Future industrial EHS management will focus on satisfying climate change expectations, preserving human and machine safety value as well as tracking and improving performance. There is a near-universal agreement among our respondents that organisations will continue to require EHS professionals who can handle a wide range of EHS functions and even allied responsibilities, despite the fact that no one can predict the long-term sustainability of safety engineers or industrial hygienists.

References

1. R.Hussain, F. M. Zakai and A. Iqbal, "Demystifying Mining Sustainability Through Efficient And Low Cost IoT Based Safety Implementations," 2022 Global Conference on Wireless and Optical Technologies (GCWOT), Malaga, Spain, 2022, pp. 1–6.

2. D. Prabha, D. B, D. M. A and S. K., "IoT application for Safety and Health Monitoring System for Construction Workers," 2021 5th International ConferenceonTrends in Electronics and Informatics (ICOEI), Tirunelveli, India, 2021, pp. 453–457.

3. S. A. Dolas, S. A. Jain and A. N. Bhute, "The Safety Management System Using Q-Learning Algorithm in IoT Environment," 2021 7th International Conference on Advanced Computing and Communication Systems (ICACCS), Coimbatore, India, 2021, pp. 1024–1028.

4. E. Tomur et al., "SoK: Investigation of Security and Functional Safety in Industrial IoT," 2021 IEEE International Conference on Cyber Security and Resilience (CSR), Rhodes, Greece, 2021, pp. 226–233.

5. V. C. Jadala, S. K. Pasupuletti, S. H. Raju, S. Kavitha, C. M. H. Sai Bhaba and B. Sreedhar, "Need of Intenet of Things, Industrial IoT, Industry 4.0 and Integration of Cloud for Industrial Revolution," 2021 Innovations in Power and Advanced Computing Technologies (i-PACT), Kuala Lumpur, Malaysia, 2021, pp. 1–5.

6. P. Karthikeyan, M. Karthik, V. Deepikapriya, S. Divya Briya, R. Dharanishwarma and S. Janakirthick, "Iot based Simulation of Robot for Pattern Painting on Walls via Android Application," 2022 2nd International Conference on Power Electronics & IoT Applications in Renewable Energy and its Control (PARC), Mathura, India, 2022, pp.1–4.

7. A. A. Shah, N. A. Bhatti, K. Dev and B. S. Chowdhry, "MUHAFIZ: IoT-Based Track Recording Vehicle for the Damage Analysis of the Railway Track," in IEEE Internet of Things Journal, vol. 8, no. 11, pp. 9397–9406, 1 June1, 2021.

8. S. Song, S. Li, H. Gao, Y. Yan, Z. Wang and W. Xu, "Research on IoT-based Operation Support System for Safety and Economic of Power Distribution Station Clusters," 2021 3rd Asia Energy and Electrical Engineering Symposium (AEEES), Chengdu, China, 2021, pp. 954–958.

9. A. K. Sutrala, M. S. Obaidat, S. Saha, A. K. Das, M. Alazab and Y. Park, "Authenticated Key Agreement Scheme With User Anonymity and Untraceability for 5G-Enabled Softwarized Industrial Cyber-Physical Systems," in IEEE Transactions on Intelligent Transportation Systems, vol. 23, no. 3, pp. 2316–2330, March 2022.

10. A. Das, A. Shukla, R. Manjunatha and E. A. Lodhi, "IoT based Solid Waste Segregation using Relative Humidity Values," 2021 Third International Conference on Intelligent Communication Technologies and Virtual Mobile Networks (ICICV), Tirunelveli, India, 2021, pp. 312–319.

11. K. Wisessing and N. Vichaidis, "IoT Based Cold Chain Logistics with Blockchain for Food Monitoring Application," 2022 7th International Conference on Business and Industrial Research (ICBIR), Bangkok, Thailand, 2022, pp. 359–363

12. P. R. M. E., H. P. S, H. L and K. E, "IoT based Industrial Automation for Various Load using ATmega328p Microcontroller," 2022 6th International Conference on Intelligent Computing and Control Systems (ICICCS), Madurai, India, 2022, pp. 471–47.

Recent Trends in Computational Intelligence and its Application – Sugumaran D. et al. (eds)
© 2023 Taylor & Francis Group, London, ISBN 978-1-032-48410-5

34

Blind Source Separation in the Presence of AWGN Using ICA-FFT Algorithms a Machine Learning Process

M. R. Ezilarasan*

Department of Electronics and Communication Engineering,
Vel Tech Rangarajan Dr. Sagunthala R & D Institute of science and technology, chennai

Rajesh G.[1]

Department of Computer Science and Engineering,
Vel Tech Rangarajan Dr. Sagunthala R & D Institute of science and technology, Chennai

Vinoth Kumar R.[2]

Department of Computer Science and Engineering,
Vel Tech Rangarajan Dr. Sagunthala R & D Institute of science and technology, Chennai

K. Aanandhasaravanan[3]

Department of Electronics and Communication Engineering,
Vel Tech Rangarajan Dr. Sagunthala R & D Institute of science and technology, Chennai

Abstract—The process of isolating sound sources in any audio environment is known as blind source separation (BSS) or audio signal separation. Applications for source signal separation include signal processing, audio processing, etc. This study investigates how to distinguish sound signals from a mixture when there is additive white Gaussian noise (AWGN). The independent component analysis (ICA) method is unable to precisely assess separated signals as a result of this additive noise. In this article, a mixing matrix is employed, which downsamples several sound sources before combining them. The mixed signal is given a variable amount of AWGN as input. The noisy mixed signal is then further denoised using a block denoising method based on the Fast Fourier Transform (FFT). The split sound signals are then reconstructed using Inverse Fast Fourier Transform (IFFT). The proposed technique is easier to remove additive noise than existing methods, and the separated sound signals have a hearing impression that is quite similar to the original signals.

Keywords—Machine learning, BSS, ICA, AWGN, FFT, IFFT

*Corresponding author: arasanezil@gmail.com
[1]rajesh702me@gmail.com, [2]vinothtechnocrat@gmail.com, [3]anand23sarvan@gmail.com

DOI: 10.1201/9781003388913-34

I. Introduction

Artificial intelligence and machine learning are techniques for imitating human intellect in machines or computers. We, therefore, want computers to be able to make decisions similar to how people do. One of a person's most important senses is hearing. Humans can quickly distinguish between various sounds, for instance, harmony and dialogue, car and truck sounds, the quality of a woman's and a man's voice, noise and valuable sound, etc., without exerting any additional effort. Machines must be able to differentiate between various noises in the same manner that humans can. This issue is also known as "machine hearing.". Regardless of the purpose, a machine learning system requires powerful and selective features that allow the machine to learn consistently and rapidly. A machine's intelligence is determined by the quantity of training it receives. A signal is represented by a feature. The issue is determining which features to extract for a machine learning algorithm to succeed. The features must be tiny in size while highlighting the signal's characteristics. The Features are packed in such a way that they considerably reduce the size of a signal while still characterizing it accurately and comprehensively. The machine learning algorithm's computational and time complexity is reduced as a result of the signal's simplified form, making them more appropriate for real-time-based applications. Thus, the process of reducing a signal's dimension so that machine learning algorithms can grasp it might be referred to as feature extraction. In this paper, only the features extracted from voice signals and musical sounds are discussed. Speech recognition, speaker recognition, blind source separation, speech enhancement, pathological speech detection, pathological speech improvement, noise reduction, noise cancellation, and other applications use voice signals. According to this paper's flow: Section 2 provides a summary of the machine learning approach. The various types of audio signals are

discussed in Section 3; the work's justification and characteristics are explained in Section 4, and the importance of the altered domain is addressed in the work that is suggested in Section 5. Final thoughts are provided in Section 6.

II. Machine Learning Process

The structure of machine learning is represented in Fig. 34.1. In the primary stage, the audio signal is pre-processed. This is where noise cancellation, quiet reduction, normalization, and other pre-processing techniques are used. The signal is then windowed to determine if the signal is non-stationary or quasi-stationary. The entire signal is inspected and evaluated by sliding the window over the complete length of the transmission. Using modern windowing techniques, the size of the window can be adjusted according to the signal's characteristics. The feature extraction and feature selection stages are then performed. The performance of the classifier is determined by these steps. The selected features are then sent into the classifier for training and testing, and the classifier's prediction is used to make a decision.

III. Audio Types

Sound signal classification is shown in Fig. 34.2. the sound can be classified into 3 categories voice, music, and environmental sound

A. Speech

In order to generate speech in humans, the activities of numerous organs, including the lungs, mouth, nose, abdomen, and brain, are combined[2]. The majority of speech is produced by the vocal tract and vocal cords. The frequency spectrum of two distinct speech transmissions is shown in Fig. 34.3. The amplitudes in this are independent. Human speech has a frequency range of 100 Hz to 17 kHz.

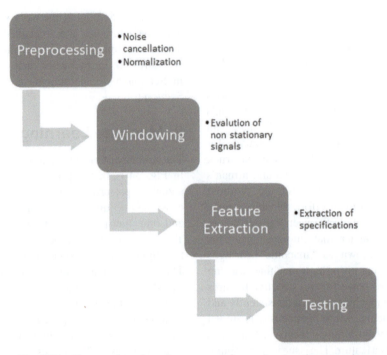

Fig. 34.1 The machine learning process for audio signal processing

Source: Author

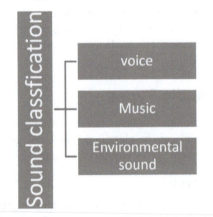

Fig. 34.2 Classification of sound signals

Source: Author

B. Musical Sounds

Humans or instruments create musical sounds to attain harmony and emotional expression. Music can be categorized using genre, mood, and acoustic characteristics [3]. Traditional classification systems have been used to group rock, jazz, classical, and pop music. The frequency range of music should ideally stretch from 40 Hz to 19.5 kHz. The amplitude of each piece of music is determined by the instruments used and can vary in pitch.

C. Environmental Sounds

There are countless environmental sounds that we hear every day, including those from cars and other moving objects, flowing water, doorbells, ringing phones, factories, animals, and so on. Throughout the whole auditory spectrum, these sounds were audible. Periodicity can be found in speech and music, but it's more difficult to find in environmental noises. The envelope of speech is smooth and continuous, whereas the envelope of guitar notes is non-continuous and has a finite duration. The sound of the fire truck is loud and looks to be noise. These three noises are distinct not only in terms of duration but also in frequency.

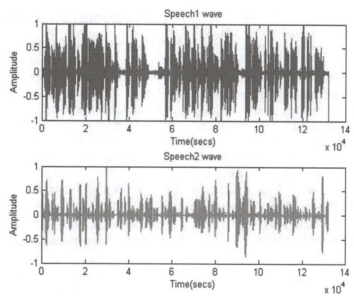

Fig. 34.3 The frequency spectrum of 2 different speech signals

Source: MATLAB

IV. Motivation for this Work

The motivation for this work is the cocktail party problem [1], [2]. The party will have multiple sound systems with many speech signals included. To listen to a specific voice or a specific instrument, As a human, he or she has a neural system in nature to differentiate the voice and the mixed signals from the sound source. But computers cannot differentiate the signals in nature. Some inputs to be faded into the system by that it can consider that input as a reference with the other source signals and recognize Separating a single source from the mixed-signal without knowing the past knowledge is called Blind source separation[3]. Many approaches are carried out in previous decades and those existing signal separation algorithms are assumed by statistical independent mixture. This assumption of statistical independence is inaccurate when many instruments take place with a common tempo [4]. The Independent component (ICA) presented in [5][6] positions the mixing and de-mixing process of BSS. But for the Demixing process ICA algorithm minimizes statistical

dependence for the Unmixed channel. Maximizing the nongaussianity [6] will also reduce the statistical redundancy which is explained in reference [2]. The separation process based on only time-frequency masking is explained in [7] where the input signals of time-dependent are treated by frequency bins. Non-negative matrix factorization [8] [9] is another separation technique where the M*N matrix is separated as M*K and K*N. the separation process is based on an adaptive filter is which has self-adjusting characteristics which are explained in [10].

A. Sound

Pitch is a parameter that can be used to define musical instruments [11]. Musical instrument sounds can be either pitched or non-pitched. Pitch is defined as "the characteristic of auditory perception that enables sound quality to be rated on a scale from low to high". This concept emphasizes that pitch is a perceptual rather than a physical quality while being unclear. Frequency (F0) is the inverse of the time period and a physical characteristic of periodic transmissions. Thus pitch is a psychophysical function of physical factors that is a psychophysical property of physical variables that are impacted by frequency F0, intensity, and duration [5]. A pitch, an onset time, and a predetermined length are all characteristics of a pitched musical instrument sound note. They can be produced by a wide range of physical systems, but they always create roughly periodic vibrations. Understanding that the sounds can be represented as a harmonic progression of sinusoidal waves has benefited in the development of analysis and synthesis systems. Fig. 34.4 shows the audio frequency characteristics, which highlight the FFT [12] characteristics of pitched instruments.

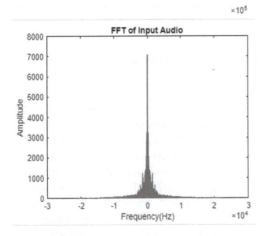

Fig. 34.4   FFT of input audio signal

Source: MATLAB

V. Proposed Architecture

The suggested architecture is displayed in Fig. 34.5. Block diagram demonstrates how audio signal samples are obtained using a mixing matrix. The mixing matrix will vary depending on the music and equipment used. white In order to represent the mixed signal in the frequency domain, Gaussian noise is added to the mixing signal. The fast Fourier transform is then applied to the mixed signal, and ICA is used to interpret the results in order to separate the components of the mixed signal.

ICA can isolate the mixed signal into individuals. And the separated signal is again applied to IFFT to convert back the frequency domain of the signal into time representation. This is explained in the below diagram

A. ICA Algorithm

For implementing the proposed module, the music system is taken as input, consider that the music system has three different sounds DrumsS1 Guitar S2 and Human voice S3 these 3 different signals are mixed with different mixing matrices. The main objective of this ICA is to find the independent components U1, U2, and U3 which is unknown based on the mixing matrix and observed signal X1, X2, and X3 which is shown in Fig. 34.6.

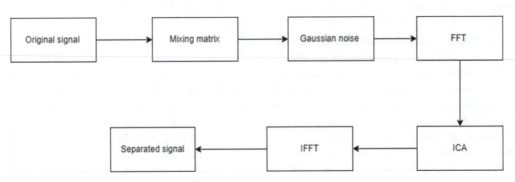

Fig. 34.5   Proposed architecture

Source: Author

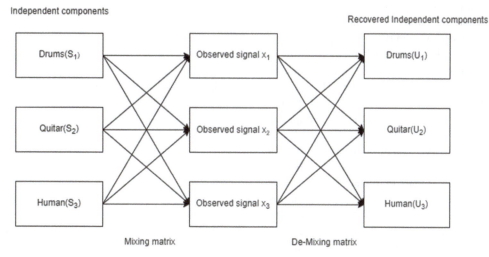

Fig. 34.6   Proposed ICA algorithm

Source: Author

B. ICA Background

This section gives a brief of the ICA algorithm used for separating Audio signal sources. Which has many applications in signal processing, speech separation, etc., and also ICA is used to focus on cocktail party problems. Consider two source signals S_1 and S_2 which changes according to time. These two source signals can be represented in a matrix as $S = \begin{bmatrix} s1 \\ s2 \end{bmatrix}$ and the mixing signal can be represented as $M = \begin{bmatrix} a11 & a12 \\ a21 & a22 \end{bmatrix}$ where $a11$, $a12$, $a21$, and $a22$ are mixing coefficients. The consequential mixed output signal is given as $X = \begin{bmatrix} a11s1 & a12s2 \\ a21s1 & a22s2 \end{bmatrix}$.

Here $X_1 = a11s1 + a12s2$ is the first mixed signals overage of the two transmitted signals, rated ($s1$ and $s2$). The distance between both the transmitted signal and the sensor device signal is adjusted to form another mixed signal arrangement ($x2$), which is computed as $x_2 = a21s1 + a22s2$. The two mixing coefficients $a11$ and $a12$ differ from the coefficients $a21$ and $a22$ because the two sensing devices used

to sense these signals are in different positions. As a result, each sensor detects various source signal combinations. Equation 1 shows the representation of the two mixed signal components. (1).

$$X = \begin{bmatrix} x1 \\ x2 \end{bmatrix} = \begin{bmatrix} a11s1 & a12s2 \\ a21s1 & a22s2 \end{bmatrix} = A.s \quad (1)$$

Because each mixing signal has varied features such as varying distance, environment, and microphone capacity, the mixing coefficient matrix A grows as the number of source signals grows. The equation gives the generic expression for mixed signal output X with n number of source signals s. (2).

$$X = \begin{bmatrix} x1 \\ \vdots \\ xn \end{bmatrix}, A = \begin{bmatrix} a11 & . & . & a1n \\ \vdots & . & . & \vdots \\ an1 & . & . & ann \end{bmatrix}, s = \begin{bmatrix} s1 \\ \vdots \\ sn \end{bmatrix}$$

$$(2)$$

C. Frequency Domain Approach

The time domain representation of an audio signal is inadequate Because it is reliant on the signal's quality and volume, Due to the simplicity of elaboration, all audio signals are processed in the frequency domain. Compared to time domain representation, signals in the frequency domain have better properties. The same presumption about signals in the frequency domain can be used to apply the ICA technique. The developed method employs self-organizing neural networks for frequency pre-processing and grouping. Using the fast Fourier transformation, the time domain has been transformed into the frequency domain (FFT). Now we can define variables for computing of this system. Let input signals be $x(t)$, which changes according to time. They are the damaged signals or mixed signals, which are separated. The separated signals are marked as $s'(t)$. in fact by applying the FFT or these signals the time variant will be converted to the frequency variant. Additionally, they are the fundamental variable and the inside (only in the system) variable is Fourier's image $X(k)$.

The Fourier's of the mixed signal is represented by $X(k)$ in the frequency domain, meanwhile the mixed signal is represented by $x(i)$ in the time domain. Although it is FFT, we also need FFT inversion (IFFT). The equation states that $s'(i)$ is the estimated signal (in the time domain) and $s'(k)$ is Fourier's image of the estimated signal (in the frequency domain).

$$f(x) = 1/2\pi \, e^{i\omega x} \, d\omega \qquad (3)$$

VI. Results and Discussion

The Software implementation of the ICA algorithm is tested using Matlab shown in the following figures. Figure 34.7 is the mixed signal of three inputs drums, guitar, and Human voice accordingly which are further demixed using IFFT. Our final experiment is a real-world problem that has been extensively examined in the ICA community and illustrates the architecture for the ICA implementation. The mixed signals are taken in a form of an audio signal, we used the .wav format for an audio signal then we used shortening the length of the audio to the nearest power of two and plotting by shifting of audio.

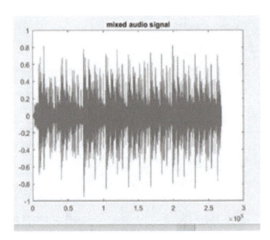

Fig. 34.7 Mixed signal of drums, guitar, and human voice

Source: MATLAB

To test the network implementation, we used real sample signals of music. Figures 34.8, 34.9, 34.10 shows the sample signals from the music recorded. Figure 34.8. Separated drum signal from mixed signal. Figure 34.9. Separated guitar signal from mixed signal. Figure 34.10 is the human voice signal all the separated signals vary according to time as all the signals are independent components.

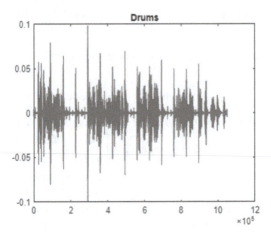

Fig. 34.8 Separated drums signal

Source: MATLAB

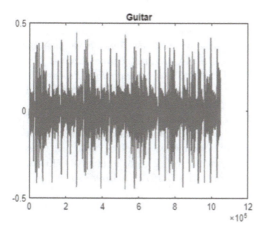

Fig. 34.9 Separated guitar signal

Source: MATLAB

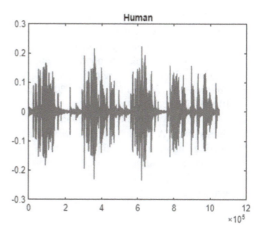

Fig. 34.10 Separated human voice signal

Source: MATLAB

VII. Conclusion

Using the ICA algorithm it is easy to separate the source signals from mixed signals, even though permutation problems will occur. To overcome this ICA algorithm is implemented with a Fast Fourier transform. By this time varients are converted into frequency variants and then implemented using the ICA algorithm. The proposed architecture is implemented using MATLAB. The simulation results show better performance compared with earlier algorithms. Typically the result connection of information signal is high in blind source separation, our proposed work concurs with the situation superior execution and sweet blind source partition calculation is prepared to do continuous signal handling.

Reference

1. A. Tharwat, "Independent component analysis: An introduction," *Appl. Comput. Informatics*, vol. 17, no. 2, pp. 222–249, 2018, doi: 10.1016/j.aci.2018.08.006.
2. Y. B. Monakhova and D. N. Rutledge, "Independent components analysis (ICA) at the 'cocktail-party' in analytical chemistry," *Talanta*, no. September, p. 120451, 2019, doi: 10.1016/j.talanta.2019.120451.
3. S. D. Jadhav and A. S. Bhalchandra, "Blind source separation: Trends of new age - A review," *IET Conf. Publ.*, no. 535 CP, pp. 251–254, 2008, doi: 10.1049/cp:20080190.
4. J. Xi, J. F. Chicharo, Ah Chung Tsoi, and Wan-Chi Siu, "On the INFOMAX algorithm for blind signal separation," no. August, pp. 425–428, 2002, doi: 10.1109/icosp.2000.894523.
5. K. Zhang and L. W. Chan, "Convolutive blind source separation by efficient blind deconvolution and minimal filter distortion," *Neurocomputing*, vol. 73, no. 13–15, pp. 2580–2588, 2010, doi: 10.1016/j.neucom.2010.05.018.
6. W. Reid, K. Huang, and D. Roberts-kedes, "Blind Audio Source Separation Pipeline and Algorithm Evaluation," pp. 1–6, 2015, [Online]. Available: http://cs229.stanford.edu/proj2015/124_report.pdf
7. D. S. Williamson and D. L. Wang, "Time-frequency masking in the complex domain for speech dereverberation and denoising," *IEEE/ACM Trans. Audio Speech Lang. Process.*, vol. 25, no. 7, pp. 1492–1501, 2017, doi: 10.1109/TASLP.2017.2696307.
8. S. Mirzaei, H. Van Hamme, and Y. Norouzi, "Blind audio source counting and separation of anechoic mixtures using the multichannel complex NMF framework," *Signal Processing*, vol. 115, pp. 27–37, 2015, doi: 10.1016/j.sigpro.2015.03.006.

9. S. Abdali and B. NaserSharif, "Non-negative matrix factorization for speech/music separation using source dependent decomposition rank, temporal continuity term and filtering," *Biomed. Signal Process. Control*, vol. 36, pp. 168–175, 2017, doi: 10.1016/j.bspc.2017.03.010.

10. A. Sayoud, M. Djendi, S. Medahi, and A. Guessoum, "A dual fast NLMS adaptive filtering algorithm for blind speech quality enhancement," *Appl. Acoust.*, vol. 135, no. April 2017, pp. 101–110, 2018, doi: 10.1016/j.apacoust.2018.02.002.

11. H. Sawada, N. Ono, H. Kameoka, D. Kitamura, and H. Saruwatari, "A review of blind source separation methods: Two converging routes to ILRMA originating from ICA and NMF," *APSIPA Trans. Signal Inf. Process.*, vol. 8, no. May, pp. 1–14, 2019, doi: 10.1017/ATSIP.2019.5.

12. Z. Albataineh and F. M. Salem, "A RobustICA-based algorithmic system for blind separation of convolutive mixtures," *Int. J. Speech Technol.*, vol. 24, no. 3, pp. 701–713, 2021, doi: 10.1007/s10772-021-09833-z.

Recent Trends in Computational Intelligence and its Application – Sugumaran D. et al. (eds)
© 2023 Taylor & Francis Group, London, ISBN 978-1-032-48410-5

35

Analysis of the System Size of a Finite Capacity M/M/C Queueing System with Synchronised Vacations and Impatient Customers

C. T. Dora Pravina*

Department of Mathematics,
Vel Tech Rangarajan Dr Sagunthala Institute of Science and Technology,
Avadi, Chennai, Tamilnadu, India

S. Sreelakshmi[1]

Department of Engineering Mathematics,
HKBK College of Engineering, Nagawara,Bengaluru, India

T. Karthikeyan[2]

PG and Research Department of Mathematics,
Ramakrishna Mission Vivekananda College, Mylapore, Chennai

Prakash M.[3]

Department of Engineering Mathematics,
HKBK College of Engineering, Nagawara, Bengaluru, India

Abstract—The paper assays Markovian multiple servers, finite capacity queue in which customers can become impatient. The servers can take a synchronous vacation, which may lead to customer impatience due to the absence of servers upon arrival. At any time when the system is void, all of the servers go on vacation. When the system is void, that is, there are no customers in the system, the servers proceed on for one more vacation; otherwise, they return to serve the queue. The governing equations are obtained for the steady-state probabilities. The probability Generating Function method is used for arriving at the balance equations. In terms of two indexes, we get the expression for system length.

Keywords—Multi-server, Finite capacity queue, Impatience customer, Synchronous vacation

I. Introduction

In numerous practical situations, queueing systems with impatient customers and synchronous vacations can occur. An example of impatient customers in a queuing system can be found in wireless networks in which data packets received after a hard deadline

*Corresponding author: doramaths@gmail.com,
[1]sree.narsh@gmail.com, [2]karthick@rkmvc.ac.in, [3]prakashm2205@gmail.com

DOI: 10.1201/9781003388913-35

at the receiver's end are useless. There are many more situations in which an impatient customer model can be applied, like inventory, telecommunication, and many other related areas. Models with impatient queues and with synchronous vacations have been studied by various authors in the past. Takacs, L, [10] Discusses various concepts in Queuing Theory and solutions to Queuing Theory problems. Levy et al. [1] analysed queueing systems where a server completes its service and there are no more customers in the waiting line, the server takes a vacation. At the end of the vacation, the server returns to the main system. A single vacation model is also being studied. Zhang, Z. G. et al. [2] enhanced the model by considering multiple servers, that is out of c servers taking synchronous vacations when d servers become idle at the service completion stage. D. Yue et al. [3] considered a queueing system with vacations and impatient customers. Wu, D. A. Takagi, H. [6] examined the $M/G/1$ queue with multiple vacations and presented a model in which the server works at various service times. Selvaraju, N et al. [5] enhanced the model by introducing impatient customers as well as a vacation policy where the server works at a lower rate during vacation instead of completely stopping service. Further, Mokhtar Kadi et al. [4] introduced a multiserver queuing system with Bernoulli feedback and impatient customers with synchronous multiple and single vacation policies. For the first time, N. Tian and Z. G. Zhang [7] analysed $G/M/1$ multi-server vacation methods for both continuous and discrete, and optimization issues were also dealt with. J.-C. Ke [8] analysed the $M^{[X]}/G/1$ system with a modified vacation policy with optional re-service. Manoharan, P. and Majid, S. [9] investigated an $M/M/c$ queue with multiple working vacations and also with impatient customers. E. Altman and U. Yechiali [11], for the first time, studied customers' impatience due to the absence of a server upon arrival. D. Yue and W. Yue [12] discuss the use of queuing theory in both wired and wireless networks.

Eitan Altman and Uri Yechiali [13] considered the system as an $M/M/\infty$ with impatient customers. Even though there are many research insights on $M/M/C$ with synchronous vacation and impatient customers, not much more work is focused on finite capacity in the system. In many situations, we found that the finite capacity of the system only creates the demand for rendering quality services. So, whenever a restriction is imposed on the capacity of the system in the queue, there is a necessity to study and analyse the queueing system. To the best of my knowledge, researchers do not include such an important factor in their work. To fill this gap, we propose and analyse the queue model with finite capacity $M/M/C$ with synchronous vacation and impatient customers. In the present paper, we extend the analysis by introducing the customer impatience phenomenon by considering multiple servers in a finite queuing system with synchronous vacation, which has not been analysed so far.

A. Application of Queueing Theory in IoT

As technology is advancing, the new buzzword Internet of Things has taken over the world currently. The IoT provides a wide range of benefits in the corporate, academic, and healthcare sectors. In most all fields, IoT has spread its wings. Markov IoT can be used to study the reliability of IoT. Queuing Theory can be applied to investigate the delay and availability of functioning things. As an example, the problem of traffic congestion can be resolved to some extent by applying queuing theory. By using this, waiting time estimation and expected queue length can be done with ease. Peter O. Peter and R. Sivasamy [14] studied the application of queuing theory in the area of health care to improve the service provided.Farhan Sufyan and Amit Banerjee [15], considered the application of queueing theory in fog computing for efficient use of the resources and to minimise the cost involved by reducing the delay. Antonio Franco et al.[16].

The paper focuses on the age of the information being transferred from IoT devices, thus improving the departure time and service time by considering $M/M/\infty$ queue.

The article is structured as follows: we outline the model in Section 2. How the governing equations are developed is explained in Section 3. We obtain the Probability generating functions for the steady-state probabilities in Section 4. In Section 5, we acquire the explicit system size expressions in terms of two indices. Section 6 deals with conclusions.

II. Model Description

The system under consideration is of finite capacity $M/M/C$ queue with impatient customers. The servers can go on synchronous vacations. According to our presumption, consumers arrive via a Poisson process with a rate of λ. The queue discipline followed is First-in, First-out (FIFO). All the servers provide the service uniformly. The service time of an individual customer follows an exponential distribution with mean $\frac{1}{\mu}$. The servers function in the following fashion: When a server completes the service, and if the system is in the void, the servers leave for vacation instantly. On the other hand, when the servers upon returning from vacation find the system to be void then, the servers immediately go for one more vacation. Upon returning from vacation, if the system is not empty, the servers are back again to serve the queue. The prolonged vacation spell follows an exponential distribution with a mean $\frac{1}{\gamma}$. At the time of the vacation spell, customers become impatient. Specifically, a customer who enters the system and notices that the server is about to go on vacation plays out an impatient timepiece that is exponentially distributed with parameters ξ. If the servers returned from their vacation well before the impatient time (and service began), the customer would remain within the

system until his service was accomplished. The customer leaves the line but never comes right back. Unfortunately, if the time has run out, the server is nevertheless on vacation.

Let X stand for the overall number of customers in the queueing system, and let S represent the server's current state. Following that, the joint process (X, S) defines a continuous-time Markov process. We define the probability as

$$\pi_{n,\,i} = \Pi[X = n, S = i \mid X = 0, S = 0],$$
$$n = 0, 1, 2, \cdots.N.; i = 0, 1.$$

III. Governing Equations

$$\lambda\pi_{0,0} = \xi\pi_{1,0} + \mu\pi_{1,1} \tag{1}$$

$$(\gamma + n\xi)\pi_{n,0} = \lambda\pi_{n-1,0} + (n + 1)\xi\pi_{n+1,0},$$
$$n = 1,2,3,...N - 1 \tag{2}$$

$$(\gamma + N\xi)\pi_{N,0} = \lambda\pi_{N-1,0} \tag{3}$$

$$(\lambda + \mu)\pi_{1,1} = \gamma\pi_{1,0} + 2\mu\pi_{2,1} \tag{4}$$

$$(\lambda + n\mu)\pi_{n,1} = \lambda\pi_{n-1,1} + \gamma\pi_{n,0} + (n + 1)\mu\pi_{n+1,1},$$
$$n = 2,3,4,...c - 1 \tag{5}$$

$$(\lambda + c\mu)\pi_{n,1} = \lambda\pi_{n-1,1} + \gamma\pi_{n,0} + c\mu\pi_{n+1,1},$$
$$n = c, c + 1, c + 2,..., N - 1 \tag{6}$$

$$c\mu\pi_{N,1} = \lambda\pi_{N-1,1} + \gamma\pi_{N,0} \tag{7}$$

IV. Steady State Solutions

Solving equations (1)–(7), we define Probability Generating Function

$$G_0(z) = \sum_{n=0}^{N} \pi_{n,0}z^n \tag{8}$$

$$G_1(z) = \sum_{n=1}^{N} \pi_{n,1}z^n \tag{9}$$

Multiplying both sides of equation (2) by z^n, summing up from $n = 1$ to $N - 1$ and using equation (1), we get

$$\xi[1 - z]G_0'(z) - [\lambda(1 - z) + \gamma]G_0(z)$$
$$= -(\mu\pi_{1,1} + \gamma\pi_{0,0} - \lambda(1 - z)\pi_{N,0}z^N \tag{10}$$

By using (6) and (5), we get

$$[(1 - z)(\lambda z - c\mu)]\,G_1(z)$$
$$= \gamma zG_0(z) - (\gamma\pi_{0,0} + \mu\pi_{1,1})z$$
$$+ \mu(1 - z)\sum_{n=1}^{c}(n - c)\pi_{n,1}z^n$$
$$+ \lambda(1 - z)\pi_{N,1})\,z^{N+1} \tag{11}$$

Solving first order differential equation (10), we get

$$G_0(z) = e^{\frac{\lambda}{\xi}z}(1-z)^{-\frac{\gamma}{\xi}}\left[-\int_0^z \frac{H}{\xi}e^{\frac{\lambda}{\xi}x}(1-x)^{\frac{\gamma}{\xi}-1}dx\right.$$

$$\left.-\int_0^z \frac{\lambda}{\xi}\pi_{N,0}x^N e^{-\frac{\lambda}{\xi}x}(1-x)^{\frac{\gamma}{\xi}}dx + G_0(0)\right]$$

(12)

where

$$H = \gamma\pi_{0,0} + \mu\pi_{1,1}$$

(13)

$$G_0(1) = e^{\frac{\lambda}{\xi}}\left[-\frac{H}{\xi}k\int_0^1 e^{-\frac{\lambda}{\xi}x}(1-x)^{\frac{\gamma}{\xi}-1}dx\right.$$

$$-\int_0^1 \frac{\lambda}{\xi}\pi_{N,0}x^N e^{-\frac{\lambda}{\xi}x}(1-x)^{\frac{\gamma}{\xi}}dx$$

$$\left.+ G_0(0)\right]\lim_{z\to 1}(1-z)^{-\frac{\gamma}{\xi}}$$

$$G_0(1) = \sum_{n=0}^N \pi_{n,0} > 0 \text{ and } \lim_{z\to 1}(1-z)^{-\frac{\gamma}{\xi}} = \infty$$

$$G_0(0) = \frac{H}{\xi}k + \frac{\lambda}{\xi}\pi_{N,0}\int_0^1 e^{-\frac{\lambda}{\xi}x}x^N(1-x)^{\frac{\gamma}{\xi}}dx \quad (14)$$

where $k = \int_0^1 e^{-\frac{\lambda}{\xi}x}(1-x)^{\frac{\gamma}{\xi}-1}dx$ substitute (14) in (12), we get

$$G_0(z) = \frac{e^{\frac{\lambda}{\xi}z}}{(1-z)^{\frac{\gamma}{\xi}}}\left[\frac{H}{\xi}k\left(1-\frac{1}{k}\int_0^z e^{-\frac{\lambda}{\xi}x}(1-x)^{\frac{\gamma}{\xi}-1}dx\right)\right]$$

$$+\left[\frac{\lambda}{\xi}\pi_{N,0}\left(\int_0^1 e^{-\frac{\lambda}{\xi}x}x^N(1-x)^{\frac{\gamma}{\xi}}dx\right.\right.$$

$$\left.\left.-\int_0^z e^{-\frac{\lambda}{\xi}x}x^N(1-x)^{\frac{\gamma}{\xi}}dx\right)\right]$$

(15)

Equation (11) well-written by

$$G_1(z) = \frac{[\gamma G_0(z) - H]z}{(1-z)(\lambda z - c\mu)} - \frac{\mu Q(z)}{(\lambda z - c\mu)} + \frac{\lambda\pi_{N,1}z^{N+1}}{(\lambda z - c\mu)}$$

(16)

Where

$$Q(z) = \sum_{n=1}^c (c-n)\pi_{n,1}z^n$$

Equations (15) and (16) express the probability generating function of all the servers on vacation and working.

V. Expected System Size

When servers are on vacation and working, the system capacity is indicated by $\mathbb{L}_0$ and $\mathbb{L}_1$, respectively. The expected system sizes are elucidated by

$$E[\mathbb{L}_0] = \sum_{n=1}^N n\pi_{n,0}$$

$$E[\mathbb{L}_1] = \sum_{n=1}^N n\pi_{n,1}$$

$$G_0'(1) = \frac{d}{dz}G_0(z)]_{z=1} = E[\mathbb{L}_0]$$

$$G_0''(1) = \frac{d^2}{dz^2}G_0(z)]_{z=1} = E[\mathbb{L}_0(\mathbb{L}_0-1)]$$

First we workout $E[\mathbb{L}_0]$: Applying L'hospital rule, we get from equation (15)

$$G_0(z) = \frac{\frac{1}{k}\frac{H}{\xi}k(1-z)^{-1} + \frac{\lambda}{\xi}\pi_{N,0}z^N}{\frac{\lambda}{\xi} + \frac{\gamma}{\xi}(1-z)^{-1}}$$

$$\gamma G_0(1) = H$$

(17)

Applying L'hospital rule, we get from equation (16)

$$G_1(1) = \lim_{z\to 1}G_1(z)$$

$$= \lim_{z\to 1}\left[\frac{[\gamma G_0(z) - H]z}{(1-z)(\lambda z - c\mu)} - \frac{\mu Q(z)}{(\lambda z - c\mu)}\right.$$

$$\left.+ \frac{\lambda\pi_{N,1}z^{N+1}}{(\lambda z - c\mu)}\right]$$

$$G_1(1) = \frac{\gamma E[\mathbb{L}_0]}{c\mu - \lambda} + \frac{\mu Q(1)}{c\mu - \lambda} - \frac{\lambda}{c\mu - \lambda}\pi_{N,1} \quad (18)$$

Using equation (10) and applying L'hospital rule to get

$$E[\mathbb{L}_0] = \lim_{z \to 1} G_0'(z) = \frac{-\lambda G_0(1) + \gamma G_0'(1) + \lambda \pi_{N,0}}{-\xi}$$

$$G_0(1) = \frac{\gamma + \xi}{\lambda} E[\mathbb{L}_0] + \pi_{N,0} \qquad (19)$$

Adopting from (18) and (19), we observing that $G_0(1) + G_1(1) = 1$

$$E[\mathbb{L}_0] = \frac{\lambda(c-\rho)(1-\pi_{N,0})}{\left[\gamma c + \xi(c-\rho)\right]} - \frac{\lambda Q(1)}{\gamma c + \xi(c-\rho)}$$

$$+ \frac{\lambda \rho \pi_{N,1}}{\left[\gamma c + \xi(c-\rho)\right]} \qquad (20)$$

where $\rho = \dfrac{\lambda}{\mu}$

Next we derive $E[\mathbb{L}_1]$:

$$E[\mathbb{L}_1] = \lim_{z \to 1} G_1'(z)$$

$$E[\mathbb{L}_1] = \frac{\gamma(c\mu - \lambda)E[\mathbb{L}_0(\mathbb{L}_0 - 1)] + 2\gamma c\mu E[\mathbb{L}_0]}{2(c\mu - \lambda)^2}$$

$$+ \frac{(c-\rho)Q'(1) + \rho Q(1)}{(c-\rho)^2}$$

$$- \left[\frac{N\rho \pi_{N,1}(c-\rho) + c\rho \pi_{N,1}}{(c-\rho)^2} \right] \qquad (21)$$

where $\rho = \dfrac{\lambda}{\mu}$

$$Q(1) = \sum_{j=1}^{c}(c-j)\pi_{j,1}$$

$$Q'(1) = \frac{d}{dz}Q(z)]_{z=1} = \sum_{j=1}^{c}j(c-j)\pi_{j,1} \quad (22)$$

To obtain, differentiate equation (10) two times.

$$\xi(1-z)G_0'''(z) = [2\xi + \lambda(1-z) + \gamma]\,G_0''(z)$$
$$- 2\lambda G_0'(z) - \lambda \pi_{N,0}\,[N(N-1)z^{N-2}$$
$$- (N+1)Nz^{N-1}] \qquad (23)$$

$$G_0''(1) = \frac{2\lambda}{\gamma + 2\xi}G_0'(1)$$

$$E[\mathbb{L}_0(\mathbb{L}_0 - 1)] = \frac{2\lambda}{\gamma + 2\xi}E[\mathbb{L}_0] \qquad (24)$$

Substitute (24) in (21) after some mathematical simplification we get $E[\mathbb{L}_1]$

$$E[\mathbb{L}_1] = \frac{\gamma \rho}{(c-\rho)}\left[\frac{1}{\gamma + 2\xi} + \frac{c}{\lambda(c-\rho)}\right]E[\mathbb{L}_0]$$

$$+ \frac{(c-\rho)Q'(1) + \rho Q(1)}{(c-\rho)^2}$$

$$- \left[\frac{N\rho \pi_{N,1}(c-\rho) + c\rho \pi_{N,1}}{(c-\rho)^2}\right] \qquad (25)$$

Therefore,the expected system capacity could be estimated as $E[\mathbb{L}_0] + E[\mathbb{L}_1]$.

where $E[\mathbb{L}_0]$ and $E[\mathbb{L}_1]$ are shown in equations (20) and (25).

VI. Conclusions

We have interpreted in this paper a queueing system of *M/M/C* with finite capacity along with impatient customers as well as with synchronous vacation of the servers. Various models for customer impatience have been studied in the past; the inception of impatience is always taken to be queue size. In this model, we show that the customers' impatience is the result of the servers being on synchronous vacation. A cognate model has not yet been investigated in the literature. Unlike others, here exists the PGF for the Steady State in terms of two indices. We have derived system sizes for the system as well.

References

1. Levy, Y. and Yechiali, U.(1976), "An Queue with Servers' Vacations", Canadian J. of Operational Research and Information Processing, Vol. 14 , pp. 153–163.
2. Zhang, Z. G. and Tian, N. (2003), "Analysis of queueing system with synchronous single vacation for some servers", Queueing Systems, Vol. 45, pp. 161–175.

3. D. Yue, W. Yue, Z. Saffer and X. Chen. (2014), "Analysis of an queueing system with impatient customers and a variant of multiple vacation policy", Journal of Industrial and Management Optimization, Vol. 10, pp. 89–112.

4. Mokhtar Kadi, Amina Angelika Bouchentouf and Lahcene Yahiaoui. (2020), "On a Multiserver Queueing System with Customers' Impatience Until the End of Service Under Single and Multiple Vacation Policies", Applications and Applied Mathematics: An International Journal (AAM) Vol. 15, Issue 2 , pp. 740–763.

5. Selvaraju, N. and Goswami, C. (2013), "Impatient customers in an M/M/1 queue with single and multiple working vacations", Comput. Ind. Eng., Vol. 65, No. 2, pp. 207–215.

6. Wu, D. A. and Takagi, H. (2006)," queue with multiple working vacations",Performance Evaluation, Vol.63, pp. 654–681.

7. N. Tian and Z. G. Zhang. (2006), "Vacation Queueing Models", Theory and Applications, Springer, New York,

8. J.-C. Ke. (2007), "Operating characteristic analysis on the system with a variant vacation policy and balking", Applied Mathematical Modelling, vol. 31, no. 7, pp. 1321–1337.

9. Manoharan, P. and Majid S. (2017), "Stationary analysis of a multiserver queue with multiple working vacation and impatient customers", Application and Applied Mathematics, Vol. 12, No. 2, pp. 658–670.

10. Takacs, L. (1962), "Introduction to the Theory of Queues," Oxford University Press, New York.

11. E. Altman and U. Yechiali, (2006), "Analysis of customers' impatience in queues with server vacations", Queueing Systems, Vol. 52, pp. 261–279.

12. D. Yue, W. Yue, (2007), "Analysis of queueing systems with balking, reneging and synchronous vacations", in: The 2nd Asia-Pacific Symposium on Queueing Theory and Network Applications, QTNA, pp. 53–62.

13. Eitan Altman and Uri Yechial. (2008),"Infinite-Server Queues With System's Additional Tasks and Impatient Customers", Probability in the Engineering and Informational Sciences, Vol.22, pp. 477–493.

14. Peter O. Peter and R. Sivasamy, (2021), "Queueing theory techniques and its real applications to health care systems – Outpatient visits", International Journal of Healthcare Management,Vol 14,Taylor Francis, pp. 114–122.

15. Farhan Sufyan and Amit Banerjee, (2021), "Computation Offloading for Smart Devices in Fog-Cloud Queuing System", IETE Journal of Research, Taylor Francis, pp. 1–13.

16. Franco, A., Landfeldt, B., Körner, U. et al., (2022), "Statistical guarantee of timeliness in networks of IoT devices", Telecommun.Syst., Vol. 80, pp. 487–496, https:doi.org/10.1007/s11235-022-00919-w.

Recent Trends in Computational Intelligence and its Application – Sugumaran D. et al. (eds)
© 2023 Taylor & Francis Group, London, ISBN 978-1-032-48410-5

36

AUBIT: An Adaptive User Behaviour Based Insider Threat Detection Technique Using LSTM-Autoencoder

Ambairam Muthu Sivakrishna*, R. Mohan[1], Krunal Randive[2]

Department of Computer Science and Engineering,
National Institute of Technology, Tiruchirappalli, India

Abstract—The most destructive cyber-attacks are often carried out by trustworthy insiders rather than suspicious outsiders or advanced persistent threats. Insiders significantly impact external factors by circumventing procedures and hiding in plain sight, leading to a substantial loss of organizational resources. Moreover, the existing approaches detect insiders with higher false alarm rates, creating chaos within an organization and demoralizing the organization's routine. This research focuses on detecting insider threats on Carnegie Mellon University's CERT version 4.2 synthetic dataset by analyzing user behavior based on their activities. AUBIT, an Autoencoders-based Long Short-Term Memory (LSTM) approach, is presented to identify insiders with improved efficacy and lower false alarm rates. When compared to other baseline methods, the proposed method has competitive results with an accuracy of (93.63%) , precision (97.19%), false alarm rate (2.5%), and F1 score (96%).

Keywords—Insiders, Insider threat, User behaviour, LSTM, Autoencoders

I. Introduction

Cyber-attacks can break and harm enterprise businesses. Every organization is prone to two kinds of attacks. One is attacks from outsiders, and the other is attacks from insiders. The attack from outsiders can be tackled easily because of the footprints left behind. Cyber security faces great difficulty detecting an attack caused by an insider, i.e., a malicious person within the organization. On November 5, 2009, A United States Army major opened re and killed around thirteen people [1]. The investigation found that radical terrorist organizations influenced him. Most of them are proven by his web browsing history and emails, where he mostly researched suicide bombers and emailed some terrorist people. In this case, this incident could have been avoided if any proactive mechanism had detected it. It can be said that insider attacks can be seen way before it happens by studying the behavior from the web browsing history and email contents.

---
*Corresponding author: amskrishna240@gmail.com
[1]rmohan@nitt.edu, [2]krunalrandive@gmail.com

DOI: 10.1201/9781003388913-36

Fig. 36.1 Types of insiders

Source: https://kratikal.com/blog/insider-threats/

In Fig. 36.1 Insiders are categorized into three types malicious, negligent, and accidental. Insider Threat is commonly connected with malicious personnel that intends to cause harm to the rm. However, in reality, the derelict and accidental cases may inadvertently cause substantial damage [2]. The damage done to an organization due to malicious insiders includes a decline in the organization's market value, impeding of regular routines, and deprivation of intellectual properties, all of which contribute to financial loss for the organization. The 2021 U.S. State of Cybercrime survey stated among a number of cyber-attacks that occurred, insiders and 30% of informants do, and 25% specified incidents from malicious activities are more costly than attacks from outside [3]. Insider threats are generally enabled by a rise in the count of users with unauthorised access to private information, a rise in the number of machines accessing sensitive data, a quick development in technological sophistication, and a lack of user awareness and training. Income, loyalties, growing frustration, vengeance, persuasion, thrills, appreciation, obsequiousness, and many other factors can motivate insiders to engage in malicious behavior, as illustrated in Fig. 36.2.

In addition, today's technological developments also paves the way for increasing malicious insiders. To increase productivity, organizations provide employees with remote access to information from almost anywhere, anytime from almost all devices, which also causes the compromise of sensitive data from insiders. This approach focuses on insider detection based on user behavior classified as either ideal or malicious based on their activity. A couple of instances are analysed for attribute selection to tamper with malicious behavior. Further, the proposed AUBIT is trained using the chosen feature vectors. To detect insiders with improved precision and a reduced false alarm rate, a hybrid deep learning technique AUBIT is proposed. AUBIT is taught using conventional usage behavior, if variation with the standard procedure is tagged as an insider while comparing the results. The article is aligned in the order as follows. A literature survey of related works is detailed in section 2, and the proposed methodology is explained in section 3 followed by the experimental setup in section 4. Results and comparisons are in section 5, and finally the conclusion about the developed model is in section 6.

II. Related Works

For many years, detecting insider threats and cyber criminals have piqued the interest of many security scientists and researchers in companies and individuals. Because of numerous issues, identifying malicious activities carried out by insiders within an organization is difficult. Firstly, an employee within the organization they are accessing can steal sensitive information without any alarm. Secondly, malicious activities are done in the long run over multiple phases. Finally, real-time data for training algorithms is very scarce because organizations in which insider threat has occurred are unwilling to share the details

Fig. 36.2 Motivation for an insider attack

Source: https://www.linkedin.com/pulse/insider-threat-internal-sabotage-cpp-ciam-fsyl-risc

for fear of damaging their reputation. Periodical data classification of user activities in different daily scenarios is represented as the feature vector. A random under-sampling technique is applied to reduce data skewness and trained over the two-layered deep autoencoder neural network [4].

In [5], a supervised learning approach is proposed where data is converted into a stream of chunks with timeslots and applied One-Class Support Vector Machine (OCSVM) algorithm to identify insider instances. A tree structure profiling approach is applied to get feature vectors as it imitates the user activities, and daily observations are compared with various anomalous metrics [6]. Dimensional attributes of a framework and its features are extracted as time series data and classified using machine learning classifiers [7]. Similarly, a deep unsupervised method used to identify malicious network instances from organization records was proposed in [8] and observed the change in user activity patterns. In [9], distance-based techniques are compared with Hidden Markov Model to identify insiders.

The restricted data consideration gave unsatisfactory results with a poor detection rate. An ensemble of deep autoencoders to detect insiders using reconstruction error obtained between initial and later data based on the features extracted from logon/logo activity records, file operations, USB device accesses, and HTTP visits are used in [10]. In [11] deep learning-based approach was proposed as an anomaly detection problem. Behavioral features of users are extracted with LSTM using a sequence of user actions. These features are generated into fixed-size matrices to feed CNN for detecting anomalies. A new framework is proposed in which frequency and sequential features are extracted and employed over machine learning classifiers to identify insiders [12]. LSTM-based autoencoder is applied over the email records of the users with timestamps as features in [13]. CNN is

used where the records of user instances are converted as color profiles [14]. Visualization plays a vital role due to unique color codes, which help in characterizing the user's activity. An unsupervised ensemble voting approach for insider identification was presented in [15] using the Mean and Median difference. Variational autoencoders are proposed where the user records are converted to integers and used as feature vectors [16]. This depicts the many ways for detecting insider threats. Few methods are quite effective, also they have flaws in terms of intricacy, and execution interpretation analysis. LSTM & Deep Auto - encoders are the most utilized algorithms because they can be employed on real-world datasets and are pretty fast and succinct.

III. Proposed Approach

LSTM becomes handy in handling sequential data and smart enough to learn features on its own. Hence, it is used to model the user behavior. In contrast, autoencoders are applied to real-world problems that are prompt & coherent. LSTM autoencoders explicitly avoid

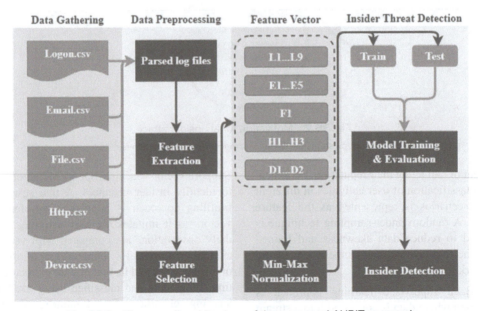

Fig. 36.3 The overall architecture of the proposed AUBIT approach

Source: Author

protracted reliance. The proposed AUBIT uses an organization's employees' resource access patterns to identify significant behavioral factors and have them as features to find patterns. Employee usage patterns on a routine basis are reflected as 1D feature vectors. An organization's usage records contain all the data needed to comprehend user behavior. Extracting influential factors from usage records yields feature vectors that represent user behavior. The odd pattern from the models is considered an anomaly, as shown in Fig. 36.3. The widely recognized CMU CERT Insider Threat dataset version 4.2 [17] is used for simulation in the first stage of AUBIT. The dataset is of a synthetic kind generated over a period of 500 days of an organization. It consists of one thousand employees' information and seventy-seven among them are intentional threat users. The second stage involves pre-processing of user access records and feature extraction. The user record files containing employees' raw resource access records in the form of rows and columns are extracted in this process of creating a feature vector with meaningful and relevant features. The dataset information is described in Table 36.1 shown below. The feature vector is framed in the third stage using a frequency-based approach [4]. These records are used to construct feature vectors, whereas every feature vector has 20 features described in Table 36.2. The features used in [18] are used in this work. The ground truth contains insight into the user, the devious events that occurred, and the date and time of the event.

Table 36.1 Contents of CMU CERT version 4.2 dataset

Dataset File	Description
logon.csv	Record of computer users signing into and out of
device.csv	Record of peripherals interactions and outages by users
http.csv	Record of users browsing history
email.csv	Record of mailing history
le.csv	Record of file usage patterns (duplicating a file to an external device)

Source: Author

Table 36.2 Feature vector description

S. no	Feature	Description
1	L1	Difference between first logon time and office start time
2	L2	Difference between last logon time and office ending time
3	L3	Avg. time gap among office start and logon times before office hours
4	L4	Avg. time gap among office end and logon times after house hours
5	L5	Total count of logins
6	L6	Count of logins after office hours
7	L7	Count of computers accessed
8	L8	Count of computers accessed after office hours
9	L9	Average session duration after office hours
10	E1	Count of emails sent to outside the organization
11	E2	Count of emails sent within the organization
12	E3	Count of attachments
13	E4	Average email size
14	E5	Count of recipients
15	F1	Count of executable(.exe) files downloaded
16	H1	Count of times websites related to Wikipedia are visited
17	H2	Count of times websites related to external sources are visited
18	H3	Count of times websites related to malicious sites are visited
19	D1	Count of drives used outside office hours
20	D2	Total count of times devices are used

For each malicious user, a separate file with details of activity that marked the user as a malicious insider, the timestamp of the event, and the log file is provided. Finally, the proposed LSTM-Autoencoder is used in the fourth stage to detect malicious insiders.

IV. Experimental Environment

The experimental environment includes a Windows operating system-based HP Workstation machine with an Intel(R) Core(TM) i7-10510U CPU running at 2.30 GHz, 32 GB of RAM, a 4GB Nvidia Quadro P520 GPU, and TensorFlow 2.6.0 with Keras 2.3.0 engine in Anaconda's Jupyter Notebook. The CERT version 4.2 dataset has 1000 users, 70 of whom are insiders. The collection comprises only 0.3% abnormal cases and 99.7% typical ones. To address the issue of class imbalance, the random over-sampling technique is applied. Duplicates of anomalous events are scattered over the dataset. For model training, validation, and testing, the dataset is separated into train, valid, and test data. The information is separated into the following proportions: 70%, 10 %, and 20%, in that order. Normal and malicious data are both included in testing data. There are a total of 3,27,70,227 logs with 15,146 trainable characteristics.

V. Results & Comparison

The model is trained with 500 Epochs, 64 batch size, 0.0001 learning rate, relu activation function for LSTM layers followed by Sigmoid for dense layer, and Adam optimizer. The input and output dimensions are intact. The deviation among their sequences is used as a loss function. Binary Cross-Entropy is used to calculate the loss function. While training AUBIT, the input is reconstructed to the output and data has been labeled 0 (normal) or 1(insider). AUBIT is analysed on both normal and malicous test samples. Insider users have a significantly larger reconstruction error than regular users. To distinguish between regular and harmful behaviour, a threshold value is determined. It is categorised as an Insider if the value is higher than the threshold; else, it is classified as Normal.

Table 36.3 depicts the performance of the AUBIT model over baseline approaches. LSTM CNN employed a feature set based on user activity. OCSVM used domain-based features, Autoencoders used user behavior-based features, Isolation Forest used psychometric results and browsing patterns, and AUBIT used a rich feature set based on shifts of use. The feature set and dataset used are two parameters that significantly impact the results. From the obtained results it can be observed that AUBIT performs well in detecting the insider threats with a lower false alarm rate which is key for the organizations to minimize the threat impact as depicted in Fig. 36.4.

Table 36.3 Performance comparison of AUBIT with standard techniques

S. no.	Algorithm	Evaluation Metrics(in %)			
		Accuracy	Precision	False alarm rate	F1 score
1	Isolation Forest [4]	79.00	82.39	NA	NA
2	OCSVM [6]	87.79	NA	NA	94.00
3	Deep autoencoders [10]	90.37	84.28	10.00	88.00
4	LSTM CNN [13]	91.85	89.50	NA	90.00
5	AUBIT	93.63	97.19	2.50	96.00

*NA- Considered metric Not Available

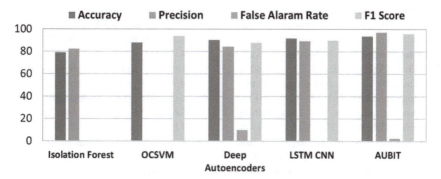

Fig. 36.4 Graph for performance comparison with standard techniques of AUBIT

Source: Authors

VI. Conclusion

It is probed that curbing the insider threat is a tricky problem. Nowadays, its remediation is attained by incorporating access control policies, user behavior surveillance, and physiological security measures. AUBIT presents a Deep Learning-based algorithm for an insider threat detection scheme. The basic aim of AUBIT is to pertain to user data sources within an entity with low computation and storage prerequisites. Furthermore, the formed system is simple and adaptable, requesting a slight knowledge base. Again, the proposed algorithm's performance was compared to other well-known techniques As per the comparison, our approach delivers pretty decent accuracy (93.89%), precision (95.43%), false alarm rate(2.5%) and F1 score (96%). We need to develop more diverse threat intelligence instances to build a reliable Insider detection system, as there aren't many available to the public threat scenarios. This need will enable us to solve insider problems with innovative thinking, good value, and correctness.

References

1. J. Jiang, J. Chen, K. K. R. Choo, K. Liu, C. Liu, M. Yu, and P. Mohapatra, Prediction and detection of malicious insiders' motivation based on sentiment prole on webpages and emails, in MILCOM 2018-2018 IEEE Military Communications Conference (MILCOM). IEEE, 2018, pp. 1–6.

2. W. Jiang, Y. Tian, W. Liu, and W. Liu, An insider threat detection method based on user behavior analysis, in International Conference on Intelligent Information Processing. Springer, 2018, pp. 421–429.

3. R. A. Alsowail and T. Al-Shehari, A multi-tiered framework for insider threat prevention, Electronics 10 (2021), p. 1005.

4. P. Chattopadhyay, L. Wang, and Y.P. Tan, Scenario-based insider threat detection from cyber activities, IEEE Transactions on Computational Social Systems 5 (2018), pp. 660–675.

5. P. Parveen, Z.R. Weger, B. Thuraisingham, K. Hamlen, and L. Khan, Supervised learning for insider threat detection using stream mining, in 2011 IEEE 23rd international conference on tools with articial intelligence. IEEE, 2011, pp. 1032–1039.

6. P.A. Legg, O. Buckley, M. Goldsmith, and S. Creese, Automated insider threat detection system using user and role-based prole assessment, IEEE Systems Journal 11 (2015), pp. 503–512.

7. A. Georgiadou, S. Mouzakitis, and D. Askounis, Detecting insider threat via a cyber-security culture framework, Journal of Computer Information Systems (2021), pp. 1–11.

8. A. Tuor, S. Kaplan, B. Hutchinson, N. Nichols, and S. Robinson, Deep learning for unsupervised insider threat detection in structured cybersecurity data streams, arXiv preprint arXiv:1710.00811 (2017).

9. O. Lo, W.J. Buchanan, P. Griths, and R. Macfarlane, Distance measurement methods for improved insider threat detection, Security and Communication Networks 2018 (2018).

10. L. Liu, O. De Vel, C. Chen, J. Zhang, and Y. Xiang, Anomaly-based insider threat detection using deep autoencoders, in 2018 IEEE International Conference on Data Mining Workshops (ICDMW). IEEE, 2018, pp. 39–48.

11. F. Yuan, Y. Cao, Y. Shang, Y. Liu, J. Tan, and B. Fang, Insider threat detection with deep neural network, in International Conference on Computational Science. Springer, 2018, pp. 43–54.

12. P. Matou†ek, V. Havlena, and L. Holk, Ecient modelling of ics communication for anomaly detection using probabilistic automata, in 2021 IFIP/IEEE International Symposium on Integrated Network Management (IM). IEEE, 2021, pp. 81–89.

13. S. Paul and S. Mishra, LAC: LSTM autoencoder with community for insider threat detection, in 2020 the 4th International Conference on Big Data Research (ICBDR'20). 2020, pp. 71–77.

14. V. Koutsouvelis, S. Shiaeles, B. Ghita, and G. Bendiab, Detection of insider threats using articial intelligence and visualisation, in 2020 6th IEEE Conference on Network Softwarization (NetSoft). IEEE, 2020, pp. 437–443.

15. D.C. Le and N. Zincir-Heywood, Anomaly detection for insider threats using unsupervised ensembles, IEEE Transactions on Network and Service Management 18 (2021), pp. 1152–1164.

16. E. Pantelidis, G. Bendiab, S. Shiaeles, and N. Kolokotronis, Insider threat detection using deep autoencoder and variational autoencoder neural networks, in 2021 IEEE International Conference on Cyber Security and Resilience (CSR). IEEE, 2021, pp. 129–134.

17. CERT, C.m.u., 2016. cmu cert synthetic insider threat test dataset. Dataset available at https://resources.sei.cmu.edu/library/assetview.cfm?assetid=508099.

18. R. Gayathri and Sajjanhar, Anomaly detection for scenario-based insider activities using cgan augmented data, in 2021 IEEE 20th International Conference on Trust, Security and Privacy in Computing and Communications (TrustCom). IEEE, 2021, pp. 718–725.

Recent Trends in Computational Intelligence and its Application – Sugumaran D. et al. (eds)
© 2023 Taylor & Francis Group, London, ISBN 978-1-032-48410-5

37

Segmentation of Brain MRI Images Using U-Net

Ramya D.*, C. Suryapunith[1], B. V. Satyendra Reddy[2], S. Mythri[3], B. Kalicharan[4]
Vel Tech Rangarajan Dr. Sagunthala R&D Institute of Science and Technology

S. K. Manigandan[5]
Vel Tech High Tech Dr. Rangarajan Dr. Sakunthala Engineering College

Abstract—A technique called U-net was primarily created for picture segmentation. Due to the significance of the U-growing net in the field of medical imaging, it is now the preferred tool for image segmentation jobs in the medical industry. Every use of U-net design, including in CT scans, MRI, X-rays, and microscopes, illustrates the efficiency of the technology. U-net is primarily thought of as a segmentation tool, but it is occasionally used for other purposes as well. It is considered the best architecture in the field of medical image segmentation. U-net strength is still increasing with so many new improvements in the architecture; this paper evaluates the growth and improvement in U-net architecture and compares it with other architectures like SegNet and LeNet. The experimental results demonstrate that the U-net model with various modifications outperforms SegNet and LeNet based approaches using the dataset BRATS2020.

Keywords—Segmentation, U-net, SegNet, LeNet, CNN, Classification

I. Introduction

Deep learning is widely used in the interpretation of medical images. During the last decade, there have been rapid developments in deep learning of computer vision. Still, it faces several difficulties in the field of image classification and segmentation using medical images. Many breakthrough strategies have been developed during past years to tackle these varied problems, and research always leads to the development of innovative and inventive methods.

U-net is an architecture that was created with picture segmentation in mind. There are two paths in a U-net architecture. The contraction path, often referred to as the encoder, is similar to conventional convolution design in that it provides classification information. The second one, known as the expansion path or decoder, consists of up-convolutions and connections with characteristics from the contraction path. Due to this broad channel, the network may now take in localised data. Furthermore, the expansion path raises the image's output resolution, which may subsequently be

*Corresponding author: raminfo84@gmail.com
[1]chithalurisuryapunith1999@gmail.com, [2]bethireddysatyendrareddy@gmail.com, [3]hyndavi678@gmail.com,
[4]charangkt@gmail.com, [5]kgmanigandan@gmail.com

DOI: 10.1201/9781003388913-37

passed to a terminal layer of CNN to receive a segmented image. The network has a u-shaped structure and is nearly symmetrical.

Most convolutional networks' principal duty is to assign the class to the entire image into a single label. U-nets create very detailed feature maps and segmentation maps using an extremely small number of samples makes it particularly helpful. The latter characteristic is particularly important in medical imaging, where exact labelled images are frequently insufficient. This is accomplished by applying changes to the data used for training at random, allowing the network to learn about the changes rather than the need for additional labelled data [1]. Finally, because of its context-based learning, compared to previous segmentation models, U-net takes significantly less time to train. In medical image processing, image segmentation is a crucial step that enables you to break an image into distinct pieces and analyse them in greater depth pathologically with varied degrees of precision and complexity. Various segmentation techniques have been proposed and utilised on brain pictures throughout the last decade.

With the use of magnetic resonance scans, Rapid advancements have been made in our comprehension of brain structure (MRI). To aid doctors in qualitative diagnosis, computerised approaches for MRI image segmentation, registration, and visualisation have been widely adopted. Because the brain tissues contain many irregularities and aberrant tissues such as tumours, brain MRI image segmentation is a very complex and demanding process. Artificial intelligence makes brain imaging analysis easier and more practical when dealing with vast amounts of data. Neural data is highly inconsistent, extremely intricate, and contains a wide range of signals. As a result, these layers improve the output resolution of the image. The contracting path consists of high-resolution characteristics which are merged with the upsampled output to localise. Based on this knowledge, a subsequent convolution layer can learn to create a more exact result.

II. Literature Review

The author Jose L Marroquin et al. [1] suggested the fully automatic 3-D segmentation method for brain MRI images. They used an efficient algorithm called expectation-maximization to find the optimal segmentation. Tohka et al. [2] introduced a framework for Image classification. A new image model has been created by combining the tissue intensities model and Markov random field. In the year 2014, B. H. Menze et al. [3] reported In collaboration with the MICCAI conferences in 2012 and 2013, the results of the Benchmark for Multimodal Brain Tumour Image Segmentation (BRATS) Author Despotic et al. [4] written the review paper for brain MRI segmentation. They discussed all the capabilities, advantages, and disadvantages of the methods. By combining adaptive mean shift based on the Bayesian formula, Q. Mahmood et alpaper .'s "Fuzzy logic with a priori spatial tissue probability maps" was published in 2015 [5]. created a technique for MRI brain image tissue segmentation that is unsupervised. Deep convolutional neural networks were utilised by Wenlu Zhang et al. [6] to segment the isotone's Newborn brain. This is very difficult as white and grey matter exhibit same level of intensity. Using CNN, they achieved a considerable improvement in segmentation. Olaf Ranneberger et al. employed U-net to segment biological images. Vijay Badrinarayanan et al. [9] [2016] developed a new architecture named SegNet for pixel-based segmentation in semantic segmentation. SegNet comprised of the encoder and decoder network after that pixel-based classification layer is there. Dong Yang [11] developed deformable models with multi-components coupled with Automated segmentation of heart walls and blood using 2D-3D U-Net. With all the limitations, this deformable model allows for the precise extraction of endo and epicardium outlines. Automated segmentation of heart walls and blood using 2D-3D U-Net. With all the limitations, this deformable model allows for the precise extraction of endo and epicardium outlines. H. Kakeya [13] proposed a paper that explains the deep learning approach Transfer learning for multi-organ CT segmentation using 3D U-JAPA-net. H. Li, A. Li proposed brain tumor segmentation architecture by modifying the existing U-Net architecture.

III. Methodology

A. U-Net Architecture

The fundamental U-net architecture (Fig. 37.1) is modified by changing the depth of the encoder and the number of convolution channels used at the encoder level of the design. By implementing these changes,

the performance of the segmentation can be improved to the default architecture. The default nnU-Net architecture has a value of 6 to a depth of the network. It is changed to the value of 7. The convolution channels used in the encoder are changed to 64, 96, 128, 194, 256, 384 and 512. The output is given in Fig. 37.2.

B. LeNet-5

LeNet-5 has a very straightforward architecture. One input layer, two pooling layers, and three dense layers make up the three convolution layers (1 output layer). It accepts a 28×28 image as input and outputs a list of probabilities for each class in the form of a 10-item list. LeNet can recognize the digits 0 to 9, as it was designed primarily to recognize handwritten

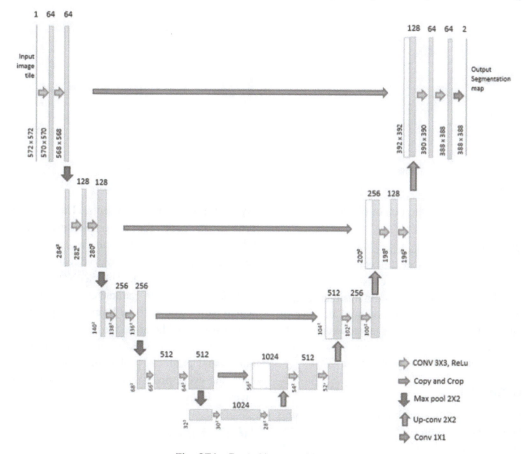

Fig. 37.1 Basic U-net architecture

Source: Author

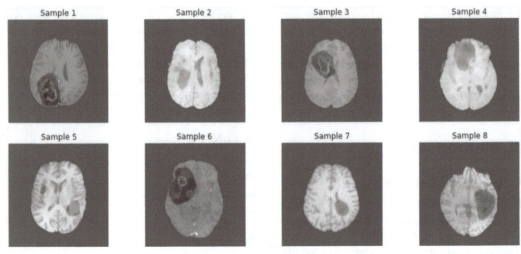

Fig. 37.2 U-Net segmented output

Source: Author

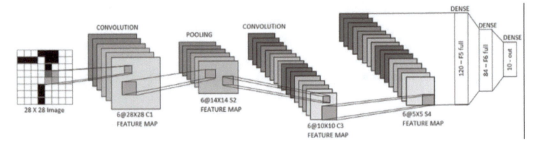

Fig. 37.3 Block diagram of convolutional neural network

Source: Author

digits. As a result, the output of the model has ten classes.

The LeNet dense block is composed of three layers, each of which has 120, 84, and 10 outputs and is fully connected. Because categorization is still being done, the 10-dimensional output layer corresponds to the potential number of output classes. It may have taken some time to completely comprehend the working of LeNet. The output of LeNet is given in Fig. 37.4.

C. SegNet

Pixelwise classification layer is carried out using Senet, which comprises an encoder network and associated decoder network. The VGG16 network's initial 13 layers are comparable to the encoder network's initial 13 convolutional

layers for categorizing objects. The training process is done for large datasets, along with weights. The fully connected layers are removed which helps to preserve feature maps with higher resolution. The SegNet encoder network uses fewer parameters as a result, which helps. The encoder layer corresponds to 13 convolutional layers in the decoder network. The soft-max classifier receives the decoder output and produces probabilities of class for each and every pixel. The segmented output is given in Fig. 37.5.

IV. Dataset

The four binary backgrounds, cerebrospinal fluid (CSF), grey matter (GM), and white matter

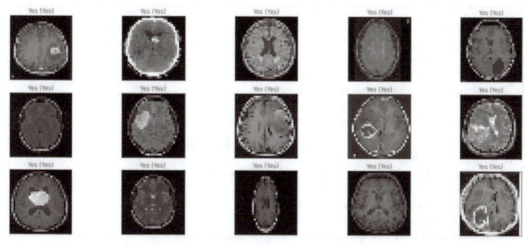

Fig. 37.4 LeNet segmented output

Source: Author

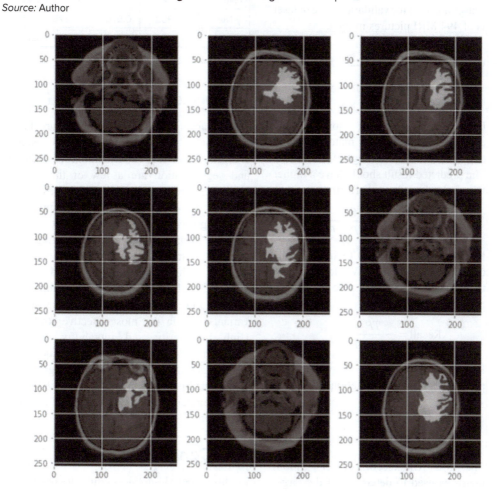

Fig. 37.5 SegNet segmented output

Source: Author

maps are the primary visual representations of the brain (WM). As a 4-channel target map, these four binary maps are taken into account before being submitted into the model with the training data. The 256×256 input image slice is divided into a number of patches, each of which is 128×128 pixels in size. For the aim of training the model, these patches are applied as inputs.

The architecture U-Net, LeNet-5 and Segnet are evaluated on Brain MRI images using the publicly available dataset BRATS 2020. The dataset contains both training and validation images which were collected from different online resources. 369 MRI images are used for training, while 125 MRI images from actual patient cases are used for validation. The dataset consists of 494 MRI pictures in total.

V. Evaluation Metrics

For evaluating the proposed method, four metrics are to be considered.

1. Dice Score measures how closely expected outcomes match actual outcomes. The value ranges from 0 to 1. The predicted result should have a higher Dice value in order to perform better.

$$\text{Dice Score} = \frac{2TP}{2TP + FP + FN}$$

2. Precision is the method to find out the correctness of the positive detections to the ground truth. The value should be high

$$\text{Recall} = \frac{TP}{TP + FP}$$

3. Recall measures how complete the positive results are in relation to the true value.

$$\text{Recall} = \frac{TP}{TP + FN}$$

4. The IoU (Intersection over Union) metric is used to determine if the target mask and the predicted output overlap.

Jaccard Index is another name for it. It is sometimes described as the proportion of common pixels between the target mask and output to all of the pixels.

$$\text{IoU} = \frac{\text{Target} \cap \text{Prediction}}{\text{Target} \cup \text{Prediction}}$$

VI. Result

Table 37.1 Comparison of U-Net with other architectures based on performance metrics

Architecture	Dice-Coefficient	Precision	Recall	Iou
U-Net	0.6141	0.9943	0.9928	0.456
LE-Net	0.4952	0.570	1.0223	0.2312
SEG-Net	0.5096	0.5098	0.9822	0.3554

Source: Author

VII. Discussion

Deep learning methods like U-net are being used more frequently in medical picture analysis. Detection, classification, localization, and segmentation are a few of the various tasks that deep learning in image processing entails. However, segmentation problems are recognised to be of great interest to those working in the field of medical imaging. This has been demonstrated by numerous research studies. This is further supported by the numerous publications on segmentation issues that have been published, here, U-net and its variants remain the most effective strategies. High modularity and mutability are two characteristics of U-net that make it a successful segmentation model. Le-Net and Segnet, two U-net versions, also perform well in the field of image segmentation. Precision, recall, and dice coefficient.

VIII. Conclusion

In this paper, U-net is mainly focussed and derived the result that it performs better than its variants if small changes are done in the basic architecture. The results are compared with the other architectures. U-Net may be compared with other variants of the architecture.

References

1. Marroquin et al., "A reliable and effective Bayesian approach for automatically segmenting the brain. MRI", IEEE Transactions on Medical Imaging, vol. 21, no. 8, 2002, pp. 934–945.
2. Tohka et al., "Magnetic resonance imaging, vol. 28, no. 4, pp. 557–573, 2010. "Brain MRI tissue categorization based on local markov random fields."
3. B. H. Menze et al., "IEEE Trans. Med. Imaging, vol. 34, no. 10, pp. 1993–2024, 2014. "The multimodal brain tumour image segmentation benchmark (BRATS)".
4. Despotovic et al., MRI segmentation of the human brain: challenges, methods, and applications," Computational and mathematical methods in medicine, vol. 2015, 2015.
5. Mahmood et al., "Automated MRI brain tissue segmentation utilising tissue probability maps and mean shift and fuzzy c-means," IRBM, 2015, vol. 36, no. 3, pp. 185–196
6. Zhang et al., "Deep convolutional neural networks for isointense newborn brain image segmentation using many modalities 2015, NeuroImage, vol. 108, p. 214–224.
7. Ronneberger et al., ""U-Net: Convolutional networks for the segmentation of biomedical imaging," 2015, CoRR, vol. abs/1505.04597
8. O. Ronneberger, P. Fischer, and T. Brox, "U-net: Convolutional networks for biomedical image segmentation," 2015, pp. 234–241.
9. Badrinarayanan et al., "SegNet: IEEE Transactions on Pattern Analysis and Machine Intelligence, vol. 39, no. 12, pp. 2481–2495, 2017. "A deep convolutional encoder-decoder architecture for picture segmentation.".
10. Q. Tong, M. Ning, W. Si, X. Liao, and J. Qin, "3D deeply-supervised U-net based whole brain segmentation," 2017, pp. 224–232.
11. D. Yang, Q. Huang, L. Axel, and D. Metaxas, "Multi-component deformable models combined with 2D–3D U-Net for automated probabilistic segmentation of heart walls and blood." 2018, pp. 479–483.
12. G. Zeng et al., "Latent3DU-net: Using radial MRI of the hip, a multi-level latent shape space limited 3D U-net automatically segments the proximal femur.," 2018, pp. 188–196.
13. H. Kakeya, T. Okada, and Y. Oshiro, "3D U-JAPA-Net: segmentation of the abdomen multi-organ CT using a combination of convolutional networks," 2018, pp. 426–433
14. S. R. Ravichandran et al., "Utilizing Computed Tomography and 3D Inception U-Net for Aorta Segmentation Angiography of the heart," 2019, pp. 1–4.
15. T. Wang et al., "MSU-Net: Multiscale Statistical U-Net for Real-time 3D Cardiac MRI Video Segmentation," 2019, pp. 614–622
16. J. Wu, Y. Zhang, and X. Tang, "Simultaneous Tissue Classification and Lateral Ventricle Segmentation via a 2D U-net Driven by a 3D Fully Convolutional Neural Network," 2019, pp. 5928–5931.
17. H. Li, A. Li, and M. Wang, "A new approach for segmenting brain tumours from beginning to end that uses enhanced fully convolutional networks," Comput. Biol. Med., vol. 108, pp. 150–160, 2019.
18. S. Bakas et al., "Improving cancer genome MRI datasets of atlas gliomas with professional segmentation labels and radiomic characteristics," Sci. Data, vol. 4, p. 170117, 2017.
19. S. Bakas et al., "Finding the most effective machine learning algorithms for the BRATS challenge's brain tumour segmentation, progression assessment, and overall survival prediction," ArXivPrepr. ArXiv181102629, 2018.
20. O. Lucena, R. Souza, L. Rittner, R. Frayne, and R. Lotufo, "Using silver standard masks, convolutional neural networks for skull-stripping in brain MR imaging AI medical art, vol. 98, pp. 48–58, 2019.
21. G. Li, L. Bai, C. Zhu, E. Wu, and R. Ma, "Convolutional Neural Networks-Based Synthetic CT Generation from MR Images is a New Approach," 2018, pp. 1–5.
22. Micha 1 Futrega, Alexandre Milesi, Micha 1 Marcinkiewicz, Pablo Ribalta, "Optimized U-Net for Brain Tumor Segmentation", arXiv:2110.03352v2, 2021

Recent Trends in Computational Intelligence and its Application – Sugumaran D. et al. (eds)
© 2023 Taylor & Francis Group, London, ISBN 978-1-032-48410-5

38

Forecasting COVID-19 Cases Using Machine Learning

Ramya D.*, Deepa J.[1], Surya Srinivasan[2], P. S. Karthik Srinath[3]
Vel Tech Rangarajan Dr. Sagunthala R&D Institute of Science and Technology

S. K. Manigandan[4]
Vel Tech High Tech Dr. Rangarajan Dr. Sakunthala Engineering College

J. Velmurugan[5]
SIMATS School of Engineering

Abstract—COVID-19 is a massive and cruel pandemic that the current generation of people experienced in their lifetime. It is a new disease that doctors cannot identify the exact cause and treatment of the disease. Many people died due to this COVID-19 disease even small children. Countries around the world faced an economically backward situation, and people have lost their growth for the past two years of their life. Researchers focus on Covid-19 disease and published many papers in journals and conferences. Machine learning algorithms and deep learning techniques were employed to find out the covid-19 cases and compare the results of the prediction around the world. Regression techniques play a crucial role in the prediction of diseases. This paper aims to predict the disease condition based on machine learning algorithms such as Support Vector Regression (SVR) and Polynomial Regression (PR) with the publicly available COVID-19 dataset.

Keywords—Machine learning, COVID-19, Support vector regression, Polynomial regression, Validation, Dataset

I. Introduction

The global spread of the COVID-19 disease has resulted in the handle of human suffering and financial misery. It has a detrimental impact on all facets of human existence and has grown to be a major burden on hospitals and medical staff. The way how the virus spreads and how many people contract the disease are important open-ended questions.

In the nation under study, there are wide variations in the trend of confirmed cases. There is a lot of variation in the response curves as well, and knowing when the virus has peaked can influence social isolation strategies. With this project, ML made progress in the field

---
*Corresponding author: raminfo84@gmail.com
[1]vasdeepa03@gmail.com, [2]suryasrinivasan76@gmail.com, [3]karthiksrinath2001@gmail.com, [4]kgmanigandan@gmail.com, [5]velmuruganj.sse@saveetha.com

DOI: 10.1201/9781003388913-38

of medicine. Large datasets are interpreted, analyzed, and their results are predicted by utilizing ML approaches. The samples were divided into treatment groups and the symptoms of the condition were identified using these ML methods. Hospitals use machine learning to manage operations and treat infectious diseases. Methods from the field of ML have been used to treat a wide range of diseases and disorders in the past, including cancer, pneumonia, diabetes, and neuromuscular disorders. They have a prediction/forecast accuracy rate of greater than 90%.

A dangerous virus that has claimed countless lives throughout the world is the COVID-19 pandemic sickness. For this infection, there is no cure. ML methods have been used to envision whether or not people are contaminated with the new virus using data from the World Health Organization and the Centers for Disease Control and Prevention. In the field of radiology, ML is also used to interpret x-ray images and make a diagnosis. A COVID-19 infection, for instance, might be identified in a pulmonary imaging study. Furthermore, the social distance may be tracked by ML using this method, allowing us to protect ourselves against COVID-19.

Predicting the infection can also assist in determining the amount of hospital equipment that will be needed. It can assist nations with future planning and resource allocation. This study attempts to forecast the confirmed infection levels five days in advance. These techniques rely heavily on presumptions about how the virus spreads, and develops in a patient, and which individuals are most vulnerable. It is preferable to avoid having such biases when researching a novel epidemic to avoid skewing the data. In order to model the transmission of illness, Dandekar and Barbastathis [2020] built a Neural Network, enabling non-linear associations to be seen, which happens when quarantine is there in a particular location. To estimate the number of confirmed illnesses in various nations, we examined three machine learning algorithms.

Future events are predicted and anticipated using a variety of ML approaches. Support Vector Machines, Linear Regression, Logistic Regression, Naive Bayes, and Decision Trees are among ML approaches for prediction.

The Moving Average, Simple Exponential Smoothing, Holt's Linear Trend Model, Holt-Winters model, Seasonal Auto-Regressive Integrated Moving Average Exogenous model (SARIMAX), and the Auto-Regressive Integrated Moving Average (ARIMA) are all Machine Learning methods that are used to make predictions about the future in comparison to the naive approach.

Each method has its own application and strengths, but they are all premised on precision. The evaluation process is where the model with the best track record for making accurate predictions or forecasts is selected. Similar to how we used the ETC to predict COVID-19 symptoms, when trying to foretell how many cases of COVID-19 would be confirmed in India, discovered that the ARIMA model gave the most reliable results.

A flowchart of the Machine Learning process Fig. 38.1 is shown in the image below. It explains how data are gathered, prepared, and split into training and test datasets for instruction and performance assessment.

II. Literature Survey

Covid-19 was spreading widely around the world from Wuhan in China. So much research works have been done based on the new virus. Our goal is to tackle the problem with new, scant, and uncertain information. The research on COVID-19 was based on the biological knowledge of SARS/MERS, however, this is not always accurate (Chinazzi et al. [2020][1], Peeri et al. [2020][2]). People are divided into four categories according to recent research (Read et al. [2020][3], Tang et al. [2020][4], and

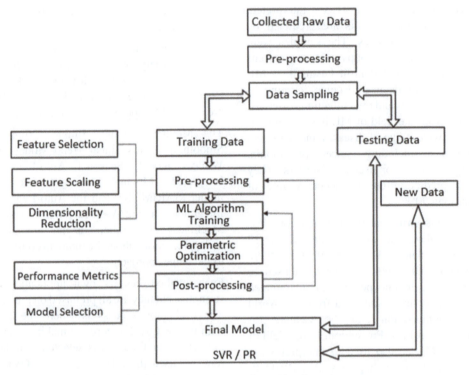

Fig. 38.1 The process of machine learning

Source: Author

Wu et al. [2020][5]), utilizing an SEIR model: susceptible, Exposed, Infected, and Resistant. At the 9th International Conference System Modeling and Advancement in Research Trends (SMART), (Akshay Kumar Siddhu) et al. [6] proposed a paper on Covid -19 by using X-ray and CT pictures from patients who had Pneumonia symptoms like Cough, cold, and fever, segregated the infected and non-infected patients. At the 2020 International Conference for Emerging Technology (INCET), the authors suggested using radiology to detect infected patients. They argued that the X-ray is the best way to determine the Covid-19 patients. Aniello Castiglion et al. [7] used convolutional neural networks to automatically detect coronavirus using CT images. They developed a new model, namely ADECO-CNN, and achieved 99.99% accuracy. Shayan Hassantabar et al. [8] developed a model named CovidDeep using Wearable Medical Sensors (WMSs) and

Deep Neural Networks (DNNs) for testing Covid-19. They obtained a test accuracy of 98.1%. Qiuwei Wan et al. [9] discussed the performance of different machine learning models in identifying disease at the 2020 International Conference on Public Health and Data Science (ICPHDS). Luzheng Bi et al. [10] used Support Vector Regression for the value of modeling human performance. They applied this algorithm to model pedestrian detection for passenger vehicles.

III. Proposed Method

The proliferation of the coronavirus has put civilization in danger of losing its social life. Moreover, it is essential to look into transmission expansion in the future and forecast transmission occurrences. One component of machine learning algorithms is regression analysis. The best machine learning algorithm is this one. Consider a straight equation line that combines any two variables, *X* and *Y*, and that can be written as:

$$Y = aX + b$$

The intercept on the y-axis a, where b is stated, is referred to as the line's slope. Therefore, a and b are however referred to as the regression analysis parameters. These criteria should be taught using appropriate teaching techniques. One or more predictor variables' values can be used in regression analysis together with a variety of machine learning techniques to forecast the value (*X*) of a continuous result variable (*Y*). The relationship between the outcome and the predictor factors is ongoing. For example, in parallel, state-of-art numerical methods are selected based on machine learning for a computer process to anticipate the transmission of the virus:

The Python library is used widely to execute machine learning and deep learning algorithms to calculate the total number of confirmed, recovered, and death cases. This forecast will enable particular decisions depending on the rate of transmission increase, such as lengthening the lockdown period, carrying out the sanitation strategy, and supplying ongoing assistance and supplies.

Arthur Samuel said that "Machine Learning (ML) is the area of research that enables machines to understand without being explicitly programmed". As a result, we could say that Machine Learning is the subset of computing that allows the creation of intelligent, self-teaching machines.

Simply said, learning occurs through experience, observations from prior work, such as samples, or teaching. This process involves searching for patterns in data and using examples to assist draw conclusions. The primary priority of ML is to automate computer learning so that it can be adjusted and used to take appropriate action.

The model is trained using historical data, and it is then tested against fresh data and utilized to make predictions. The effectiveness of the

trained ML model is assessed using a piece of historical data. The validation process is the general term for this. The ML model is assessed in this procedure based on performance metrics like accuracy. Measurement of a Machine Learning model's precision is achieved by the division of a number of features by the number of features that were properly predicted.

A. Support Vector Regression

The widely-used machine learning technique known as Support Vector Machine (SVM) is put to use for this purpose. Like SVM, SVR applies the central tenet to a specific problem domain; in this case, regression. SVMs frequently employed to solve grouping and sorting issues. Regression using SVMs is not always well-documented. SVRs are these models. When utilizing regression, it is challenging to find a function that roughly maps the input domain to the actual number in the training sample. SVR focuses primarily on taking into account factors within a decision's parameters. With a fair error margin and an acceptance adjustment that is larger than the needed error rate, the popular SVR approach is versatile and lets you choose the degree of error tolerance. Like Linear Regression, Support Vector Regression assumes that the world is a straight line with the equation $y = wx + b$.

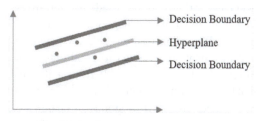

Fig. 38.2 Support vector regression
Source: Author

In SVR, this is called a hyperplane. The closest data points on either side of a hyperplane are used as support vectors to represent the boundary line. The objective of the SVR, in comparison to other regression models, is not to reduce the difference in the middle of the actual and predicted values but rather to match up the best line within a threshold value. (Separation of the hyperplane from the boundary line). Therefore, this $y = wx + b$ is a goal of the SVR model. The locations along this boundary were used to determine the value. In SVR, this is called a hyperplane. The closest data points on either side of a hyperplane are used as support vectors to represent the boundary line. In contrast to other regression models, the goal of the SVR is to fit the best line within a threshold value, rather than to minimize the difference between the actual and predicted values (Separation of the hyperplane from the boundary line in Fig. 38.2).

B. Polynomial Regression

By fitting a polynomial equation to the data with a curvilinear link between the selected variable and the independent variables, it may perform polynomial regression, a particular case of linear regression. When a connection is curvilinear, the predictor's value affects the target variable in a non-linear way. With regard to the unique instance of Multiple Linear Regression in ML, the equation for multiple linear regression is changed to the equation for polynomial regression by adding certain polynomial terms. It is indeed a linear model with various adjustments made to improve accuracy. ML is often used in industries, including medical, to diagnose and forecast disease. Regarding medicine, timely diagnosis and treatment are essential to good outcomes. Several fatalities might result from a therapy that has a high mistake rate. Therefore, scientists have started using AI programs in healthcare. Because it's a question of life or death, the researchers' choice of the best equipment makes the work challenging. The equation for polynomial regression is as follows:

$$y = b_0 + b_1 x + b_2 x^2 + \ldots\ldots b_n x^n$$

For polynomial regression, a non-linear dataset is used for training (Fig. 38.3). Fitting complex, non-linear functions and data sets with a linear regression model.

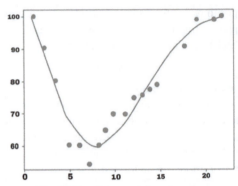

Fig. 38.3 Polynomial regression

Source: Author

Table 38.1 MAE and MSE values of two algorithms

Algorithm	MAE	MSE
Polynomial Regression	0.2153264485	0.4693112637
Support Vector Regression	0.3152274485	0.5683102611

Source: Author

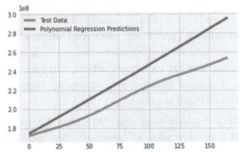

Fig. 38.4 Prediction of polynomial regression vs. test data

Source: Author

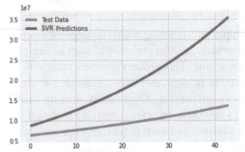

Fig. 38.5 Prediction of support vector regression vs. test data

Source: Author

IV. Result

The loss function like Mean Absolute Error (MAE) and Mean Squared Error (MSE) are calculated for Polynomial regression and Support Vector Regression. Both functions represent the prediction ability of the algorithm. They are the performance metrics for regression problems. Polynomial regression performs better when compared with Support vector regression.

V. Conclusion

The COVID-19 pandemic has now spread to every continent on the planet. It will have spread to over 85 countries by April 2020. According to their respective governments, the USA and India claim that vaccines had developed and are currently being tested as a result of scientists' tireless efforts to discover solutions. Scientific research has made extensive use of computers for early forecasting. To counteract COVID-19, ML is the subject of extensive investigation. The stability and expansion of COVID-19 were particularly predicted by the chapter using the ARIMA time technique. Due to COVID-19, there have been several nations with significant mortality tolls. It is hypothesized that by providing the model with access to more data, its performance can be enhanced and it can produce more reliable results. It is clear that by providing the model with access to more data, its performance can be enhanced and it can produce more reliable results. Based on the information provided by health authorities, the model generates conclusions. As a result, predicting may not be totally accurate, but it may still be utilized as a corrective action. Research development may be improved further by mixing additional factors and algorithms with ARIMA to provide findings that are more precise.

References

1. Dianbo Liu; Leonardo Clemente; Matteo Chinazzi, "A machine learning methodology for real-time forecasting of the 2019–2020 COVID-19 outbreak using Internet searches, news alerts, and estimates from mechanistic models" arXiv:2004.04019v1 [stat.OT] 8 Apr 2020.

2. Noah C Peeri; Nistha Shrestha; Md Siddikur Rahman, "The SARS, MERS and novel coronavirus (COVID-19) epidemics, the newest and biggest global health threats: what lessons have we learned" International Journal of Epidemiology, Volume 49, Issue 3, June 2020.

3. Jonathan M. Read; Jessica R.E. Bridgen; Derek A.T. Cummings, "Novel coronavirus 2019-nCoV: early estimation of epidemiological parameters and epidemic predictions" Version 2. Updated 27 Jan 2020.

4. BiaoTang; Nicola Luigi Bragazzi; Qian Li; Sanyi Tang, "An updated estimation of the risk of transmission of the novel coronavirus (2019-nCov)" Infectious Disease Modelling Volume 5-2020.

5. Chaomin Wu; Xiaoyan Chen; Yanping Cai, "Risk Factors Associated With Acute Respiratory Distress Syndrome and Death in Patients With Coronavirus Disease 2019 Pneumonia in Wuhan, China" JAMA Intern Med. 2020;180(7): 934–943. doi:10.1001/jamainternmed.2020.0994.

6. Akshay Kumar Siddhu; Ashok Kumar; Shakti Kundu, "Review Paper for Detection of COVID-19 from Medical Images and Symptoms of Patient using Machine Learning Approaches" 2020 9th International Conference System Modeling and Advancement in Research Trends (SMART), Proceedings of the SMART–2020, IEEE Conference ID: 50582.

7. Aniello Castiglione; Pandi Vijayakumar; Michele Nappi; Saima Sadiq; Muhammad Umer, "COVID-19: Automatic Detection of the Novel Coronavirus Disease from CT Images Using an Optimized Convolutional Neural Network", IEEE Transactions on Industrial Informatics, Volume: 17, Issue: 9, Sept. 2021.

8. Shayan Hassantabar; Novati Stefano; Vishweshwar Ghanakota, "CovidDeep: SARS-CoV-2/COVID-19 Test Based on Wearable Medical Sensors and Efficient Neural Networks", IEEE Transactions on Consumer Electronics, Volume: 67, Issue: 4, Nov. 2021.

9. Qiuwei Wang, "Performance of Different Models of Machine Learning in Predicting the COVID-19 Pandemic", 2020 International Conference on Public Health and Data Science (ICPHDS)" 20-22 November 2020.

10. Luzheng Bi; Omer Tsimhoni; Yili Liu, "Using the Support Vector Regression Approach to Model Human Performance," IEEE Transactions on Systems, Man, and Cybernetics - Part A: Systems and Humans (Volume: 41, Issue: 3, May 2011).

Recent Trends in Computational Intelligence and its Application – Sugumaran D. et al. (eds)
© 2023 Taylor & Francis Group, London, ISBN 978-1-032-48410-5

39

Efficient Pollutants Assessment of Larger Water-Bodies using Sentinel-2 MSI

Kumaresan M.*

Junior Research Fellow, Centre for Autonomous System Research,
Vel Tech Rangarajan Dr. Sagunthala R&D Institute of Science and Technology,
Avadi, Chennai, India

E. Balasubramanian[1]

Professor, Department of Mechanical Engineering,
Vel Tech Rangarajan Dr. Sagunthala R&D Institute of Science and Technology, Chennai, India

S. Vishnu Kumar[2]

Assistant Professor, Department of Electronics & Communication Engineering,
Vel Tech Rangarajan Dr. Sagunthala R&D Institute of Science and Technology, Chennai, India

Ch Nirmal Prabhath[3]

PhD Research Scholar, Department of Computer Science and Engineering,
Vel Tech Rangarajan Dr. Sagunthala R&D Institute of Science and Technology,
Avadi, Chennai, India

Abstract—Water bodies are considered a living heritage from which our ancestors retained a healthy environment. However, rapid urbanization and anthropogenic activities such as industrial waste discharge, sewage, and other household garbage directly into freshwater began to have an impact on the biological health of wetland ecosystems. In terms of analyzing water quality, satellite data products-based water quality monitoring is extremely helpful in drawing conclusions about water quality based on its spectral reflectance behavior. The current research focuses on developing a method for indirectly quantifying the presence of pollutants in larger bodies of water. In multispectral remote sensing, nitrate, phosphates, and ammonia are considered optically inactive parameters. Chlorophyll-a (CHL-a) and turbidity, on the other hand, are regarded as optically active components with adequate spectral reflectance properties. The water quality was estimated by establishing an empirical relationship between optically active and inactive parameters of the water-body. A linear relationship is established based on the CHL-a concentration using a two-band ratio model CHL - an a (VNIR/red) and a three-band ratio model CHL - an a [(1/red- 1/VNIR) *VNIR]. As a result, the relevant in-situ samples for quantifying the pollutants are collected and measured. According to the findings, multispectral remote sensing data for monitoring CHL-a can be used to indirectly quantify the pollutants like nitrate, phosphates, and ammonia.

Keywords—Sentinel-2, Multispectral remote sensing, Chlorophyll-a, Water quality, Two band algorithm, Three band algorithm

*Corresponding author: kumaresh_m@outlook.com
[1]esak.bala@gmail.com, [2]cmail2jitvishnu@gmail.com, [3]dnirmalprabhath@outlook.in
DOI: 10.1201/9781003388913-39

I. Introduction

Water pollution is a major concern due to its serious harmful effects on human health (Dokulil et al., n.d.; Mokaya et al., 2004). The various anthropogenic activities across the water bodies also affect the habitats of the aquatic systems. Rapid urbanization and industrial growth have increased the release of toxic-biological, organic, and chemical compounds getting discharged into water-bodies. Thus, it is essential to perform water quality inspections at regular intervals over water reservoirs for efficient water resource management to ensure a healthy ecosystem and for sustainable water supply for human beings (Bai et al., n.d.).

Water quality was measured with respect to its biological, chemical and physical properties. The traditional way of measuring the quality of water is by collecting water samples from the field randomly and performing analysis in the laboratory it is time-consuming and tedious.

Conventional point sampling analysis is not an efficient method to understand the quality of the water across larger lakes, rivers and other water bodies (Patra et al., n.d.). The larger water bodies are difficult to access for in-situ measurements, and the accuracy and precision of collected data may not be accurate. But with the advent of remote sensing capabilities and increased computing abilities, large water bodies can be effectively monitored. It gives a synoptic and complete view of water bodies covering the vast regions (Ansper & Alikas, n.d.). It also provides a comprehensive history of various parameters affecting water quality with respect to spatial and temporal variation that can be effectively captured (Soomets et al., 2020).

From the literature, it is evident that, spectral reflectance characteristics from sentinel satellites are predominantly used in satellite data products in the field of water quality aspects with respect to the various spectral band ratios (Grendaitė et al., n.d.). CHL-a has a dominant effect in the visible spectral regions, especially in the blue and green spectral ranges. In order to estimate the CHL-a content using sentinel data, an empirical approach-band ratio is utilized (Sent et al., n.d.).

This paper is further formulated as follows: section two defines the methodology and materials employed. Section three discourse the results obtained. Section four emphasis the author's conclusion and the future line of study.

II. Materials & Methodology

A. Satellite Data

As the authors (Spoto et al., 2012), indicated, the Sentinel-2 is a satellite with Multispectral Imager (MSI) which has three different spatial resolutions (10 m, 20 m, and 60 m) and works with thirteen spectral bands as described in Table 39.1. It also has a radiometric resolution of twelve bits.

Table 39.1 Sentinel-2: Spatial resolution and their respective bands

Spatial Resolution	Spectral Bands (nm)
10 m	B2:490, B3:560, B4:665 and B8:842
20 m	B5:705, B6:740, B7:783, B8a:865, B11:1610 and B12:2190
60 m	B1:443, B9:940 and B10:1375

Source: https://sentinels.copernicus.eu/web/sentinel/user-guides/sentinel-2-msi/resolutions/spatial

B. Data Product Used

S2A_MSIL1C_20220614T045711_N0400_R119_T44PMV_20220614T065511 is downloaded from https://scihub.copernicus.eu and used.

C. Study Area

The Kollimedu lake is roughly 175 acres in extent and is located at 13.163676° latitude and 80.104406° longitude as represented in Fig. 39.1. The lake is surrounded by a sparsely colonized residential area due to which, waste

Fig. 39.1 Kollimedu lake with marked sample points

Source: Google earth imagery 2022

materials are being disposed of near the lake's eastern and western parts of lake sewage inflow was observed. Human defecation was also observed in the foreshore areas and it was also observed the water is used by locals for bathing and fishing. The water was also seen to be more turbid in nature.

D. Procedure

Water samples were collected using a random sampling method. The assessment of pollutants was planned spatially within the lake's extent. CHL-a concentration was used as an indicator to establish the expected relationship between pollutants, the level of CHL-a was estimated using a remote sensing method (A. A. Gitelson et al., 2007). The values in remote sensing were derived from the object's spectral reflectance properties. Atmospheric effects are important in this reflection. Because of the lower reflectance

of the water surface (Mobley et al., 2016), atmospheric scattered radiance accounts for 90% of the reflectance reaching the sensor. Thus, sen2cor atmospheric correction was used to obtain an atmospherically corrected reflectance (Hadjimitsis & Clayton, n.d.). Water masking was applied for 10 m resampled imagery to avoid unwanted scattering from land features. Normalized differential water index was calculated and the water pixel alone was isolated to demarcate the water spread area. The band combination ratio such as the two-band and three-band algorithms were used as defined in equations 1 and 2 for obtaining the concentration of CHL-a. As mentioned by authors (A. Gitelson, n.d.), an initial absorption peak near to blue region (450 nm) and a second absorption peak near to red region (665 nm) were observed in surface water reflectance where chlorophyll is present. Since the lake under study was visually turbid and nutrient-rich, scattering of the medium was reduced by using a third band of 740 nm which is less sensitive to CHL-a pigment. The CHL-a pigment concentration was calculated by multiplying the inverse difference of the NIR-Red ratio by the third band.

$$CHL\text{-}a \propto K_{n/r} \qquad (1)$$

$$CHL\text{-}a \propto (M * VNIR) \qquad (2)$$

where, $K_{n/r} = \dfrac{VNIR}{Red}$ and $M_{n/r} = \dfrac{1}{VNIR} - \dfrac{1}{Red}$

As communicated by authors (Ogashawara et al., n.d.; Sent et al., 2021), the empirical quantification of CHL-a $\left(\dfrac{mg}{m^3}\right)$ concentration was calculated using Equation 3 and Equation 4,

$$CHL\text{-}a = [61.32 * K_{n/r}] - 37.94 \qquad (3)$$

$$CHL\text{-}a = [56.31 * M_{n/r} * VNIR] - 19.89 \qquad (4)$$

Physical samples were taken from the sampling points, in total five samples were collected and analyzed in the laboratory for chemical characteristics. The Jal-TARA Water Testing Kit (*Jal-TARA Water Testing Kit*, n.d.) was used to conduct laboratory experiments. Ammonia, Phosphate, and Nitrate were estimated using chemical reagents as illustrated in Fig. 39.2.

III. Results and Discussion

The CHL-a was estimated using 2-band and 3-band images as illustrated in 3 and Fig. 39.4 respectively.

The two-band model has the highest R-square value of 0.933 for Ammonia as illustrated in Fig. 39.6, whereas, the three-band model has 0.901 as illustrated in Fig. 39.7.

Similarly, for phosphate the R square value was obtained as 0.899 in the two-band and 0.895 in the three-band, which are illustrated in Fig. 39.8 and Fig. 39.9 respectively.

Under laboratory testing, the presence of Nitrate was estimated as 10mg/l for all 5 samples and the linear relation cannot be established because of the uncertainties in the instrument as shown in Fig. 39.10, and Fig. 39.11.

Overall, it was inferred that, low nitrate content was present in the lake. It was also observed from these maps that comparatively out of five sample points, points 1 and 2 have higher CHL-a concentration and these two points lie on the foreshore area of the lake where runoff

Fig. 39.2 Sample collection and clinical testing

Source: Author

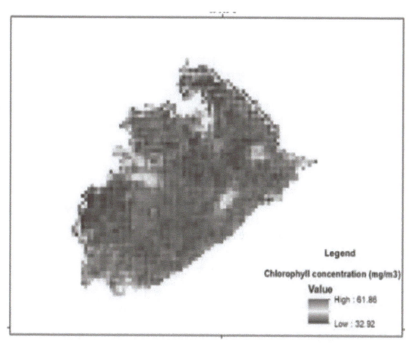

Fig. 39.3   Sentinel-2 imagery: CHL-a concentration - two-band ratio

Source: Author

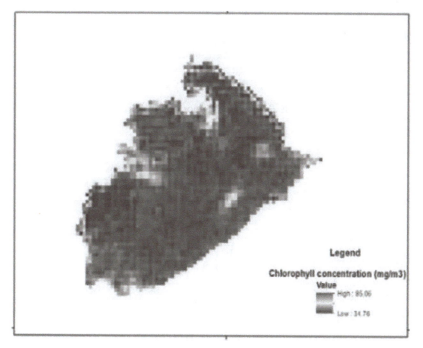

Fig. 39.4   Sentinel-2 imagery: CHL-a concentration - three-band ratio

Source: Author

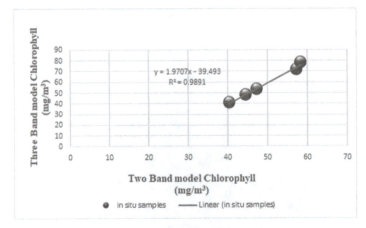

Fig. 39.5 Correlation among 2 and 3 bands

Source: Author

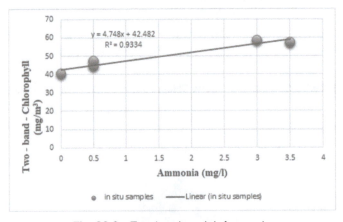

Fig. 39.6 Two-band model: Ammonia

Source: Author

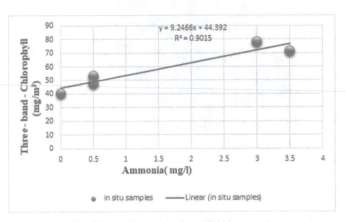

Fig. 39.7 Three-band model: Ammonia

Source: Author

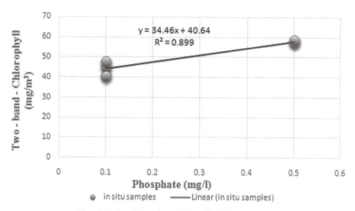

Fig. 39.8   Two-band model: Phosphate

Source: Author

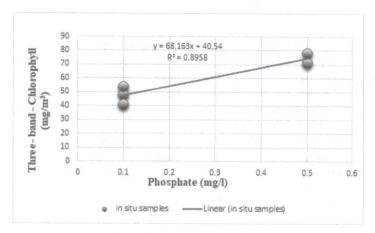

Fig. 39.9   Three-band model: Phosphate

Source: Author

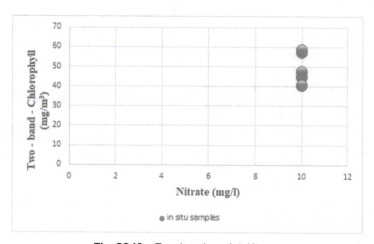

Fig. 39.10   Two-band model: Nitrate

Source: Author

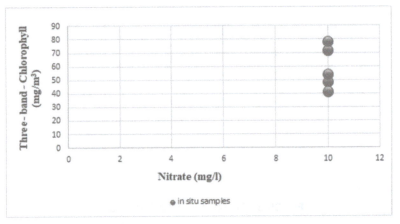

Fig. 39.11 Three-band model: Nitrate

Source: Author

enters the lake. Moreover, the GPS point of the sample point-1 and point-2 was very much closer to sewage inflow form the household, and sample point-2 was engulfed in an island-like shape where nutrients get stagnated. The rest of the points were located downstream of the lake where the concentration of CHL-a was seen to be slightly reduced. It was evident that the above-mentioned two-band and three-band model was used to derive notable inferences to map pollutant levels in water by using CHL-a concentration as a proxy.

IV. Conclusion

Inland water bodies have to be regularly monitored for addressing water quality issues and also to minimize degradation. In this study, the Remote Sensing method was utilized to estimate the chlorophyll concentration on Kollimedu lake using Sentinel-2 data by using two band model (CHL-a a 704/664) and a three band model (CHL-a a [1/664-1/704] × 740). Both the models were compared and they obtained R^2 value of 0.989 which confirms that they were fit for further analysis. In addition, the Chlorophyll map was obtained from the reflectance properties of water using satellite data. The Chlorophyll map derived has strong linear relation between ammonia and

phosphate. The two-band model has achieved R^2 values of 0.93 and 0.89 and the three-band model obtained R^2 values of 0.90 and 0.89 respectively. It was observed that ammonia and phosphate concentration was higher at the foreshore area (sample point 1) which is near to sewage inlet and lower near to bund (sample point 5) which was near to command area. Similarly, by considering other significant sampling points of interest of the lake, the chlorophyll concentration can be estimated and it will be related to obtaining the concentration of ammonia and phosphate. This study used limited samples and single season data. In the future, pre- and post-monsoon data will be considered and a greater number of in-situ samples to estimate the biological and chemical parameters of this lake will be performed.

References

1. Ansper, A., &Alikas, K. (n.d.). Retrieval of chlorophyll a from Sentinel-2 MSI data for the European Union water framework directive reporting purposes. *Remote Sensing, 11*(1), 64.
2. Bai, J., Chen, X., Li, J., Yang, L., & Fang, H. (n.d.). Changes in the area of inland lakes in arid regions of central Asia during the past 30 years. *Environmental Monitoring and Assessment, 178*(1), 247–256.

3. Dokulil, M., Chen, W., & Cai, Q. (n.d.). Anthropogenic impacts to large lakes in China: The Tai Hu example. *Aquatic Ecosystem Health & Management*, *3*(1), 81–94.

4. Gitelson, A. (n.d.). The peak near 700 nm on radiance spectra of algae and water: relationships of its magnitude and position with chlorophyll concentration. *International Journal of Remote Sensing*, *13*(17), 3367–3373.

5. Gitelson, A. A., Schalles, J. F., &Hladik, C. M. (2007). Remote chlorophyll-a retrieval in turbid, productive estuaries: Chesapeake Bay case study. *Remote Sensing of Environment*, *109*(4), 464–472. https://doi.org/10.1016/J.RSE.2007.01.016

6. Grendaitė, D., Stonevičius, E., Karosienė, J., Savadova, K., &Kasperovičienė, J. (n.d.). Chlorophyll-a concentration retrieval in eutrophic lakes in Lithuania from Sentinel-2 data. *Geologija. Geografija*, *4*(1).

7. Hadjimitsis, D. G., & Clayton, C. (n.d.). Assessment of temporal variations of water quality in inland water bodies using atmospheric corrected satellite remotely sensed image data. *Environmental Monitoring and Assessment*, *159*(1), 281–292.

8. *Jal-TARA Water Testing Kit*. (n.d.). Retrieved June 30, 2022, from https://www.techxlab.org/solutions/taraenviro-jal-tara-water-testing-kit

9. Mobley, C. D., Werdell, J., Franz, B., Ahmad, Z., &Bailey, S. (2016). *Atmospheric Correction for Satellite Ocean Color Radiometry*. http://www.sti.nasa.gov

10. Mokaya, S. K., Mathooko, J. M., &Leichtfried, M. (2004). Influence of anthropogenic activities on water quality of a tropical stream ecosystem. *African Journal of Ecology*, *42*(4), 281–288. https://doi.org/10.1111/J.1365-2028.2004.00521.X

11. Ogashawara, I., Kiel, C., Jechow, A., Kohnert, K., Ruhtz, T., Grossart, H. P., &Wollrab, S. (n.d.). The use of Sentinel-2 for chlorophyll-a spatial dynamics assessment: A comparative study on different lakes in northern Germany. *Remote Sensing*, *13*(8), 1542.

12. Patra, P., Dubey, S. K., Trivedi, R. K., Sahu, S. K., & Rout, S. K. (n.d.). *Estimation of chlorophyll-a concentration and trophic states for an inland lake from Landsat 8 OLI data: a case of Nalban Lake of East Kolkata Wetland*.

13. Sent, G., Biguino, B., Favareto, L., Cruz, J., Sá, C., Dogliotti, A. I., & Brito, A. C. (n.d.). Deriving water quality parameters using Sentinel-2 imagery: A case study in the Sado estuary, Portugal. *Remote Sensing*, *13*(5), 1043.

14. Sent, G., Biguino, B., Favareto, L., Cruz, J., Sá, C., Dogliotti, A. I., Palma, C., Brotas, V., & Brito, A. C. (2021). Deriving Water Quality Parameters Using Sentinel-2 Imagery: A Case Study in the Sado Estuary, Portugal. *Remote Sensing 2021, Vol. 13, Page 1043*, *13*(5), 1043. https://doi.org/10.3390/RS13051043

15. Soomets, T., Uudeberg, K., Kangro, K., Jakovels, D., Brauns, A., Toming, K., Zagars, M., & Kutser, T. (2020). Spatio-Temporal Variability of Phytoplankton Primary Production in Baltic Lakes Using Sentinel-3 OLCI Data. *Remote Sensing 2020, Vol. 12, Page 2415*, *12*(15), 2415. https://doi.org/10.3390/RS12152415

16. Spoto, F., Sy, O., Laberinti, P., Martimort, P., Fernandez, V., Colin, O., Hoersch, B., &Meygret,

17. A. (2012). Overview of sentinel-2. *International Geoscience and Remote Sensing Symposium (IGARSS)*, 1707–1710. https://doi.org/10.1109/IGARSS.2012.6351195

Recent Trends in Computational Intelligence and its Application – Sugumaran D. et al. (eds)
© 2023 Taylor & Francis Group, London, ISBN 978-1-032-48410-5

40

Safety and Legal Measures to Protect Women from Cyber Crimes

Sridevi J.[1]
Head - Department of Commerce and Business Administration,
Vel Tech Rangarajan Dr. Sagunthala R&D Institute of Science and Technology, Chennai

M. S. R. Mariyappan[2]
Dean-School of Management,
Vel Tech Rangarajan Dr. Sagunthala R&D Institute of Science and Technology, Chennai

Eskandhaa K. Vel[3]
Student, Department of Management Studies, SRM Institute of Science and Technology

Abstract—Role of Technology is playing a vital role in the current scenario. People may now interact with social media on cellphones and laptops thanks to technological improvements. On the dark side, strangers making threats, bullying, harassing, and stalking others online are the most commonly reported and seen crimes, especially against women on social media. Any criminal behaviour involving a computer, a networked device, or a network is considered cybercrime. Cyber-stalking, cyber-defamation, cyber-sex, dissemination of obscene material, and trespassing into one's private domain are all crimes that are specifically targeted at women. This type of crime induces the women to take the wrong decision 'suicide'. It is also possible to attribute the main cause of the rise in cybercrimes against women in India to a lack of legal protection. Making strict laws is urgently necessary, and their proper application must also be guaranteed. The main objective of the paper is to study the cyber problems against women and the legal measures available to protect them. The target for the study is 200 women between the ages of 10 to 30. But 132 were only able to complete the questionnaire becomes a sample size. A convenient sampling method is used; due to the large population and being a sensitive issue, some are not willing to reveal their opinion. The Likert 5-point scale is used to measure the responses. The percentage analysis method and Ranking method is used as statistical tool for the analysis. The legal measures information are taken from the cyber law and Information Technology Act 2000. The result of the study shows that it creates some awareness about the legal measures to protect women. Women should be more courageous and self-disciplined to lead happy life.

Keywords—Social media, Women, Cybercrimes, Legal measures, Happy life

[1]sridevij@veltech.edu.in, [2]Deansomgt@veltech.edu.in, [3]kvdkishorekumarrc@gmail.com

DOI: 10.1201/9781003388913-40

I. Introduction

Cybercrime is an unethical one, creates mental pressure, a serious threat worldwide. It is a computer crime. Many young girls have been affected between the ages of nine to fourteen by their addicted to social media. National Crime Records Bureau (NCRB) clearly says that nearly a thousand women are affected every year due to this cybercrime. Crime Rate is increasing day by day.[1] As per National Crime Records Bureau, a total of 18,657 cases have been registered under Cyber Crimes, representing a 0.8% increase over 2019. (18,500 cases). The rate of cybercrime increased from 16.2 in 2019 to 16.4 in 2020. Computer Related Offenses (section 66 of the IT Act) (11,356 cases) constituted the greatest number of Cyber Crimes in 2020, accounting for 60.9% of all Cyber Crimes. In terms of women's safety, because of the rise in cybercrime, Females are negatively impacted and become victims of crimes like pornography and cyberstalking, which can lead to significant violent actions like serious physical harm or harassment, or any form of sexual abuse.

There are four key factors that contribute to cybercrime are:

- Mobile devices are the cause of a breach. Mobile devices were deemed safe in 2015 since their contamination rate was less than 1%. But now it's totally different.
- Malware Inclusion in Trustworthy Applications.
- Making Use of Illegal Products.
- Internet access without limits.

The study focuses on women's perception of cybercrimes and the safety and legal measures taken by the Law. Descriptive type of research is undertaken. 200 samples were collected but 132 were absolute. Convenient sampling is chosen since the subject was sensitive and emotional. The respondents are only women between the ages from 10 to 30. Most of the women are shy and insecure, not willing to disclose their feelings. Moreover, they are not aware of the safety and remedial measures taken by the Government. The study creates awareness among women about cybercrimes and protects them legally.

A. Statement of the Problem

The study aims to create awareness about cybercrimes against women and the legal measures available in the law. It concentrates on various cybercrimes against women and how to overcome those issues.

B. Objectives of the Study

- To analyze the women's perception of cyber crimes.
- To identify the legal measures taken by the Government and the laws to protect women.

C. Scope of the Study

The study covers the women's perception of cybercrimes in Chennai only. The scope of the study can be extended to various parts of the state. The outcome of the study may help women, how to respond to crimes.

D. Limitations of the Study

The study is confined to Chennai, the mixed age group of 10 to 30 and the duration of the study is one month. There may be a personal bias among the respondents which may affect the results of the study.

II. Review of Literature

[2] A. Kumari, K. Sharma, and M. Sharma, (2015), have presented a predictive analysis of cybercrimes against women in India and laws that prevent cyber victimization in general and women especially. Through the study, authors have predicted the effectiveness of laws that provides protection available to women victims of cyber-crimes like stalking harassment, threatening, and blackmailing defamation in cyber space.

III. Research Methodology

Descriptive type of study is adopted. The study undertook the factors such as harassment through emails, cyberstalking, cyber defamation, cyber bullying, cyber grooming, and child pornography are cyber crimes which help to know the opinion of women about cybercrimes.

A. Data Instrument

The questionnaire was developed for the data collection. It investigated the women's perception of their cybercrime data of theirs. The first part of the questionnaire collected demographic data of the women participants only such as age and student status. The second part is on women's perceptions of the solicited responses of the seriousness of cybercrimes which creates more stress and affects their both physical and mental health.

B. Data Collection

Data for the women's survey was collected from different schools, colleges, offices, and companies. Both direct and online questionnaire was used for data collection. The data collection exercise lasted a week, out of 200 collected data, 132 samples were taken for data analysis, as the remaining data are incomplete. For the 10-14 age group the questionnaire was explained in detail. The secondary sources of information were collected from the Research gate database, NCERB statistics, and the website resources.

C. Sampling Method

Convenient sampling a Nonprobability sampling method is adopted. The issue was very sensitive some may be not willing even to give their opinions. The data collection was carried out in different places like schools, colleges, and companies convenient sampling was found more suitable than any other type of sampling. The percentage analysis method and Ranking method is used as statistical tool for the analysis.

IV. Result Analysis

Out of 132 women participants, the majority of them belonged to the age groups of 10-14 years (33.33%)and 15-20 years (44.7%) (Table 40.1)

Table 40.1 Demographic profile of the respondents

Particulars	Frequency	Percentage
Age		
10-14	44	33.3
15-20	59	44.7
21-25	20	15.2
26-30	9	6.8
Total	132	100

Source: Primary Data

Interpretation: Table 1reveals the demographic characteristics of the respondents. Majority of the respondents are 10 to 14 age group and the minimum number of participants belongs to the age group between 26 to 30.

Table 40.2 In your opinion, which one is considered most serious crime?

Crimes	1	2	3	4	5
Child pornography					
Harassment through emails					
Cyber grooming					
Cyber stalking					
Cyber defamation					
Cyber bullying					

Note: 1. Not serious, 2. Slightly serious, 3. Fairly serious, 4. Serious, 5. Very serious.

Source: Primary Data

Table 40.3 Women's perception of the seriousness of cybercrimes (N = 132)

Rank	Cyber crimes	Mean (1-5)	SD
1	Cyber grooming	3.99	1.043
2	Cyber stalking	3.94	0.990
3	Cyber bullying	3.88	1.0990
4	Cyber defamation	3.86	1.011
5	Child pornography	3.74	0.960
6	Harassment through e mails	3.47	0.972

Source: Primary Data

The women were asked about their opinion on the seriousness of different cybercrimes which affects their both health and mental. In Table 40.2 a set of cybercrimes were provided in the questionnaire. A 5-point Likert scale was used to measure the responses, where 1 represented 'not serious' and 5'very serious. As shown in Table 3, the women felt that the top crime is cyber grooming (mean 3.99). The next crime that hurts women is cyberstalking (mean 3.94), and the third rank goes to cyberbullying (mean 3.88). However, it was found that the mean scores of all the cybercrimes fell in a narrow range of 3.47 to 3.99, reflecting women's agreement that all the cybercrimes listed in the questionnaire were serious only.

Table 40.4 Reasons for increasing cyber crimes

Rank	Particulars	Percent (N = 132)
1	Lack of self-control	47.7
2	More time spending on internet	46.2
3	Non-Awareness of the Government provisions and safety measures	28

Source: Primary Data

The women participating in this study were asked to identify the reasons for increasing cybercrimes. As shown in Table 5, the top reason was a lack of self-control (47.7%), more time spent on the internet (46.2%), and the remaining is not willing to complain or non-awareness of the Government provisions and safety measures.

V. Discussion

If a woman faces cybercrimes, she can file a complaint in any cybercrime unit irrespective of where the incident took place. Cybercrimes are established to provide redressal of grievances to the woman victims. These cells operate as a part of the criminal investigation department for offenses related to criminal activity on the internet. If there is no cyber cell near the vicinity, the woman can also file an FIR at the local police station.

This study investigated women's perception of the seriousness of cyber crimes such as cyber grooming, cyberstalking, cyberbullying, etc. Also, the overwhelming majority of the respondents felt that cyber grooming (mean 3.99) is the most serious cybercrime which upsets the growth of women. And the reason for increasing cybercrimes is a lack of self-control (47.7%). Educational institutions and employers can take adequate measures to safeguard women. A shared responsibility approach is likely to be more effective and successful.

A. Suggestions to Preventing Cybercrime

- Be wary of meaningless or fraudulent phone or email messages.
- Emails requesting personal information should not be answered.
- Be wary of fraudulent websites that attempt to obtain your personal information.
- Pay close attention to the privacy policies that come with the software and are available on websites.
- Check the security of your own email address.
- Make use of Secure Passwords.
- Victim of cybercrime should contact their local police station or cyber cell.
- An anonymous complaint can also be submitted through the National Cybercrime Reporting Portal.

VI. Conclusion

Cybercrimes have always been a serious and continuous one for women. Metaverse is again a milestone in information technology, but this type of cybercrime brings the world into

Table 40.5 Provisions governing on cyber crimes

Crimes	Meaning	Section	Code/Act	Measures	Fine	Imprisonment
Defamation	The action of damaging the good reputation of someone	499	Indian penal code	FIR	depends	depends
		500	Indian penal code	FIR	Fine-depends	Up to 3 years
		503		FIR	depends	depends
		66A	IT Act 2000	FIR	depends	depends
Cyber bullying	The use of electronic communication to bully a person, typically by sending messages of an intimidating or threatening nature.	507	Indian Penal Code	FIR	-	Up to 2 years
		66 E	IT Act 2000	FIR	3 lakhs	Up to 3 years
Cyber grooming	When someone (often an adult) befriends a child online and builds an emotional connection with future intentions of sexual abuse, sexual exploitation or trafficking.	67 B	IT Act 2000	FIR		imprisonment
		66 D	IT Act 2000	FIR		Imprisonment
			IT rules, 2021	Robust grievance redressal mechanism	depends	depends
		14C	Ministry of affairs	Indian cybercrime coordination centre	depends	depends
		-	Govt Nirbhaya fund.	Cybercrime prevention against women and children	depends	depends
		Portal	www.cybercrime.gov.in	FIR	depends	depends
		Toll free number	1930	Lodging complaint	depends	depends
		CBSE	Guidelines on 18.08.2017	Safe and secure use of internet/	depends	depends
		Meity	ISEA Dedicated website http://www.infosecawareness.in	Awareness material	depends	depends
		Govt	Orders to internet service providers	To implement Internet watch Foundation and block access	depends	depends

Crimes	Meaning	Section	Code/Act	Measures	Fine	Imprisonment
		Dept of tele commu-nications	Request to IPSs	To use parental control filters	depends	depends
		Protec-tion of children from sexual offences POSCO Act	Effective provisions	FIR	depends	depends
Cyber stalking	(1) Email Stalking: sending hate, obscene, or threatening emails, or sending viruses and spam. (2) Internet Stalking: spreading rumors or tracking victims on the web. (3) Computer Stalking: hacking into a victims computer and taking control of it.	Sec 67	IT ACT 2000	FIR	Fine-depends	Up to 3 years
		Sec 67A		FIR	5 lakhs	Up to 5 years
		Sec 354D	Indian Penal Code 1860	FIR	Fine-depends	Up to 3 years

Source: Primary data

an older one. Women can avoid disclosing information on social media. They can focus on their studies rather than on Instagram and WhatsApp. Courage and self-discipline are more important to get peace in life. Parents, employers, employees, and the public should help and support women against cybercrimes, Moreover, women with confidence, good attributes, attitudes, ethics, and behaviour can challenge any situation and lead a happy life. Also, the legal system has passed a number of laws to address cybercrime against women. To ensure that technology advances in a healthy manner and is used for legal and ethical economic growth rather than illegal activities, rulers and legislators must work constantly to achieve this goal.

References

1. NCERB statistics Crime in India 2020 Statistics Volume –I,Ministry of Home affairs.
2. https://ijcrt.org/papers/IJCRT2203318.pdf
3. NSPCC,"Grooming," NSPCC, WWW.nspcc. org.uk/prevening-abuse/child-abuse-and-neglect/grooming/.
4. https://www.legalserviceindia.com/legal/article-4998-cyber-crime-in-india-an-overview.html
5. https://cybercrime.gov.in
6. https://www.thehindu.com/news/national/india-reported-118-rise-in-cyber-crime-in-2020-578-incidents-of-fake-news-on-social-media-data/article36480525.ece
7. https://infosecawareness.in/cyber-laws-of-india

Recent Trends in Computational Intelligence and its Application – Sugumaran D. et al. (eds)
© 2023 Taylor & Francis Group, London, ISBN 978-1-032-48410-5

41

Swin Transformer Based COVID-19 Identification and its Severity Quantification

Deepa J.*, Ramya D.[1], K. S. Rishab[2], Sachi Shome[3]
Vel Tech Rangarajan Dr. Sagunthala R&D Institute of Science and Technology

S. K. Manigandan[4]
Vel Tech High Tech Dr. Rangarajan Dr. Sakunthala Engineering College

J. Velmurugan[5]
SIMATS School of Engineering

Abstract—The entire world suffers from a health crisis in the midst of a global COVID-19 pandemic. Researchers are trying to address their concerns by providing remedies to save lives and halt the pandemic outbreak. As early detection and diagnosis of the cases slows down the spread of this disease among the people, there is a need of an automatic detection of COVID-19. Creating a workable algorithm to detect and assess COVID-19 severity using the chest X-ray (CXR) obliges a vast amount of carefully chosen COVID-19 datasets, which are hard to obtain. This circumstance is appropriate for the Swin Transformer architecture because it enables the self-attention mechanism to utilize a significant amount of unlabeled information through structural modeling. The backbone network's inherent attributes are employed as inputs in Transformer architecture for COVID-19 identification and to estimate severity. To assess our model's generalization ability, we use a variety of test datasets. The testing results demonstrate that the proposed approach obtained the best results in both detection and diagnosis tasks and yields an average precision value of 0.93 as a result.

Keywords—Swin transformer, CXR, Coronavirus, Segmentation, Classification, Severity quantification

I. Introduction

The unique coronavirus disease of 2019 (COVID-19), caused by the severe acute respiratory illness coronavirus 2, is the lethal virus of the century (SARS-CoV2) [18]. Given the risks associated with COVID-19 [17] disease, early diagnosis is crucial to prevent future disease transmission and lessen the burden on an already overburdened healthcare system. A Swin model [4] for COVID-19 is demonstrated by combining an existing large-scale dataset with a low-level CXR feature corpus that includes depictions for common CXR [1], [12]

*Corresponding Author: vasdeepa03@gmail.com
[1]raminfo84@gmail.com, [2]rishabjain2030@gmail.com, [3]sachishome26@gmail.com, [4]kgmanigandan@gmail.com, [5]velmuruganj.sse@saveetha.com

DOI: 10.1201/9781003388913-41

results. Furthermore, by carefully analyzing low-level features like structure, repetition, position, and dispersion, our technique mimics the clinical specialists who make the final CXR diagnosis. One of the most common strategies for addressing this obstacle is to build a vigorous model with a large training data, however given the current pandemic crisis, creating a large dataset with labeled COVID-19 [17] examples is hard. As a result, many solutions have been created to address the issue, including domain adaptation, weak reinforcement learning, and outlier detection, but their effectiveness is still lacking. Convolutional neural network (CNN) models [7], [8], [10], which were not explicitly designed for COVID-19 appearances, were extensively used in previous studies. Although CNN architecture [15] has demonstrated higher performance in many visual applications, high-level CXR illness classification [6] problems may not be the ideal fit. The reduced CXR feature corpora are used to create a unique ViT model for COVID-19, which incorporates depiction for frequent CXR [2] finds and a pre-built huge dataset. We expanded the scope of our model beyond diagnosis to assess severity in order to provide doctors with the clinical standards they need to choose the best course of action. Our method outperforms current ones. Using previously unexplored data, experiments reveal that transformer-based models adapt better than CNN-based systems.

A. Motivation and Justification

To improve the accuracy, time taken and performance of other existing models, we employed swin transformer. We developed our model beyond diagnosis to evaluate severity in order to offer doctors clinical criteria for determining treatment options. When compared to other vision models, Swin Transformers accomplish a better speed-accuracy trade-off. Swin integrates inductive assumptions including transfer function, locality, and hierarchical recognition algorithms, making it a versatile foundation for a range of images depending on access.

B. Outline of the Proposed Work

The proposed model has three parts. 1. Backbone Network pre-training for Low-level Feature Corpus. 2. Swin Transformer for Classification. 3. Swin Transformer for Severity Quantification.

Backbone Network pre-training for Low-level Feature Corpus:

With multiple low-level discoveries, the common embedding performed prior to the PCAM technique and the feature mappings in each layer may serve as templates for future Transformer feature embedding. To get the optimum intermediate-level embedding, the PCAM approach for particular low-level CXR abnormalities has proven to be essential since it aligns these characteristics for better classification outcomes.

Swin Transformer for Classification:

Swin Transformer adds transformer encoder layers with the input embedding as the transformer encoder uses a fixed vector for dimensions. By including an additional linear classifier as the classifier head, we may use the input CXR picture to obtain the diagnostic result.

Swin Transformer for Severity Quantification:

Transformer provides feature embedding at every component position which will be helpful in determining severity. The severity map is then produced using ROI max-pooling (RMP) [17]. As a result, the map heading network can be trained, and therefore damaged and healthy lung areas can be precisely segmented.

II. Related Work

A. Vision Transformer

The ViT[14] was the first substantial attempt to standardize the method of applying a straight transform directly to the image, which implies it can entirely replace old convolution approaches by performing as well as SOTA. However, the studies showed that there is a significant computational cost associated with training a basic ViT model. As a result, the authors propose a hybrid framework that combines transformer [9], [11], and a CNN backbone (such as ResNet). The transformer may concentrate on utilizing the ResNet feature to mimic global attention [13]. According to the results of the experiments, the hybrid technique achieved greater performance with fewer calculations. Since the debut of ViT, the usage of transformers in computer vision has increased in popularity, leading to the creation of many ViT variant models.

B. Probabilistic Class Activation Map Pooling

A type of class-oriented feature map called a class activation map (CAM) is created by assessing how much a particular visual segment contributes to network prediction. The capacity to use image-level supervision to identify the critical area with just subpar labeling is the most helpful feature of CAM. They developed Probabilistic-CAM (PCAM) pooling, a revolutionary method for global pooling that uses the CAM in a probabilistic way. The normalized attention weights derived from the output probabilities were utilized to generate weighted feature maps that represented each class more accurately. They won the 2019 CheXpert Challenge by demonstrating that PCAM pooling may increase the localization as well as the diagnostic performance of the model.

C. Severity Quantification for COVID-19

Lesion segmentation labels and other frame annotations can provide a wealth of information for building an automated method for calculating severity. Because such labeling approaches take time, a global COVID-19 epidemic would make compiling enormous amounts of data labeled at the pixel level impracticable. Score-based and array-based severity annotation approaches have been developed as solutions to the problem in order to design an automated process for severity assessment. Both lungs are divided into six regions, and each section is given a score of 0 or 1 considering whether COVID-19-based involvement is present. This results in a severity range of 0-6. An array-based annotation method [19] is used to determine COVID-19 severity.

III. Proposed Methodology

A PCAM approach was used to pre-train the dataset in order to extract the backbone of CXR results. Patching and positional embedding will be used to feed the extracted output into the Swin Transformer. In a Transformer encoder, the backbone is divided into numerous layers, including multilayer perceptrons, multi-head self-attention, residual connections, and layer normalization at each block. The output of the transformer is a group of tokens. We'll focus on a certain aspect of classification and severity assessment. Classification is accomplished by merely adding a classifier, whereas severity quantification is accomplished by ROI-max pooling. As a result, it generates two outputs: accuracy-based categorization and segmented bounding box severity quantification. Fig. 41.1 shows the process layout of the proposed approach.

A. Swin Transformer

Vision transformers include Swin Transformers (swifted windows transformers). It produces hierarchical feature mapping by combining image patches into deeper layers and has a linear computing cost proportionate to the size of the input image because self-attention processing only takes place inside each local window. Transformer is a deep neural network that provides very large receptive fields using a self-attention technique. It was initially created for NLP. Vision Transformer was the first serious endeavor to apply the Transformer directly to images, which implies it can entirely replace current convolution approaches by

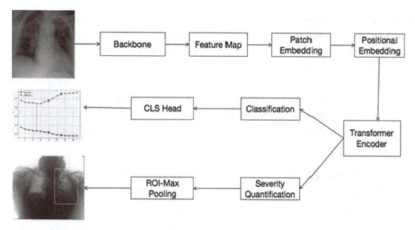

Fig. 41.1 Process layout

Source: Author

performing as well as SOTA. However, the experiments revealed that there was a significant computational cost involved with training a simple ViT model. As a consequence, the authors created a hybrid architecture comprised of Transformer and a CNN backbone (such as ResNet). The Transformer may concentrate on utilizing the ResNet feature to mimic global attention [13]. According to the results of the experiments, the hybrid technique achieved greater performance with fewer calculations.

B. Backbone Network pre-training for Low-level Feature Corpus

The backbone network was initially pre-trained to recognize ten different types of data. Using a sizable public dataset, we found a lot of low-level discoveries that were consistent. We found that the most crucial information is contained in the conventional embedding technique employed before the PCAM procedure, and that feature mappings in every layer could be possibilities for future Transformer feature embedding. The

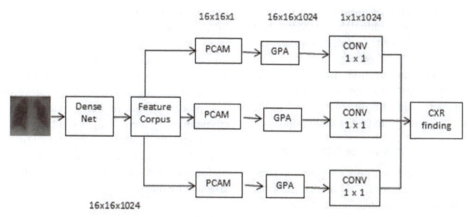

- PCAM-Probabilistic Class Activation Map
- Gap-Global Average Pooling

Fig. 41.2 The backbone network pre-training for low-level feature corpus

Source: Author

PCAM procedure, which aligns these features for better classification outcomes, has been shown to be essential for attaining the best transitional embedding for a specific low-level chest X-ray image abnormalities (such as lung opacity, consolidation, etc.). As shown in Fig. 41.2 each layer's feature maps might be candidates for useful feature embedding in ensuing transformer.

C. PCAM (Probabilistic Class Activation Map)

Calculating the input of a precise part of an image to network prediction yields a saliency map known as a class activation map (CAM). CAM's most useful feature is its ability to apply image-level supervision to identify the important area with just poor labeling. CAM, despite its excellent localization capabilities, was mostly used to create heatmaps for lesion identification and visualization during interpretation. To make better use of the CAM's localization capacity, a recent study used it during training in the phase of CXR classification as well as localization tasks. They developed Probabilistic-CAM (PCAM) pooling, a game-changing global pooling solution that influences the CAM in a probabilistic way. Both classification and severity forecasting operations benefit from PCAM operation, although categorization benefits more. These findings support the idea that the PCAM process may be more beneficial in classification jobs when the strong representations for numerous low-level characteristics are directly related to the ultimate diagnosis of the chest radiograph.

D. Swin Transformer for Classification

As the transformer encoder employs a fixed vector for dimensions, Swin Transformer adds transformer-encoded layers to an input embedding. The low-level chest X-ray images characteristic of the transformer is built using these feature vectors. The projected feature tensor is represented sequentially using the patch and positional embedding. Our model uses recurrent layers of multi-head self-attention, multilayer perceptron, layer normalization, and residual connections as encoder layers, just like the classic Transformer. With respect to the [class] token, the Transformer attended the feature vector. As a result, we can get the diagnostic outcome of the input CXR image by adding an extra linear classifier as the classifier head.

E. Severity Quantification Using Swin Transformer

While taking into consideration long-range interactions between the blocks, the remaining transformer output embeds features at every block location. As a result, this information will be useful in establishing how serious the issue is. The severity map is obtained by pixel-wise multiplying the network output by the segmentation mask, and then ROI max-pooling (RMP) [16] is used. The map head network is trained to minimize the projected severity array's inaccuracy in comparison to the poorly annotated severity [3] label. This novel method makes possible the exact segmentation of damaged and healthy lung tissue.

F. ROI max-pooling (RMP)

For a given input image, find all feasible locations where items may be found. This stage should produce a list of bounding boxes for potential object placements. Decide if each area suggestion from the previous step belongs to one of the target classes or the background. The process of the region of interest pooling (sometimes called RoI pooling) [16] is commonly employed in object identification tasks. Its goal is to execute max pooling on non-uniformly sized inputs in order to create fixed-size feature maps.

IV. Result

As a result, we can get the diagnostic outcome of the input CXR image by adding an extra linear classifier as the classifier head. And

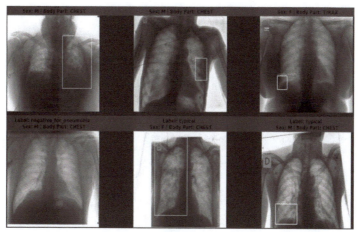

Fig. 41.3 Output

Source: Compiled by Author

this information will be useful in establishing how serious the issue is. The severity map is obtained by pixel-wise multiplying the network output by the segmentation mask, and then ROI max-pooling is used. The map head network can be trained to minimize the inaccuracy of the projected severity array relative to the poorly annotated severity label. This exact segmentation of damaged and healthy lung tissue is made possible by this novel method as shown in Fig. 41.3. The publicly available COVID-19 Chest X-Ray Dataset Initiative [5] is used to measure the performance of our proposed model.

The training accuracy is calculated to be 0.932 and the training loss is assessed to be 0.2778. The training graphs are shown in Fig. 41.4. The valid precision is determined to be 0.806. The valid specification is computed as 0.8744. The model is tested in stages to guarantee that it can forecast properly. The first stage is to make predictions on the testing set. As a result, the average valid accuracy of all modules is 0.806. The proposed model is compared with the other state-of-arts and its result is shown in Table 41.1.

V. Conclusion

This study presents a deep learning-based approach for COVID-19 diagnosis and severity assessment using images from chest X-rays.

Table 41.1 Comparison of CNN with other transformer-based models with respect to precision metric score

Methods	Average	Normal	Others	COVID
CNN-Based				
ResNet-50	0.842	0.726	0.882	0.921
ResNet-152	0.822	0.754	0.892	0.894
DenseNet-121	0.853	0.741	0.906	0.909
Transformer Based				
ViT	0.884	0.882	0.878	0.895
Swin	**0.932**	**0.938**	**0.926**	**0.932**

Source: Author

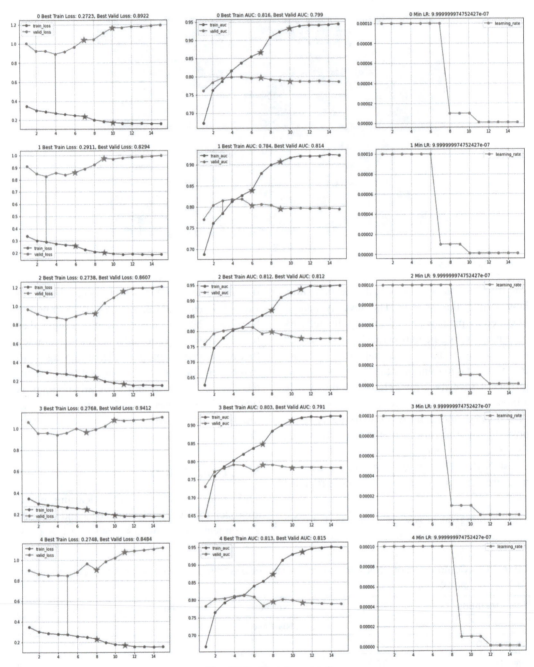

Fig. 41.4 Training result

Source: Compiled by Author

Beyond diagnosis, we developed our model to compute severity in order to give clinical guidelines to clinicians when evaluating alternative therapeutic options. According to experiments, our approach performs better than the current transformer-based and CNN-based models when it comes to generalizing to new data.

Finally, because newly emerging pathogens are expected to share low-level Chest X-ray image features with established diseases, the idea of creating better diagnoses by computing low-level feature corpus, that is easily accessible with already available data sources, could be implemented to create a functional algorithm at odds with them quickly.

References

1. Le Dinh T, Lee S-H, Kwon S-G, Kwon K-R. COVID-19 Chest X-ray Classification and Severity Assessment Using Convolutional and Transformer Neural Networks. Applied Sciences. 2022; 12(10):4861. https://doi.org/10.3390/app12104861.

2. Borghesi, A., Maroldi, R., 2020. COVID-19 outbreak in italy: experimental chest x-ray scoring system for quantifying and monitoring disease progression. La radiologia medica 125, 509–513.

3. Xie, F., Huang, Z., Shi, Z., Wang, T., Song, G., Wang, B., & Liu, Z. (2021). DUDA-Net: a double U-shaped dilated attention network for automatic infection area segmentation in COVID-19 lung CT images. International journal of computer assisted radiology and surgery, 16(9), 1425–1434. https://doi.org/10.1007/s11548-021-02418-w.

4. Weiwei Sun, Jungang Chen, Li Yan, Jinzhao Lin, Yu Pang, Guo Zhang 2022. COVID-19 CT image segmentation method based on swin transformer. Medical Physics and Imaging. 2022; https://doi.org/10.3389/fphys.2022.981463.

5. COVID-19 Chest X-Ray Dataset Initiative. https://github.com/agchung/Figure1-COVID-chestxray-dataset. Accessed 2 July 2020.

6. Wang, X., Deng, X., Fu, Q., Zhou, Q., Feng, J., Ma, H., Liu, W., Zheng, C., 2020b. A weakly-supervised framework for COVID-19 classification and lesion localization from chest CT. IEEE transactions on medical imaging 39, 2615–2625.

7. Zheng, C., Deng, X., Fu, Q., Zhou, Q., Feng, J., Ma, H., Liu, W., Wang, X., 2020a. Deep learning-based detection for COVID-19 from chest CT using weak label. MedRxiv.

8. Zhang, J., Xie, Y., Li, Y., Shen, C., Xia, Y., 2020. COVID-19 screening on chest x-ray images using deep learning based anomaly detection. arXiv preprint arXiv:2003.12338 .

9. Zheng, S., Lu, J., Zhao, H., Zhu, X., Luo, Z., Wang, Y., Fu, Y., Feng, J., Xiang, T., Torr, P.H., et al., 2020b. Rethinking semantic segmentation from a sequence-to-sequence perspective with transformers. arXiv preprint arXiv:2012.15840.

10. Narin, A., Kaya, C., Pamuk, Z., 2020. Automatic detection of coronavirus disease (COVID-19) using x-ray images and deep convolutional neural networks. arXiv preprint arXiv:2003.10849

11. Devlin, J., Chang, M.W., Lee, K., Toutanova, K., 2018. Bert: Pre-training of deep bidirectional transformers for language understanding. arXiv preprint arXiv:1810.04805 .

12. Cozzi, D., Albanesi, M., Cavigli, E., Moroni, C., Bindi, A., Luvara, S., Lu-ᶜ carini, S., Busoni, S., Mazzoni, L.N., Miele, V., 2020. Chest x-ray in new coronavirus disease 2019 (COVID-19) infection: findings and correlation with clinical outcome. La radiologia medica 125, 730–737.

13. Kaiming He, Xiangyu Zhang, Shaoqing Ren, Jian Sun, 2015 Deep Residual Learning for Image Recognition arXiv:1512.03385.

14. Alexey Dosovitskiy, Lucas Beyer, Alexander Kolesnikov, Dirk Weissenborn, Xiaohua Zhai, Thomas Unterthiner, Mostafa Dehghani, Matthias Minderer, Georg Heigold, Sylvain Gelly, Jakob Uszkoreit, Neil Houlsby, 2021An Image is Worth 16x16 Words: Transformers for Image Recognition at Scale, arXiv:2010.11929v2.

15. Bastiaan S. Veeling, Jasper Linmans, Jim Winkens, Taco Cohen, Max Welling, 2018 Rotation Equivariant CNNs for Digital Pathology, arXiv:1806.03962v1

16. Ross Girshick, Jeff Donahue, Trevor Darrell, Jitendra Malik, 2014 Rich feature hierarchies for accurate object detection and semantic segmentation arXiv:1311.2524v5.

17. Shi, H., Han, X., Jiang, N., Cao, Y., Alwalid, O., Gu, J., Fan, Y., Zheng, C., 2020. Radiological findings from 81 patients with COVID-19 pneumonia in wuhan, china: a descriptive study. The Lancet infectious diseases 20, 425–434.

18. Ng, K., Poon, B. H., Kiat Puar, T. H., Shan Quah, J. L., Loh, W. J., Wong, Y. J., Tan, T. Y., Raghuram, J., 2020. COVID-19 and the risk to health care workers: a case report. Annals of internal medicine 172, 766–767.

19. Toussie, D., Voutsinas, N., Finkelstein, M., Cedillo, M.A., Manna, S., Maron, S.Z., Jacobi, A., Chung, M., Bernheim, A., Eber, C., et al., 2020. Clinical and chest radiography features determine patient outcomes in young and middle-aged adults with COVID-19. Radiology 297, E197–E206.

Recent Trends in Computational Intelligence and its Application – Sugumaran D. et al. (eds)
© 2023 Taylor & Francis Group, London, ISBN 978-1-032-48410-5

42

Dissertation on Password Space of Disparate Graphical User Authentication Systems

Shanthalakshmi M.[1]
Research Scholar, SRM Institute of Science and Technology, Chennai, India

Ponmagal R. S.[2]
Associate Professor, SRM Institute of Science and Technology, Chennai, India

Abstract—User authentication is a critical component of data security. A password is a key that is used to verify user identity. The most frequent and commonly used method of user authentication is text-based passwords. The alphanumeric modes of authentication, on the other hand, have a number of drawbacks. For example, it is predictable and memorable easily. Passwords that are difficult to remember are often forgotten. Passwords that are hard to know or break, on the other hand, are typically harder to recall. Because users can only remember a limited number of passwords, research shows that individuals tend to write passwords down or reuse passwords throughout many accounts. These drawbacks are the main reason for the introduction of graphical passwords. Humans can remember and recall images easily rather than text-based passwords. This paper reviews the graphical password authentication systems in terms of usability and security measures.

Keywords—User authentication, Text-based passwords, Graphical passwords, Usability

I. Introduction

Passwords are extremely important in determining the users' identities. Alphanumeric usernames and passwords are the most widely used computer authentication technique. Textual passwords are insecure because they are recognizable (weak passwords include names, birthdates, and phone numbers), and attackers may readily guess them. Strong passwords, on the other hand, are not only tough to guess or crack, but also to remember. There is a potential that the user will forget their password if they do not use it regularly. Various approaches such as brute force assault, shoulder surfing, and spyware attacks may readily guess text passwords. To address these flaws, graphical user authentication has been introduced. Humans recall visuals more often than words, according to studies. These passwords are considered to be stronger and easier to remember. In recent years, mobile devices have provided humans with enormous convenience by allowing users to access a variety of apps

[1]sm9257@srmist.edu.in, [2]ponmagas@srmist.edu.in

DOI: 10.1201/9781003388913-42

such as online shopping, banking, navigation, and mobile media at any time and from any place. While people love the convenience and flexibility of the "Go Mobile" movement, their important personal information (such as their name and credit card number) may be compromised on mobile devices. As a result, mobile phones should be protected using strong user authentication [25] [27].

Graphical User authentication can be done in various ways. For example, Recognition based systems (Pass Faces, Photographic authentication, and Picture based authentication), Pure Recall based systems (DAS, Pass Doodle, Pass Shapes), Cued Recall based systems (Pass points, Pass Faces, Pass logix), Biometric based authentication, Captcha and QR based authentication [28]. This paper reviews various types of graphical user authentication. Researchers have devised authentication systems that employ images as passwords to tackle the problem of low security.

II. Recognition and Recall-Based Authentication Systems

Many Recognition and recall-based techniques are available. Generally, graphical user authentication invokes the processing of images, pass faces, pass points, etc. Fig. 42.1 illustrates the systematic process of graphical password authentication. Bhaveer Bhana et al. [1] followed a method to evaluate the effectiveness of a two level user authentication solution that uses passcodes and keystroke dynamics. The password strength passes, and keystroke dynamics were calculated using the Shannon Entropy theory and Chunking theory. Password policies are guidelines that a system administrator imposes on a user when they create a password. The keystroke model was used to evaluate passwords. Here the entropy was calculated for a two-tier authentication method using passphrases and keystroke dynamics. A comparison of the entropy for various authentication methods has been given.

Much text-based and graphical authentication systems have been proposed. However, the vast majority of them are susceptible to shoulder surfing attacks or must make a trade-off in terms of usability. Driven by this challenge Teng Zhou et al.[2] proposed a pass grid-graph supplemented method with the use of a camera by using a sequence of one-time login indications and cyclic moveable blocks containing textual components. Users simply need to remember one set of passwords to decrease their strain. According to our user research, Pass Grid can achieve high security

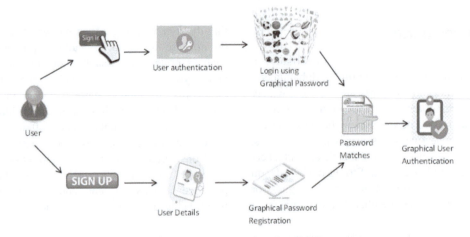

Fig. 42.1 Graphical password authentication system

Source: Author

and usability performance, with an average login time of 22 seconds and a short password length [29].

A novel graphical authentication method for shoulder surfing resistance is proposed by Misbah Urrahman Siddiqui et al. [3].Here R*R grid is filled with various images which represent alphabets, numbers and special characters. The password can be submitted via images that are displayed randomly. The sequence of the images will also be varied [22]. So the shoulder surfers find it very difficult to hack the password. Noor Ashitah Abu Othman et al. [4] proposed a directional based graphical authentication method where the image will be selected from 9 images that are displayed from different folders [26]. Next user needs to select the proper direction. If both match, the authentication can be done successfully. Since the images are varying for each session and direction, the shoulder surfing attack is very difficult. Usability suffers when we place a higher attention on security. Motivated by this challenge, Gi-Chul Yang et al. [5] proposed a user friendly graphical pass position password technique. The user will choose 3 points and corresponding (x,y) coordinates. By using this pass point, R-String (RD, LD) can be generated that holds the relative positions of the selected positions. No need to select the same points as in the registration phase. Authentication will be done whenever the generated R-String matches. Mudassar Ali Khan et al. [6] [25] proposed gRAT - A novel randomized graphical authentication method against shoulder surfing attacks. Here the pattern lock application contains various categories for images. The user needs to pick the category and select the password from the 3*3 matrix. Then by swiping the images, draw the pattern. If the user draws the same pattern that has been selected in the previous phase, authentication will be done [24].

Security can be enhanced through multi-level authentication [21]. Xian Chu et al. [7] proposed Pass Page, a two-level graphical password authentication appropriate for online authentication with better security. It makes advantage of implicit memory depending on the user's online surfing history. During the login phase, the server provides 9 tiny pages as a challenge and requires the user to pick all the sites the user has viewed in addition to entering a text password. The user test is conducted on 12 participants. The findings indicated that when users are frequently using the login process, the average success rate is consistently above 80%, and this success rate did not drop significantly after 6 days [30].

Pass points can also be used in different ways. TrustRatchasan et al. [8] proposed a method using haptic points. The aforementioned issues may be avoided by providing enhanced pass points as haptic points. Haptic Points improve Pass Points by delivering haptic feedback randomly. Pass points may also be applied in conjunction with the shapes. Lissette Suárez-Plasencia et al. [9] proposed a method that uses the number of sides of polygons used as the pass point. The ability of this feature to identify weak passwords composed of clustered dots is assessed. In some cases, passwords can be generated dynamically to increase security. Ramsha Fatima et al. [18] proposed a method. Even though the real user password is the same, the password given at the moment of login will be varied for each session under this technique. Here the operator is selected during the registration. The corresponding operation will be done with the random number which is generated at each session. The user authentication will be done successfully when the password matches. Brute force attacks, Dictionary attacks, Spyware attacks, and Shoulder surfing attacks are eliminated.

III. Authentication Based on Captcha/QR Code

Graphical user authentication can also be done with Captcha and QR Code for enhancing security. Altaf Khan et al. [10] proposed a scheme that integrates recall-based n*n grid points and captcha-based password that uses Click Symbols, Alphanumeric, and Visual symbols shortly known as the CS-AV method. DivyansMahansaria et al. [11] proposed a unique one-time QR - Quick Response code in order to avoid different types of attacks. Here the use of a QR code will make context-based authentication easier. The QR code stores certain information that varies with each login process authentication.

Digitally signed QR codes can also be used for secure authentication. Hamdi Abdurrahman Ahmed et al. [12] proposed an authentication method that uses a QR code with a digital signature. A digital signature is applied for the student data for example Student Name, Course name, GPA, etc. that are signed by the Higher Educational Institute. The digitally signed certificate-generating process can be done in offline mode. An android application is used to verify the QR code with the digital signature signed. The certificate will be scanned, verified, and authenticated without contacting the certificate-issuing organization or getting access to the user's security credentials. In order to increase the security level, some other techniques can also be used in addition to QR codes. In the same way, Jun Zhao et al. [13] proposed spatial watermarking and QR code which is implemented on paper-cut images.

IV. Authentication Based on Hybrid-Textual and Graphical Passwords

In order to improve security, we can integrate both textual and graphical passwords. In this way, Lauren N. Tiller et al. [14] proposed an EaN - Explore a Nation technique. The EaN method presents users with a fixed representation of a map that conceals an icon passcode route amid other distracter icons. Users may create a strong password by following the icon route.

By using this Over Shoulder Attacks (OSA) can be avoided. High usability can be achieved. Shailja Varshney et al. [15] proposed a hybrid scheme that uses both textual and graphical passwords. To log in, the user must click the chosen colour immediately when it appears on the relevant picture in combination with dynamic images [23]. This system is protected against brute force assaults, guessing, and shoulder surfing. As Fallback authentication, if you lose your password, security questions are a common way to get back in which is easily guessed. In order to address this issue, Joon Kuy Han et al. [16] proposed the Pass Tag technique Pass Tag includes deceptive images that make it more difficult to identify the real ones that the user has uploaded. Pass Tag is much stronger against close opponents and easy to remember (92.6 percent to 95.0 percent) after one, two, and three months. Pass Tag is a potential option for fallback authentication, according to the findings of this longitudinal study. It's still a big challenge to come up with an authentication technique that's both secure and easy to remember and use. To address this challenge, Shah Zaman Nizamani et al. [17] proposed a hybrid authentication mechanism that uses the dynamic mode of password entry based on the environment. A draw metric method is used to set the password. One time password mechanism is used for randomness in the selection and arrangement of password components. Random numbers are assigned to passwords for increasing security. This technique is against brute force attacks, dictionary attacks, guessing attacks, shoulder surfing attacks, keystroke loggers and various recording attacks.

V. Password Space

For attaining security during the authentication, the password should not be predictable by anyone. To ensure this we are using the password space and password entropy as the measure. Password space is a metric of how many random attempts are required during a brute force attack to locate any password therein and is thus a measure of password security and susceptibility. With the high values of password space, the security level of the authentication method will also be high. The comparison of Password Length and Password Space for various authentication methods is given in chart Fig. 42.2 and Fig. 42.3.

VI. Results and Discussion

In any authentication method, there is a possibility of various attacks. By using any of these attacks, hackers will try to crack the system. The possibility of the most common attacks like guessing attacks in various authentication methods is mentioned in Table 42.2. In this table, the rating is given as 1, 2, and 3 which represents low, medium, and high-security level. The security level for various authentication methods has been mentioned as very weak, weak, reasonable, and strong in Table 42.1. From this, we can say that the two-level authentication scheme which includes passcodes and keystroke dynamics has the highest password space which leads to a strong security authentication method.

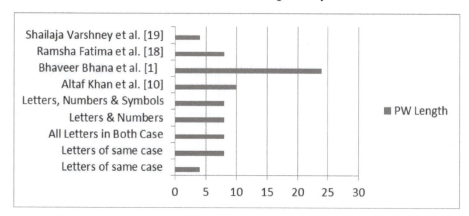

Fig. 42.2 Comparison of password length for various authentications

Source: Author

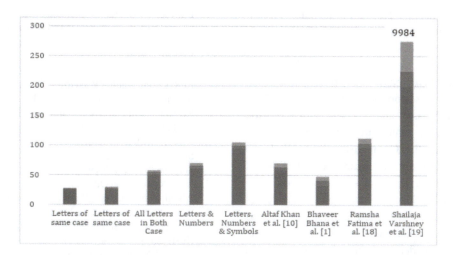

Fig. 42.3 Comparison of password space for various authentication methods

Source: Author

Table 42.1 Comparison of password space and password entropy for various authentications

Password contains	Length	Possible symbol	Possible combinations	Strength
Letters of same case	4	26	456,976	Very weak
Letters of same case	8	26	2.09E+11	Weak
All Letters	8	52	5.35E+13	Reasonable
Letters & Numbers	8	62	2.18E+14	Reasonable
Letters, Numbers and Symbols	8	95	6.63E+15	Reasonable
Altaf Khan et al. [10]	10	58	4.31E+17	Reasonable
BhaveerBhana et al. [1]	16+keystroke	26+(keystroke)	2.22E+25	Strong
Ramsha Fatima et al. [18]	8	96	7.21E+15	Reasonable
Shailaja Varshney et al. [19]	4	52*16*12	9.94E+15	Reasonable

Source: Author

Table 42.2 Comparison of various attacks in authentication methods

State	Shoulder suffering	Hacking	Guessing
Text password	1	1	2
Hardware based	3	3	2
Bhaveer Bhana et al. [1]	1	2	2
Altaf Khan et al. [10]	2	3	2
Ramsha Fatima et al. [18]	1	3	2
Shailaja Varshney et al. [19]	1	3	2

Source: Author

VII. Conclusion

In this paper, Recognition, and Recall-based authentication systems, Captcha/QR-based authentication systems, and Hybrid authentication systems are reviewed. Security measures like password space have been discussed. Password space has been calculated for some basic and specific authentication methods and values are mentioned in Table 42.1. From the Comparison Table 42.2, we may look at the vulnerabilities of different authentication techniques and see how they might be exploited. Even though researchers have come up with new methods to address this issue. An additional factor that researchers should consider is the tradeoff between graphical password usability and security, and how to maintain a balance between the two.

References

1. Bhaveer Bhana, Stephen Flowerday, "Passphrase and keystroke dynamics authentication: Usable security", Computers & Security, 2020.
2. Teng Zhou, Liang Liu, Haifeng Wang, Wenjuan Li & Chong Jiang, "PassGrid: Towards Graph-Supplemented Textual Shoulder Surfing Resistant Authentication", Springer, 2019.
3. Misbah Urrahman Siddiqui; Mohd. Sarosh Umar; Miftah Siddiqui,"A Novel Shoulder-Surfing Resistant Graphical Authentication Scheme ", IEEE, 2019.
4. Noor Ashitah Abu Othman; Muhammad Akmal Abdul Rahman; Anis Shobirin Abdullah Sani "Directional Based Graphical Authentication Method with Shoulder Surfing Resistant", IEEE, 2018.
5. Gi-Chul Yang, "PassPositions: A Secure and User-Friendly Graphical Password Scheme", IEEE, 2018.

6. Muddasar Ali Khan,Ikram Ud Din,Sultan Ullah Jadoon,Muhammad Kurrahm Khan," gRAT| A Novel Graphical Randomized Authentication Technique for Consumer devices", IEEE, 2019.

7. Xian Chu, Huiping Sun & Zhong Chen, "PassPage: Graphical Password Authentication Scheme Based on Web Browsing Records", Springer, 2020.

8. Trust Ratchasan & Rungrat Wiangsripanawan, "HapticPoints: The Extended Pass Points Graphical Password", Springer, 2019.

9. Lisset Suárez-Plasencia, Joaquín A. Herrera-Macías, Carlos M. Legón-Pérez, Raisa Socorro-LLanes, Omar Rojas, "Analysis of the Number of Sides of Voronoi Polygons in PassPoint", Springer, 2021.

10. Altaf Khan; Alexander G. Chefranov "A Captcha-Based Graphical Password With Strong Password Space and Usability Study", IEEE, 2020.

11. Divyans Mahansaria & Uttam Kumar Roy "Secure Authentication Using One Time Contextual QR Code", Springer, 2020.

12. Hamdi Abdurhman Ahmed,Jong wook Jang "Document Certificate Authentication System Using Digitally Signed QR Code Tag", IEEE, 2018.

13. Jun Zhao,Guangyong Gao,Jianhua Cheng "Identity Authentication Protection for "Spatial Watermarking + QR Code" Paper-cutting Digital Image", ACM, 2017.

14. Lauren N. Tiller, Catherine A. Angelini, Sarah C. Leibner & Jeremiah D. Still "Explore-a-Nation: Combining Graphical and Alphanumeric Authentication", Springer, 2019.

15. Swaleha Saeed, Mansur Umar "A hybrid graphical user authentication scheme", Springer, 2020.

16. Joon Kuy Han,Xiaojun Bi,Hyoungshick Kim "PassTag: A Graphical-Textual Hybrid Fallback Authentication System", ACM, 2020.

17. K. Vijay, R. Vijayakumar, P. Sivaranjani, and R. Logeshwari, "Scratch detection in cars using mask region convolution neural networks," Adv. Parallel Comput., vol. 37, pp. 575–581, 2020.

18. Ramsha Fatima, Nadia Siddiqui, M. Sarosh Umar & M. H. Khan "A Novel Text-Based User Authentication Scheme Using Pseudo-dynamic Password", Springer, 2018.

19. Shailja Varshney, Mohammad Sarosh Umar & Afrah Nazir "A Secure Shoulder Surfing Resistant Hybrid Graphical User Authentication Scheme", Springer, 2020.

20. Kapil Juneja "An XML transformed method to improve effectiveness of graphical password authentication", Science Direct, 2020.

21. Leonardo dos Santos Dourado, Edison Ishikawa, "Graphical Semantic Authentication", IEEE, 2020.

22. Dr. Uday Pratap Singh, Siddharth Singh Chouhan, Sanjeev Jain, "Images as graphical password: verification and analysis using non-regular low-density parity check coding", Springer, 2020.

23. Saleha Ahmad,Muhammad Faisal Hayat, Muhammad Ali Qureshi, Shahzad Asef, Yasir Saleem," Enhanced halftone-based secure and improved visual cryptography scheme for colour/binary Images", Multimedia Tools and Applications", 2021.

24. Esra Alkhamis, Helen Petrie & Karen Renaud," KidsDoodlePass: An Exploratory Study of an Authentication Mechanism for Young Children", Springer, 2020.

25. Yean Li Ho, Michael Teck Hong Gan, Siong Hoe Lau & Afizan Azman, "Pilot Evaluation of BlindLoginV2 Graphical Password System for the Blind and Visually Impaired", Springer, 2021.

26. R. Vijayakumar, K. Vijay, P. Sivaranjani, and V. Priya, "Detection of network attacks based on multiprocessing and trace back methods," Adv. Parallel Comput., vol. 38, pp. 608–613, 2021.

27. Pradheeba U., Bhavani M. B., Yuvaraj B. R., Krithika V, "Prospective classification over various handwritten character recognition algorithms—A Survey", Adv. Parallel Computing, 2021.

28. S. Sreesubha, R. Babu, R. Vijayakumar, and K. Vijay, "An efficient data hiding approach on digital color image," Adv. Parallel Comput., vol. 37, pp. 552–557, 2020.

29. V. R, R. S, V. K and S. P, "Integrated Communal Attentive & Warning System via Cellular

Systems," 2022 International Conference on Electronic Systems and Intelligent Computing (ICESIC), 2022, pp. 370–375, doi: 10.1109/ICESIC53714.2022.9783481.

30. V. K, K. Jayashree, V. R and B. Rajendiran, "Forecasting Methods and Computational Complexity for the Sport Result Prediction," 2022 International Conference on Electronic Systems and Intelligent Computing (ICESIC), 2022, pp. 364–369, doi: 10.1109/ICESIC53714.2022.9783514.

Recent Trends in Computational Intelligence and its Application – Sugumaran D. et al. (eds)
© 2023 Taylor & Francis Group, London, ISBN 978-1-032-48410-5

43

Convolutional Neural Network Based Image Steganography

Lijetha C. Jaffrin[1], P. N. Karthikayan[2]
Assistant Professor, Department of IT,
Vel Tech Rangarajan Dr. Sagunthala R&D Institute of Science and Technology, India

Soumik Datta[3], Arjit Majumdar[4]
UG Student, Department of IT,
Vel Tech Rangarajan Dr. Sagunthala R&D Institute of Science and Technology, India

Abstract—In cryptography, image steganography is the activity of hiding message into an image by pixel value alteration depending upon the encryption algorithm. However, hiding images in a cover image is a great provocation today. Here Variational Auto-Encoder architecture, a type of artificial neural network is chosen which learns the maximum from the minimum dimension possible. It consists of three parts- encoder, latent space representation(compression), and decoder. It does image concealing by minimizing the loss and removing noise from the image. The secret image is first routinely embedded into the cover image by the encoding network. Then decoding network is utilized to recover the concealed image. Image Steganography using CNN has good compatibility, accuracy, and minimum loss and can be done on dissimilar data types like remote sensing images and aerial images. This model has good training and loss reduction capabilities depending upon its hyperparameters. This model is also used to train other image datasets that are diverse from this ImageNet dataset, such as remote sensing images and aerial images. This model also understands concealing and extraction ensuring secure steganography.

Keywords—Deep learning, Image steganography, Variational auto-encoder architecture, Convolution neural network, Cryptography

I. Introduction

End-to-end encryption and privacy protection of communication are very essential to maintain confidentiality and integrity. Thus, information hiding plays an important role in various sectors to maintain secure communication. Image Steganography is a vital factor in information concealing. The source conceals the hidden image inside a cover image and sends it to the recipient. The recipient receives the image and reconstructs it to get the secret image. Image

[1]lijethacjaffrin@veltech.edu.in, [2]pnkarthikayan@veltech.edu.in, [3]vtu11765@veltech.edu.in, [4]vtu11667@veltech.edu.in

DOI: 10.1201/9781003388913-43

is a main form of data carrier thus secure retention and transmission has always been an issue. The use of CNN in image segmentation, classification and concealing has been a vital topic in recent times.

The use of Variational Autoencoder architecture and end-to-end training of decoder and encoder is one of the most important works in image steganography. The safety of image steganography relies on two factors: first, the amount of data to be concealed and second, the presence of the cover image itself and the amount of change in basic statistics after message embedding. Perfect image steganography depends upon the maximum amount of information hidden, while stego-image and basic statistical data persist. Depending upon the issues faced due to the problem, an information concealing scheme based on deep learning has been designed which concentrates on how successfully the image steganography can be balanced which is also a challenge to overcome in this research work. A steganography model based on a deep convolutional neural network that can efficiently rise the steganography capacity in the minimum dimension possible ensuring steganography security has been proposed in this paper. However, this is not viable, because the appearance of the cover image and the amount of embedded information hold an aggressive association. More the embedded messages, the greater the degree of image distortion, which lowers the safety and vice versa.

The main objective of this paper is briefed as follows: A CNN-based model is proposed and the network is prepared to acquire maximum features of images under minimum dimensionality and to use them in a data-driven way. It effectually realizes the union of features of the cover image and the original concealed image, so as to understand the embedding and extraction of the secret image. It produces high-volume image steganography. The ratio of the secret image to be hidden in the cover image is 1:1. Depending on its hyperparameters,

this model can do well in training and loss reduction. This model is also used to train other picture datasets that differ from the ImageNet dataset, like aerial photos and remote sensing images, which also comprehend concealment and extraction to provide secure steganography.

A. Literature Survey

Deshpande Neeta et al. [1] proposed a Least Significant Bit(LSB) embedding technique where data could be hidden in the least significant bits of the cover image and humans could not find the concealed image in the cover file. It will help people to make their data more secure. It is the simplest one to understand, easy to implement and results in stego images that contain embedded data. But it is vulnerable to steganalysis and is not secure at all

Hiroshi Naito et al. [2] used GAN to make huge unique and ordinary images. Really, training of GAN consequence in two neural networks namely the generator and the discriminator. The generator created virtual images, and the discriminator assessed the reality of virtual images. The advantage here was that GAN got an insight into data and could simply understand various versions thus helpful in the machine learning activity. But producing outcomes from text or speech was complex.

Alexander G. et al. [3] combined the CNN-two-stage procedure into a joint training system. They verified their process on semantic image segmentation and showed improved outcomes on PASCAL VOC 2012 dataset. Here authors proposed a deep learning-based generic decoder and encoder architecture for image steganography. They introduced a new loss function which automatically detected important features without human aid.

Bharath Hariharan et al. [4] detected all occurrences in an image type and, for each occurrence, the pixels that fit it were noticeable and called that task Simultaneous Detection and Segmentation (SDS). Unlike normal bounding box detection, SDS required segmentation.

Unlike traditional semantic segmentation, all object occurrences were required. The advantage was that it can handle a large amount of dataset at a time. But many different sources can be combined into one solution.

Neena Aloysius et al. [5] said that Convolutional Neural Networks (CNNs) have provided an alternate process for routinely understanding domain-specific features. Now issues in the wider domain of computer vision were re-reviewed from a new methodology perspective. The advantage was that it automatically detected vital features without human intervention.

Yinyin Peng et al. [6] did a rescindable image verification technique that could improve the accuracy of tamper detection and the quality of watermarking images. Here, rescindable data concealing was applied with two undistinguishable host images, where the hidden data was embedded in one host image and misrepresentation data was embedded in another image. The advantage was that when the original image was separated into chunks, the image would be covered with a chaotic watermark and none could recuperate the original image without the key. But some losses would take place while recovering the original image.

L. M. Marvel et al. [7] presented a digital steganography method referred to as spread spectrum image steganography (SSIS). This system concealed and recuperated digital imagery-based messages of substantial length while preserving original image dimensions and dynamic range. The concealed message could be recovered using appropriate keys without knowing about the original image. The advantage was that SSIS used digital imagery as a cover signal. But here also data loss took place while decoding the process.

Xin Liao et al. [8] explained classical steganographic systems where RGB three-channel payloads were allotted likewise in a true colour image. In fact, the security of colour image steganography is linked not only to data- embedding procedures but also

to diverse payload partitions. The advantage was the colour image-hiding process. But the colours may change or fade while recovering the original image.

Pin Wu et al. [9] combined the latest deep convolutional neural network approaches with image-into-image steganography. The main contribution was to reduce the loss which was already done. It automatically perceived the important features. But large training dataset was needed and did not encode the position and orientation of the object.

Zihan Wang et al. [10] research worked on a high-volume image steganographic model termed Hiding GAN. The proposed model employed a secret of hiding data within an ordinary file. In this way, hiding GAN avoids the detection of human eyes. The advantage was that GAN explored into details of data and could simply infer various versions of helping machine learning tasks.

B. Motivation and Justification of Work

Nowadays many people, organizations and industries are very much concerned with the privacy of their data. To keep this confidential and protected this proposed idea of image steganography using CNN can be used. This cryptographic process will help users hide their confidential and private data and help them transmit it without any breach of privacy or any kind of unusual attacks for the stealth of data. The main motivation for doing this work are as follows: (i) To keep someone's secret image safe from any third-party attacks (ii) To decode the secret message and to establish the authenticity of a piece of information

This work is justified using Image steganography and by this process many people, organisations or industries can communicate with each other privately. For example, intelligence agencies can use them for communication. Moreover, cryptocurrencies also have reconstructed uses of image steganography.

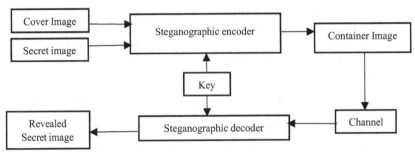

Fig. 43.1 Overview of methodology

Source: Authors

C. Contributions

The contributions of this work include a comprehensive review and perceptive analysis of deep learning segmentation algorithms such as CNN covering training data, network architecture selection, loss functions, training methods, and their major contributions.

D. Outline of the Paper

Figure 43.1 shows the overview of the proposed methodology. The embedding of a secret image inside the cover image and the extraction of the embedded secret image from the stego image are both possible using a lightweight and straightforward deep convolutional autoencoder architecture. Utilizing ImageNet datasets, the proposed technique is assessed.

E. Organization of the Paper

The paper is systematized as below: In the abstract, the idea, approach of implementation and how it is useful and can be used in different industry sectors side by side are discussed. Sect. 1 represents the introduction, which discussed the problems faced, the innovative approach, and how the implementation is done. Sect. 2 proposes the system methodology which discussed the architecture model and algorithms used, Sect. 3 represents the experimental design, Sect. 4 proposes performance evaluation, Sect. 4.1 identifies the proposed metrics which discussed the metrics used to measure the degree of correctness of the model, Sect. 4.2 identifies the performance analysis and the final

Sect. 5 includes results and discussion Sect. 6 proposes the conclusion with references.

II. Proposed Method

CNN based architecture model for image steganography includes variational autoencoders which compare each neural network with another giving the best output possible. The encoder receives the cover image and the secret image as inputs, creating the stego image as an output, and the decoder receives the stego image as input, producing the secret image. While changes in the convolutional layer and pooling layer are predicted, the ways in which the input cover picture and the secret image are combined vary as well. Different methods utilise different numbers of filters, strides, filter sizes, activation functions, and loss functions. The size of the cover image and the secret image must be the same (64*64), which means that each pixel from the secret image must be evenly distributed throughout the cover image.

A. CNN Concept

CNN refers to the Convolutional Neural Network. It is a multi-layered neural network. It is especially employed in image processing and image designing. It is comprised of two main blocks: The first block acts as a convolution filtering operation (feature extractor). The first layer returns the feature map by applying a feature extractor. This process will take place several times. The output of the first block and the input of the second block is created when the value of the final feature map is concatenated into a vector. The second block is used for classification. It is located at the conclusion of every neural network. It takes the vector as input which comes from the first block and distributes all elements of the vector as 0 and 1 and the sum of all is worth 1. Basically, CNN is a multi-layered neural network so it has 4 layers. These are the Convolutional Layer, Pulling Layer, ReLU Connection Layer, and fully connected Layer. The purpose of the convolutional layer is to detect the existence of a set of features in an image received as input. The pooling technique involves shrinking the size of the image while maintaining its crucial elements. All negative values that are received as inputs are replaced by zero via the ReLU connection layer. It serves as an activating mechanism. A new output vector is created by the fully connected layer after receiving an input vector. In this layer, we can also calculate the loss function of the input image and output image. The kernel convolution is given as follows:

$$G[m, n] = (f * h)[m, n]$$
$$= \sum_j \sum_k h\,j, k\,f[m - j, n - k] \qquad [1]$$

The Pixel position can be given by

$$n = (f - 1)/2 \qquad [2]$$

(a) Challenges

Some challenges that appear while doing this work were-

1. Adjusting the size of the secret image and cover image to merge both images.
2. Implementing the network in such a way that overfitting/underfitting doesn't occur.

(b) Key Features

CNN is composed of 2 key features. That is Feature extrusion and classification. Multiple convolutional layers, max-pooling, and an active function are all included in feature extraction. The classifier often has layers that are fully coupled.

(c) Merits

Main advantage of CNN is that it automatically recognizes the crucial characteristics without any human supervision. It is little dependent on pre-processing and decreases the need for human effort to develop its functionality. The best thing is that it is easy to recognize and fast to deploy and accuracy is very high to predict the images. Error back propagation function has been introduced to hurry up training tasks.

The technical novelty in this work comes up with a novel concept of image steganography that has high efficiency, easy implementation and minimal software/hardware requirements. From the existing scientific knowledge, this work is developed in an innovative manner. It is planned to develop a new approach to exploit the existing scientific knowledge.

(d) Algorithm

In this research work, CNN and Variational Autoencoder have been used for image steganography [11]. A sort of artificial neural network called a variational autoencoder is used to learn effective data in an unsupervised setting [12]. By teaching the network to disregard signal noise, the goal is to learn a representation for a set of data, often for dimensionality reduction.

Step 1: Dataset Acquisition, preparation and image pre-processing

Step 2: Loading dataset into the program. Normalize the image vectors ranging from 0 to 255. Divide it by 255 so that the input image pixel intensity ranges between 0 to 1.

Step 3: Divide the prepared dataset for training, testing and validation

Step 4: Preparation of Loss model for full model

Step 5: Preparation of Keras model consisting of Preparation Network, Hiding Network and Reveal Network.

Step 6: Training the model with appropriate hyper-parameters and their values

Step 7: Loading and configuring the model to retrieve decoded predictions.

Step 8: Display the result.

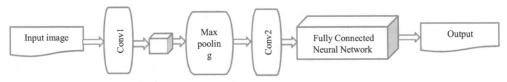

Fig. 43.2 Architecture of CNN

Source: Authors

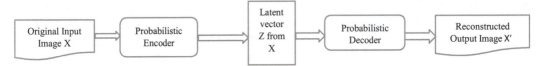

Fig. 43.3 Variational autoencoder architecture and CNN

Source: Authors

III. Experimental Design

Experimental Design consists of the overall model preparation design and architecture and how the algorithms are being used in the model. To understand this better a diagrammatic explanation is required.

The Figure 43.2 and 43.3 show the overall model architecture of Variational Autoencoder and CNN.

Figures 43.2 shows the CNN architecture. A Convolutional Neural Network (ConvNet/CNN) is a Deep Learning procedure that considers an input image, allot weights and biases to various aspects of the image, and can distinguish one from the other [13] [14] [15]. It consists of:

1. *Convolution Layer:* The convolution layer is basically a filter that is used to show the result of an input in a map of activation. When this application of the same filter is used repeatedly that is called a feature map. In a detected feature of input, it indicates the strengths and locations like an image.

2. *Padding Layer:* In this layer, all images are padding together to upgrade their edge pixels. Here advanced filtering techniques are used to provide a border to every image for giving space for image boundary or to write notations.

3. *Pooling Layer:* Pooling layers are especially used in feature maps to adjust the proportions of an image. As a result, it decreases the number of constraints to learn and the quantity of network processing. These layers do nothing but all things of the feature map which is already done by the convolution layer.

4. *Flattening Layer:* Data is flattened when it is made into a 1-dimensional array for input into the following layer. We flatten the convolutional layer output to produce a solitary, lengthy feature vector. Additionally, it is linked to the last classification model, sometimes known as a completely connected layer.

Figure 43.3 is the Variational Autoencoder Architecture. An artificial neural network called an autoencoder is used for unsupervised learning and effective data coding [16] [17]. An autoencoder trains the network to ignore signal "noise" with the goal of learning a representation for a set of data, generally for dimensionality reduction [18].

A. Applications

1. *Dimensionality Reduction:* The premise behind dimension reduction techniques [19] is that the data's intrinsic dimension is substantially lower than its artificially inflated dimension. The size of the hidden layer decreases as the number of layers

in an autoencoder increases. When the concealed layer's size shrinks below the data's inherent dimension, information is lost [20].

2. *Image Generation:* Variational Autoencoder (VAE), a form of autoencoder that is a generative model and is used to create images [21]. The concept is that the system will produce comparable images when given input images, such as images of a person or scenery. The purpose is to create new animated characters and to create counterfeit human images.

3. *Image Denoising:* An image is referred to as being noisy when it becomes corrupted or has some noise. Image denoising [22] is what we need to get accurate information about an image's content. We design our autoencoder to reduce the majority, if not all, of the image noise.

4. *Sequence to Sequence Prediction:* [23] Machine translation issues can be solved using the Encoder-Decoder Model, which can capture temporal structure, such as LSTM-based autoencoders. This can be used to generate phony videos by foretelling the following frame of a video.

In this research work, Sequence to Sequence Prediction implementation is used.

Figure 43.4 here stacking one layer upon another is performed. From the above snippet it is seen that three convolution layers x3, x4 and x5 with different input and kernel size

take input from input_S and after that feature extraction it is sent to the concatenation layer. Then again, this x(concatenate)layer is given as input to the convolution layers and their outputs are again concatenated.

IV. Performance Evaluation

The next phase is to use test datasets to determine how well the model predicts or states output based on some metric after performing the customary Feature Extraction, Selection, and Model Implementation. Different algorithms are assessed using various performance indicators. Classification performance metrics such as Log-Loss, Accuracy, AUC (Area under Curve) were used.

A. Performance Metrics

1. *Loss function for reveal network:* Here we have loss function for reveal network to determine how far or close our reveal network is from its goal [24]. (Loss for reveal network is: beta * |S-S'|)

2. *Loss function for full model:* This takes care of defining the loss function for full model. (Loss for the full model is: |C-C'| + beta * |S-S'|)

B. Performance Analysis

Depending upon the performance metrics (loss function) we see that our model has a good efficiency as shown below in the loss history graph (Fig. 43.5). This means that our data is well trained and has an appropriate accuracy.

```
def make_encoder(input_size):
    input_S = Input(shape=(input_size))
    input_C= Input(shape=(input_size))

    # Preparation Network
    x3 = Conv2D(50, (3, 3), strides = (1, 1), padding='same', activation='relu', name='conv_prep0_3x3')(input_S)
    x4 = Conv2D(10, (4, 4), strides = (1, 1), padding='same', activation='relu', name='conv_prep0_4x4')(input_S)
    x5 = Conv2D(5, (5, 5), strides = (1, 1), padding='same', activation='relu', name='conv_prep0_5x5')(input_S)
    x = concatenate([x3, x4, x5])
```

Fig. 43.4 Code snippet of preparation network

Source: Authors

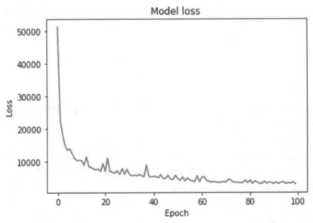

Fig. 43.5 Model loss history

Source: Authors

Table 43.1 CNN preparation network structure

Operation Layer		Number of Filters	Kernel Size	Stride Value	Padding Value
Input- input_S		-	-	-	-
Convolution Layer- x3	Convolution ReLU(two times)	50	3 x 3	1 x 1	same
Convolution Layer- x4	Convolution ReLU(two times)	10	4 x 4	1 x 1	same
Convolution Layer- x5	Convolution ReLU(two times)	5	5 x 5	1 x 1	same
Concatenation Layer (two times)x = (x3, x4, x5) x = (input_C, x)		-	-	-	-

Source: Authors

Table 43.2 CNN Hiding network structure

Operation Layer		Number of Filters	Kernel Size	Stride Value	Padding Value
Input-x		-	-	-	-
Convolution Layer- x3	Convolution ReLU(five times)	50	3 x 3	1 x 1	same
Convolution Layer- x4	Convolution ReLU(five times)	10	4 x 4	1 x 1	same
Convolution Layer- x5	Convolution ReLU(four times)	5	5 x 5	1 x 1	same
Concatenation Layer (five times) x = (x3, x4, x5) x 5		-	-	-	-
Convolution Layer-	Convolution ReLU output_Cprime	3	3 x 3	1 x 1	same

Source: Authors

Table 43.3 CNN Reveal network structure

Operation Layer		Number of Filters	Kernel Size	Stride Value	Padding Value
Input-x, input_with_noise		-	-	-	-
Convolution Layer- x3	Convolution ReLU(five times) input_with_noise	50	3 x 3	1 x 1	same
Convolution Layer- x4	Convolution ReLU (five times)	10	4 x 4	1 x 1	same
Convolution Layer- x5	Convolution ReLU(four times)	5	5 x 5	1 x 1	same
Concatenation Layer (five times) x = (x3, x4, x5) x 5		-	-	-	-
Convolution Layer-	Convolution ReLU output_Sprime	3	3 x 3	1 x 1	same

Source: Authors

Table 43.1, Table 43.2, Table 43.3 above depicts the CNN preparation network structure, CNN Hiding network structure, and CNN Reveal network structure which shows the various operation layer, number of filters used, Stride Value, kernel size, and Padding Value respectively.

Table 43.4 Model training history

Batch Size = 32		
Epochs	Average Auto-Encoder Loss	Reveal Network Loss
1 - 10	18,742.01	10,283.09
11 - 20	7,406.12	5,146.36
21 - 30	5,902.70	3,925.42
31 - 40	4,789.81	2,958.23
41 - 50	4,293.11	3,038.43
51 - 60	3,785.99	2,208.71
61 - 70	3,775.60	2,084.58
71 - 80	4,091.47	2,534.59
81 - 90	3,093.25	1,825.58
91 - 100	2,956.57	1,791.61
Average	5,883.66	3,579.66

Source: Authors

Table 43.4 above depicts the model training history which includes Epochs, average auto-encoder loss, reveal network loss. The optimal number of epochs should be used to train the model. After each training period, the model's performance is checked using a portion of the training data set kept aside for model validation.

V. Results and Discussion

Significant results and unique findings have been established through this research work. After taking the results into consideration it is seen that the model works properly to perform image steganography. In this work the dataset used is Tiny ImageNet Visual Recognition Challenge. Our training set is made of a random subset of images from all 200 classes. We have taken 10 images from each class to train (200*10=2000 images) and the testing set contains a total of 500 images. The training set is split into two halves, the first half for secret images and the second half for cover images.

After model has been trained with a large dataset from Fig. 43.6 it is seen that the model has successfully established our idea and main

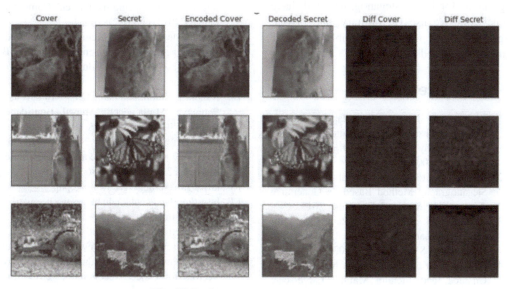

Fig. 43.6   Image steganography results

Source: Authors

motive of the research work. It shows six cover pairs from the dataset on which image steganography is performed [25] and it also shows the difference between predictions and ground truth.

VI. Conclusion

This research work proposes an application of CNN based on image steganography. One secret image will be concealed in one cover image as part of this research work. Concatenating the two images in the encoder allows the sender to obtain the stego-image, which is then transmitted to the receiver. The receiver then receives the stego-image from the decoder. The necessary secret image is then rebuilt. In addition, the trained model works well for virtually any color image steganography, including images from remote sensing and aerial photography. The work that has been done includes- defining and understanding the problem, design and model selection, preparation of dataset and understanding the concept of variational autoencoder architecture, training, implementing and then observing the output. This unique idea of using deep learning algorithms for image steganography will have some extensive uses in military, and private organizations as well as the public for their authentication and privacy.

References

1. Deshpande Neeta, Kamalapur Snehal, Daisy Jacobs: "Implementation of LSB Steganography and Its Evaluation for Various Bits", June 2007

2. Hiroshi Naito, Qiangfu Zaho: "A New Steganography Method Based on Generative Adversarial Networks", December 2019.

3. Alexander G. Schwing and Raquel Urtasun: "Imagenet Large Scale Visual Recognition Challenge(ILSVRC)", June 2003

4. Bharath Hariharan, Pablo Arbel´aez, Ross Girshick, and Jitendra Malik: "Large-scale CelebFaces Attributes (CelebA) Dataset" July 2020

5. Neena Aloysius, M. Geetha: "Designing steganographic distortion using directional filters", January 2013

6. Yinyin Peng, Xuejie Niu, Zhaoxia Yin: "Image authentication scheme based on reversible fragile watermarking with two images", Volume 40, June 2018, pp. 236–246

7. L. M. Marvel, C. G. Boncelet, C. T, Retter: "Spread spectrum image steganography", Volume 8, August 1999, pp. 1075–1083

8. Xin Liao, Yingbo Yu, Bin Li, Zhongpeng Li, Zheng Qin: "A New Payload Partition Strategy in Color Image Steganography", Volume 30, Mach 2020, pp. 685–696

9. Pin Wu, Yang Yang, Xiaoqiang Li: "StegNet: Mega Image Steganography Capacity with Deep Convolutional Network", June 2018, pp. 401–405

10. Zihang Wang, Neng Gao, Xin Wang, Ji Xiang: "HidingGAN: High Capacity Information Hiding with Generative Adversarial Network", October 2019, pp. 393–401

11. Haichao Shi, Xiao-Yu Zhang, Shupeng Wang, Ge Fu, and Jianqi Tang: "Synchronized detection and recovery of steganographic messages with adversarial learning" International Conference on Computational Science, June 2019, pp. 31–43.

12. N. F. Johnson and S. Jajodia. "Exploring steganography: Seeing the unseen".Computer, January 2012, pp. 31(2): 26–34

13. Deshpande Neeta, Kamalapur Snehal, Daisy Jacobs: "Implementation of LSB Steganography and Its Evaluation for Various Bits", June 2007

14. Nandhini Subhramanian, Omar Elharrouss, Soumaya Al-ma'adeed, Ahmed Bouridane: "Image Steganography: A Review of the Recent Advances", January 2021, pp. 315–336

15. https://www.sciencedirect.com/science/article/abs/pii/S2214212617305586

16. https://image-net.org/download.php

17. http://mmlab.ie.cuhk.edu.hk/research works/ CelebA.html

18. https://arxiv.org/abs/1806.06357

19. R. K. Keser and B. U. Töreyin, "Autoencoder Based Dimensionality Reduction of Feature Vectors for Object Recognition," 2019 15th International Conference on Signal-Image Technology & Internet-Based Systems

(SITIS), 2019, pp. 577–584, doi: 10.1109/ SITIS.2019.00097.

20. W. Wang, Y. Huang, Y. Wang and L. Wang, "Generalized Autoencoder: A Neural Network Framework for Dimensionality Reduction," *2014 IEEE Conference on Computer Vision and Pattern Recognition Workshops*, 2014, pp. 496–503, doi: 10.1109/CVPRW.2014.79.

21. P. Cristovao, H. Nakada, Y. Tanimura and H. Asoh, "Generating In-Between Images Through Learned Latent Space Representation Using Variational Autoencoders," in *IEEE Access*, vol. 8, pp. 149456–149467, 2020, doi: 10.1109/ACCESS.2020.3016313.

22. A. Pawar, "Noise reduction in images using autoencoders," *2020 3rd International Conference on Intelligent Sustainable Systems (ICISS)*, 2020, pp. 987–990, doi: 10.1109/ ICISS49785.2020.9315908.

23. H. Jain and G. Harit, "An Unsupervised Sequence-to-Sequence Autoencoder Based Human Action Scoring Model," *2019 IEEE Global Conference on Signal and Information Processing (GlobalSIP)*, 2019, pp. 1–5, doi: 10.1109/GlobalSIP45357.2019.8969424.

24. T. Dung Pham, V. Cuong Ta, T. Thanh Thuy Pham and T. Ha Le, "Reducing Blocking Artifacts in CNN-Based Image Steganography by Additional Loss Functions," *2020 12th International Conference on Knowledge and Systems Engineering (KSE)*, 2020, pp. 61–66, doi: 10.1109/KSE50997.2020.9287408.

25. M. Sharifzadeh, C. Agarwal, M. Aloraini and D. Schonfeld, "Convolutional neural network steganalysis's application to steganography," *2017 IEEE Visual Communications and Image Processing (VCIP)*, 2017, pp. 1–4, doi: 10.1109/VCIP.2017.8305045.

Recent Trends in Computational Intelligence and its Application – Sugumaran D. et al. (eds)
© 2023 Taylor & Francis Group, London, ISBN 978-1-032-48410-5

44

Smart Digilocker Using IoT

S. Shanmuga Priya[1]
Department of Artificial Intelligence and Data Science,
Saveetha Engineering College, Chennai

R. Umamaheswari[2]
Department of Computer Science and Engineering,
Vel Tech Rangarajan Dr Sagunthala R&D Institute of Science and Technology, Chennai

S. Uma[3]
Department of Computer Science and Engineering,
Rajalakshmi Institute of Technology, Chennai

R. Ganesan[4]
Department of Computer Science and Engineering,
Vel Tech Rangarajan Dr Sagunthala R&D Institute of Science and Technology, Chennai

Abstract—Development of technologies, and smart devices are very common in everyday life. Smart designs have benefited from the development of gadgets that can connect to the Internet and transmit data. The main constraint with lockers has forced them to remember locker passwords or misplacing locker keys is a widespread issue at workplaces and colleges. Also, the pre-assignment of lockers can be a waste of space for those who might not use them that often. So, we have designed Smart DigiLocker that can be accessed using an RFID card, which can be incorporated into the ID card itself. A green LED on the locker would indicate vacancy and a red LED occupancy. On the back end, we store locker usage information. As an added layer of security, the intruder details are sent to the locker occupant via SMS and email. With this Smart DigiLocker, anyone can use any unoccupied locker safely. Also, one common master card shall be possessed by the authority in case one loses their ID card. Moreover, it is quite cost-efficient.

Keywords—Smart DigiLocker, RFID card, IoT, LED, SMS, Internet

I. Introduction

Today everything is smart. Let it be a smart watch, smart TV, smart washing machine, or smart air conditioner. Then why not make our lockers smart too? During university examinations, students don't have any safe place to keep their belongings and when it comes to manual lock and key systems, the common problem with it is the hassle of remembering

[1]spriyasenthil2009@gmail.com, [2]uma2007ap83@gmail.com, [3]dewuma@gmail.com, [4]ganeshitlect@gmail.com

DOI: 10.1201/9781003388913-44

locker passwords or forgetting locker keys. This makes it a tedious task to remember the password, whereas in our daily life we already have to remember bank passwords, email id passwords, and social media accounts passwords since this RFID card would also be the individuals' identity card that they carry to their offices or Universities daily without fail, a separate key need not be issued for accessing lockers. Moreover, assigning lockers to each individual can be a waste of space. There might be some people who never use a locker. So, to overcome that issue, this paper presents an idea where any user can use any unoccupied locker, thus the space is not wasted.

II. Literature Review

Alqahtani et al. [1] proposed an Automated smart locker that relies on Bluetooth and can be controlled remotely via a mobile application aimed at improving convenience and security. A keypad can be used as an alternative method to open the lock. If Mobile isn't available at the time and the user has forgotten their password. So, we avoided the hassle of remembering passwords or pins by using an RFID card. Anusha et al. [2] proposed a Locker security system utilizing identification by facial recognition. When the user accesses the locker, they literally put their face in front of a camera and enter the OTP supplied by the locker. The OTP (One Time Password) technique connects the system to the user's email or mobile number. When the camera is linked to the locker and the PIN is genuine, the Eigen face and PCA (principles components analysis) algorithms are used to compare and identify the user's image. It is a very cost-effective locker yet extremely secure. So, we try to design a cost-efficient locker that can be accessed using an RFID card. Mohammed et al. [3] proposed methods for constructing a security door lock system. An RFID card and keypad-based Secured Locker system have been proposed by the author. Instead of a keypad, we made some

changes like an added layer of security, and intruder details are sent to the locker occupant via SMS and email. Gupta et al. [4] proposed a combination of the tri securities methods such as the sequence of RFID, Password, and Biometrics. It proposes multi-layer security. These all modules are managed through a microcontroller. This system application is used where a large amount of protection is required, but it is a very complex system. We try to make it simple and cost-efficient. Mahendra et al. [5] focused on enhancing the traditional security system based on the IoT platform, which has suitable for communicating in real-time with devices. The system contains cameras, speech or voice sensors\microphones, moving sensors, and an LTE\WIFI MODULE which is an interface and essential part of the system. Here the main drawback is whenever the power is cut the Wi-Fi will not work. Instead of Wi-Fi in our model, we used a battery.

Kabir et al. [6[used six Tier Multipurpose Security Locker systems depending on the Arduino concept, the authors have proposed a six-layer locker system that consists of a biometric sensor, password keypad, GPS, GSM, and PIR sensor. Although it is a highly protected locker system, it exceeds the budget when we need to implement it on a larger scale. Hence, this paper proposes to design a system that has a simple UI (user interface) and is affordable. Jester et al. [7] proposed a Two-point Security system for doors/Lockers. In that, the authors developed a system that protects lockers using biometric sensors and human face recognition using machine learning (ML) techniques and the Internet of Things (IoT). For the database, they used Google Firebase. The drawback of this database is that the Google Firebase has limited querying capabilities and which makes it quite complicated to make complex queries.

Knight et al. [8] elaborate on the Lock take in the time of IoT. If not properly protected, the proposed usage of a third-party back-end service to process data and provide critical functionalities can result in major risks. As a result, we suggest a secure back-end service in this article. The Super Secure Door Lock System for critical zones authors proposed a two-factor authentication system. They have used RFID cards along with OTP matching for securing lockers. Moreover, the RFID card's code is encrypted using four different encryption algorithms. Mathew et al. [9] discussed increasing bank security systems by using Facial Recognition when the bank manager/employee wants to open the locker just show the face in front of the camera, Iris Scanner which is used to scan the object perfectly and accurately and Palm Vein Technology is a kind of biometric method, the authors have presented a secure locker system that uses face recognition, Palm Vein technology, and iris scanner used to verify the user. But cost-wise more expensive. Also, we have used chosen this paper as a part of the future scope for our proposed Smart Locker. Overall, in the Existing Model, some of the limitations are identified as

- There is always the hassle of carrying the key to the access locker.
- If any theft occurs, the occupant cannot be intimated.
- Wastage of space due to the pre-allocation of lockers

III. Proposed System

The proposed Smart DigiLocker using IoT is an intelligent and affordable locker system. The major goal of our suggested method is to eliminate carrying keys around all the time (let's say the user forgot the locker key, in which case it would be quite challenging for the user to unlock the locker again) and having to remember passwords (in-case if the user is unable to remember the password of the locker). [10] The Smart DigiLocker can be opened by just scanning the RFID card. This RFID card is actually the identity card itself which everybody is obliged to carry daily. People can access any vacant locker by just

scanning their identity card or RFID card. The vacancy of the locker is shown with a green LED, and if the locker is full or occupied it is indicated by a red LED. If someone tries to scan their ID on an occupied locker, the information of the unauthorized person will be stored in the database on a separate sheet. Moreover, there is always a chance that one may lose their ID, in such cases, one master card is provided to the locker maintenance department which can access all the lockers with that single card.

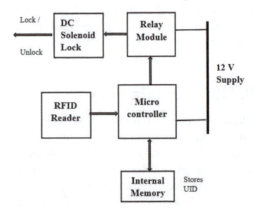

Fig. 44.1 System architecture
Source: Authors

The Smart DigiLocker system's IoT architecture is seen in Fig. 44.1. The 12V DC electrically actuated solenoid lock, relay, Node MCU, RFID reader, and internal memory are the main parts of our Architecture.

Components needed for implementing the work

DC Solenoid Lock

A solenoid lock is generally also called a small electromagnetic lock. It works on the principle of electromagnetic force. This lock has copper wire winding which provides a voltage that generates an electromagnetic field. So due to this force will move the solenoid lock.

Relay

The relay is also known as an electromagnetic switch, which is used for turning on and turning off a circuit where many circuits are controlled by a single signal or by-passing low signal power to the circuit. By turning ON the switch control with the help of the magnetic field current is generated as passed through the coil so, that armature is attracted and the load circuit is closed.

Node MCU

One of the available open-source software and hardware is Node MCU. It is based on the ESP-12 module which is developed as an inexpensive chip ESP8266 or System-on-a-Chip (SoC).

RFID Reader

Digital data is used by Radio-Frequency Identification (RFID) and is encoded in smart labels or Radio waves have been used by a reader to pick up RFID tags. An interrogator is the brain of the RFID system as it is a necessary functioning of the system. The interrogators are devices used to transmit and receive radio waves so that we can able to communicate with RFID tags.

Internal Memory

The internal memory stores the details of the users when they scan their RFID tag on the RFID card reader.

Different levels of the work process

(a) RC522 RFID Reader Module

The technology known as radio-frequency identification (RFID) is used to encode digital data onto RFID tags or smart labels. With the help of the reader and by using the radio waves these smart labels are captured. Interrogators in the brain of the RFID system which is one of the necessary functioning parts of the system. [11] The interrogators are the device used for the transmission and receiving of radio waves so that we can able to communicate with RFID tags. In this module, we are using an RC522 RFID reader and two RFID cards for demo purposes. The RFID card reader is first used by the user to scan their RFID card. The microcontroller receives the data from the RC522 RFID card reader after it has read the card's information. [12] The microcontroller now checks this data

with the existing data present in the internal memory. If both the card data match then the solenoid lock is triggered. If the locker is empty then the data of the new RFID card is stored in an excel sheet. If any intruder scans their RFID card, those card details are stored in a separate intruder excel sheet.

(b) ESP8266 Wi-Fi 802.11 Band Module

In this module, a transmitter is used for connecting with the admin's device and this transmitter provides a hotspot for the remote locker. We are using the ESP8266 Wi-Fi 802.11 band for this purpose. The hotspot name and IP address of ESP8266 can be specified in the Arduino code and can be changed at our convenience. Once the remote locker is connected to this WiFi, we can read and store the data of all the scanned cards in the internal memory. The ESP8266 can be controlled from our local Wi-Fi network or from the internet.

(c) 12V DC Electrically Actuated Solenoid Lock Module

A solenoid lock (12V) has a slug in form of a slanted cut and it is also called an electromagnetic lock. It is designed for cabinets, safes, or doors basically When 9- 12VDC is applied, the slug is pulled in so it won't stick out so that we can able to open the door. In this module whenever a trigger is received from the ESP8266 Wi-Fi 802.11 band module, the lock gets opened. And in all other cases, i.e., when there is a mismatch of the RFID card data with the data present in the database, the lock remains closed.

(d) Database Storage

The database is an integral part of any real-time application to store data. In our system, we are using Microsoft Excel for storing data. Whenever a user tries to scan their card for the first time, details such as date, RFID number, locker open time, locker close time, and duration of usage are stored in an excel sheet. If an unauthorized person tries to open the locker with their RFID card, they will be considered an intruder, and data such as date, intruder

RFID number, and access time is stored in a separate excel sheet.

Overall, in our proposed model, we are having some pros such as:

- Anti-theft
- Encrypted
- Theft detection
- Cost-effective
- Robust.

IV. Results and Discussion

IoT and RFID-based Smart DigiLocker implementation proved successful. A relay, which is nothing more than a power relay module—an electrical switch that is powered by an electromagnet—is present on our circuit board for turning modules on and off. When the relay is re-energized, the closed sets of contacts open and break the connection, and if the contacts were open, the connection is broken. A latch for electrical locking and unlocking is referred to as a **12V DC solenoid lock**.

Figure 44.2 Represents the DC solenoid Lock and Fig:3 represents the overall circuit board with the relay switch.

Fig. 44.3 Overall circuit board

Source: Authors

Figure 44.4 represents the top & Front view of the locker When an authentic user scans their card, the lock opens as demonstrated below

Front View

Fig. 44.2 DC solenoid lock

Source: Authors

Top View

Fig. 44.4 Top and front view of the locker

Source: Authors

	A	B	C	D	E
1	Date	Accessed By	Locker Opened at	Locker Closed at	Duration of Usage
2	31-3-2021	65 C5 0B 2C	13:15:20	13:15:29	0:0:9
3	31-3-2021	65 C5 0B 2C	13:15:40	13:20:24	0:4:44
4					
5					
6					
7					
8					
9					
10					
11					
12					
13					
14					
15					
16					

Fig. 44.5 Data set of authentic users

Source: Authors

A. Data Set of Authentic Users

Whenever a user scans their card to access the locker, the following details are stored in an excel sheet.

B. Data set of Intruders

Whenever an intruder tries to access an occupied locker, they will be considered intruders and their data will be stored in a separate excel sheet.

V. Conclusion

In this article, we introduced the IoT-based Smart DigiLocker with an RFID card approach using a low-cost budget. It will be more beneficial for little businesses and organizations. We need to combine different units into one unit in order to get the output needed to secure lockers. The theoretical and technical application of initiative of engineering practice to execute. Thus, this device has been initiated to provide

	A	B	C	D	E
1	Date	Intruder ID	Access Time		
2	31-3-2021	5D AF 5B 7C	13:20:20		
3					
4					
5					
6					
7					
8					
9					
10					
11					
12					
13					
14					
15					
16					

Fig. 44.6 Intruders data

Source: Authors

security by using the RFID technology which is implemented in the locker. It allows only authorized people to access the system. It is a demonstration model but it can be implemented in the field for lockers security purposes in schools, colleges, and offices. In this model, advancements can be done and it can be made more innovative by providing multiple verification options so that this project can be used in various other applications such as in banks, military, hospitals, etc., where a higher level of security is required. In future work, we will consider the influence of more security layers and face Recognization techniques for capturing intruders and will send them back to the authorized person and administrator.

References

1. Alqahtani, Hanan F., Jeehan A. Albuainain, Badriayh G. Almutiri, Shahad K. Alansari, Ghaliah B. AL-awwa Nada N Alqahtani, Samia M. Masaad, and Rania A. Tabeidi. "Automated Smart Locker for College." In 2020 3rd International Conference on Computer Applications & Information Security (ICCAIS), pp. 1–6. IEEE, 2020.

2. Anusha, N., A. Darshan Sai, and B. Srikar. "Locker security system using facial recognition and One Time Password (OTP)." In 2017 International Conference on Wireless Communications, Signal Processing and Networking (WiSPNET), pp. 812–815. IEEE, 2017.

3. Mohammed, Salma, and Abdul Hakim Alkeelani. "Locker Security System Using Keypad and RFID." In 2019 International Conference of Computer Science and Renewable Energies (ICCSRE), pp. 1–5. IEEE, 2019.

4. Gupta, Ashutosh, Prerna Medhi, Sujata Pandey, Pradeep Kumar, Saket Kumar, and H. P. Singh. "An efficient multistage security system for user authentication." In 2016 International Conference on Electrical, Electronics, and Optimization Techniques (ICEEOT), pp. 3194–3197. IEEE, 2016.

5. Mahendra, Santosh, Mithila, Sathiyanarayanan and Rajesh Babu Vasu. "Smart Security System for Businesses using Internet of Things (IoT) IEEE, pp. 424–429. 2018.

6. Kabir, AZM Tahmidul, Nirmal Deb Nath, Utshawafin Akther, Fukrul Hasan, and Tawsif Ibne Alam. "Six Tier Multipurpose Security Locker System Based on Arduino." In 2019 1st International Conference on Advances in Science, Engineering and Robotics Technology (ICASERT), pp. 1–5, IEEE, 2019.

7. Jeste, Manasi, Paresh Gokhale, Shrawani Tare, Yutika Chougule, and Archana Chaudhari. "Two-point security system for doors/lockers using Machine learning and Internet of Things." In 2020 Fourth International Conference on Inventive Systems and Control (ICISC), pp. 740–744. IEEE, 2020.

8. Knight, Edward, Sam Lord, and Budi Arief. "Lock Picking in the Era of Internet of Things." In 2019 18th IEEE International Conference on Trust, Security and Privacy in Computing and Communications/13th IEEE International Conference on Big Data Science and Engineering (TrustCom/BigDataSE), IEEE, pp. 835–842, 2019.

9. Mathew, Meera, and R. S. Divya. "Super secure door lock system for critical zones." In 2017 International Conference on Networks & Advances in Computational Technologies (NetAct), pp. 242–245. IEEE, 2017.

10. Gusain, Raj, Hemant Jain, and Shivendra Pratap. "Enhancing bank security system using Face Recognition, Iris Scanner, and Palm Vein Technology." In 2018 3rd International Conference on Internet of Things: Smart Innovation and Usages (IoT-SIU), pp. 1–5. IEEE, 2018.

11. Sangiampak, J., C. Hirankanokkul, Y. Sunthornyotin, J. Mingmongkolmitr, S. Thunprateep, N. Rojsrikul, T. Tantipiwatanaskul et al. "Locker Swarm: An IoT-based Smart Locker System with Access Sharing." In 2019 IEEE International Smart Cities Conference (ISC2), pp. 587–592. IEEE, 2019.

12. Singla, Muskan. "Smart Lightning and Security System." In 2019 4th International Conference on Internet of Things: Smart Innovation and Usages (IoT-SIU), pp. 1–6. IEEE, 2019.

Recent Trends in Computational Intelligence and its Application – Sugumaran D. et al. (eds)
© 2023 Taylor & Francis Group, London, ISBN 978-1-032-48410-5

45

Cybersecurity and Artificial Intelligence: A Systematic Literature Review

R. Srinivasan[1], M. Kavitha[2], R. Kavitha[3]
Professor, Department of Computer Science and Engineering,
VelTech Rangarajan Dr. Sagunthala R&D Institute of Science and Technology, Chennai, Tamil Nādu
Uma S.[4]
Assistant Professor, Rajalakshmi Institute of Technology, Chennai, Tamil Nādu

Abstract—In an era when technological advancement is so fast, with connected devices and the Internet of things, the experts of Cyber Security in today's world are facing a number of challenges. It is necessary to provide them with the resources they need to prevent cybercrime. There are organizations today that are more connected than ever before, and they have to face cyber-attacks, data breaches, as well as a lot more threats in the cyber area that are becoming increasingly difficult for humans to defend against. It is however challenging to create a standard logic for a software system that is able to effectively defend against cyberattacks. On the other hand, cyber security issues can be effectively addressed by AI-based strategies that use machine learning. Cybersecurity threats can increase as technology develops and changes rapidly. With an emphasis on institutional digital transformation, cybersecurity issues and ways of enhancing and developing them became more important. In literature, it has been noted that traditional computer algorithms for cybersecurity may not be able to counter the creativity and development capabilities of hackers and saboteurs. Therefore, Artificial Intelligence (AI) techniques must be utilized to enhance cyber security. In this paper, we examine how artificial intelligence is applied to strengthen and improve cyber security, as well as how it benefits from artificial intelligence applications. The importance of utilizing Artificial Intelligence techniques to enhance cybersecurity was discussed utilising an analytical descriptive technique of earlier writings, then the utmost vital applications of AI that increase cybersecurity.

Keywords—Artificial intelligence, Cyber-security, Expert system, Intelligent agent

I. Introduction

Artificial Intelligence in the business world is becoming increasingly integrated with organizational processes and systems [8]. As a general rule, a majority of AI companies recognize that AI is the greatest threat to cybersecurity [8, 9]. In terms of cybersecurity, AI is a technology, and because of its all-purpose, dual-use character, it consumes the possibility

[1]srinivasanrajkumar28@gmail.com, [2]kavitha@veltech.edu.in, [3]rkavitha@veltech.edu.in, [4]dewuma@gmail.com

DOI: 10.1201/9781003388913-45

towards remain mutually a gift and also a burden. AI can be used both as a sword and a shield when it comes to reducing cybersecurity risks, as has been proven [9]. Cybersecurity is concerned with safeguarding data, networks, systems, etc., from being attacked by such threats. The following are some of the tools that are used in order to protect computers against such attacks: cryptography, antivirus software, and Intrusion Detection System (IDS). These are tools that have historically prevented a lot of attacks and malware. These methods use signatures of previous attacks to detect such attacks using the traditional methods of security. These databases contain information about known attacks and malware types, which are compared with the traffic on a daily basis. Any unusual activity, matching one of the signatures in the database, will trigger an alarm in the event that it is detected. Cyberattacks continue to improve and methods such as these can be of value to us by protecting us from attacks that we already know about. However, the trend currently envisaged is that cyberattacks will continue to improve [10]. Rapid advances in information and communication technology, including the Internet, have had positive effects on organizations and society. Networking and communication are enabled by the Internet. People need to share knowledge [1] and have social interactions [2]. Figure 45.1 shows the areas and various applications of AI. Using shield methods, data, applications, and networks are protected against unauthorized

access, modification, or vandalism [3]. A variety of strategies is also offered for protecting the integrity of system networks, applications, and stored data from damage, illegal access, and cyberpunk assaults [4]. With the advancement of ICT, new threats are emerging and are changing at a rapid rate. The field of artificial intelligence, which is concerned with making robots intelligent [5], continues to grow rapidly and has an impact on every aspect of our lives and business [6]. Artificial intelligence is providing benefits to industries such as gaming, manufacturing, health care, education, and natural language processing. In the cyber world, Artificial Intelligence (AI) is being used for both offense and defense, which ultimately translates to the improvement of cyber security. The x-axis in Figure 45.2 shows temporal information, while the y-axis represents popularity variables for data science, machine learning, and cybersecurity. By 2019, the popularity index values for these places will exceed 70, more than double the trend from 2014

It is becoming increasingly apparent that the fields of cyber security and artificial intelligence, initially considered to be distinct, are increasingly being integrated as attacks become much more sophisticated. Some of these areas include developing software to fix data leaks and to improve the security of computer systems, since attackers are increasingly imitating user behavior [7]. It is the purpose of

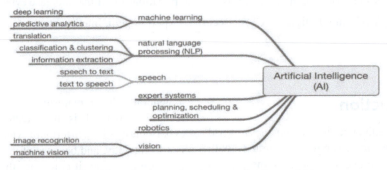

Fig. 45.1 Areas and various applications of AI

Source: Author

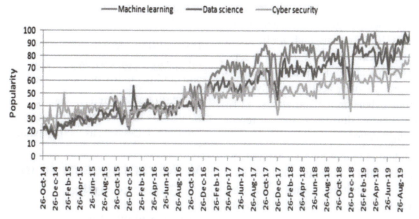

Fig. 45.2 Data science, machine learning, and cybersecurity popularity patterns across time [12]

this paper to describe how artificial intelligence can be applied to cybersecurity.

II. Problem Statement

This paper aims to explore how artificial intelligence is applied to cybersecurity, as it is one of the major concerns regarding this field. In the past decade, cybersecurity has dominated the headlines as a result of the increased risk and the efforts by cybercriminals must keep one step ahead of law enforcement. Hacking goals have mainly stayed consistent throughout time, while hackers have gotten more sophisticated [19].By using traditional methods, consumers can protect themselves from certain types of cyberattacks by using traditional methods. Organizations have a challenging time managing and prioritizing the many new vulnerabilities that are announced every day. It is common for vulnerability management solutions to respond to proceedings individuals later the susceptibility takes remained broken [20]. Here section III explains the elements of cyber security, and the advantages of cyber security explains in section IV the different types of cyber security applications will explain in section V.

III. Motivation

By identifying patterns of behaviour that indicate unusual behaviour, Artificial Intelligence can prevent cybercrime. Cybercriminals will typically attack systems that need to cope with so many events happenings every second using AI. AI's predictive capabilities make it so useful, which is why so many businesses will invest in these solutions in the coming year. Cybercriminals are also aware of the benefits of AI, and new attacks relying on technology such as machine learning to circumvent cyber security safeguards are emerging. As a result, AI becomes even more critical since it is the only way to prevent AI-powered cyber-attacks.

IV. Cybersecurity

This section's major goal is to offer an overview of the many sorts of cybersecurity events and the technology that supports them. Information security (InfoSec) is considered by several researchers to be of utmost importance [11]. Information and communication technologies (ICT) have developed tremendously over the past half century and are now omnipresent as well as deeply integrated into the society in which we live. In recent years, policymakers have become increasingly concerned about the security of ICT systems and applications against cyber-attacks [13]. In recent years, cyber security has become a term used to describe the process of protecting ICT systems against various cyber threats and attacks [14]. Figure 45.3 below illustrates the different aspects of cyber security.

- A confidential property is defined as a property that prevents unauthorized individuals, entities, and systems from having access to and disclosing information.

- The integrity of your information refers to your ability to prevent any unauthorized changes or destruction of your data.

- Information assets and systems can be accessed by authorized entities based on their availability.

Fig. 45.3 Shows the different properties of cyber security

Source: Author

Applications of cybersecurity range from businesses to mobile computing, and can be classified according to their context.

Network Security: A cybersecurity approach that focuses on the protection of computer networks against cyber threats is termed a cyber security approach.

Application security: In terms of application security, you want to ensure that your software

and devices are protected against cyber threats or risks.

Information security: The primary concern of information security is to ensure the security and confidentiality of the related information.

Operational security: The term operational security is used to refer to processes related to securing data assets as well as handling data.

V. Advantages of AI in Cybersecurity

Accuracy of humans

Using artificial intelligence, tens of thousands or even millions of data points can be analysed to identify patterns and anomalies that are otherwise invisible to humans.

AI Works 24/7

The AI systems do not have to sleep or take time off from work to spend time with their families. Business can function at peak performance 24 hours a day, 7 days a week, making this an ideal solution for anyone who runs a business.

Detect and alert changes

It is possible to monitor your network for physical and digital changes with artificial intelligence. A second set of eyes watches over your system or a hundred pairs of eyes if you prefer.

VI. Applications of AI in Cybersecurity

In a lot of interdisciplinary fields, cyber security and artificial intelligence are interwoven together. In order to sense and stop cybercrime, expert systems, artificial intellect, neural networks, intelligent agents, machine learning, data mining and so on [15] are all becoming increasingly important. Artificial intellect can stand used to recover the effectiveness of cyber security solutions. It is possible for companies to implement artificial intelligence to improve their cybersecurity systems 16][17]. As illustrated in Fig. 45.4, there are various applications AI has in terms of cyber security.

- Defense by automation
- Development of biometric-based login methods that are more accurate
- Predictive analytics is a powerful tool for detecting threats and malicious activities
- Using natural language processing to enhance learning and analysis
- Security for authentication and access to conditional access
- Using this technology to automate mundane security tasks

Cyber security uses expert systems: An expert system is a software package that contains knowledge that can be provided to a customer or another software package in order for it to function correctly. Knowledge is embedded in this system, and it is crucial to surround it with expert assistance to make it effective and efficient [20].

Deep learning applications in cybersecurity: One of the challenges in cybersecurity research can be attributed to the lack of disaggregated data. Oftentimes the lack of machine learning capabilities has been attributed to secrecy reasons furthermore, recent experiences have shown that, behind closed doors of huge corporations with significant internal expertise,

threat data may be turned into categorise data that machine learning can exploit. In my opinion there are a number of reasons why this is happening, which include the large quantity of large and unbalanced datasets, the difficulties in conducting manual categorizations, and the complexity of such disciplines as semantic categorization, which makes it more difficult to relate technological expertise to mathematical modelling [21].

Machine learning applications for cybersecurity: Cybersecurity threats evolve continuously, requiring the need for immediate responses. Machine learning techniques, particularly deep learning algorithms that do not require prior expertise or expert categorization, may be very useful for implementing AI technology into cybersecurity [23]. Machine learning approaches have been investigated in this study in relation to cybersecurity. During the study, machine learning was used to detect intrusions, spam, and malicious software. This debate focuses mainly on the efficiency of computer-based technologies and their drawbacks, which make them less suited to the direct application of cybersecurity and machine-learning approaches [22].

Data mining and cyber security: Through data mining, patterns and trends that cannot be identified through other means are analysed. Using data mining, we can uncover

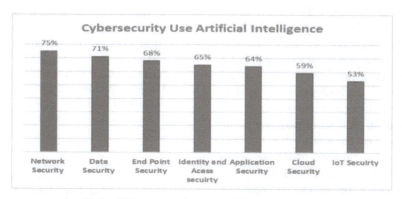

Fig. 45.4 Different application of AI with cyber security

Source: Capgemini Research Institute

hidden patterns in large datasets and gain valuable insights [24]. Machine learning requires "databases, analytics, expert systems, visualisation, high-performance computation, rough sets, and neural networks." Data mining necessitates the use of a host capable of obtaining data using a number of methods, including clustering, grouping, and association analysis [25].

VII. Results and Discussion

Every company's first priority is always data privacy and security. Digital or cyber information rules today's globe. On social networking sites, users may interact with their friends and family in a safe environment. Cybercriminals will continue to target social networking sites in order to steal personal information from end users. All required security precautions must be followed when participating in social networking and financial transactions.

VII. Conclusion

In this paper, we look at the usage of artificial intelligence in cybersecurity from the standpoint of artificial intelligence. According to one study, AI is fast becoming one of the most essential technologies for improving the efficacy of information security teams. In today's highly connected world, artificial intelligence (AI) offers this much-needed threat assessment and discovery that safety authorities can draw upon to decrease the risks of breaches and then advance their administration's overall security posture. Technology will gradually become more incorporated into our daily lives, which will lead to artificial intelligence having a greater impact on the way we live our lives. Furthermore, AI might stand used to recognize and prioritise hazards, direct incident response, besides detecting cyber-attacks in enterprises before they happen. As a consequence, despite its inherent drawbacks, artificial intelligence

may help businesses improve their security posture and cybersecurity.

References

1. F. N. Koranteng and I. Wiafe, ``Factors that promote knowledge sharing on academic social networking sites: An empirical study,'' Edu. Inf. Tech-nol., vol. 24, no. 2, pp. 1211–1236, Mar. 2019, doi: 10.1007/s10639-018-9825-0.

2. F. N. Koranteng, I. Wiafe, and E. Kuada, "An empirical study of the relationship between social networking sites and students' engagement in higher education," J. Educ. Comput. Res., vol. 57, no. 5, pp. 1131–1159, Sep. 2019, doi: 10.1177/0735633118787528.

3. T.S. Tuang. Diep. Q. B, and Zelinka. I, Artificial Intelligence in the Cyber Domain: Offense and Defense: Symmetry, 2020, 12, 410 available: www.mdp.com/journal/symmetry on [assessed Apr. 20, 2020.

4. A. M. Shamiulla, Role of Artificial Intelligence in Cyber Security, International Journal of Innovative Technology and Exploring Engineering, vol. 9 issue 1 pp. 4628–4630, November 2019.

5. P. Pranav, "Artificial Intelligence in cyber security", International Journal of Research in Computer Applications & Robotics, vol 4, 1, pp. 1–5, May 2016.

6. T.S. Tuang. Diep.Q. B, and Zelinka. I, Artificial Intelligence in the Cyber Domain: Offense and Defense: Symmetry, 2020, 12,410 available: www.mdp.com/journal/symmetry on [assessed Apr. 20, 2020.

7. S. Bhutada and P. Bhutada, Application of Artificial Intelligence in Cyber Security: in IJERCSE, 5(4): 214–219, 2018.

8. Ishaq Azhar Mohammed, "Artificial Intelligence For Cybersecurity: A Systematic Mapping Of Literature ", International Journal of Innovations In Engineering Research And Technology, vol.7, Issues 9, pp. 172–176, 2020.

9. C. Oancea, "Artificial Intelligence Role in Cybersecurity Infrastructures", International Journal of Information Security and Cybercrime, vol. 4, no. 1, pp. 59–62, 2015.

10. Harsh Chaudhary, Ankit Detroja, Priteshkumar Prajapati and Dr. Parth Shah. "A review of

various challenges in cybersecurity using Artificial Intelligence", Proceedings of the Third International Conference on Intelligent Sustainable Systems ICISS 2020, pp. 829–836, 2020.

11. G. Wang, X. Zhang, S. Tang, C. Wilson, H. Zheng, and B. Y. Zhao, ``Clickstream User Behavior Models," *ACM Trans. Web*, vol. 11, no. 4, pp. 1–37, 2017.

12. Iqbal H. Sarker, A. S. M. Kayes, Shahriar Badsha, Hamed Alqahtani, Paul Watters ,Alex Ng, "Cybersecurity data science: an overview from machine learning perspective", Journal of Big Data, pp. 2–29, 2020.

13. Rainie L, Anderson J, Connolly J. Cyber attacks likely to increase. Digital Life in. 2014, vol. 2025.

14. Fischer EA. Cybersecurity issues and challenges: In brief. Congressional Research Service (2014)

15. S. Dilek, H. Çakır, and M. Aydın, "Applications of artificial intelligence techniques to combating cyber crimes: A review," International Journal of Artificial Intelligence & Applications, vol. 6, no. 1, January 2015, pp. 21–39.

16. C. Crane, "Artificial intelligence in cyber security: The savior or enemy of your business?" July 2019, https://www.thesslstore.com/blog/artificial-intelligence-in-cyber-security-the-savior-or-enemy-of-your-business/

17. "The role of AI in cyber security," https://blog.eccouncil.org/the-role-of-ai-in-cybersecurity/

18. S. Rubin, "Knowledge-Based Programming for the Cybersecurity Solution", The Open Artificial Intelligence Journal, vol. 5, no. 1, pp. 1–13, 2018.

19. T. Tagarev, "Intelligence, Crime and Cybersecurity", Information & Security: An International Journal, vol. 31, pp. 05–06, 2014.

20. Pandey, M., "Artificial Intelligence in Cyber Security. On Emerging Trends In Information Technology"- 'The Changing Landscape Of Cyber Security: Challenges, 66, 2018.

21. Anagnostopoulos, C. Weakly Supervised Learning: How to Engineer Labels for Machine Learning in Cyber- Security. Data Science for Cyber-security, 3, 195, 2018.

22. Apruzzese, G., Colajanni, M., Ferretti, L., Guido, A., & Marchetti, M. (2018, May). On the effectiveness of machine and deep learning for cybersecurity. In 2018 10th International Conference on Cyber Conflict (CyCon) (pp. 371–390). IEEE, 2018.

23. Katoua, H. S., Exploiting the Data Mining Methodology for Cyber Security. Egyptian Computer Science Journal, 37(6), 2013.

24. Ashwini Sheth, Sachin Bhosale, Adnan Bukhari, "A Survey On Cyber Security", Contemporary Research In India, Special Issue, April, 2021.

25. Yuchong Li, Qinghui Liu, "A comprehensive review study of cyber-attacks and cyber security; Emerging trends and recent developments", Energy Reports, Volume 7, pp. 8176–8186, November 2021.

46

Weighted Support Vector Machine Classification by using Confidence Parameters for Unbalanced Sampling Target Data Sets

M. Premalatha[1]

Department of Mathematics, Veltech Dr. Rangarajan Dr. Sagunthala R & D Institute of Science and Technology, Avadi, Chennai, Tamil Nadu, India

C. Vijayalakshmi[2]

Department of Statistics and Applied Mathematics, Central University of Tamil Nadu, Thiruvarur, India

Abstract—Artificial Intelligent plays a vital role in Machine Learning. ML is one of the subfields of AI. In Support Vector Machine algorithm developed based on nonlinear SV in the period of sixties in Russia is called Generalized Algorithm of SVM. SVM first founded by Vapnik. In a classification of SVM for unbalanced datasets perform in original SVM and Weighted Support Vector Machine using Target data set (WSVMA). SVMs give effective solutions for datasets that are balanced, in a similar way they give effectiveness to unbalanced datasets to create sub-optimal models. When creating the unbalanced training sets, where the data of the Target class are exceeded by the data present in the class of the Non–Target class the accomplishment of the SVM classifier is so enhanced. In sequence to balance the distribution of the target class, our algorithm study is based on density information in training sets to eliminate unwanted data of the non-target class and bring about new artificial data of the target class. The main idea is for the unbalanced datasets to assign the different sets of weighted values to different sets of data points such that they called the Weighted Support Vector Machine algorithm for the training data set (WSVM) by using the "Confidence parameter, C". In this training, the algorithm finds out the decision surface according to the correlative set of data points in the training phase.

Keywords—Machine learning, Classification, SVM classifier, Weighted support vector machine, Hyperplane, Training data

I. Introduction

SVM algorithm works in the size of margin bounds on unseen data set of hyperplane. It's mainly associated with the process of calculation based on the mathematical procedure and memory space [1]. One of the main phases of the algorithm farmed by SVM is

[1]drmpremalatha@veltech.edu.in, [2]vijusesha2010@gmail.com

DOI: 10.1201/9781003388913-46

to find a decision boundary of a hyperplane that gives the best splits of the data according to the ratio in the target and non-target data set into two phases of classes. The split is made soft margin that allows some misclassified points [2]. The margin allows the non-target class on unbalanced datasets also it can be updated to each class to improve the working process of the algorithm on datasets with oblique distributions of the two classes. This new modification gives the weights of SVM data that weight the margin corresponding to the class. The class distribution of the test data set to split into simulated proportions for the target example to the non-target example. Adapt the proportion of 100 for the target class (Maximum) and 1 for non-target class (Minimum).

II. Support Vector Machine Methodology

SVMs play an important and effective role in the nonlinear machine learning algorithm. The margin that optimizes the distance between the decision boundary and the nearest instances from each of the two classes determines the shape of the hyperplane [3-5].

$$\underset{W,\xi}{\text{Min}} \frac{1}{2}\|W\|^2 + C \sum_{i=1}^{n} \xi_i$$

subject to

$$y_i(x^T w + b) \geq 1 - \xi_i$$

$$\xi_i \geq 0$$

Parameter $C > 0$

Using a kernel to define a linear hyperplane to separate the classes in the transformed space (feature), which is comparable to a nonlinear class boundary in the original feature space, becomes a method of data transformation. Minimizing the bound and misclassification error is possible in VC dimension. A set of data includes both updated data and previously used SVM training SVs [6-7]. Hyperplane $f(x)$ is classified into two parts by the intersection of the two classes of SVM maximization which is

related to w and b.

Ratio Margin

The ratio margin method is similar to the Max-Min Margin method. The *relative* sizes of −1 and +1 are the largest (Fig. 46.1). The Max-Min and Ratio methods still work on without the constraint on the modulus of the training feature vector phase.

III. SVM for Unbalanced Minority Classification of Data set

The hyperplane margin (soft) is to permit various points to emerge on the wrong side of the decision limits. A normalization confidence parameter known as the soft-margin parameter ("C"), which controls the trade-off between minimizing the number of misclassified instances and maximizing the separation margin between classes, controls the margin (Fig. 46.1) [8].

The unbalanced datasets make them more effective for modified SVM algorithms and they are sensitive to the unbalance in the datasets and execute the optimal models.

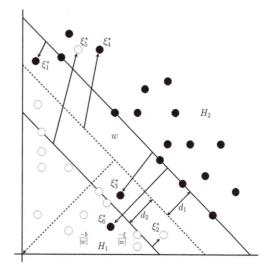

Fig. 46.1 Linear and non-linear classifier
Source: Author

Given more data sets in the non-target class than the target class, the soft margin concept gives the advantage of the decision boundary an act performed good well to the non-target class. The learning algorithm will support them in non-target classes it gives a better exchange between classification error and margin maximization. But it works at the rate of the target class, particularly for unbalance ratio is high, to get better results to omit the target class data set and then will reach the optimum.

The main concept of weighted SVM is an assign different weighted values to a different sets of data points otherwise it is called as Cost-Sensitive SVM of different weighted values [9].

IV. Difference between Original SVM and WSVM for Unbalanced Data Set

A. Original SVM

SVM: Input set

1. Training data sets are represented by $S = ((x_1, y_1) \dots (x_m, y_m))$, y_j in -1 to $+1$ as base classified (weak) the number of iteration T.

2. *Initialization:* Weighted Distribution $W(D_i)$ over the data such that $W[D_1(j)] = 1/m$.

3. Train the weighted training set on a weak base.

   Weak hypothesis $h_t = X \to \{-1, +1\}$, $t = 1, 2, 3 \dots T$

4. Error Calculation of Training set

$$h_j : e_t = \sum_{j=1}^{m} D_1(j) \neq h_t(x_j)$$

   Calculating

$$d_t = \frac{1}{2} \ln\left(\frac{1-e_t}{e_t}\right)$$

5. Updated Weighted
   Weight of training sample data

$$D_{t+1}(j) = \frac{D_t(j) \exp(-d_t y_j h_t(x_j))}{Z_t}$$

Where Z_t is a constant representing for Normalization of the distribution and

$$\sum_{j=1}^{m} D_{t+1}(j) = 1$$

6. *Output:* Based on Final postulation

$$H(x) = sign\left(\sum_{j=1}^{T} d_t h_t(x_j)\right)$$

7. To define "Confidence Parameter, C"

$$C = \frac{2}{lk} \sum_{j=1}^{l} \min(k_j^-, k_j^+)$$

Here k represents an integer for both the class belonging to -1 to $+1$

B. Weighted Support Vector Machine (WSVM)

Modified SVM learning

1. Training data sets are represented by $S = ((x_1, y_1) \dots (x_m, y_m))$, y_j in -1 to $+1$ as base classified (weak) the number of iteration T.

2. *Initialization:* Weighted Distribution $W(D_i)$ over the data such that $W[D_1(j)] = 1/m$.

   To create N synthetic (Numerical) data from the target class by using modified distribution Dt .

3. Error Calculation of Training set

$$h_j : e_t = \sum_{j=1}^{m} D_t(j); y_j \neq h_j(x_j)$$

   Calculating

$$d_t = \frac{1}{2} \ln\left(\frac{1-e_t}{e_t}\right)$$

4. Calculation based on Updated Weighted

$$D_{t+1}(j) = \frac{D_t(j) \exp(-d_t y_j h_j(x_j))}{Z_t}$$

Where Z_t is a constant representing for Normalization of the distribution and

5. *Output:* Based on Final postulation

$$H(x) = \text{sign}\left(\sum_{j=1}^{T} d_t h_j(x_j)\right)$$

6. To define "Confidence Parameter, C"

$$C = \frac{2}{lk} \sum_{j=1}^{l} \min(k_j^-, k_j^+)$$

7. $\text{Min } C = \begin{cases} 1 & \text{if } k_j^- = k_j^+ \\ 0 & \text{if } (k_j^-, k_j^+) = 0 \end{cases}$

V. Confidence Parameter

A value of C indicates a strong margin with no chance of error in the margin. While high integer numbers, like 1, 10, and 100, allow for

a much softer margin, small positive values, like 1 and 2, allow for some traduce. The number and severity of transgress to the margin of the n observations are determined by [C]. The model's penalty for finding the decision boundary is the C confidence parameter. The weighting for each class is the same, and the margin is symmetrical.

Table 46.1 Comparison result for each Misclassification Error Rate of SVM and WSVMA using different splitting ratios on the unbalanced data set

No	Splitting ratio	SVM	WSVM
1	95:5	10.9413	7.8910
2	90:10	10.8300	8.1221
3	80:20	11.0004	9.2441
4	75:25	12.9967	11.1008
5	50:50	11.7523	10.5920

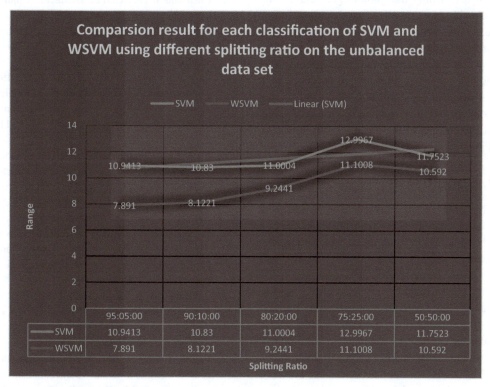

Fig. 46.2 The comparison result of SVM and WSVM for unbalanced data set

Source: Author

The modified SVM for unbalanced classification imputes a proportional weight to each class based on the C value. In SVMs, an instance-level weighted modification is applied directly to the weight values based on the misclassification ratio between classes or the confidence factor for each individual instance. Each example in the training data set contains a unique penalty term (C value) that is used in the SVM model's margin computation. A weighting of the global C-value that is defined proportionally to the class distribution can be used to determine the C-value of an example [10–11].

$$C_i = \text{weight}_i \times C$$

Small Weight: Higher penalty for incorrectly labeled data set, smaller C value.

Larger Weight: Lower penalty for incorrectly labeled data set, larger C value.

Examples from the non-target class are misclassification on the corresponding target side. As the target class data set is assigned with a rate greater of misclassification, the modified SVM algorithm would not tend to bias the hyperplane that separates toward the target and class instances to minimize and control the overall misclassification rate represent in Table 46.1 and corresponding linear (SVM) represent in Fig. 46.2 [12-13].

VI. Conclusion

Each pair of classes has a classifier designed for it, and the classifier with the most control is the final one. We applied non-parametric machine learning algorithmic approaches that were tractable for large datasets with high dimensional.In classification tasks, generalization performance is implemented by maximizing the margin, which is a canonical framework related to minimizing the weight vector. SVMs produce sub-optimal models whenever the datasets are severely unbalanced, but they do provide effective solutions whenever the datasets are balanced.

The approach of assigning various weights to various sets of data can also be used to compute the confidence factor. SVM learning problems maintain a strong approximation solution in a faster technique, minimize the time provided, as well as provide great speed. SVM makes it simple to execute mathematical calculations and to achieve the scope.

References

1. V. N. Vapnik, The Nature of Statistical Learning Theory (Springer, New York, 1995).
2. Cerri and A. de Carvalho., (2010). New top-down methods using SVMs f or hierarchical multi label classification problems. In IJCNN 2010, pages 1–8. IEEE Computer Society.
3. Colin Campbell, Yiming Ying., (2011). Learning with Support Vector Machines Synthesis Lectures on Artificial Intelligence and Machine Learning 5: 1, 1–95.
4. Chen and J. Hu., (2012). Hierarchical multi-label classification based on over-sampling and hierarchy constraint for gene function prediction. IEEE Transactions on Electrical and Electronic Engineering, 7: 183–189.
5. Pragya Sur and Emmanuel J Cand`es., (2019). A modern maximum-likelihood theory for high dimensional logistic regression. Proceedings of the National Academy of Sciences, 116 (29): 14516–14525.
6. Yang Liu, Gareth Pender., (2015). A flood inundation modelling using v-support vector machine regression model. Engineering Applications of Artificial Intelligence 46223-231.
7. Q. Wu and W. Wang, "Piecewise-Smooth Support Vector machine for Classification," Hindawi Publishing Corporation Mathematical Problems in Enginering, 2013.
8. Z. Nazari, M. Nazari, M. S. S. Danish, and D. Kang. Evaluation of Class Noise Impact on Performance of Machine Algorithms, Int. J. Comput. Sci. Netw. Secur, Vol.8, 2018.
9. F. Zhu, J. Yang, J. Gao, and C. Xu. Extended nearest neighbor chain induced instance-weights for SVMs, Pattern Recognition, Vol. 60, pp. 863–874, 2016.

10. C. Liu, W. Wang, M. Wang, F. Lv, and M. Konan. An efficient instance selection algorithm to reconstruct training set for support vector machine, Knowledge-Based Systems, Vol.116, pp. 58–73, 2017.

11. Y. Haowen and G. Rumbe, "Comparative Study of Classification Techniques on Breast Cancer,"International Journal of Artificial Intelligence and Interactive Multimedia , vol. 1, no. 3, pp. 5–12, 2010.

12. S. W. Purnami, V. Chosuvivatwong, H. Sriplung, M. R. Dewi and E. Suryanto, "Comparison of Piecewise Polynomial Smooth Support Vector Machine to Classify Diagnosis of Cervical Cancer,"International Journal of Applied Mathematics and Statistics, vol. 53, no. 6, pp. 159–166, 2015.

13. M. Premalatha, C. Vijayalakshmi, "Using Optimization Methodologies to find the solution of Support Vector Machine with Maximum Accuracy", Pensee Journal, vol 76, issue 4, pp. 197–206, 2014.

Recent Trends in Computational Intelligence and its Application – Sugumaran D. et al. (eds)
© 2023 Taylor & Francis Group, London, ISBN 978-1-032-48410-5

47

Regional Stores Products Recommender Using Attention-Based Long-Term Short-Term Memory

P. N. Karthikayan[1], Lijetha C. Jaffrin[2]
Assistant Professor, Department of IT,
Vel Tech Rangarajan Dr. Sagunthala R&D Institute of Science and Technology, Chennai

R. Saravana Perumal[3], Bhavishya P.[4], Yugandhar Surya[5]
UG Student, Department of IT,
Vel Tech Rangarajan Dr. Sagunthala R&D Institute of Science and Technology, Chennai

Abstract—People across the world use the internet to access information. No longer is it necessary to leave the house to go shopping, consult a phone book for contacts, or utilize a map for directions. These characteristics have rendered many previously necessary items outdated and can now be found more quickly online. Despite the fact that consumers have had access to online shopping for a while now, they still prefer to shop at reputable local establishments. Shopping in-person is still very advantageous and has a significant role in the retail industry. A few elements, including physically checking a goods and pricing, are unique to an in-character experience. When a user searches for local items at provincial marketplaces, it is typically difficult to identify the precise location where the user could purchase the item. Additionally, there aren't enough facilities for local shop owners to advertise and show the goods they sell. To solve these issues, an application is being developed that will help an individual to identify the correct shop, visit and buy before they depart from their home as per their requirement. Additionally, this application includes a feature that helps store owners increase sales. An individual will save time and energy using this application without spending any money. Additionally, it would simplify the user's search for nearby products and services based on their preferences and present location. This application is a product recommendation system implemented with ALSTM which uses the previous search and purchase history of a user. People would benefit from it by saving time, effort, and money. Additionally, it will aid neighbourhood merchants in growing their sales. By promoting their stores, this application will also assist those who want to become sellers on this platform. The rate of new customers would increase as a result of this.

Keywords—Flutter, Barcode, MLkit, E-commerce, Dart, Firebase, ALSTM, RNN model, Attention, Location-based recommendation, Regional retail stores

[1]pnkarthikayan@veltech.edu.in, [2]lijethacjaffrin@veltech.edu.in, [3]vtu16828@veltech.edu.in, [4]vtu12362@veltech.edu.in, [5]vtu12278@veltech.edu.in

DOI: 10.1201/9781003388913-47

I. Introduction

Online shopping is growing, which supports the theory that traditional stores would eventually close. Even though the demand for internet shopping is growing, customers still prefer visiting physical businesses. Customers can look at the products to aid in making decisions [1] [2]. A physical store also gives the shopkeepers a sense of security. Online retailers cannot duplicate the experience that customers get in physical ones. Many consumers prefer to personally visit stores to browse new merchandise. This increases customers' trust in these stores, which boosts sales [3]. Because of how things are displayed in stores, customers may end up purchasing more than they intended. A physical store's sales improve because it exposes buyers to a wide range of products. In spite of the availability of multiple E-commerce applications, there is an enormous dependence on regional markets and supplies.

Now, the problem is how and on what basis a person should choose the exact shop which is near them in a particular area *i.e.* in a fence, particularly when you are going to a new city or town. Because of the easy accessibility of the internet nowadays, consumers don't need to struggle to find what they want. Using our application, in order to find the specific product a customer needs, it is now quick and simple to browse among hundreds of products in a couple of seconds. By looking at these problems, a solution is being provided by which a person can reduce the wastage of time and energy. Geofencing and geo-sorting technology are implemented to create a fence and to check whether the user is present within that fence or not. If the user is present within a fence, our app will allow an individual to find the right destination along with its route access. With the addition of search options on our platform, users can now discover the stores that have the specific product in stock without browsing all the potential locations that have the product they desire. The user can get a more realistic idea of the goods themselves from the product descriptions of the items they are purchasing. Users can learn more about the goods, their quality, and their price by reading reviews of the retailer to learn about actual customers' experiences. This can assure buyers that the product is of good quality. A supreme feature is included for a user that will make sure that a user is getting a more precise and effective product on their feed and to achieve this feature, different algorithms were implemented like cosine similarity, collaborative filtering, association rule mining, min hash, and slope one. These algorithms measure the resemblance between two vectors by estimating the cosine angle between the two similar users' preferences and recommend products based on purchase history and rating consistency provided by the other two algorithms, but these algorithms are not that efficient to implement on a big scale. After analyzing the results of the above-mentioned algorithms, an analysis was made on 10 different RNN-based algorithms for a recommendation system, and among them, ALSTM is the most efficient algorithm [4], to recommend the correct user's choice product because the features of this algorithm are that it uses a huge dataset collection from amazon and Instacart. It also varies with user preference. This algorithm accurately predicts the next product that a user can buy based on the user's past search and purchase history. This provides an incentive to buyers relying only on the product's quality and appropriateness. Many small-scale retailers struggle as they are deprived of the facilities and platforms that the internet provides. So, the local retail store dealers might be benefited from more exposure to their store with separate access to the application than the customers to maintain and update the inventory of a shopkeeper on our application. For the seller's convenience, a barcode scanner is implemented in the application that can process 30-50 pics per min with 100% accuracy and it is also cost-effective and easy to implement on an android device

[5]. It can scan the multi-format 1D/2D barcode as well as also be able to decode any type of barcode [6] [7] [8]. This process will save the time and energy of the seller that is going to be wasted in updating the database. And due to geofence, it is going to give a better experience to the buyers to access the nearest shop with the cheapest product.

II. Motivation and Justification of Work

A. Motivation

Google Analytics report states that an individual loses 45-60 minutes (depending upon age) in a day for finding the appropriate product on online platforms like Flipkart, Amazon, etc. Most of the time users do not get a particular product. Nearly all grocery delivery platforms have an order limit on free delivery and typically human beings do not opt to shop online because of Quality and trust difficulties. These issues lead us to analyse and develop the best suitable application which can triumph over those issues. A startup known as Local Banya got here with the answer to fixing those troubles. However, because of mismanagement and lack of the right guidance, they failed in 2015. So, the intrinsic motivation is to deliver a competent, reliable, and scalable solution to the retail industry.

B. Justification

A solution needs to be developed that makes a huge difference in the life of an individual by helping them in finding the correct shop to go to before they depart from their home as per their requirement. Thus, the plan is to develop a platform wherein customers can decide on the product and retail store before leaving their houses. It would help in saving essential time and energy in today's fast-paced world. It would additionally give an incentive for sellers to enlarge their trade. This platform will aid the individuals who want to sign up as sellers, by facilitating them to exhibit their services online and therefore strengthen their sales by raising the inflow of purchasers.

C. Related Work

Product recommender systems have received a lot of attention in terms of research and application. Many algorithms have proven high performance on product recommendation, including Collaborative Filtering (CF), Recurrent Neural Networks (RNN), and Long Short-Term Memory (LSTM). However, while dealing with enormous amounts of data, various flaws were discovered. These algorithms are capable of accurately predicting the subsequent word in a list of words. The vanishing gradient problem is a major issue that makes training RNNs highly sluggish and inefficient for use. Small gradients continue to reduce by considerable margins at every layer as the weights are allocated based on the preceding layer until they reach a point where they are extremely near to zero. As a result, learning slows over the first few layers, and overall effective training slows. And thus, due to diminishing gradients, RNNs do not learn long-range dependencies consistently over time steps. This implies that even if the previous tokens in a sequence are critical to the overall context, they will not be given much significance. This failure to learn long sequences results in short-term memory. Several studies have shown that when the length of the input sentence increases, the performance of the neural network diminishes rapidly. LSTM was intended to address the long-term dependency concerns that regular RNNs encountered. As the default behaviour of LSTM is to remember long-term information, it overcomes the short-term memory issue. However, there is no option to prioritize some input terms over others while translating the statement.

III. Proposed Model

The recommendation system plays a major role in the phase where the progress is from online-to-offline retail. Here, the attention-based Long-

Short Term Memory Model (ALSTMM) is being used for building up a product recommendation system which reduces the shortcomings in the past research on recommendation algorithms. The attention system examines the user's temporary preferences and assigns unique weights corresponding to the priority of the items. Here, attention is paid to the input array of items corresponding to the output sequence. It identifies differing choices of shoppers over the period and comprehends more patterns based on past purchases. Recommendations are enhanced by considering them. Users' choice keeps on changing at different times. At one time, the consumer may have some other preference. This affects future recommendations on items [9] [10] [11]. So, this ends up with exceptional effects on the forecast of the subsequent items for diverse users. This is also a location recommendation system that displays products from retail stores near the user's current location. This application would be implemented using geofencing and sorting techniques and thus help the user in finding the right destination along with its map route access. It gets the user's current location and longitude and latitude and creates a geofence [12] [13]. Then the distance from the user to each of the POIs (point of interest) is determined and sort the list of POIs based on distances. QR codes, and barcodes can be scanned by the retailer using the smartphone camera for storing the product details in the database. Automated data collection through barcodes will help in saving sellers time and reduce data entry errors. The data will be stored in Firebase and MongoDB. The compact UI simplifies product searching and thus helps in saving money, time, and energy.

Previous methods and difficulties were outlined before focusing our attention on the ALSTM. This model has the potential to significantly improve efficiency. This is a simple proposal that recommends assigning proportional importance to each of the input words. In contrast to the traditional approaches, the attention mechanism allows for the examination of all hidden states from the encoder sequence in order to make predictions. As a result, whenever the suggested model generates a phrase, it looks for a collection of points in the encoder's hidden states that contain the most relevant information.

A. Implementing Product Recommendations Using ALSTM

The purpose of ALSTM is to enhance the accuracy of the present benchmark model (Long-Short Term Memory LSTM) supported by the previous purchase history. The user's short-term intent for recommendations is tracked via the attention method. The long-term Preference Layer and consequently the Attention-based short-term Intent Layer are the following layers in the ASLTM paradigm. The long-term Preference Layer is responsible for a crucial function. Prior to inspecting the current session at the start, itself, it gives the Attention-based short-term Intent Layer the records of the user's long-term choices. Considering the user's long-term preference and conveying it to the attention-based short-term Intent Layer are the tasks of this layer.

B. Geofencing and Location Recommendation Through Geo-Sorting

Geofencing tells about the user's current location. It recognizes the places that may be of the user's interest. When a device moves into or out of the created geofence, the application gets notified of it. To figure out the distance between the user and the POIs, POI Latinos need to be defined as LatLon objects. Firstly, get the user's position from the geolocation and add it to the LatLon object. Now, determine the distance from the user with the JavaScript array sort method by navigating over the list of POIs. Then it is passed to a custom comparator function that takes two location elements as parameters and determines which location is closer to the user.

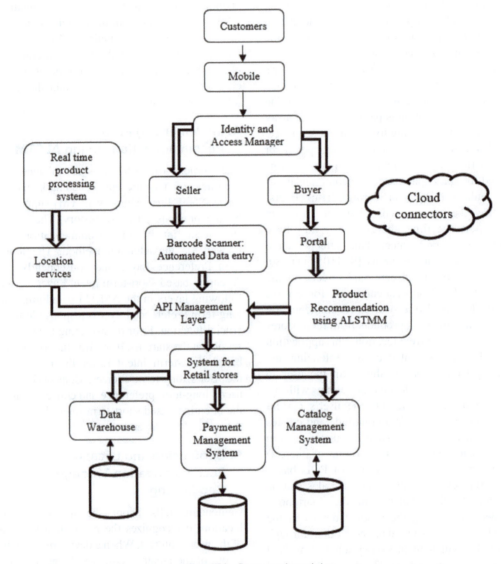

Fig. 47.1 Proposed model

Source: Authors

C. Challenges

The challenges of this project are to overcome the restrictions of the usual offline market. In offline retail, all data is not accessible to us. Digitization is not easy in online e-commerce. Getting reviews and ratings of regional retail stores is a profoundly challenging task as compared to the online market [14]. Thus, it becomes crucial to infer the user preferences from the experiences and data that has customer purchase feedback.

D. Merits

This platform is helpful towards the energy and time savings of people. This would bring trade expansion opportunities for the local traders

Fig. 47.2 UI design

Source: Authors

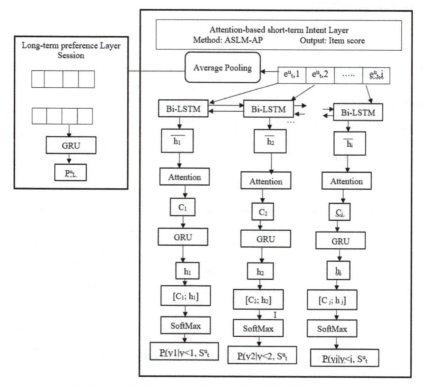

Fig. 47.3 Architecture diagram of LSTM with attention

Source: Authors

Table 47.1 Comparison of the functionality of proposed methodology with its predecessors

Measures	RNN	LSTM	ALSTM (proposed)
Architecture	RNN feeds its output value to itself while having hidden layers.	It has a cell that remembers values. The input gate, output gate, and forget gate regulate the flow of information into and out of the cell.	Attention-based encoder-decoder architecture. **One Attention layer** (the Attention gate) is used.
Functioning	An RNN passes its output to itself at subsequent time-step. It remembers the previous inputs stored in the memory.	Handles long-range dependency better than RNN. Removes vanishing gradient problem.	It searches for the most important information in the encoder's hidden states where attention has to be placed.
Training	Due to the memory-bandwidth-bound computation they demand, RNN and LSTM are challenging to train.		Attention weights are calculated as context vectors and passed to softmax.

Source: Authors

as it would help local traders increase their sales. The search process of the user will be now easier as they can decide the product and destination to buy the product beforehand.

IV. Comparative Functionality of the Algorithm

See Table 47.1

V. Performance Evaluation

A. Performance Metrics

Various metrics such as Mean Absolute Error (MAE) and Mean Squared Error (MSE) Fig. 47.5 were computed to assess the accuracy

and model performance in the recommendation system, which was evaluated on previous algorithms. These statistics are based on 512 features that the model was trained on. The MSE of Attention Based LSTM was 0.0014, while it gave the MAE: 0.0303 and RMSE: 0.0387, with a validation loss of 0.0012.

Accuracy

All three networks appear to function similarly. It is revealed that their performance is comparable when measuring how effectively they produce new predictions based on past forecasts. In terms of accuracy, however, ALSTM has a little edge. Despite being the slowest to train, they offer the advantage of evaluating extended sequences of input data

Table 47.2 Performance evaluation

Precision	Recall	Mean Reciprocal Rank	Mean Average Precision	Coverage	Model
0.0276	0.0565	0.0733	0.0345	0.06963	LSTM with "Description" + Attention Layer
0.0266	0.0553	0.0707	0.0336	0.7006	ALSTM with description feature bucketized 100 bins
0.0224	0.045	0.0527	0.0242	0.4296	RNN recommendation system
0.0213	0.0434	0.0525	0.0248	0.8606	LSTM (only description/no attention layer
0.0184	0.0339	0.0415	0.0228	0.9691	Last Purchased product

Source: Author

without expanding the network size. After 100 epochs of training, loss:

0.0012 - val loss: 0.0015 is evaluated. This loss is quite minimal when compared to other methods, resulting in improved accuracy and resilience. The Accuracy against various parameters is plotted in Fig. 47.4, Fig. 47.5, Fig. 47.6, Fig. 47.7, and Fig. 47.8.

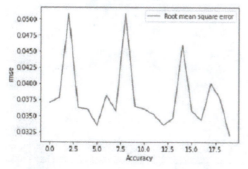

Fig. 47.4   Accuracy vs root mean square error
Source: Authors

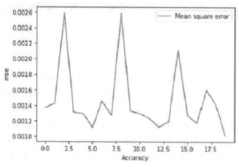

Fig. 47.5   Accuracy vs mean square error
Source: Authors

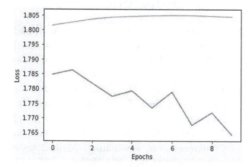

Fig. 47.6   Epochs vs loss
Source: Authors

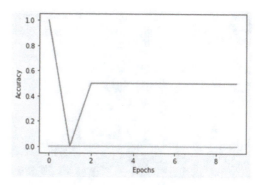

Fig. 47.7   Epochs vs loss
Source: Author

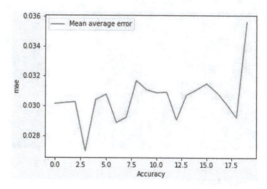

Fig. 47.8   Accuracy vs mean average error
Source: Author

B. Performance Analysis

With the help of this application, millions of people who shop every day across the nation will be connected to the dispersed vendor-client community. By developing a platform that enables sellers to list their services, manage interactions with buyers, and offer a range of services to buyers, primarily to help them choose the location of a retail shop for the goods they want, information will become more readily available. This application will target common people across cities, small towns, and remote areas who often lacked the patience for buying basic resources. Moreover, people who lack time and hesitate to visit multiple retail stores for finding a product will benefit from it.

Fig. 47.9 Application output

Source: Authors

VI. Conclusion

With the use of an application created with Flutter, users will be able to search for nearby, regional reputable stores to find the products they're looking for. listing all the retail establishments inside that geofence using the flutter geofencing approach Customers can search for the goods they require and receive suggestions for shops close to their present or ideal location. Dijkstra's algorithm [15] will be used to implement the order of shortest distance for these regional stores. This project efficiently recommends the products to the users taking into context their previous purchase history, pricing, and other features, using the Attention Based Long Short-Term Memory Model (ALSTM). Also, the buyers will be benefitted as they get a platform to display the variety of products they have. This increases sales. This application also has the facility of automated data collection through barcodes. This will reduce manual data entry errors and delays and thus result in higher productivity and efficiency. In short, this application will bring ease to the sellers' buyers by saving their time, money, and energy.

References

1. Abdulhamid, M., & Matongo, A. Mobile Phone, and Barcode Scanner. The Scientific Bulletin of Electrical Engineering Faculty, 20(1), 25–28.
2. Visa, M., & Patel, D. (2021, April). Attention-based Long-Short Term Memory Model for Product Recommendations with Multiple Timesteps. In 2021 5th International Conference on Computing Methodologies and Communication (ICCMC) (pp. 605–612). IEEE.

3. Chincholkar, Snehal. (2018). Consumer Behaviour towards Online Grocery RetailStore "Localbanya.com" in Mumbai Region.

4. David Zhan Liu, & Gurbir Singh (2016). A recurrent neural network-based recommendation system. tech. Rep Stanford University.

5. T. Sangkharat and J. La-or, "Application of Smart Phone for Industrial Barcode Scanner," 2021 7th International Conference on Engineering, Applied Sciences and Technology (ICEAST), 2021, pp. 9–12.

6. Sampada S. Upasani, et.al. (2016) A Robust Algorithm for Developing a Barcode Recognition System using the Web-cam.

7. Abdulhamid, M., & Matongo, A. Mobile Phone, and Barcode Scanner. The Scientific Bulletin of Electrical Engineering Faculty, 20(1), 25–28.

8. Ohbuchi, E., Hanaizumi, H., & Hock, L. A. (2004, November). Barcode readers using the camera device in mobile phones. In 2004 International Conference on Cyberworlds (pp. 260–265). IEEE.

9. Suja Panikar, Prayag Mane, et.al. (2018) Online Grocery Recommendation System.

10. Satheesan, P., Haddela, P. S., & Alosius, J. (2020, December). Product Recommendation System for Supermarket. In 2020 19th IEEE International Conference on Machine Learning and Applications (ICMLA) (pp. 930–935). IEEE.

11. Yeo, S. F., Tan, C. L., Lim, K. B., & Wan, J. H. (2020). To Buy or Not to Buy: Factors that Influence Consumers' Intention to Purchase Grocery Online. Management & Accounting Review, 19(3).

12. R., A. A. R., R., L., G., H. and N., L. (2020), "Tracking the Covid zones through geo-fencing technique", International Journal of Pervasive Computing and Communications, Vol. 16 No. 5, pp. 409–417.

13. Rahimi, S. M., Far, B. & Wang, X.(2020) Contextual location recommendation for location-based social networks by learning user intentions and contextual triggers.

14. Shetty, K. S., & Singh, S. (2018). Cloud-based application development for accessing restaurant information on mobile devices using LBS. arXiv preprint arXiv:1111.1894.

15. Dian Rachmawati and Lysander Gustin 2020, "Analysis of Dijkstra's Algorithm and A* Algorithm in Shortest Path Problem" J. Phys.: Conf. Ser. 1566 012061.

Recent Trends in Computational Intelligence and its Application – Sugumaran D. et al. (eds)
© 2023 Taylor & Francis Group, London, ISBN 978-1-032-48410-5

48

Single Product Inventory System Serving Israeli Queued Customers with Batch Service and Instantaneous Replenishment Policy

T. Karthikeyan*

Department of Mathematics, Ramakrishna Mission Vivekananda College, Chennai, India

J. Viswanath[1], C. T. Dora Pravina[2]

Department of Mathematics,
Vel Tech Rangarajan Dr. Sagunthala R & D Institute of Science and Technology, Chennai, India

J. Vijayarangam[3]

Department of Mathematics, Sri Venkateswara College of Engineering, Chennai, India

Abstract—We considered a stochastic model of continuous review, Non-perishable single commodity inventory of items. We are channelizing the arriving demands via the server facility for the smooth functioning of the system for customer satisfaction. Israeli Queue discipline is followed in rendering service to the customer in batch mode. A product for each Israeli group is served. No lead time is taken into account at the time of reorder and an instantaneous replenishment policy is adopted whenever the items in the system reach zero. It is restricted that only Israeli groups are allowed with an unrestricted number of customers in each Israeli group. We develop an integrated stochastic model of an inventory system with Israeli Queue and solved by matrix method with MATLAB code. Key performance measures are derived and validated with numerical illustrations.

Keywords—Israeli queue, Non-perishable items, Batch service, Instantaneous replenishment

I. Introduction

Quite extensive works have been carried out by several researchers in the recent past on continuous review inventory systems with the consideration of uncertainties as probability measures. In these articles, Queuing systems are modulated by the inventory system. When we come across the situations like physical waiting queues in either buying or booking tickets at the counters, if an arriving customer who needs the tickets may find someone in the queue who is known to him and already waiting for booking the ticket then immediately arriving customer

---
*Corresponding Author: karthick@rkmvc.ac.in
[1]mathsviswa@gmail.com, [2]doramaths@gmail.com, [3]jvijayarangam75@gmail.com,

DOI: 10.1201/9781003388913-48

will hand over his job to him. Otherwise, he will join the queue as the last person. In the prior case, the accepted person will be considered a leader and representative of the group if many arriving customers are known to him. In such cases, we assume that the service time of the whole batch will not affect the others. It is restricted only to N-such groups who are allowed to get service from the server facility. The influence of the number of customers in each group to get service is negligible. As a result, the customers in each group get served at a time or at one time period. Such a queue discipline is termed as Israeli Queue. Boxma, van der Wal and Yechiali [3] have identified Israeli queue in their work infinite batch size.

Several researchers have studied inventory models with queues upon service for efficient usage of shared resources not only limited to the business environment but also includes manpower modelling, machinery and highly costly equipment used in Agriculture, the Medical field, Social services and so on. In the field of agriculture, there are certain machines and equipment with advanced futures that cost major expenses to the customer. Such items are essential but not in continuous usage. Instead of purchasing such an item as a single individual, it will be appropriate to purchase by a team of members who are all in need of the item during some partial period of their work. For example, Purchasing huge expensive Medical devices like CT and CAT scans used for tomography tests, Magnetic Resonance Imaging Machines, Surgical Robots, Ultrasound Machines, PET (Positron Emission Tomography), and more expensive and commonly used software [Unreal Engine is a gaming engine developed by Epic Games, Autodesk Maya, Adobe Acrobat Capture 3, Core Impact PRO- security testing and measurement of software products and services, Softimage Face Robot, Source Engine is a 3D game engine] the customers forming as a group among themselves for purchase of single commodity in partnership. Therefore the service is not depending on the group of people who

want to purchase a single product for common usage in different slots. A mathematical theory of the production inventory system was first registered in the literature by Arrow et.al, [1]. Berman et.al, [2] made a study on inventory control at service facilities using the Markov renewal theory. Many researchers have carried out their research on Inventory modulated with queuing systems see ref [5-7][9][10][12]. Perel and Yechiali [8] have considered a pre-emptive priority queuing system served by a single server with higher and lower priority customers. For more work on batch services refer to [13-14]. A batch service queueing model with constrained waiting for a server to start service after getting the desired number of customers in a batch is studied by [15] and framed as a rule for the flexibility of the batch.

Chakkravarthy [4] has done a study on steady-state analysis of customer loss probability upon arrivals with batch demand. Recently, Viswanath et.al.,[11] have proposed and discussed a model for a single product inventory system and demand served by Israeli Queue fashion, hence obtaining the various performance measures by considering lead time as a random variable. Instantaneous replenishment appears in many real-life situations such as medical emergencies in hospitals like ventilation facilities for Covid affected patients, High power electricity generators in hospitals, and air conditioning high-scale types of machinery etc. Deal with such kind of significant factors in different fields triggers the necessity of constructing a model of an inventory system that serves Israeli disciplined customers with instantaneous replenishment in order to fill this gap. The flow of the article is as follows: In Section 2 model assumption and description are given, and Section 3 is devoted to the governing equations and transition rate diagram of the model. In Section 4, all necessary performance measures are discussed and Section 5 covers sensitivity analysis, and Section 6 is devoted to the Conclusion.

II. Model Assumptions and Description

A single product inventory system is considered with Maximum storage capacity S. Whenever the inventory level reaches zero, an adjustable reorder is placed and instantaneous replenishment is allowed. Service rendered to the Israeli type queued customers via a single server facility. Only Israeli groups are allowed. Customer arrival and service follow respectively Poisson and exponential distributions with parameters λ and μ. Assumed that batch size never influences the service time. If an arriving customer finds that there are already N groups in the system then he will check with all groups and confirm his familiarity. If no group members are familiar to him he will join the Nth group else he will join the group in which the member is familiar to him. Let be the probability for any existing group leader to be known to the arriving customer else the probability is $q(p + q = 1)$. Two cases arise upon the arrival of a new customer to the system: Case:1 If there are groups (which includes the group which gets service) in the system while a customer arrives in the system then the customer either joins the ith group if the leader of the ith group is known to him with probability $q^{i-1}p$ where $i \leq k$ or he may initiate a new group at the end of the queue with probability q^k. Case:2 There are already N Israeli groups in the system at the time of arrival of a new customer who knows non in the queue then with probability q^{N-1}, he is forced to join in N^{th} group.

III. Flow Balance Equations and Solution by Matrix Method

We identified $S(t) = \{X(t), Y(t)\}$ as a two-dimensional stochastic process with state space $\Omega = \{(l,k), l = 1,2,3,4... N; k = 1,2,3,4, ...S\}$. Where $X(t)$ and $Y(t)$ are respectively represents a number of Israeli groups in the system and the number of items loaded in the inventory at the time. Then the flow of transition of this model is given in Fig. 48.1.

Define the conditional probability:

$$Pr(l, k; t) = Pr[X(t) = l, Y(t) = k|X(0) = 0, \\ Y(0) = S] \tag{1}$$

The following Integral equations are arrived at by using probability law for mutually exclusive and exhaustive cases.

$$Pr(0, S; t) = \mu Pr(1, 1; t)\Theta e^{-\lambda t} \tag{2}$$

$$Pr(0, k; t) = \mu Pr(1, k + 1; t)\Theta e^{-\lambda t}), \\ k = 1, 2, 3, ..., S - 1 \tag{3}$$

$$Pr(l, S; t) = [\lambda q^{l-1} Pr(l - 1, S; t) \\ + \mu Pr(l + 1, 1; t)] \Theta e^{-\{\mu + q^l\lambda\}t}, \\ l = 1, 2, 3, ..., N - 1 \tag{4}$$

$$Pr(N, k; t) = \lambda q^{N-1} Pr(N - 1, k; t)\Theta e^{-\mu t}, \\ k = 1, 2, 3, ..., S \tag{5}$$

$$Pr(l, k; t) = [\lambda q^{l-1} Pr(l - 1, k; t) \\ + \mu Pr(l + 1, k + 1; t)]\Theta e^{-\{\mu + q^l\lambda\}t}, \\ l = 1, 2, 3, ..., N - 1, k = 1, 2, 3, ..., \\ S - 1 \tag{6}$$

Applying Laplace Transform on both sides of the equations (2) to (6) and using final value theorem we get the following balance equations for steady state

$$\lambda\pi(0, S) = \mu\pi(1, 1) \tag{7}$$

$$\lambda\pi(0, k) = \mu\pi(1, k + 1), k = 1, 2, 3, ..., S - 1 \tag{8}$$

$$\{\mu + q^l\lambda\}\pi(l, S) = \lambda q^{l-1} \pi(l - 1, S) + \mu\pi(l + 1, 1), \\ l = 1, 2, 3, ..., N - 1 \tag{9}$$

$$\mu\pi(N, k) = \lambda q^{N-1} \pi(N - 1, k), k = 1, 2, 3, ..., S \tag{10}$$

$$\{\mu + q^l\lambda\}\pi(l, k) = \lambda q^{l-1}\pi(l - 1, k) \\ + \mu\pi(l + 1, k + 1) \tag{11}$$

$$l = 1, 2, 3, ..., N - 1, k = 1, 2, 3, ..., S - 1$$

By total probability law, we arrive

$$\sum_{i=0}^{N}\sum_{j=1}^{S}\pi(l,k) = 1 \tag{12}$$

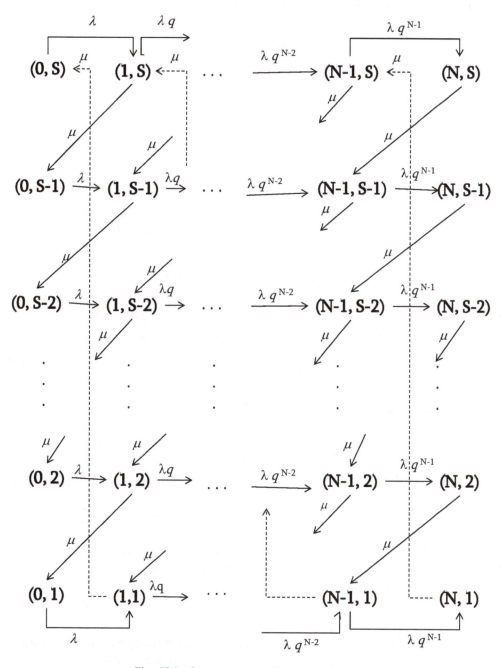

Fig. 48.1 State transition diagram of the model

Source: Author

Rate matrix of the identified stochastic processes as Markov processes is

$$
Q = \begin{pmatrix}
H_0 & U_0 & 0 & 0 & 0 & . & . & . & 0 & 0 & 0 \\
L_1 & H_1 & U_1 & 0 & 0 & . & . & . & 0 & 0 & 0 \\
0 & L_1 & H_2 & U_2 & 0 & . & . & . & 0 & 0 & 0 \\
0 & 0 & L_1 & H_3 & U_3 & . & . & . & 0 & 0 & 0 \\
0 & 0 & 0 & L_1 & H_4 & . & . & . & 0 & 0 & 0 \\
. & . & . & . & . & & & & . & . & . \\
. & . & . & . & . & & & & . & . & . \\
. & . & . & . & . & & & . & & & . \\
0 & 0 & 0 & 0 & 0 & . & . & . & H_{N-2} & U_{N-2} & 0 \\
0 & 0 & 0 & 0 & 0 & . & . & . & L_1 & H_{N-1} & U_{N-1} \\
0 & 0 & 0 & 0 & 0 & . & . & . & 0 & L_1 & H_N
\end{pmatrix}
$$

where $L_1 = \mu \begin{pmatrix} 0 & 1 \\ I' & 0 \end{pmatrix}$, I' is unit matrix with order $S - 1$, $H_0 = -\lambda I$, $H_N = -\mu I$ and $H_n = -\{\lambda q^n + \mu\} I$, $U_n = \lambda q^n I$. Here I is a unit matrix of order $(S \times S)$. Define the vector $P = [P_0 P_1 P_2 \dots P_N]$.

Where $P_\tau = [\pi(\tau, 1)\ \pi(\tau, 2)\ \pi(\tau, 3)\ \dots\ \pi(\tau, S)]$, $\tau = 0, 1, 2, \dots, N$

Since $QP = 0$, we get the following equations

$$P_0 H_0 + P_1 L_1 = 0 \tag{13}$$

$$P_{n-1} U_{n-1} + P_n H_n + P_{n+1} A_1$$
$$= 0, n = 1, 2, 3, \dots, N - 1 \tag{14}$$

$$P_{N-1} U_{N-1} + P_N H_N = 0 \tag{15}$$

From (15), we get

$$P_N = P_{N-1} X_N, \tag{16}$$

Where $X_N = -\lambda q^{N-1} H_N^{-1}$.

By Substituting $n = N - 1$ in (14), we get

$$P_{N-1} = P_{N-2} X_{N-1}, \tag{17}$$

Where $X_{N-1} = -U_{N-2}(H_{N-1} + X_N L_1)^{-1}$

For $n = N - 2$ in (14), we get

$$P_{N-2} = P_{N-3} X_{N-2}, \tag{18}$$

Where $X_{N-2} = -U_{N-3}(H_{N-2} + X_{N-1} L_1)^{-1}$

For $n = N - 3$ in (14), we get

$$P_{N-3} = P_{N-4} X_{N-3}, \tag{19}$$

Where $X_{N-2} = -U_{N-4}(H_{N-3} + X_{N-2} L_1)_{-1}$

Similarly for $n = 1$ in (14), we get

$$P_1 = P_0 X_1, \tag{20}$$

Where $X_1 = -U_0(H_1 + X_2 L_1)^{-1}$

Equation (17)-(20) yield

$$P_n = P_{n-1} X_n, \tag{21}$$

Where $X_n = -U_{n-1}(H_n + X_{n+1} L_1)^{-1}$, $n = 1, 2, 3, \dots, N - 1$. Then from (12) we get P_0.

Then all steady state probabilities are derived.

IV. Measures of the Stystem

A. Mean (Average) Number of Replenishments

$$E(r) = \sum_{l=1}^{N} \pi(l, 1)\mu \tag{22}$$

B. Mean (Average) Number of Israeli Groups in the System

$$M(g) = \sum_{l=1}^{N} \sum_{k=1}^{S} l\pi(l, k) \tag{23}$$

C. System Throughput with Respect to the Inventory Level

$$T = \left[1 - \sum_{k=1}^{S} \pi(0, k) \right] \mu, \qquad (24)$$

D. Mean (Average) Number of Bypassed Israeli Groups by the Arriving Customer

Let B_P be the number of bypassed Israeli groups while the arrival of new customers. Which is considered a random variable denoted as $B_P, 0 \leq B_P \leq N - 1$.

$$E[B_P] = \sum_{h=1}^{N-1} h \sum_{l=h+1}^{N} \sum_{k=1}^{S} pq^{l-h-1} \pi(l, k) \quad (25)$$

V. Sensitivity Analysis

By varying the parameters λ, μ and p, we have arrived Mean stationary rate of replenishments, average Israeli groups in the system, system throughput and $E[B_P]$ for both the cases $N < S$ and $N > S$ and listed out the values in the Table 48.1. We can observe that from the first part of Table 48.1 that irrespective of the cases either $N < S$ or $N > S$ all the measures $E(r)$, $M(g)$, T and $E[B_P]$ increase while increasing λ. We can observe that from the second part of Table 48.1 irrespective of the cases either $N < S$ or $N > S$ the measures $E(r)$, T are increased and the measures $M(g)$, $E[B_P]$ are decreased as increasing the rate of service μ. Decrease the number of Israeli groups in the system, since more Israeli groups get service as the service rate increases; also decrease the mean number of bypassed customers by the arriving customer, since fewer Israeli groups will be there in the system due to the high service rate. As a result, there is less chance for the arriving customer to bypass. We can observe that from the third part of Table 48.1 irrespective of the cases either $N < S$ or $N > S$ the measures $E(r)$, $M(g)$, T are decreased and the measures $E[B_P]$ alone increased while increasing the probability p of

an arriving customer to know a person in any of the existing Israeli groups. Since increasing p this causes naturally the following: A decrease in the number of replenishment events, a decrease in the number of Israeli groups in the system, since the increase in p restricting the generation of more Israeli groups in the system, a decrease in the value of systems throughput since by having less Israeli groups in the system but increase the mean number of bypassed customers by the arriving customer, since if p the value is high it leads that the chance for an arriving customer to know any one of the members in the Israeli groups, as a result, a chance for bypassing is more.

6. Conclusion and Future Directions

In this article single product inventory system is analyzed with the consideration of server facility to serve the customers. Steady-state probabilities are arrived at by the matrix method. Derived effective performance measures the mean number of demands satisfied, replenishment events and bypassed Israeli groups. Moreover, the system throughput is measured. All performance measures are analyzed for the effect of variation of changes in parameters and the same is listed in tables and figures. Moreover, the model may provoke interest among the researchers to develop this model further in the future by considering other factors like consideration of perishable items, prioritized customers,and environmental influences with the inclusion of the occurrence of catastrophe.

References

1. Arrow, K. J., S. Karlin, and H. E. Scarf, (1958) Studies in the Mathematical Theory of Inventory and Production. Stanford University Press, Stanford, California.
2. Berman, O., K. P. Sapna.,(2000). Inventory management at service facilities for

Table 48.1 Performance measures by varying λ, μ and p for the cases $N < S$ and $N > S$

Vary	[For N = 3 and S = 5]↓				[For N = 5 and S = 3]↓			
	Fix μ = 5, p = 0.2				Fix μ = 5, p = 0.2			
(λ)	$E(r)$	$M(g)$	T	$E[B_p]$	$E(r)$	$M(g)$	T	$E[B_p]$
1	0.1910	0.2235	0.9550	0.0070	0.3188	0.2249	0.9565	0.0074
1.5	0.2784	0.3503	1.3920	0.0160	0.4661	0.3568	1.3982	0.0179
2	0.3593	0.4833	1.7964	0.0282	0.6041	0.5015	1.8124	0.0337
2.5	0.4331	0.6190	2.1655	0.0430	0.7323	0.6583	2.1969	0.0552
3	0.4996	0.7544	2.4982	0.0598	0.8500	0.8258	2.5501	0.0827
3.5	0.5590	0.8868	2.7950	0.0779	0.9570	1.0018	2.8711	0.1159
4	0.6115	1.0141	3.0576	0.0968	1.0532	1.1837	3.1595	0.1544
4.5	0.6577	1.1350	3.2884	0.1159	1.1387	1.3688	3.4160	0.1973
(μ)	Fix λ = 2, p = 0.2				Fix λ = 2, p = 0.2			
3	0.3240	0.8431	1.6200	0.0718	0.5535	0.9423	1.6606	0.1042
3.5	0.3370	0.7159	1.6848	0.0548	0.5722	0.7770	1.7167	0.0743
4	0.3465	0.6190	1.7324	0.0430	0.5858	0.6583	1.7575	0.0552
4.5	0.3537	0.5435	1.7685	0.0345	0.5961	0.5698	1.7884	0.0425
5	0.3593	0.4833	1.7964	0.0282	0.6041	0.5015	1.8124	0.0337
5.5	0.3637	0.4344	1.8187	0.0234	0.6105	0.4474	1.8316	0.0273
6	0.3674	0.3941	1.8368	0.0197	0.6158	0.4036	1.8474	0.0225
6.5	0.3703	0.3604	1.8517	0.0168	0.6201	0.3675	1.8604	0.0189
(p)	Fix λ = 2, μ = 5				Fix λ = 2, μ = 5			
0.1	0.3713	0.5205	1.8566	0.0176	0.6301	0.5599	1.8902	0.0241
0.15	0.3652	0.5014	1.8260	0.0236	0.6165	0.5284	1.8494	0.0300
0.2	0.3593	0.4833	1.7964	0.0282	0.6041	0.5015	1.8124	0.0337
0.25	0.3536	0.4661	1.7679	0.0314	0.5929	0.4781	1.7786	0.0357
0.3	0.3481	0.4497	1.7404	0.0335	0.5825	0.4575	1.7474	0.0367
0.35	0.3428	0.4342	1.7139	0.0346	0.5728	0.4392	1.7184	0.0369
0.4	0.3377	0.4196	1.6884	0.0350	0.5637	0.4226	1.6911	0.0365
0.45	0.3327	0.4057	1.6637	0.0346	0.5551	0.4074	1.6902	0.0355

Source: Author

systems with arbitrarily distributed service times. *Communications in Statistics. Stochastic Models*, 16 (3–4) 343–360. doi: 10.1080/15326340008807592.

3. Boxma, O. J., J.van der Wal, and U. Yechiali., (2008). Polling with batch service. *Stochastic Models*, 24 (4) 604–625. doi: 10.1080/15326340802427497.

4. Chakravarthy, S. R., (2020). Queueing-Inventory Models with Batch Demands and Positive Service Times. *Autom Remote Control*, 81 (4) 713–730. doi: 10.1134/S0005117920040128.

5. Hanukov, G., T. Avinadav, T. Chernonog, and U. Yechiali., (2019). A multi-server queueing-inventory system with stock-dependent demand, *IFAC-PaperOnline*, 52 (13) 671–676.

6. Jiang, T., Xingzheng Lu., Lu Liu., Jun Lv., and Xudong Chai., (2020). Strategic Behavior of Customers and Optimal Control for Batch Service Polling Systems with Priorities. *Complexity*, 2020, Article ID 6015372, 19 pages. doi:10.1155/2020/6015372.

7. Krishnamoorthy, A., and K. P. Jose, (2007). Comparison of Inventory Systems with Service, Positive Lead-Time, Loss, and Retrial of Customers. *International Journal of Stochastic Analysis*, 2007, Article ID 037848, 23 pages.doi.org/10.1155/2007/37848.

8. Perel, N ., and Yechiali, U., (2013). The Israeli queue with priorities. *Stochastic Model*, 29 (3) 353-379. doi:10.1080/15326349.2013.808911.

9. QI-Ming Hh and E. M. Jewkes, (2000). Performance measures of a make-to-order inventory-production system. *IIE Transaction*, 32 (5) 409–419. doi: 10.1080/07408170008963917.

10. Sophia Lawrence., B. Sivakumar, and G. Arivarignan., (2013). A perishable inventory system with service facility and finite source. *Applied Mathematics Modelling*, 37 4771–4786.

11. Viswanath, J., C. T. Dorapravina., T. Karthikeyan., and A. Stanley Raj., (2021). Serving Israeli Queue on Single Product Inventory System with Lead Time for Replenishment. *Mathematical Analysis and Computing,Springer Proceedings in Mathematics & Statistics*, 344 .doi: 10.1007/978-981-33-4646-8_14.

12. Zipkin,P.(1986) Stochastic lead-times in continuous-time inventorr models, Naval Research Logistics Quarterly, vol. 33 pp. 763–774.

13. Rajat Deb and Richard F. Serfozo, (1973). Optimal control of batch service queues. *Advances in Applied Probability*, 5 (2) 340–360. doi: 10.1017/S0001867800039215.

14. Jens Baetens, Bart Steyaert, Dieter Claeys, Herwig Bruneel, (2019). Analysis of a batch-service queue with variable service capacity, correlated customer types and generally distributed class-dependent service times. *Performance Evaluation*, 135, 102012. doi: 10.1016/j.peva.2019.102012.

15. Deena Merit C. K., Haridass M., (2021). Analysis of Flexible Batch Service Queueing System to Constrict Waiting Time of Customers. *Mathematical Problems in Engineering*, 2021, Article ID 3601085, 16 pages. doi: 10.1155/2021/3601085.

Recent Trends in Computational Intelligence and its Application – Sugumaran D. et al. (eds)
© 2023 Taylor & Francis Group, London, ISBN 978-1-032-48410-5

49

Solving Assignment Problem and Game Theory Using Graph Theory

N. Kalaivani*[1]

Department of Mathematics,
Vel Tech Rangarajan Dr. Sagunthala R& D Institute of Science and Technology, Chennai, India

E. Mona Visalakshidevi[2]

Department of Mathematics,
Misrimal Navajee Munoth Jain Engineering College, Chennai, India

Abstract—Graph theory aids as a mathematical model to signify any scheme confining a binary relation. In this paper, the Assignment problem and Game theory are solved in symmetric and non-symmetric problems using the Graph Theory approach. The method proposed in this work is a methodical process, apparent to implement for resolving assignment problems and Game theory problems in which so many routes of the path are possible to find the optimal solution in the directed graph and undirected graph in Graph theory. In this proposed work, the main graph is used to frame three other sub-graphs. Four routes of the path are taken in different ways in the graph. For each route, there is an in-degree and out-degree which frames the matrix representation. A numerical specimen is examined by means of a new technique in addition, the solution is computed by the existing two methods. Also, a performance comparison is made for the optimum solutions amongst this new method besides the two prevailing methods by means of Matlab. Using the graph theory, the solution to the assignment problem and game theory can be determined. The present work finally concludes the frame of the matrix of indegree and outdegree and also the results of the Assignment problem and game theory can be obtained.

Keywords—Graph, Directed graph, Simple graph, Undirected graph, Indegree-outdegree, Assignment problem, Game theory, Matlab

I. Introduction

Assignment problem is generally selected such that it exhibits the lowest cost and highest profit (in terms of the time of distance) while assigning that i-th person allocated to the j-th job. The issue is to find a task (which job should be dispensed to which person one-on-one basis) such that the entire charge of the execution of all jobs is lowest, problem of this kind is

---
*Corresponding author: kalaivani.rajam@gmail.com
[1]drkalaivani@veltech.edu.in; [2]monashravanthi@gmail.com, [2]monavisalakshidevie.mat@mnmjec.ac.in

DOI: 10.1201/9781003388913-49

identified as an assignment problem. This paper presents useful thoughts from graph theory and displays how assignment problems can be solved optimally [10]. Many of the Indian industries are using the routes to reach the place in the shortest path and shortest duration as discussed in [5, 6]. These possible routes are used in operation research for finding the optimal solution.

II. Motivation and Justification

A different strategy of Game theory is played in this present work to solve the assignment problem. Whereas Assignment problems from various applications predominantly prominence on investigational study and execution of the graph theory procedures. Graph portrayal [1] is a vital theme in the execution viewpoint because the programmed generation of portrayal graph has significant applications in vital computer science knowledge such as database strategy, software manufacturing, automated circuit conniving, circuit automatic verification [7], computer network designing, and visual interfaces. Predominantly, deprived of knowing the ideas of graphs [11,12] in our everyday life span, graph theory techniques are being applied. Here this problem from the outlook of graph theory can be formulated by considering the two places as vertices besides roads as per edges. If the direction of transit is measured, then the graph must be directed. Correspondingly, these perceptions of graph theory are used in numerous circumstances. A graph can be practiced to extant nearly any corporeal state connecting separate and association among them [11]. Graphs afford an expedient mode to signify numerous categories of scientific substances as cast-off in the toil stated in [14,7]. Graphs give us many methods and litheness [7] while defining and answering a real-life problem.

III. Application in Deep Learning

As a distinct case of task disputes, the linear sum task problem is a traditional combinatorial optimization task and can be extensively launched in many wireless transmission structures, for instance, mode assortment for gadget-to-gadget transportations, joint source distribution in numerous-input and numerous-output structures. Specified a problem portrayal, deep learning can be applied to find nearby best heuristics by means of trifling human input [2-4]. The credit assignment problem is related to training/learning and optimization in Deep learning. There is a key role in graph philosophy in computer science especially Deep learning. There are innumerable procedures ensuing from graph models explicitly unswerving path procedure in a network, catching the least straddling tree, detecting grid planarity, procedures to treasure contiguity matrices, procedures to discover the connectedness, procedures to invent the rotations in a graph, etc. [8-10].

Graphs are availed to express the stream of computation besides utilized to characterize networks of communication. Graphs epitomized in the work described in [11-15] are handled to characterize data organization. Graph transformation systems toil on rule-based in-memory handling of graphs. Graph concept is applied to invent an unswerving track in a boulevard or else a network. In Google Maps, several positions are signified as vertices or nodes, and the streets are signified as edges, and the graph concept is utilized to discover the paths amid two nodes [7].

IV. Description of Proposed Technique for Solving Assignment Problem

In the present work, a new graph theory approach has been used in Operations Research

Table 49.1 Approach of assignment problem [4]

$$
\begin{array}{cc}
 & \begin{array}{ccccccc} & & \textit{Activity} & & & & \textit{Available} \\ & 1 & 2 & 3 & . & j & r & 1 \end{array} \\
\text{Resource} \begin{array}{c} 1 \\ 2 \\ 3 \\ \\ i \\ r \\ required \end{array} &
\left[\begin{array}{ccccccc}
h_{11} & h_{12} & h_{13} & . & h_{ij} & h_{1r} & 1 \\
h_{21} & h_{22} & h_{23} & . & h_{2j} & h_{2r} & 1 \\
h_{31} & h_{32} & h_{33} & . & h_{3j} & h_{3r} & 1 \\
 & & & & & & \\
h_{i1} & h_{i2} & h_{i3} & . & h_{ij} & h_{im} & 1 \\
h_{r1} & h_{r2} & h_{r3} & . & h_{rj} & h_{rr} & 1 \\
1 & 1 & 1 & & 1 & 1 &
\end{array}\right]
\end{array}
$$

by applying it to an assignment problem. The terminologies and theorems used are designated first following which the proposed graph technique is elucidated. In a graph a vertex v through null in-degree is termed a vertex (source) besides a vertex through null out-degree is entitled a degree (sink). Then we started the $r \times r$ matrix showing the assignment of resources to activity in the following Table 49.1.

V. Innovative Procedure Intended for Explaining Assignment Problem and Game Theory

Now we can apply the innovative procedure for finding the Assignment problem and game theory with the help of direct graph and undirected graph. A matrix is framed with the help of graph theory-finding indegree and outdegree but diverse after them. Also, the

problem is solved by this technique and the result is in matrix form. Here and now, we consider the assignment problem and game theory in matrix form, where h_{ij} is the charge of assigning i^{th} items as vertices to j^{th} receiver indegree and outdegree.

A. Proposed Method: Subtract Row and Add One-Assignment Method

B. Algorthim for Game Theoy in Dominance Rule and Graphical Method

C. Algorithm for Directed Graph

D. Problem Based New Technique used in Indegree and outdegree apply Game Theory and Assignment Problem

First we can draw a graph and find in-degree and out-degree of the main graph. Then in the main graph, we can take three subgraphs to find the indegree and outdegree. Now we can collect four indegree graphs and four outdegree graphs and frame a new matrix. Solve the problem based on in-degree and out-degree in Assignment Problem and game theory problem.

By the four routes (see next page) we can find the different direction ways indegree and outdegree of the routes. Based on this we can frame the matrix.

Table 49.2 Approach of game theory [4]

$$
\begin{array}{c|cccccc}
 & 1 & 2 & 3 & . & j & . & r \\
\hline
1 & h_{11} & h_{12} & h_{13} & . & h_{ij} & . & h_{1r} \\
2 & h_{21} & h_{22} & h_{23} & . & h_{2j} & . & h_{2r} \\
 & . & . & . & . & & & . \\
i & h_{i1} & h_{i2} & h_{i3} & . & h_{ij} & . & h_{im} \\
r & hr1 & h_{r2} & h_{r3} & . & h_{rj} & . & h_{rr}
\end{array}
$$

Main Graph-Directed graph	Subgraph-1	Subgraph-2	Subgraph-3
Fig. 49.1(a)	**Fig. 49.1(b)**	**Fig. 49.1(c)**	**Fig. 49.1(d)**
	we can find the route of the graph to the above Fig. 49.1(b), k to l, l to m, m to n, n to k and m to k.	we can find the route to the graph to the above Fig. 49.1(c), k to l, l to m, m to n and n to k.	we can find the route to the graph to the above Fig. 49.1(d), k to l, l ton, m to k and n to l.

Main Graph - indegree and outdegree.	Subgraph-1: output of indegree and outdegree	Subgraph-2: output of indegree and outdegree	Subgraph-3:output of indegree and outdegree
$\deg^-(k) = 5, \deg^+(k) = 2$	$\deg^-(k) = 2, \deg^+(k) = 1$	$\deg^-(k) = 1, \deg^+(k) = 1$	$\deg^-(k) = 2, \deg^+(k) = 0$
$\deg^-(l) = 3, \deg^+(l) = 3$	$\deg^-(l) = 1, \deg^+(l) = 1$	$\deg^-(l) = 2, \deg^+(l) = 0$	$\deg^-(l) = 0, \deg^+(l) = 2$
$\deg^-(m) = 1, \deg^+(m) = 6$	$\deg^-(m) = 1, \deg^+(m) = 2$	$\deg^-(m) = 0, \deg^+(m) = 2$	$\deg^-(m) = 0, \deg^+(m) = 2$
$\deg^-(n) = 4, \deg^+(n) = 2$	$\deg^-(n) = 1, \deg^+(n) = 1$	$\deg^-(n) = 1, \deg^+(n) = 1$	$\deg^-(n) = 2, \deg^+(n) = 0$

Indegree Graph					Outdegree Graph				
indegree					*outdegree*				
vertices	1	2	3	4	*vertices*	1	2	3	4
k	1	1	0	2	k	2	1	2	5
l	1	0	2	3	l	1	2	0	3
m	2	2	2	6	m	1	0	0	1
n	1	1	0	2	n	1	1	2	4

Explaining the indegree matrix using Assignment Problems and Game Theory.

Vertices/in degree	1	2	3	4
k	1	1	0	2
l	1	0	2	3
m	2	2	2	6
n	1	1	0	2

Solution: Vertices and indegree = 4. The optimal solution using Assignment Problem is $u \to 3, v \to 2, w \to 1, x \to 4, 0 + 0 + 2 + 2 = 4$. Total cost = 4. Also, by using Game theory, the value of the game is 2

Solve the outdegree matrix using Assignment Problem and Game Theory

Vertices/ indegree	1	2	3	4
K	2	1	2	5
L	1	2	0	3
M	1	0	0	1
N	1	1	2	4

Solution: Optimal Assignment is $u \to 2, v \to 3, w \to 4, x \to 1, 1 + 0 + 1 + 1 = 3$. Cost Total = 3. Also, by using Game theory, the value of the game is 1.33333.

E. Solve the Problem Based on the Undirected Graph

First, we can find the undirected graph based on degree. In the main graph we can take three subgraphs and frame the 4X4 matrix. Then we can take the value of the degree. We can frame the 4X4 matrix of the degree of the route to find the solution of assignment problem and Game theory.

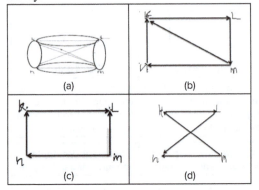

(a)	(b)
(c)	(d)

Fig. 49.1 Graphs explaining different directions

Source: Authors

By the above graphs of the Figures, we can find the different directions ways degrees of the routes.

$\deg(k) = 7$ $\quad \deg(k) = 3$ $\quad \deg(k) = 2,$ $\quad \deg(k) = 2,$

$\deg(l) = 6,$ $\quad \deg(l) = 2$ $\quad \deg(l) = 2,$ $\quad \deg(l) = 2,$

$\deg(m) = 7,$ $\quad \deg(m) = 3$ $\quad \deg(m) = 2,$ $\quad \deg(m) = 2,$

$\deg(n) = 6$ $\quad \deg(n) = 2$ $\quad \deg(n) = 2,$ $\quad \deg(n) = 2$

Solution of Assignment problem based on Undirected graph and Game theory

Vertices\degree	1	2	3	4
K	7	3	2	2
L	6	2	2	2
M	7	3	2	2
N	6	2	2	2

Solution: The number of rows = 4 and columns = 4. Optimal solution is $u \rightarrow 3, v \rightarrow 1, w \rightarrow 4, x \rightarrow 2,$ cost = 2 + 6 + 2 + 2 = 12. Value of the game theory is 2

Table 49.3 Comparison for indegree and outdegree of the directed graph

Graph	Directed graph frame the matrix		Game theory		Assignment problem	
	Indegree-Matrix	Outdegree-Matrix	Indegree-optimal value	Outdegree-optimal value	Indegree-optimal value	Outdegree-optimal value
	A $\begin{pmatrix} 1 & 0 & 1 & 1 \\ 2 & 1 & 1 & 1 \\ 1 & 1 & 0 & 1 \\ 1 & 1 & 2 & 1 \end{pmatrix}$ B C D	A $\begin{pmatrix} 1 & 0 & 1 & 1 \\ 2 & 1 & 1 & 1 \\ 1 & 1 & 0 & 1 \\ 1 & 1 & 2 & 1 \end{pmatrix}$ B C D	2	1.33	4	3
	A $\begin{pmatrix} 0 & 1 & 2 & 1 & 4 \\ 1 & 1 & 0 & 2 & 4 \\ 0 & 0 & 0 & 2 & 2 \\ 2 & 1 & 1 & 1 & 5 \\ 0 & 1 & 1 & 1 & 2 \\ 1 & 2 & 1 & 0 & 6 \end{pmatrix}$ B C D E F	A $\begin{pmatrix} 2 & 1 & 0 & 1 & 4 \\ 1 & 1 & 2 & 0 & 4 \\ 2 & 2 & 2 & 0 & 6 \\ 0 & 1 & 1 & 1 & 3 \\ 2 & 1 & 1 & 1 & 6 \\ 1 & 0 & 1 & 2 & 2 \end{pmatrix}$ B C D E F	1.1	1	2	3
	A $\begin{pmatrix} 1 & 0 & 1 & 1 \\ 2 & 1 & 1 & 1 \\ 1 & 1 & 0 & 1 \\ 1 & 1 & 2 & 1 \end{pmatrix}$ B C D	A $\begin{pmatrix} 1 & 1 & 1 & 0 \\ 0 & 1 & 1 & 1 \\ 1 & 1 & 2 & 1 \\ 1 & 1 & 0 & 2 \end{pmatrix}$ B C D	1	1	2	1
	A $\begin{pmatrix} 0 & 2 & 1 & 3 \\ 2 & 1 & 1 & 5 \\ 1 & 1 & 0 & 5 \\ 1 & 1 & 1 & 3 \\ 1 & 0 & 0 & 4 \end{pmatrix}$ B C D E	A $\begin{pmatrix} 2 & 0 & 1 & 5 \\ 0 & 1 & 1 & 3 \\ 1 & 1 & 2 & 3 \\ 1 & 1 & 1 & 5 \\ 1 & 2 & 2 & 4 \end{pmatrix}$ B C D E	1	1.33	3	4

Source: Authors

Table 49.4 Comparison for indegree and outdegree of the undirected graph

Graph	Directed graph frame the matrix	Game theory	Assignment problem
	degree	Optimal value	Optimal cost
(graph)	$\begin{array}{l}A \\ B \\ C \\ D\end{array}\begin{pmatrix} 7 & 3 & 2 & 2 \\ 6 & 2 & 2 & 2 \\ 7 & 3 & 2 & 2 \\ 6 & 2 & 2 & 2 \end{pmatrix}$	2	12
(graph)	$\begin{array}{l}A \\ B \\ C \\ D \\ E \\ F\end{array}\begin{pmatrix} 4 & 4 \\ 4 & 4 \\ 2 & 6 \\ 5 & 3 \\ 2 & 6 \\ 6 & 2 \end{pmatrix}$	4	4
(graph)	$\begin{array}{l}A \\ B \\ C \\ D\end{array}\begin{pmatrix} 3 & 3 \\ 5 & 3 \\ 4 & 5 \\ 5 & 4 \end{pmatrix}$	4.5	6
(graph)	$\begin{array}{l}A \\ B \\ C \\ D \\ E\end{array}\begin{pmatrix} 3 & 5 \\ 5 & 3 \\ 5 & 3 \\ 3 & 5 \\ 4 & 4 \end{pmatrix}$	4	6

Source: Authors

VI. Algorithmic Simulation— Through Matlab Encryption

Through the directly above the counter, we contract the outcome of symmetric in addition non-symmetric delinquent in a directed graph and undirected graph: The constraint morals are taken as 4 costs, 2 costs, 2 costs, 3 costs as indegree and 3 costs, 3 costs, 1 costs, 4 costs as outdegree of assignment problem, 2,1.1,1,1 as indegree and 1.33,1,1,1.33 as outdegree of Game Theory using for directed graph in Grid (i).The constraint morals are taken as 12 costs, 4 costs, 6 costs, 6 costs as task problem,

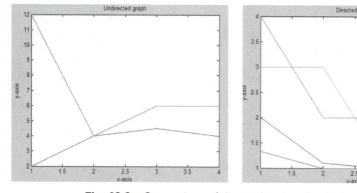

Fig. 49.2 Comparison of directed and undirected graphs

Source: MATLAB

2,4,4.5,4 as Game Theory-using for undirected Grid(ii).In the relative learning, concerning numerous complications, when devoted by means of "Assignment problem and Game theory", the occasioned morals are portrayed in the subsequent Line-chart in Grid(i)&Grid(ii).

Grid of comparison Table 49.3 & Table 49.4 using Matlab the grid as a numerical value of the results.

VII. Conclusion

We can find the optimal solution based on the assignment problem and game theory using a directed graph and an undirected graph. Comparing the indegree and outdegree matrix we can find the numerical result which is the nearest to optimal cost. So, we can get the result of the in-degree graph routes which is greater than the out-degree graph routes based on that we can frame the matrix using a directed graph. Same way by comparing the undirected graph and the directed graph, we can find the result of the degree. So, the conclusion of this paper is that the degree of the directed graph is less than that of the undirected graph. Using Matlab coding we can draw the graph of a directed graph and an undirected graph.

──────── References ────────

1. S. G. Shrinivas, S. Vetrivel and G. M. Elango, Application of graph theory of computer science an Overview, International journal of Engineering Science and Technology, vol. 2 (9) 2010. 4610–4621.

2. Bronstein, M. M., Bruna, J., LeCun, Y., Szlam, A., and Vandergheynst, P. (2017). "Geometric Deep Learning: Going Beyond Euclidean Data". In: IEEE Signal Processing Magazine 34.4, pp. 18–42.

3. Mengyuan Lee, Yuanhao Xiong, Guanding Yu, and Geoffrey Ye Li, Deep Neural Networks for Linear Sum Assignment Problems, IEEE Wireless Communications Letters,

4. Afroz, Humayra Dil, and Mohammad Anwar Hossen. Divide column and subtract one assignment method for solving assignment problem. *American Scientific Research Journal for Engineering, Technology, and Sciences (ASRJETS)* 32,no.1(2017): 289–297.

5. Kumar Bisen, Sanjay. Application of Graph Theory in Transportation Networks. *International Journal of Scientific research and Management* 5, no. 7 (2017): 6197–6201

6. Eneh, A. H., and U. C. Arinze. Comparative analysis and implementation of Dijkstra's shortest path algorithm for emergency response and logistic planning.*Nigerian Journal of Technology* 36, no. 3 (2017): 876–888.

7. Zoutendijk M, Mitici M. Probabilistic Flight Delay Predictions Using Machine Learning and Applications to the Flight-to-Gate Assignment Problem. *Aerospace*. 2021; 8(6): 152. https://doi.org/10.3390/aerospace8060152

8. Amous Gilat, MATLAB: An Introduction with Applications, John Wiley & Sons, Inc. (2004).

9. Stormy Attaway, MATLAB: A Practical Introduction to Programming and Problem Solving, Elsevier, Inc (2009).

10. Burkard, R. E. "Optimierung und Kontrolle: Selected Topics on Assignment Problems." *Karl-Franzens-Universitat Graz & Technische Universitat Graz* (1999).

11. Balakrishnan, Rangaswami, and Kanna Ranganathan. *A textbook of graph theory.* Springer Science & Business Media, 2012.

12. Harinarayanan, C. V. R., and S. Lakshmi. Applications of graph theory in job scheduling and postman's problem, *Advances and Applications in Mathematical Sciences* (2019).

13. Eneh, A. H., and U. C. Arinze. Comparative analysis and implementation of dijkstra's shortest path algorithm for emergency response and logistic planning., *Nigerian Journal of Technology* 36, no. 3 (2017): 876–888.

14. Sporns, Olaf. "Graph theory methods: applications in brain networks." *Dialogues in clinical neuroscience* (2022).

15. Fornito, Alex, Andrew Zalesky, and Edward Bullmore. *Fundamentals of brain network analysis.* Academic Press (2016).

Recent Trends in Computational Intelligence and its Application – Sugumaran D. et al. (eds)
© 2023 Taylor & Francis Group, London, ISBN 978-1-032-48410-5

50

Real Time Video Instance Segmentation Using Mask RCNN

R. Hariharan*, M. Dhilsath Fathima, B. Prakash,
S. Abi Sundar, E. Punith, M. Mohammed Muzameel Hussain
Department of Information technology,
Vel Tech Rangarajan Dr. Sagunthala R&D Institute of Science and Technology, Chennai, India

Abstract—In the real world, object recognition in digital photos and videos is crucial. Using Mask R-CNN and PixelLib, we provide a simple, versatile, and general architecture for Real-Time video instance segmentation. This paper came up with the idea of accurately recognizing objects and also the meantime generates a greater segmentation mask for every instance. Mask R-CNN extends Faster R-CNN by combining the existing bounding box identification branch with the real-world object mask prediction branch PixelLib is a library that makes segmenting objects and instances in real-time applications straightforward. Mask R-CNN is easy to understand and adds only a less amount of advanced methods to Faster R-CNN, which works at 5 fps. Objects of the same class will be assigned a distinct instance in the instance segmentation. Mask R-CNN paired with PixelLib outperforms all prior single-model entrants because we can compute a mask (pixel level) for each and every object in the input instance.

Keywords—Mask RCNN, Real time video, Segmentation, Faster R CNN

I. Introduction

We propose Real-Time Video Instance Segmentation Using Mask R-CNN and PixelLib. Given a pre-trained dataset including many image frames as input, the algorithm continuously generates the mask for each instance in the Real-Time video. Each instance's output sequence is termed as an instance sequence. These approaches are intuitive from a conceptual standpoint, and they provide

flexibility and resilience, as well as quick training and inference times. The objective of this paper is to improve a paradigm for example segmentation that is both enabling and comparable. Instance segmentation is difficult as it requires exact detection of all elements of an instance and precise segmentation of each instance. It takes features from traditional computer vision tasks such as object detection, where the objective is to categorize and localize individual objects using box boundaries as well

*Corresponding author: hharanbtech@gmail.com

DOI: 10.1201/9781003388913-50

as RoIAlign class. PixelLib is a library designed to make object segmentation in real-time as simple as possible and allows developers to easily create semantic segmentation and also instance segmentation.

We display, that a deceptively easy, adaptable together with quick approach can outperform prior state instance segmentation findings. Mask R-CNN with PixelLib is a way to improve Faster R-CNN by providing branches for prediction and training. Classification of box boundaries which is already available with that segmentation mask is applied to each ROI. Mask R-CNN is easy to set up and train than the Faster R-CNN architecture that allows for a wide range of customizable architecture configurations. The mask branch also adds only a small amount of computational cost, allowing for a speedy solution and efficient experimentation. Although the Principle of Mask R-CNN is a simple modification of Faster R-CNN, appropriately designing the mask branch is crucial for good outcomes. Faster RCNN, for example, is not created to match network deliverables pixel-by-pixel. The way RoIPool, the basic function for attending to instances, implements coarse spatial quantization for extracting features is a good example of this.

We offer RoIAlign, a simple, arrangement layer that accurately preserves specified geographic coordinates to correct the misalignment. Despite the fact that it appears to be a little modify, RoIAlign has a considerable influence on mask accuracy, improving it by 10 to 50 percent compared to tighter localization measurements. Second, we noticed that separating mask with class prediction was crucial; we predict a binary mask for each class and depend on the RoI classifier branch of the network to predict the group independently while competing with other classes. FCNs, on the other hand, often execute per-pixel for multi-class classification, in that segmentation and classification combine and, according to our tests, perform badly

for instance segmentation. On the COCO video instance segmentation problem, Mask R-CNN exceeds all prior form solitary results without any bells and whistles. In a single 8-GPU computer, our prototype can operate at roughly 200ms per frame, and COCO training involves approximately two days. We believe that the framework's versatility and accuracy, as well as the quick train and test speeds, will gain and clarify future video instance segmentation studies. Ultimately, we use the COCO key point dataset to perform human posture estimation, demonstrating the generality of our method. Each key point can be perceived as a binary mask that can be modified with relatively little effort. The Mask R-CNN is used to identify instance-specific postures while also running at a rate of 5 fps. As a result, Mask R-CNN and Pixellib can be thought of as a flexible approach for instance-level recognizing that might easily be applied to much more sophisticated applications.

A. Motivation

This proposed model aims to create an improved real-time video segmentation using mask RCNN. We tackle this task using a Real-Time video instance segmentation (Mask R-CNN) by modelling it with PixelLib as a direct parallel sequence prediction problem. Mask R-CNN algorithm along with Pixellib can be used to mask and detect the objects in the respective frame.

2. Related Work

In the paper, Ross Girshick suggested a method that employs a selective search algorithm[1] Instead of working on a large number of areas, it only works on 2000 regions while extracting 2000 areas out of an image (region of proposals). It has some disadvantages such as the time it takes to train and classify 2000 region proposals in a single image. There is no scope for improvement with the selective search approach because it is a predefined algorithm.

Then Ross Girshick presented a Region-based convolution neural network to detect segments of an object with more accurate results [2]. To solve the problems of R-CNN, they developed Faster R-CNN. Instead of providing a region of proposals to CNN, they provide an input picture for CNN to produce a feature map, from which we determine the region of proposals and convert them to a fixed size using the ROI polling layer before feeding them to the fully connected network. The softmax layer is used to predict the bounding box class and offset value. R-CNN is slower than fast R-CNN since 2000 regions of proposals must be provided to CNN every time, whereas fast R-CNN generates a feature map by performing convolution operations once per image.

Instead of using selective search algorithms like R-CNN and fast R-CNN, Shaoqing Ren et al introduced a method called Faster R-CNN[3], which allows a separate network to forecast regions by region proposals algorithm instead of using R-CNN and fast R-CNN. The algorithm introduces a Region Proposal Network (RPN) which integrates complete convolutional features with the learning algorithm, allowing for nearly free region proposals. An RPN simultaneously detects the objects and convolutional network fully at each point.

Saining Xie et al proposed a correct residual transformation for deep neural networks[4], which reveals in addition to the dimensions of layer thickness, a new dimension termed cardinality (the size of the set of transformations) is an important feature.Increased cardinality can enhance classification accuracy even under the constrained condition of retaining complexity, according to empirical evidence. Haochen Wang proposed SwiftNet for real-time semi-supervised video object segmentation (One-Shot VOS) [5], which claims 77.8% J&F and 70 FPS on the DAVIS 2017 validation dataset. speed. To do this, we use pixel adaptive memory (PAM) to compress the spatiotemporal redundancy of the correlated VOS.

III. Outline of the Proposed Work

We use PixelLib and Mask R-CNN to model the real-time video instance segmentation challenge. The Mask-RCNN produces the sequences of masks for every instance in the given Real-Time instance with multiple instances as input. To do this, we monitor and segment instances just at the sequence level using the Mask-R CNN algorithm in accordance with the PixelLib.

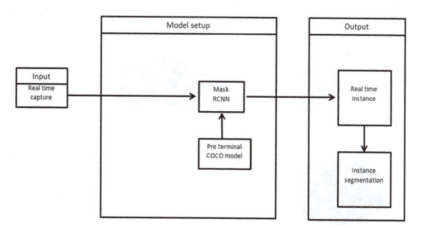

Fig. 50.1 Architecture of real-time video instance segmentation using Mask-RCNN

Source: Author

A. Architecture

In this proposed model Fig. 50.1 the Faster R-CNN produces two branches, class names, and bounding boxes. Mask R-CNN adds an additional branch to the mask as well as the class name and bounding box. According to Pixellib, it is a library meant to make real-time object segmentation as simple as possible and allow developers to easily perform semantic segmentation and instance segmentation.

B. Faster R-CNN

First, we'll go through the Faster R-CNN classifier. There are two phases in the faster R-CNN. The first level is a Region Proposal Network (RPN) which explains potential predicted bounding boxes.

Fast R-CNN is a second level in which each potential box uses features from RoLPooL before classification. In the same way, bounding box regression is also done by RoLPooL. Features used in both phases can be shared to speed up inference Fig. 50.2.

In object recognition tasks requiring convolutional neural networks, region of interest pooling (also known as RoI pooling) is a typical procedure. For example, recognizing many autos and pedestrians in a single shot. It is anticipated that the image's dominant object, as well as all of the objects' positions, be accurately labelled.

C. Mask R-CNN

Mask R-CNN is basically an easy concept to understand, but Faster R-CNN provides multiple outputs for each instance element and bounding box offset.. It contains a third branch called Object Mask. In this, we include a third branch called object mask which requires a much more precise spatial configuration of an object to be extracted.

After that, the basic components of Mask R-CNN are introduced, including feature map positioning, one of the major missing elements in Fast/Faster R-CNN.

Mask R-CNN offers further features including instance segmentation and masking. Each item is separated from its background by Mask R-CNN, which also creates segments on every pixel of the image Fig. 50.3.

Mask R-CNN consists of two stages:

By the given instances are scanned in the first stage to creating the proposals. The proposals are classified in the second stage, which generates a mask and builds the bounding box.

D. Working Principle of Mask R-CNN:

The Mask R-CNN is given input instances. Which persisted model will then produce a feature map for the RPN. The RPN (Regional Proposal Network) takes CNN's output and uses a light-weight classifier model to generate

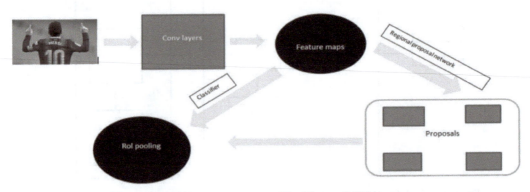

Fig. 50.2 Working model of Faster-RCNN

Source: Author

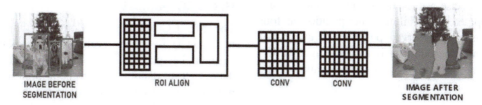

Fig. 50.3 Working model of Mask R-CNN

Source: Author

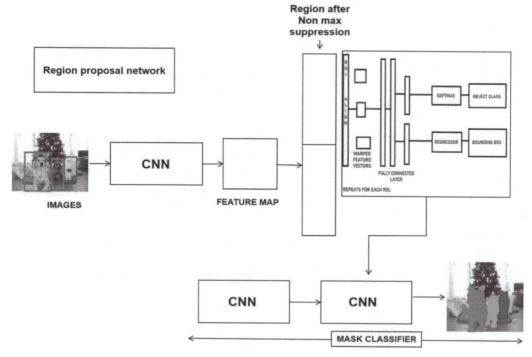

Fig. 50.4 systematic working of Mask R-CNN

Source: Author

numerous regions of interest. By wrapping multiple bounding boxes to fixed dimensions, the Align ROI layer receives the feature map along with the image ROI and generates a fixed-size feature map. The ROI Align feature map is then delivered to the fully connected layers(FCN) and Mask Branch as a fixed-size feature map Fig. 50.4.

The fully connected Layer will use bounded boxes to categorize the objects in the instances. Then the mask branch then produces a segmentation mask for each instance's ROI. This will create instances with bounding boxes

over the objects and a mask for them. Using anchor boxes, Mask R-CNN recognizes various objects, objects of varied sizes, and overlapped objects in a picture. As a result, object detection becomes faster and more efficient.

E. ROI Allign

Region of Interest Align, or RoIAlign, is a technique used for the extraction of a tiny feature map of each individual RoI in detection and segmentation applications. RoI Pool's severe quantization is reduced, next, the retrieved features are aligned with the input[10].

The precise values of the input features are computed using bilinear interpolation at four consistently observed positions in every RoI bin, and the outcome is then averaged (using max or average), preventing any scaling of the RoI boundaries or bins.

F. Pixellib

PixelLib is a computer language framework based on Python. PixelLib is a versatile toolkit that enables easy object segmentation in pictures and videos with just only a few lines of Python code. Instance segmentation is implemented using Mask R-CNN. PixelLib trains a custom model using the Mask R-CNN model with the coco dataset and transfer learning, which may be used to solve a variety of machine learning issues [6]. It allows anyone with a basic understanding of programming and little or no experience with machine learning to properly implement instance segmentation.

G. Training the Model

When we need to send data between databases that are used for various purposes yet need to exchange certain data between development and production databases, being capable to exports and importing data is useful. The dataset we used is called COCO Model (Pre-Trained) dataset [7]. therefore no need for training.

H. COCO Dataset

The MS COCO dataset (Common Objects in Context) is a large-scale object tracking, segment, key-point recognition, and captioning dataset. The dataset, which is a pretrained model dataset used in real-time video instance segmentation, has 328 K pictures. Machine learning and computer vision scientists use the COCO dataset for a number of computer vision tasks. Computer vision aims to understand visual scenes by recognizing what items are visible in their relationships. As a result, item detection and classification algorithms can be trained using the dataset, localizing them in two and three dimensions, establishing their attributes, and defining them [8][9].

I. Input

As Input will be taken using a webcam in Real-Time, no specific Image or video is given in Fig. 50.5. Through the input, real-time instances are masked with class names and bounding boxes. As a result, we get a segmented output with accuracy.

IV. Segmented Output

A. Result

As Proposed Our method of using Mask R-CNN and PixelLib together proves to

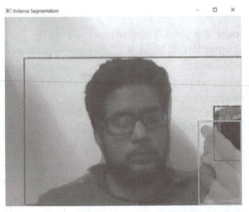

Fig. 50.5 Segmented output

Source: Author

Table 50.1 Evaluation results

S. No	Class Names	AP	S. No	Class Names	AP	S.No	Class Names	AP
1	Person	1.00	11	Tie	1.00	21	Teddy bear	1.00
2	Bicycle	1.00	12	chair	1.00	22	Tooth brush	1.00
3	Car	0.95	13	cake	1.00	23	Dining Table	1.00
4	MotorCycle	0.97	14	suitcase	1.00	24	Potted Plant	1.00
5	Tv	1.00	15	Handbag	1.00	25	Spoon	1.00
6	Laptop	1.00	16	cellphone	1.00	26	Bowl	1.00
7	Pizza	1.00	17	cup	1.00	27	Sofa	1.00
8	Knife	1.00	18	book	1.00	28	microwave	1.00
9	Glasses	1.00	19	clock	1.00	29	bed	1.00
10	Umberlla	1.00	20	vase	1.00	30	Fan	1.00
Mean Average Precision(mAP)						**96.6**		

Source: Author

produce better outputs than any other models. When compared to earlier weights, the new MS COCO-trained weights improve accuracy. These are the evaluation results in Table 50.1 [1] and a comparison with other models is given in Table 50.2.

B. Comparison with Other Models

See Table 50.2

V. Conclusion

In this experiment, we presented a novel library that uses fast deep-learning architectures to conduct image classification, video segmentation, and background editing in photos and videos with only a few lines of Python code. In accordance with Mask R-CNN, we displayed the library's results. The proposed approach will

Table 50.2 Comparison with other model

S.No	Model	Algorithm	Dataset	mAP
1	Rich feature hierarchies for accurate object detection and semantic segmentation[1]	Selective Search Algorithm	PASCAL VOC Dataset	53.3
2	Region based convolution neural networks for accurate object detection and segmentation[2]	Scalable Detection Algorithm	ILSVRC2013 Detection Dataset	62.4
3	Faster R-CNN towards real time object detection with region proposals network[3]	Region Proposal Algorithm	PASCAL VOC Dataset	75.9
4	Aggregated residual transformation for deep neural networks[4]	Cardinality Method(the size of the set of transformations)	Imagenet-1k Dataset	81.9
5	Real-time Video Object Segmentation[5]	PAM(Pixel Adaptive Memory) method	DAVIS2017 Dataset	77.8
6	Real-time Video Instance Segmentation Using Mask-RCNN	Mask- RCNN Algorithm	MS COCO Dataset	96.6

Source: Author

overcome the limitations of object detection. Fast R-CNN and Faster RNCN were previous models that could only accomplish semantic segmentation and not instance segmentation. To perform instance segmentation, the proposed model Mask RCNN can be employed. Auto cars and counting the number of persons or objects in a photograph are only two examples of how instance segmentation is used. It aids in the separation of foreground and background objects. When compared to earlier models, this has a high level of accuracy.

References

1. Girshick, R., Donahue, J., Darrell, T. and Malik, J., 2014. Rich feature hierarchies for accurate object detection and semantic segmentation. In Proceedings of the IEEE conference on computer vision and pattern recognition (pp. 580–587).
2. Girshick, R., Donahue, J., Darrell, T. and Malik, J., 2015. Region-based convolutional networks for accurate object detection and segmentation. IEEE transactions on pattern analysis and machine intelligence, 38(1), pp. 142–158.
3. Ren, S., He, K., Girshick, R. and Sun, J., 2015. Faster r-cnn: Towards real-time object detection with region proposal networks. Advances in neural information processing systems, 28.
4. Xie, S., Girshick, R., Dollár, P., Tu, Z. and He, K., 2017. Aggregated residual transformations for deep neural networks. In Proceedings of the IEEE conference on computer vision and pattern recognition (pp. 1492–1500).
5. Wang, H., Jiang, X., Ren, H., Hu, Y. and Bai, S., 2021. Swiftnet: Real-time video object segmentation. In Proceedings of the IEEE/CVF Conference on Computer Vision and Pattern Recognition (pp. 1296–1305).
6. Krizhevsky, A., Sutskever, I. and Hinton, G.E.H., 2017. ImageNet classification with deep convolutional neural networks Communications of the ACM. 60 (6): 84–90.
7. He, K., Gkioxari, G., Dollár, P. and Girshick, R., 2017. Mask r-cnn. In Proceedings of the IEEE international conference on computer vision (pp. 2961–2969).
8. Farhat, H., Sakr, G.E. and Kilany, R., 2020. Deep learning applications in pulmonary medical imaging: recent updates and insights on COVID-19. Machine vision and applications, 31(6), pp. 1–42.
9. Li, Y., Qi, H., Dai, J., Ji, X. and Wei, Y., 2017. Fully convolutional instance-aware semantic segmentation. In Proceedings of the IEEE conference on computer vision and pattern recognition (pp. 2359–2367).
10. Chen, Y., Pont-Tuset, J., Montes, A. and Van Gool, L., 2018. Blazingly fast video object segmentation with pixel-wise metric learning. In Proceedings of the IEEE conference on computer vision and pattern recognition (pp. 1189–1198).

Recent Trends in Computational Intelligence and its Application – Sugumaran D. et al. (eds)
© 2023 Taylor & Francis Group, London, ISBN 978-1-032-48410-5

51

Ingenious Concepts of Machine Learning Towards Furtherance

Bhavani M[1]

Research Scholar Department of Computer Science and engineering,
SRMIST, Vadapalani, Chennai

M. Durgadevi[2]

Assistant Professor, Department of Computer Science and engineering,
SRMIST, Vadapalani, Chennai

Abstract—Image process is a very helpful technology and in recent times it's experienced dramatic steady growth. It relates to multiple and wide variety of platforms from medicine to statistics of mathematics. It had been well known for this purpose and serves as an area that attracts researchers and various students to make more findings. Computer-aided diagnosis has already become an unavoidable part of providing treatments. Traditionally, image processing that makes use of system getting to know seemed inside the Nineteen Sixties as an attempt to simulate the human imaginative and prescient device and automatize the photograph evaluation method. Due to the era improvement and improvement, answers for precise duties started out to seem. Due to the era improvement and improvement, answers for precise duties started out to appear. Medical imaging is growing efficaciously because of traits in image processing strategies collectively with image recognition, evaluation, and enhancement. Recently, many useful libraries are created which will be used to solve image processing problems with machine learning. Similarly, Medical image process tools were additionally very important. Employing a medical image process tool, its potential to accelerate and enhance the operation of the analysis of the medical image. In major ideal of this paper is to bandy about the lately developed tools and packages. Additionally, medical image processing techniques and applications were conjointly enclosed within the study.

Keywords—Computer-aided diagnostic, Machine learning, Medical image processing, Image processing, Image recognition

I. Introduction

An imaging method is also a form of knowledge method anywhere the data and yield are pictures, almost like photos or casings of video. Image process from time to time strategy pictures as second signals, and apply normal sign method strategies to them. For

[1]bm6010@srmist.edu.in, [2]durgadem@srmist.edu.in

DOI: 10.1201/9781003388913-51

the foremost half, image methods are usually separated into advanced image method and clinical image method. This paper will target clinical image method apparatuses. In clinical fields, clinical imaging and process instruments are collaborating in crucial jobs in very few applications. Such applications occur in the course of the clinical track of occasions; not solely among demonstrative settings, in any case conspicuously among the domain of designing, finishing, and examination before careful tasks, during this method.

The medical specialty image method has learned sensational development Associate in Nursing has been a knowledge base examination field drawing in expertise from science, computer sciences, designing, insights, physical science, and medicine. Computer helped the analytic method has effectively become an important region of clinical everyday follow. Within the interior of a surge of the foremost recent improvement of innovation, and utilization of different imaging modalities, a ton of difficulties emerge[2].

II. Objective and Motivation

In major ideal of this paper is to bat about the recently developed tools and packages. also, medical image processing ways and operations were jointly enclosed within the study. The appetite to work with medical image processing and to gain the implicit knowledge from the images led to this detailed analysis of tools and packages available for the processing of data.

III. Methodology

Two assortments have strategies applied for photo processing are Analog and Digital Image handling. Simple or visible strategies of a image interplay are commonly applied for the debilitating duplicates like printouts and pics. Picture examiners make use of a form of fundamentals of an information though, using those visible techniques. The image instruction

is not simply limited to a area that must be examined.

However, on facts of an expert. Association becomes moreover taken into consideration as some other good sized tool in a photo method via visible procedures. In this way, investigators observe a combination of character facts, and coverage records to an photo method. The three standard tiers that everyone's assortments of facts want to endure though, using a automatic method is a pre-method, an improvement, and a showcase, facts extraction.

A. Medical Imaging

In this process captured images from human parts for the purpose of diagnosis and treatment. The following are some of the medical imaging technique [9].

- X-ray radiography
- Fluoroscopy
- MRI
- Endoscopy
- Elastography
- Tactile imaging
- Thermography

B. Machine Learning Technique in Medical Image Processing

The following are the major machine learning techniques employed in processing medical image:

Classification: It's thought of to be a big space wherever machine learning techniques was utilized for Diagnostic purpose. This method includes classifying pictures diagnosed in previous, during this process each diagnosed example is additionally a sample.

Detection: Anatomical object localization like organs is taken under consideration to be one among the mandatory preprocessing a section of segmentation. Localization of object gift throughout an image wants 3D parsing, many algorithms are projected to convert 3D area as the composition of second orthogonal planes.

Segmentation: The segmentation of human's organs and alternative subparts in pictures used for medical diagnosis permits qualitative inquiry related to features. Segmentation of lesion integrates the challenge in detection of objects and organ and subpart segmentation[3].

Registration: Referred as spacial alignment is a conventional image analysis process throughout that coordinate transforms is obtained by using a image, later it will be converted to a singular image. Though lesion detection and object segmentation are popeyed as the main use of DL rules.

III. Tools and Libraries

The following are the tools and libraries commonly used in medical image processing

OpenCV: OpenCV can be a cross-stage library abuse that we are going to generally be ready to foster in progress computer vision applications. Abuse this library, you will get, pack, improve, restore, and extricate information from images. PyTorch: PyTorch is that the partner open-source structure essentially created by the Facebook AI science laboratory employed in profound learning.

MATLAB: matrix laboratory. It's the call of each desired platform for decision clinical and mathematical finding, and collectively programing language. This platform encompasses an Image process toolbox (IPT) that has more than one algorithm and development programs for technique, analyzing, and visualizing, collectively for growing algorithms.

Google Co laboratory: it is in any other case referred to as Colab, A loose cloud provider that may be used for constructing deep getting to know programs from scratch. Colab makes it less complicated to apply great libraries like OpenCV, Keras related TensorFlow as soon as growing an AI-primarily based totally application.

GCV (Google Cloud Vision): It could also be the vicinity of GCP (google cloud platform, enables the use of API for human action decisions prefer to image annotation, classification, object localization, perception.

TensorFlow: it's an associate open-source code package library for machine learning, created to unravel problems with constructing and together, training a neural network with the

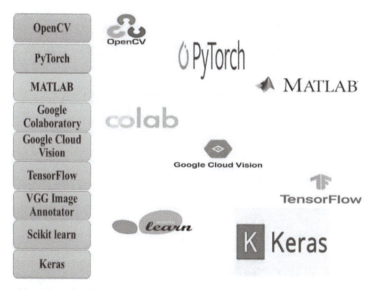

Fig. 51.1 Commonly used frameworks and libraries in recent times

Source: Author

Table 51.1   Comparison of tools

Tools	ALGORITHMS	OPEN SOURCE
OpenCV	• Boosting • Decision tree learning • Gradient boosting trees • Expectation-maximization algorithm • k-nearest neighbour algorithm • Naive Bayes classifier • Artificial neural networks • Random forest • Support vector machine (SVM) • Deep neural networks (DNN)	Yes
PyTorch	• Neural networks	Yes
MATLAB	• Recurrent Neural Networks (RNN) • Stacked Autoencoders (SAE) • Convolutional Neural Networks (CNN) • Regions with CNN (R-CNN) • Fast R-CNN / Faster R-CNN • Directed Acyclic Graph Networks (DAG) • Long Short-Term Memory Networks (LSTM) • Semantic Segmentation	No
Google Colaboratory	• Decision Trees • Boosted Trees • Random Forest • Support Vector Machines • Neural Networks	Yes
Google Cloud Vision	• Linear learner • Wide and deep • TabNet • XGBoost • Image classification • Object detection	Yes
TensorFlow	• Linear regression • Classification • Deep learning classification • Deep learning wipe and deep Booster tree regression • Boosted tree classification	Yes
VGG Image Annotator	• Object detection • supported region shapes: rectangle, circle, ellipse, polygon, point and polyline	Yes
scikit learn	• K-Nearest Neighbors • Support Vector Machines • Decision Tree Classifiers/Random Forests • Naive Bayes • Linear Discriminant Analysis • Logistic Regression	Yes
Keras	• Auto-Encoders • Convolution Neural Nets • Recurrent Neural Nets • Long Short-Term Memory Nets • Deep Boltzmann Machine (DBM) • Deep Belief Nets (DBN)	Yes

Source: Author

Table 51.2 Some of the tools and reference links

Tools	links
OpenCV	https://opencv.org/
PyTorch	http://torch.ch/
MATLAB	https://www.mathworks.com/products/matlab.html
Google Colaboratory	https://colab.research.google.com/notebooks/intro.ipynb
GCV (Google Cloud Vision)	https://cloud.google.com/vision
TensorFlow	https://www.tensorflow.org/
VIA (VGG Image Annotator)	https://www.robots.ox.ac.uk/~vgg/software/via/
scikit learn	https://scikit-learn.org/stable/
Keras	https://github.com/EderSantana/keras

Source: Author

training of automatically finding and classifying pictures.

VIA (VGG Image Annotator): VIA is put in a browser directly. The most reason for this application is Annotation. Every one of the perceived articles contained inside image, sound, and video records are clarified.

Scikit learn: it's additionally referred to as sklearn. Sklearn is an open-source, the foremost normally used package for ML. It permits the choice of efficient tools for machine learning.

Keras: It is an open-source library that enables a python interface for an artificial neural network. Keras has many implementations of neural-network building blocks, that are commonly used such as layers and also tools to work with an image. Activation functions, optimizers, tools to work with image and text data which simplifies the coding essential for writing deep neural network code.

The preceding Fig. 51.2 O'Reilly survey on machine learning tools used by the organization contains the result of a survey conducted by a notable American learning company, that publishes books, produces technical school conferences, and provides a web learning platform within the year 2019 within the month of December.

The survey relies on the AI and machine learning tools employed by the organizations. The resultant graph clearly depicts that, there is a huge need in using the tools we discussed previously and there is a need to use these tools in their organizations

The following were the most familiar languages that are used to implement machine learning Problems namely,

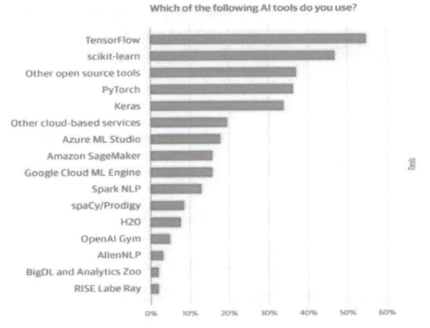

Fig. 51.2 O'Reilly survey on machine learning tools used by the organizations

Source: Author

- Python
- C++
- R

Among these programming languages, python was preferred by most of the researchers in recent times.

IV. Conclusion

With the assistance of tools, techniques, and algorithms, machines are often educated to ascertain and interpret pictures within the method needed for selected a process. Progression inside, execution of the image processing is doing good, it brought about an enormous change of chances in fluctuated fields from medications and agribusiness to retail and requirement. The success rate, power of execution relies upon a few variables, anyway choosing the right devices is one of the chief indispensable errands, it licenses you to extensively save time and assets and tracks down the best outcomes. By abuse the information of

machine learning tools to a picture interaction, one will address different assortments of issues of an machine learning drawback in a simpler, quicker, and extra conservative technique.

References

1. A. Karpathy, L, "Deep visual-semantic alignments for generating image descriptions", *Proc. IEEE Conf.Comput. Vis. Pattern Recognit.*, pp. 3128–3137, June 2015.
2. Adithya.A K, Kavya Ramesh and Hemalatha N, Machine Learning in Image Processing, International Journal of Latest Trends in Engineering and Technology, 2016.
3. Anjna, and Rajandeep Kaur, "Review of Image Segmentation Technique, "*International Journal,* volume 8, 36–39, 2017.
4. J. Chen, L. Yang et.al., "Combining fully convolutional and recurrent neural networks for 3D biomedical image segmentation", pp. 3036–3044, 2016.
5. Jiang Feng et. al., "Medical image semantic segmentation based on deep learning", Neural Computing and Applications, 1–9, 2017.

6. V. K, K. Jayashree, V. R and B. Rajendiran, "Forecasting Methods and Computational Complexity for the Sport Result Prediction," 2022 International Conference on Electronic Systems and Intelligent Computing (ICESIC), 2022, pp. 364–369, doi: 10.1109/ICESIC53714.2022.9783514.

7. Kannan, E., Ravikumar, S., Anitha, A. et al. Analyzing uncertainty in cardiotocogram data for the prediction of fetal risks based on machine learning techniques using rough set. J Ambient Intell Human Comput (2021). https://doi.org/10.1007/s12652-020-02803-4.

8. Kemal Polat, et.al., "Histo-gram-based automatic segmentation of images." *Neural Computing and Applications,* Volume 27, pp. 1445–1450, 2016.

9. Ker, Wang et. al., "Deep Learning Applications Medical Image Analysis," in IEEE Access,volume 6, pp. 9375–9389, 2018.

10. Olivier l'ezoray, christophe charrier, hubert, a survey on tools used for machine learning, International journal of engineering applied sciences and technology, 2020

11. S. Sreesubha, R. Babu, R. Vijayakumar, and K. Vijay, "An efficient data hiding approach on digital color image," Adv. Parallel Comput., vol. 37, pp. 552–557, 2020.

12. Prospects of deep learning for medical imaging, Precis Future Med. 2018; 37–52.

13. Qayyum Adnan et.al., "Medical image retrieval using deep convo- lutional neural network.", *Neurocomputing,* Volume 266, pp. 8–20, 2017.

14. Ravikumar, S. and Kannan, E., 2021. Machine Learning Techniques for Identifying Fetal Risk During Pregnancy. International Journal of Image and Graphics, p. 2250045.

15. Kirubasri, G. (2020). A Machine Learning Model for Improved Prediction of Alzheimer's Progression, International Journal of Advanced Science and Technology, 29(6), 4204–4215.

16. Shital Patil and Surendra Bhosale, Machine learning applications in medical image analysis, 2019.

17. Solomon, Chris, and Toby Breckon. Fundamentals of digital image processing:, 2011.

18. Srinivasa reddy et. al., "Generalised rough intuitionistic fuzzy c-means for magnetic resonance brain image segmentation.", *IET,* Volume 11, 777–785, 2017.

19. Wang S, Summers RM. Machine learning and radiology. Med image Anal 2012; 16: 933–51.

20. Wernick et. al., Machine learning in medical imaging, IEEE, 2010.

Recent Trends in Computational Intelligence and its Application – Sugumaran D. et al. (eds)
© 2023 Taylor & Francis Group, London, ISBN 978-1-032-48410-5

52

Converter with Analog MPPT for a Photovoltaic Water Pumping System: High Efficiency and Low Cost

K. Prabu[1], S. Irudayaraj[2], P. Chandrasekar[3]

Dept. of Electrical and Electronics Engineering,
Vel Tech Rangarajan Dr. Sakunthala R&D Institute of Science and Technology,
#42, Avadi Veltech Raod, Avadi, Chennai, Tamilnadu, India

Abstract—Renewable electricity sources are sporadic; this system takes advantage of photovoltaic technologies to carry out continuous irrigation without requiring back control. We're using the photovoltaic system for complete power point monitoring here. Directly coupled 3-phase induction motor to drive the photovoltaic mechanism. Most present engines end in a smoother display because of maintenance-free induction motors. Perturbed & Observe algorithm The company is considering an economically feasible and market-friendly approach. With a Three-motor Voltage Source Inverter and Reverberating Inductor Boost Converter combination, the percentage of efficiency at rated power is 97%; estimated lifetime costs for the proposed device are likely to be heavy, given the use of electrolytic capacitors.

Keywords—Photovoltaic solar energy, Perturb & Observe algorithm, Boost converter, Maximum Power Point Tracking (MPPT)

1. Introduction

The application of modern and improved irrigation methods for agriculture enabled the implementation of such a new and highly effective solar technology. Currently, over 800 million people don't have water to drink [1-3]. Of the number, much of it is found in locations where only rainwater is used as a water source. Due to power shortages, traditional water treatment methods are unusable in this location. One way of solving this will be using photovoltaic (P.V.) pumping and treatment systems, which have been used for decades [4-5]. However, until recently, most commercial water pumps were powered by electric motors or batteries. The batteries provide the device with a steady power supply despite no sunlight. This will allow for the electrical connection of the pump motor to the solar panel. In this device, the battery usually has a short life cycle of two years, and the P.V.

[1]prabuk@veltech.edu.in, [2]drirudayaraj@veltech.edu.in, [3]drchandrasekar@veltech.edu.in

DOI: 10.1201/9781003388913-52

module has a 15-year usable life span [6-8]. Further, in addition, installing and maintaining these devices can be costly. Battery loss also leads to complete device failure. This D.C. motor usually has no boost stage between the P.V. module and the motor. Unfortunately, D.C. motors are not appropriate for the project [9]. This paper recommends using a three-phase induction motor for applications like this due to its greater strength, reduced maintenance, and lower cost [10]. Designing a direct-flow motor drive system that uses photovoltaic power supplies involves innovative ways to contend with supply constraints and produces as much electricity as a good as the standard solar panels for 15 years is a challenge. This paper devises a simple means of harvesting solar energy and filtering it [11].

A. Scope of the Papers

This paper's primary goal is to provide clean and renewable energy in rural areas

The solar water pumping system is a renewable energy source that requires little maintenance.

II. Block Diagram

A proposed method for a new overall device includes three SEPIC converters, D.C. Rectifier, SPWM Rectifier, and a pulse generator taken from ultra-low power control and coupled at 2.5 kHz from the input. In this case, the SEPIC converter extracts power from the solar information. It feeds it into a galvanic current converter, while a direct current supply provides the load power requirement through an inverter and a boost converter [12-13]. The Perturb and Observe method was used to monitor a P.V. cell from an MPPT. The SEPIC converter and DC-DC converter receive D.C. input. Unwanted signals and harmonics are removed utilizing a snubber circuit. The P.I. Controller will produce the triggering pulses. To transform from D.C. to A.C., a three-phase inverter is used. The term "resistive load" refers to using a resistive load.

A. Photovoltaic Cell Operating Principle

This is the base of all photovoltaic panels. Most commonly, silicon is used. They use the photoelectric effect. The solar cell's current is created by separating the charged particles through the design below. A p-n junction is a silicon doped with a minimal amount of anion of donor atoms; in the player's case, one has more electrons (valence), and one has fewer (P.N.). When the two layers are joined, the n-side and the donors are neutralized at the interface; free electrons can diffuse through the p-side. Similarly, the free sites on the n-side leave behind an area adversely paid by the acceptors. This is an obstacle to flowing further. The electron and hole equilibrium is balanced as the potential barrier is crossed. Holes will flow in one direction while electrons flow in the other. A representation of the described electric field, as seen in Fig. 52.2.

To absorb the electrons and the holes, metallic contacts are positioned on both sides. In addition, metallic connections must be provided for the n-layer, which is exposed to sunlight.

B. Equivalent Solar Cell Circuit

The solar cell is seen in Fig. 52.2 It currently has the following voltage:

$$I = I_L - I_0 \left\{ \exp\left[\frac{q(V + IR_S)}{nkT} \right] - 1 \right\} - \frac{V + IR_S}{R_{SH}}$$

For solar cells, I is the current (Luminescent), V is the voltage (Dark Saturation), and I0 is the dark current; where $R_{.S.}$ is the diode efficiency factor, k is the Boltzmann constant, and T is the absolute temperature. $R_{.S.}$ is the resistance in the contacts and the bulk of the solar cell. The resistance's source is impossible to pinpoint. The accumulation of impurities on the surface of the p-n junction does this. When R_{SH} is infinite, $R_{.S.} = 0$. However, manufacturers strive to remove the resistances to boost their materials.

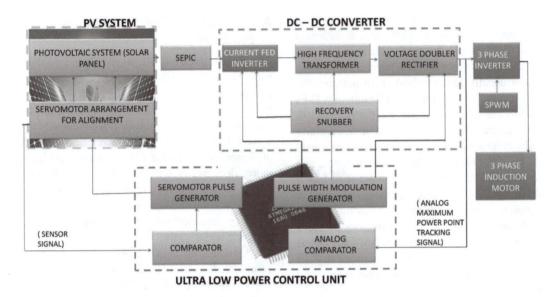

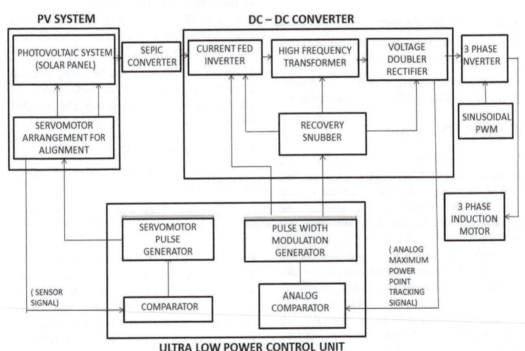

Fig. 52.1 System block diagram, the proposed method

Source: Author

In other cases, the resistance is infinite, so the last term is usually ignored when building the model.

Since each panel has several solar cells attached in series and parallel, the output voltage and current are both high enough for the grid or the

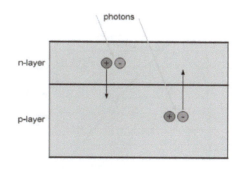

Fig. 52.2 Sun cell

Source: Author

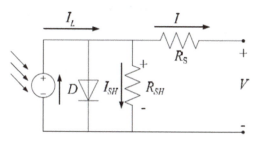

Fig. 52.3 Equivalent circuit of a P.V. cell

Source: Author

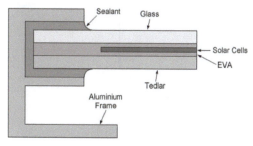

Fig. 52.4 P.V. module typical construction

Source: Author

relevant devices. Because of the simplification described above, the P.V. output voltage is given by the equation, where np and ns are the number of solar cells in parallel and series.

III. Algorithms for Maximum Power Point Tracking

An illustrated, Maximum Power point Tracking algorithms are required in **photovoltaic** applications because of the irradiance and temperature. A variety of approaches to locate

the MPP have been used for decades. This process differs significantly in many ways, such as sensor requirements, complexity, speed of integration, accuracy when irradiation or temperature changes, and popularization. This can be done in about 19 different ways. Among these techniques, P&O and In Cond are the most common. This technique offers a lot of benefits, but also some disadvantages, as you can see. Additional methods include fuzzy logic control, neural network, fractional open voltage, short circuit, or something else. Most methods yield a maximal local voltage and a fractional value, which means it is an approximation. It isn't easy to establish a single limit for the V-P curve. However, under partial sunlight, the P.V. system has multiple optima. To handle this problem, some algorithms have been created. Next are the most common methodologies

A. Perturb and Watch

Based on how it is used, the P&O algorithm is also known as "hilly-climbing." altering the power converter's service cycle and lowering the D.C. connection's operating voltage between the P.V. array and the power converter, Either altering the D.C. contact voltage or the P.V. duty cycle will change the power converter's duty cycle. The next perturbation is identified using the perturbation symbol and the most recent power sign rise. The Fig. 52.5 demonstrates that increasing the voltage causes the current to grow while reducing it causes it to go down.

When the power increases, the perturbation should decrease; when the power decreases, the partition should increase. The details corroborate the algorithm. Before the MPPT is found Oscillates around the MPP also said, this is an issue Fig. 52.6 of the algorithm is shown here

B. Incremental Conductance

As in Fig. 52.5, the curve slopes up (down) to the MPP (powers) and to the left (as shown) and to the right (impeded ate), the P.V. current flows

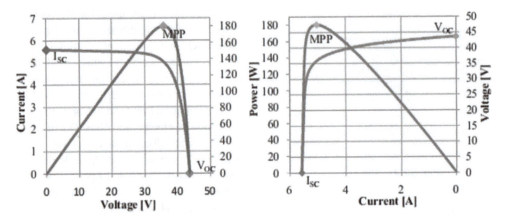

Fig. 52.5 PV panel characteristic curves

Source: Author

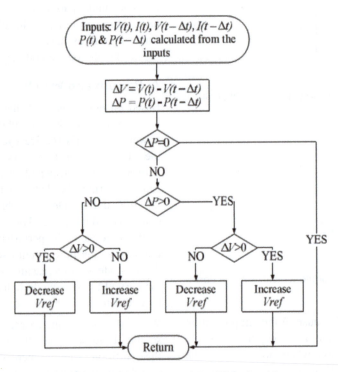

Fig. 52.6 The flowchart of the P&O algorithm

Source: Author

in the opposite direction (to the impediment), as is shown. Increasing the reference voltage quickly determines how quickly the MPP is achieved.

Many of these strategies have pitfalls. The primary issue is that they are unable to keep track of the MPP Changes in phase can be tracked accurately since the MPP remains constant. However, as the slope of the signal varies, the algorithm will be dependent on the voltage perturbation only since it is not possible for the algorithms to decide if the power transition is a voltage increment or a power surge. Additionally, in the steady state, the voltage and current oscillate about the MPP. This is because the control signal is discrete, and the current and voltage aren't continuously fluctuating around the MPP. The amplitude of the voltage fluctuations depends on the magnitude of the applied voltage. The more intense, the more violent. However, the MPP's speed is equal to this change and is inversely proportional to the voltage change in magnitude. When the amplitude of the oscillations is minimal, a compromise is necessary.

For some alternatives, recently, workarounds have been proposed. Regarding the rapid voltage and current transition, P&O's "Sera" released an improved approach called "dP-P&O." By using this approach, the influence of the perturbation (changing the voltage) and the MPPT algorithm itself can be measured in three consecutive cycles and the total effect (the difference in voltage times) can be estimated. In relationship to the MPP, Zhang et al. suggested a constant perturbation for the P and O algorithm. In this thesis, there are so many approaches

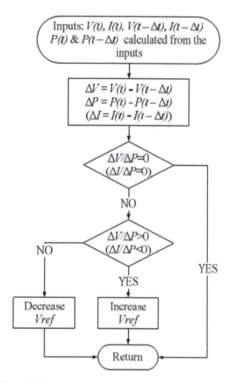

Fig. 52.7 Incremental conductance algorithm

Source: Author

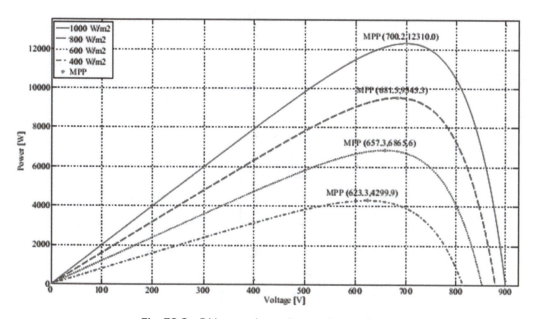

Fig. 52.8 P-V curve depending on the irradiation

Source: Author

available that it is almost impossible to cover any of including them. Notwithstanding these facts, in everyone's efforts, no viable alternative has been found, and the proposed MPPT methods have not been put to the test. These cloud profilers model environmental changes. Maximizing MPP in these instances requires monitoring the MPP.

C. Operation of SEPIC Converter

The single-ended primary-inductor converter (SEPIC) enables a voltage to be greater than, equal to, or lower than its D.C. voltage supply. The SEP is regulated by the service cycle of the turn. The method uses two diodes and two capacitors and is thus a fourth-order nonlinear system.

D. Basic Three-phase Voltage Source Inverter

The dc to ac converters, also known as VSI, depend on the form of the source and topology of the power circuit (CSIs). The single-phase

inverters and switching patterns were covered in chapter two, so you can read about three-phase inverters here.

$$S11 + S12 = 1 \tag{1}$$
$$S21 + S22 = 1 \tag{2}$$
$$S31 + S32 = 1 \tag{3}$$

Eight switching states mentioned in Table 52.1 produce no output voltage. In this case, the ac current flows up or down the armature windings. In the remaining states, zero out to produce a given voltage waveform, and the device goes through different states. Thus, the resulting ac voltages being isolated values are -V, -0, and 0 V

IV. Simulation and Result

Simulation has been extremely useful in the workplace and academia, too. It is now a must for an electrical engineer to have a fundamental understanding of simulation and recognize its implementation in different they go hand in hand. They complement each other perfectly:

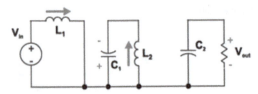

Fig. 52.9 SEPIC converter when the switch is turned on

Source: Author

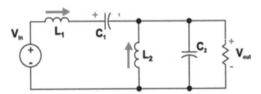

Fig. 52.10 SEPIC converter when the switch is turned off

Source: Author

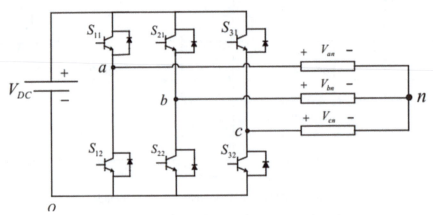

Fig. 52.11 Basic three-phase voltage source inverter

Source: Author

Table 52.1 The switching states in a three-phase inverter

S_{11}	S_{12}	S_{31}	V_{ab}	V_{bc}	V_{ca}
0	0	0	0	0	0
0	0	1	0	$-V_{DC}$	V_{DC}
0	1	0	$-V_{DC}$	V_{DC}	0
0	1	1	$-V_{DC}$	0	$-V_{DC}$
1	0	0	V_{DC}	0	$-V_{DC}$
1	0	1	V_{DC}	$-V_{DC}$	0
1	1	0	0	V_{DC}	$-V_{DC}$
1	1	1	0	0	0

Source: Author

prototyping in the lab is a valuable research aid to computer simulation. A hardware experiment must not be used as a replacement for a computer simulation. The aim of this chapter is to show how to manage the open-loop SEP, and PID controller Laboratory is called MATLAB. The very first MATLAB, created in the early and late 70s for Matrix, Linear Algebra, and Numerical Analysis, is still being used today. With the inclusion of toolboxes, MATLAB has vastly improved.

A. Simulation Block Diagram

See Fig. 52.12

B. DC-DC Converter

See Fig. 52.13

C. DC AC Inverter

See Fig. 52.14

D. Simulation Result

See Fig. 52.15

V. Conclusion

In this process, the aim is to use as much power as possible, regardless of the irradiance and

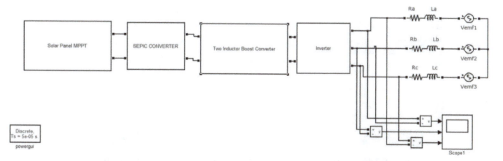

Fig. 52.12 Simulink model for photovoltaic water pumping system

Source: Author

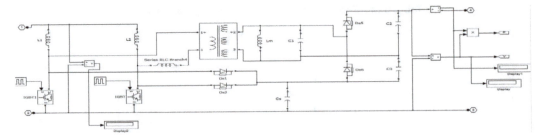

Fig. 52.13   Simulink model for two inductor boost converter

Source: Author

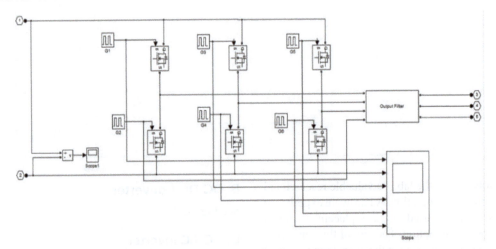

Fig. 52.14   Simulink model for three-phase inverter

Source: Author

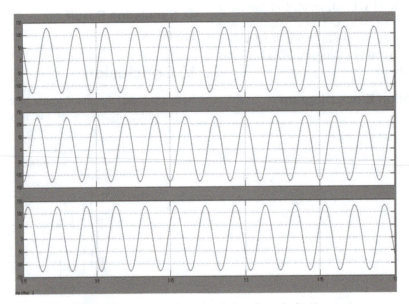

Fig. 52.15   Output waveform for three phase induction motor

Source: Author

temperature. A number of control schemes have already demonstrated problems, necessitating the development of new methods to improve performance. An MPPT controller has been obtained; it makes it easier to locate the pinpoint of power in shorter runs than the well-known P&O controller. Using the fuzzy and Neuro-fuzzy controller, we achieve a steady output voltage. Using the above techniques and simulation data, the P.V. maxim voltage-sensing model is a safer alternative. It can be estimated from the simulations that the Neuro-fuzzy model delivers greater ripple current ripple. Neuro-fuzzy voltage balancer is more suited to balancing more data and will boost the efficiency of the Neuro-fuzzy model It was shown that a non-storage photovoltaic pumping and treatment system could be used. Three-phase PV power converters are considered commercially feasible because they are low cost, high performance, and durable. It was diagrammed out, calculated, and designed. The proposed solution can prove to be a workable solution after more durability tests are conducted to ensure its strength.

References

1. Hahn, A. Technical maturity and relia ability of photovoltaic pumping systems. In: 13th European Photovoltaic Solar Energy Conference, Nice France, pp. 1783-1786. 1995
2. Vitorino, M.A.; Correa, M.B.R., "High-performance photovoltaic pumping system using induction motor," Power Electronics Conference, 2009. COBEP '09. Brazilia an, vol., no., pp. 797–804, Sept. 27 2009-Oct. 1 2009
3. Mr. Prabu. K, Mr. Vinoth John Prakash. S, Mr. Barathi. K, Mr.Gopikaramanan R, Renewable Energy Based Torque Control of Sqim Based onVoltage Angle, International Journal of Mechanical Engineering and Technology 8(8),2017, pp. 1287–1293.
4. P. J. Wolfs, "A Current-Sourced Dc-Dc Converter Derived via the Duality Principle from the Half-Bridge Converter," IEEE Trans. Industrial Electronics, Vol. 40, No. 1, pp. 139–144, Feb. 1993.
5. Meikandan, M; Karthick, M; Natrayan, L; Patil, Pravin P; Sekar, S; Rao, Y Sesha; Bayu, Melkamu Beyene; Experimental Investigation on Tribological Behaviour of Various Processes of Anodized Coated Piston for Engine Application, Journal of Nanomaterials, 2022, Hindawi

Recent Trends in Computational Intelligence and its Application – Sugumaran D. et al. (eds)
© 2023 Taylor & Francis Group, London, ISBN 978-1-032-48410-5

53

Prediction of Pediatric Cardiomyopathy Genes Associations Using N2V- LSTM

K Jayanthi*, C. Mahesh[1], D Sugumaran[2]
Department of Information Technology,
Vel Tech Rangarajan Dr. Sagunthala R&D Institute of Science and Technology, India

S. Sithsabesan[3]
Department of Mathematics,
Vel Tech Multitech Dr. Rangarajan Dr. Sakunthala Engineering College, India

Abstract—For the condition to be caught early, it is essential to find the genes linked to pediatric cardiomyopathy. In recent years, machine learning approaches have been applied to predict disease-related genes in various ways. It is necessary to increase the effectiveness of current techniques for predicting disease genes. Furthermore, the bulk of machine learning-based illness gene prediction algorithms are unable to identify indirect correlations between gene features. While understanding the relationship between the genes can improve the diagnosis of disease at the earliest. We propose a deep learning framework in this paper, named N2V- LSTM (Node2Vec – Long Short-Term Memory) to predict genes associated with pediatric cardiomyopathy disease. The Microarray gene data was converted to a graph and information related to genes was obtained using the Node2Vec method. The Embedded Vector was generated with learned features for each gene trained with LSTM. N2V-LSTM framework outperforms existing methods significantly.

Keywords—Graph embedding, Node2vec, LSTM, Disease gene prediction, Random walk, Pediatric cardiomyopathy disease

I. Introduction

In that they both include a chain of proteins known as bases, DNA and RNA have similar features. Adenine, Cytosine, Guanine, and Thymine are the bases, and its respective first letters are abbreviated as A, C, G, and T. Both DNA and RNA have four nucleotides in common. Uracil has been used to treat thymine RA. The genetic building blocks that are passed down from one generation to the next are called genes. DNA contains information about protein synthesis in the form of genes. Protein serves as a cellular building block. A genetic disease occurs when one of the foregoing pathways fails. Changes in cell DNA content, on the other

---
*Corresponding Author: jayanthi2contact@gmail.com
[1]chimahesh@gmail.com, [2]dsugumaran@veltech.edu.in4, [3]sithsabesan@veltechmultitech.org

DOI: 10.1201/9781003388913-53

hand, will alter genes. Inconsistencies in protein production are caused by changes in gene mutation. Because the irregular protein never acts correctly, a hereditary disease develops. It is difficult to forecast disease using genes. To identify disease from genes, researchers have presented algorithms [1], [2], and [3. Traditional classification algorithms place a strong emphasis on annotation, incorporate structures involved in biological processes or molecular activities, and then compare those annotations to those of known disease genes.

The approaches based on annotated data [3] are limited and never capture indirect gene connection. The shared behaviour or function of genes is never utilised to detect diseases. The ontology-based disease gene similarity network method was used to prioritise genes [4] and it found the association between cellular, molecular, protein, and microarray data when compared to annotation-based genes [3, 4]. This study suggests using genetic testing to diagnose pediatric cardiomyopathy. Pediatric cardiomyopathy is a cardiac muscle condition that affects children. Cardiomyopathy fatalities and impairments have increased dramatically in modern society as a result of genetic illnesses. In the pediatric cardiomyopathy registry, cardiomyopathy is diagnosed in 1.1 to out of every 100,000 children under the age of 18. Approximately half of disease cardiomyopathies patients died abruptly in childhood or during heart transplantation. Children with cardiomyopathies account for roughly 17.5 million deaths, or 31% of all deaths worldwide. By 2030, the death toll may reach 23.6 million. 1.7 to 2.0 million Indians suffer from heart disease. In Kerala, around (187 - >350 men/100,000 men/year) suffer from heart disease. South India is responsible for one-fifth of all causation in India. The Indian Medical Board predicts that the number of people suffering from cardiomyopathies will reach 17.9 million by 2030. According to the Global Society of Heart and Lung Transplant in Prague, 4% of patients experienced infections

following their transplant, while roughly 25% of patients need heart transplants due to cardiomyopathies (2012). No signs or symptoms of cardiomyopathies are seen in the affected individuals. The indications and symptoms of cardiomyopathies treatment fluctuate with age. Cardiomyopathy in children must be accurately diagnosed, which is very important. It is, however, a difficult assignment for researchers. A microarray is a method for determining whether a specific individual's DNA includes a gene mutation. A specific gene's DNA may contribute to a disease. Microarray technology is used in a number of research to simultaneously measure the expression levels of thousands of genes. Microarray datasets have low statistical power to detect GES because to the small sample size (Gene Expression Signature).

When microarray datasets from multiple studies are combined, the sample size increases, making the results more credible. However, the number of features in a microarray experiment is substantially greater than the number of samples, and supervised Machine Learning algorithms are insufficient when applied to high- dimensional datasets. Reduce the number of features before utilising machine learning methods to overcome the problem of too many dimensions. Supervised machine learning methods were utilised to identify genes that are expressed in diverse ways. Using these methods, a prediction model that can be used to categorise fresh samples can be produced. The t-test and other statistics tests are often used. These methods choose which gene features to use based on arbitrary thresholds, and there is no consensus on the appropriate values. These methods struggle to find non-linear structures within datasets and restore the original structure when the datasets are highly skewed in a high-dimensional environment. In this stage, we will learn low-dimensional features with Node2Vec. According to research, most diseases are caused by more than one gene. This suggests that most diseases are caused by more than one gene. According to gene-based research publications, 1269 genes have been associated to cardiomyopathy disease. Finding the genes associated with Pediatric Cardiomyopathy will allow us to learn more about the disease, find out how it operates at the molecular level, and detect it early. So, devising a method to determine which genes are associated to pediatric cardiomyopathy is a fantastic idea.

This research presents a deep learning model named N2V- LSTM (Node2Vec - Long Short-Term Memory) to predict the genes associated with the illness juvenile cardiomyopathy. Our work has contributed the following things:

- N2V – LSTM can capture microarray disease information from a gene using the Node2Vec method.
- Embedded Vector generated with learned features for each gene trained with LSTM.
- N2V – LSTM uses a proposed deep learning framework model to assign each gene a low-dimensional representation. N2V- LSTM outperforms existing methods significantly.

II. Related Works

According to (Cáceres and Paccanaro) [5, the Network approach to prioritising gene-disease connections can anticipate genes for conditions without a recognised molecular cause.They used a measure of how similar the symptoms of various diseases are to one another as well as semi-supervised learning. The Source of Asian PDIs platform, created by Shivakumar and Keerthikumar [6] include molecular details. These include microarray analyses of all known PID genes, protein-protein interaction networks, and mice studies. Using this material, create an algorithm that can find important PID genes. Utilizing the support vector machine learning method, they discovered 1442 candidate PID genes. 148 known PID genes and 3162 non-PID genes were used as training data, totaling

69 binary characteristics. Recent research has revealed that six of the predicted genes are PID genes, demonstrating the utility of this technique. From a large collection of candidate genes, (Rosario M. Piro and Ferdinando Di Cunto) [7] used a binary classification algorithm to identify illness genes. Many times, hundreds of positional candidates are selected using the association analysis and association research methods. Shannon Kelly and Weidong Mao [8] Proposed approach for discovering Parkinson's disease genes discovered utilising an N-sembles method (N-semble). Filters, wrappers, and embedding techniques for gene prediction feature selection were investigated in [9] by E. Neelima and M.S. Prasad. These strategies could make a considerable difference in performance quality and outcome. Muhammad Asif and others [10] Machine learning classifiers trained on gene functional similarity may help researchers identify genes implicated in complicated disorders more quickly. A quantitative estimate of gene functional similarity can be obtained by integrating various semantic similarity measurements. To discover previously unknown functional commonalities between genes. (ParkChihyun and others) [11] Using large-scale gene expression and DNA methylation data, a deep learning algorithm was able to predict illnesses. Conventional machine learning methods are outperformed by a prediction model based on deep neural networks. In Vietnamese, this would be "Duc-Hau Le." [3] Classification methods based on networks were discussed. In a binary classification system using positive and negative training samples, disease and non-disease genes are employed as training samples. Sikadar and associates (et. al.) As an illustration, [12] provided the support for prior knowledge to identify gene-disease associations' basic network properties of topological features, as well as gene sequence information or biological features that significantly restrict the process of choosing a single gene-disease link.

III. Methodology

Three steps are involved in N2V - LSTM. The vector representation of each gene is extracted from the biological process, molecular function, and protein interactions in the first stage using Node2Vec. LSTM is used to lower the dimension of the produced vector in the second stage. Finally, we compared the results of our predictions to the existing machine learning technique. The architecture of the proposed deep learning framework is depicted in Fig. 53.4.

A. Vector Representation

Molecular information is supplied by Shivakumar and Keerthikumar [6]. These consist of mouse studies, protein-protein interaction networks, and microarray investigations of all known PID genes. Make an algorithm that can locate significant PID genes using the information provided. Using a method called support vector machine learning, they discovered 1442 candidate PID genes. They employed 3162 non-PID genes and 69 binary characteristics from 148 known PID genes as training data. Six of the anticipated genes actually PID genes, according to recent research, proving the technique's usefulness. $\pi_{vx} = \alpha_{pq}(t, x) \cdot w_{vx}$ $d_{tx} = 2$ where d_{tx} is the shortest distance between nodes t and x and must be either 0 or 1. During the random walk, the variable p decides whether a node can be visited again. When p is large, previously visited nodes are hardly sampled. In 2-hop sampling, this method allows for considerable exploration while avoiding repetition. If p is small, the walk will backtrack one step (Fig. 53.1), keeping the walk "local" to the starting node u. The q option instructs the search to distinguish between "local" and "global" nodes. Fig. 53.1 shows that if q is greater than one, the random walk has a larger likelihood of selecting nodes surrounding node v. These walks can provide a more detailed representation of the underlying graph. BFS gathers nodes from a limited geographic area. If q is bigger than one, the

random walk will move further from v, allowing it to collect more global feature information. As a result, the distance between the given source node u and the sample node is not strictly increased. The measurement, on the other hand, benefits from the pre-processing and the higher sampling efficiency of the random walk. In this study, we use the node2vec method to create 512-dimensional vector representations of each gene in the PPI network.

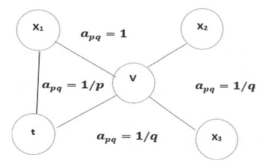

Fig. 53.1 Illustration diagram of Node2Vec model

Source: Author

B. Reducing Dimension Using LSTM

A deep RNN solution that uses LSTM units is the LSTM [13] network. As was previously said, an RNN is a deep learning network that contains intrinsic neuronal feedback. These internal feedbacks support the retention and assimilation of important past events. RNN, as opposed to a fully connected feedforward network, shares parameters across all model components, allowing it to generalise to sequence lengths that weren't encountered during training. The RNN architecture shown in Fig. 53.2 includes recurrent connections between Sustainability hidden neurons and generates an output [14] at each time step.

The Simple RNN is shown in Fig. 53.2. Weight matrices U and W connect the input to the hidden, while weight matrices V connect the hidden to the output. Forward propagation defines the hidden unit j_0 starting state. The following update equations are then applied to

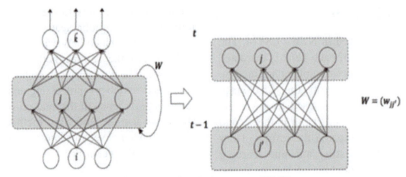

Fig. 53.2 Block diagram of RNN

Source: Author

each time step starting with j_0. At time t, the hidden neuron h input value is:

$$h^t = \sum_j i\, u_{ji}\, x^t + \sum_i j'\, w_{jj}\, z^{t-1} \qquad (2)$$

where u_{ji} weight of input i over hidden j and x_t is the at time t, input value

The concealed neuron's transfer function is denoted by f, and its output is denoted by

$$z_j^t = f(h^t) \qquad (3)$$

Finally, the hidden layer z output value is fed into output neuron k, and the output layer output value is given.

$$s_k^t = \sum_j v_j z^t \qquad (4)$$

where v_j is the weighted average of the hidden and output neurons

The acquisition of complex time-dependencies that span more than a few time steps is difficult for RNN, though. The information from previous occurrences significantly vanishes as the number of time steps to be considered increases. The use of LSTM is made to solve the problem of long-term dependence. In LSTM networks, data from over a thousand prior time steps can be stored. [15] LSTM can scale to far longer sequences than ordinary RNN, which has limitations like vanishing and exploding gradients. The use of LSTM in numerous sequential modelling applications, including as speech recognition, motion detection, and natural language processing, has increased dramatically during the past few years. A block diagram for the LSTM is shown in Fig. 53.3. The following describes the

computation process inside an LSTM block. The input value cannot be stored in the cell's state unless the input gate permits it.

Its input value and potential memory cell value, C_t, are calculated at each time step, t, as follows:

$$i_t = \sigma(W_i x_t + U_i h_{t-1} + b_i)$$
$$\tilde{C}t = \tanh (W_c x_t + U_c h_{t-1} + b_c) \qquad (5)$$

where W, U, b represents the weight matrices and bias, respectively.

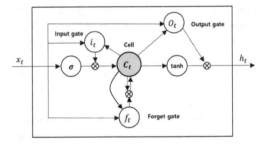

Fig. 53.3 LSTM cell unit gates

Source: Author

Figure 53.3 depicts the LSTM gate. A memory cell, three multiplicative gating units, an input, an output, and a forget gate are all part of the LSTM block. The cells are related by recurrent connections, and each gate provides continuous operations to the cells. Each gate handles the cells' write, read, and reset operations, while the cell is in charge of sending state values over arbitrary time intervals. The forget gate manages the weight of the state unit, and its value is calculated as

$$f_t = \sigma(W_f x_t + U_f h_{t-1} + b_f) \qquad (6)$$

The new state of the memory cell is updated as a result of this process.

$$C_t = i_t \times Ct + f_t \times C_{t-1} \qquad (7)$$

Using the memory cell's new state, the following formula is used to determine the gate's output value:

$$o_t = \sigma(W_o x_t + U_0 h_{t-1} + V_o C_t + b_o) \qquad (8)$$

The value of cell's task depends is described as

$$h_t = o_t \times \tanh(C_t) \qquad (9)$$

All gates use sigmoid nonlinearity, and the state unit can act as an additional input to other gating units. The LSTM architecture can address the problem of long-term dependencies with this method at a low processing cost.

IV. Results and Discussion

We compare the performance of four N2V approaches on predicting genes associated with pediatric cardiomyopathy disease in this section: LSTM, RWR, and Shortest Path. RWR is a popular method for predicting disease genes in networks. Path-based disease gene prediction employs the ED and SPL techniques. We also investigate the effect ofthe N2V- LSTM algorithm's various steps and parameters on its performance. Finally, we use N2V- LSTM to find new genes associated with pediatric cardiomyopathy. The findings show that some of the genes identified by the N2V- LSTM algorithm are supported by existing methods.

In order to demonstrate the model's durability, the suggested method is compared to different methodologies: N2V-LSTM. The accuracy of the N2V-LSTM is 93.16 and 93.13 percent, respectively. The six separate classes' classification accuracy is attained. The performance analysis models are trained to categories literature into up to three groups. The proposed model for binary classification has an accuracy of 88.99 percent.

V. Conclusion

The dataset comprises 66,153 inputs with a training, validation, and testing ratio of 70%, 10%, and 20%, respectively. There are 46307 samples in the training set, 6615 samples in the validation set, and 13231 samples in the testing set. A vocabulary can have up to 8972 characters, whereas a sequence can have up to 2000 characters. The binary cross-entropy function is employed as the loss function during the training phase. The difference between the actual output and the objective label is calculated using this loss function, which is used to train and update the weights. By adjusting the settings of many hyperparameters, including filters, filter size, layer count, and embedding dimension, we put the LSTM to the test. The number of filters in each layer is the ideal value in all three models (128, 64, and 32). The filter contains 32 embedding dimensions, 100 training batches, 10 training epochs, and a per size of 6.

Table 53.1 Performance metrics of N2V – LSTM

Method	Metrics	Samples											
		1	2	3	4	5	6	7	8	9	10	11	12
N2V- LSTM	Accuracy	0.93	0.99	0.92	0.99	1	0.99	0.93	0.99	0.93	0.99	0.99	0.99
	Precision	0.89	0.98	0.82	0.99	1	0.99	0.89	0.94	0.93	0.98	0.99	0.99
	F1	0.94	0.95	0.49	0.99	1	0.99	0.94	0.95	0.48	0.99	0.99	0.99
	Sensitivity	0.98	0.93	0.35	0.99	1	0.99	0.99	0.96	0.32	0.99	0.99	0.99
	Specificity	0.85	0.99	0.99	0.99	1	0.99	0.84	0.99	0.99	0.99	0.99	0.99

Source: Author

References

1. A. Vasighizaker, A. Sharma, and A. Dehzangi, "A novel one-class classification approach to accurately predict disease-gene association in acute myeloid leukemia cancer," PLoS One, vol. 14, no. 12, pp. 1–12, 2019, doi: 10.1371/journal.pone.0226115.

2. P. Luo, Y. Li, L. P. Tian, and F. X. Wu, "Enhancing the prediction of disease-gene associations with multimodal deep learning," Bioinformatics, vol. 35, no. 19, pp. 3735–3742, 2019, doi: 10.1093/bioinformatics/btz155.

3. D. H. Le, "Machine learning-based approaches for disease gene prediction," Brief. Funct. Genomics, vol. 19, no. 5–6, pp. 1–14, 2020, doi: 10.1093/bfgp/elaa013.

4. D.-H. Le and V.-T. Dang, "Ontology-based disease similarity network for disease gene prediction," Vietnam J. Comput. Sci., vol. 3, no. 3, pp. 197–205, 2016, doi: 10.1007/s40595-016-0063-3.

5. J. J. Cáceres and A. Paccanaro, "Disease gene prediction for molecularly uncharacterized diseases," PLoS Comput. Biol., vol. 15, no. 7, pp. 1–14, 2019, doi: 10.1371/journal.pcbi.1007078.

6. S. Keerthikumar et al., "Prediction of candidate primary immunodeficiency disease genes using a support vector machine learning approach," DNA Res., vol. 16, no. 6, pp. 345–351, 2009, doi: 10.1093/dnares/dsp019.

7. R. M. Piro and F. Di Cunto, "Computational approaches to disease-gene prediction: Rationale, classification and successes," FEBS J., vol. 279, no. 5, pp. 678–696, 2012, doi: 10.1111/j.1742-4658.2012.08471.x.

8. M. Weidong and S. Kelly, "An optimum random forest model for prediction of genetic susceptibility to complex diseases," Lect. Notes Comput. Sci. (including Subser. Lect. Notes Artif. Intell. Lect. Notes Bioinformatics), vol. 4426 LNAI, pp. 193–204, 2007, doi: 10.1007/978-3-540-71701-0_21.

9. E. Neelima and M. S. Prasad Babu, "A comparative Study of Machine Learning Classifiers over Gene expressions towards Cardio Vascular Diseases Prediction," Int. J. Comput. Intell. Res., vol. 13, no. 3, pp. 403–424, 2017, [Online]. Available: http://www.ripublication.com.

10. M. Asif, H. F. M. C. M. Martiniano, A. M. Vicente, and F. M. Couto, "Identifying disease genes using machine learning and gene functional simse Gene Association Using Machine Learning," IEEE Access, vol. 8, pp. 160616–160626, 2020, doi: 10.1109/ACCESS.2020.3020592.

Recent Trends in Computational Intelligence and its Application – Sugumaran D. et al. (eds)
© 2023 Taylor & Francis Group, London, ISBN 978-1-032-48410-5

54

Video Surveillance for Intruder Detection and Tracking

Vijayakumar R.[1], Vijay K.[2], S. Muruganandam[3]
Assistant Professor, Department of Computer Science and Engineering,
Rajalakshmi Engineering College, Chennai, India

S. Ravikumar[4]
Department of CSE, Vel Tech Rangarajan Dr. Sagunthala R&D
Institute of Science and Technology, Chennai, India

Kirubasri G.[5]
Assistant Professor, Department of Computer Science and Engineering,
Sona College of Technology, Salem, India

Abstract—In order to monitor security-sensitive locations, such as retail centres, educational facilities, highways, congested public spaces, and borders, video surveillance systems have been used. As the amount of crime, theft, and robbery is increasing every day, it is necessary to provide proper security in the above-mentioned places using CCTV. Intruder detection, categorization, tracking, and analysis algorithms need to be effective and quick when creating video surveillance systems. There are methods implemented for this purpose ranging from statistical coherence based on Bayesian rule, background subtraction, temporal frame differencing for object detection, Point tracking, Silhouette tracking, Kernel tracking for object tracking, but the problem of occlusion during detection and tracking is still a problem. The proposed system helps to remove this problem during intruder detection and tracking using background subtraction with trimmed median filtering algorithm with edge histogram method. In this paper we used different methods such as Voila Jones and Background Subtraction Algorithms, Linear Discrimination Analysis (LDA), Trimmed Median Filtering with Edge Detection, Kalman Filter and Edge Histogram.

Keywords—Intruder detection, Tracking, Occlusion, CCTV, Background Subtraction, Trimmed Median Filtering, Linear Discrimination Algorithm, Viola Jones, Security, Edge Histogram

I. Introduction

Artificial intelligence is done through machines, which is also called as "machine intelligence", which uses natural language described by humans. Video surveillance is very interesting area in artificial intelligence mainly for security purposes. CCTV (Closed-

[1]r.vijayakumar91@gmail.com, [2]vijayk.btech@gmail.com, [3]murugan4004@gmail.com, [4]ravikumars@veltech.edu.in, [5]kirubasri.cse@sonatech.ac.in

DOI: 10.1201/9781003388913-54

Circuit Television) is used for security in public and crowded areas. There are three main steps for this purpose starting from object detection, object classification to object tracking and analysis. Detection of moving intruders is a difficult task for any video surveillance system. Classification of objects involves classifying based on shape and motion features [13, 17]. Object tracking is involves tracking the moving objects by comparing multiple frames and tracks the path through which an objects move. The main objective is to monitor, detect and track intruders to identify theft or any suspicious activity in security sensitive and crowded areas like banks, shopping malls, colleges using CCTV [6] [7]. Intruder detection, classification, tracking, and activity analysis are used to construct smart video surveillance systems that accomplish this. The main development in this paper is that the intruder detection and tracking is done without overlapping that is occlusion. The working is done by first inputting a video sample recorded in the CCTV initially the intruders are detected and classified using Pre-processing steps and converted into hundreds of frames of images [8] [9]. For a one second video it gets separated into at least of 100 frames. From which required frame is selected and the intruders face is detected using Voila John's Algorithm, Linear discrimination is used to identify the faces with samples [10] [11]. The necessary face samples are cropped and collected into a folder giving a unique identifier for each face. The intruders are then tracked by Kalman Filter and Edge Histogram. The result is enhanced by using Trimmed Median Filter with Edge Detection [12]. The output of this gives whether the suspected intruder is present in that frame and where the intruder was found.

II. Related Works

They have demonstrated in the suggested study [1] that moving object recognition is handled by the use of an adaptive background-subtraction technique that is effective in both indoor and outdoor environments. For comparison of performance and quality measures, they discussed two more effective object detection techniques, temporal differencing and adaptive background mixing models. In this study, the adaptive background method was used to initialise a reference background with the system's first few video frames and update it to adapt to both short-term and long-term dynamic scene changes. This paper [2] showed an experimental result on the efficiency and accuracy of Viola-Jones Algorithm under three background conditions – black, white, and noisy background. And proved that Voila-Jones Algorithm works well in these backgrounds with reasonable efficiency. The three methods for object detection suggested in the paper [2] are Model-Based System, Image Invariance Method, and Example-based Learning Method. Moving intruders are detected using Temporal differencing, Optical Flow, Background Subtraction. In this paper they used different features of an object for tracking which used foreground objects between a series of frames like velocity, color and texture. They used two approaches for object tracking where first approach is based on correspondence matching and the other depends on distinct tracking.

The basic features of an intelligent video surveillance system based on moving object recognition and tracking are demonstrated in this study [3] and they are analysed. The intelligent video surveillance system is capable of reliably and automatically detecting suspicious activity. The three major techniques for detecting moving objects are the optical flow method, the back difference method, and the frame difference approach [15, 16]. In this survey research [4], video samples were compressed by reducing spatial and temporal repetitions in the sample and by using the 2D discrete cosine transform method. By extracting the alterations in the object boundaries, the moving objects are recognised and detected. Utilizing the Horn-Schnuck algorithm for moving object detection, the optimal flow vector is computed. The maximum likelihood densities over the complete set of separated frames were created using the peak correlation coefficient, which was also utilised to calculate posterior probabilities. The probability points in the chosen image are localised using the Bayesian rule. This paper [5] proposed a tracking algorithm focusing on occlusion with multiple backgrounds. They have showed that Mean Shift algorithm works for occlusion free tracking and Kalman filter is good in the case of similarity measures [14]. This paper gives the comparison on the working of Mean Shift, Kalman Filter, and Edge Histogram algorithms.

III. Existing System

There are many methods implemented in this video surveillance object detection and tracking to improve performance and efficiency. In the existing system, moving object detection used background subtraction and the detect object is then tracked using Euclidean distance calculation between the present frame and the previous frames. But this method is not efficient because when two persons overlap or cross each

other, they get misguided or their identifiers get exchanged. In the proposed idea the detection is based on complete object and face detection and given a unique identification number, for that we have used both background subtraction method for object detection along with trimmed median filtering with edge detection for removing foreground noise for each frame and Voila Jones for the face detection and after detection the tracking is done using Edge histogram algorithm which works well for occlusion. Another add-on to our proposed idea is during tracking it experimentally verifies detected face database (i.e., each intruder face is given an identification number) through checking the each and every frame of the video and it concludes the presence of the intruder in that frame and gives the location of the intruder.

IV. Framework

The proposed framework takes recorded video as the input, which was in the format of .mp4. It was infused as input to the preprocessing, where

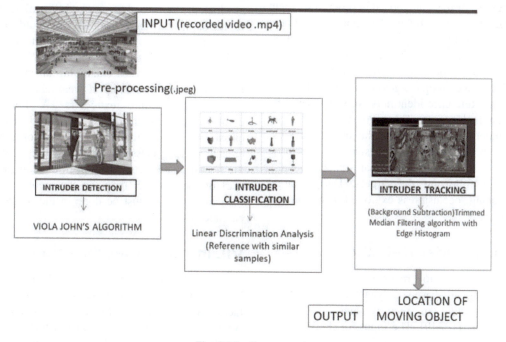

Fig. 54.1   Framework

Source: Authors

noise has been removed and transform the video into images frames; Then for detecting intruder over every image, VIOLA JOHN'S algorithm has been applied.

Once object has been detected, those detected object been given input to the classification. Here classification task has been done with help of Linear Discrimination Analysis (LDA). Once it was classified, intrusion has been identified.

V. Proposed System

The proposed system aims to aid in the monitoring, tracking, and detection of moving intruders. To find intruders, the Viola-Jones method is used. The key benefit of employing this method is that while detection appears to be quick, training is slow. Intruder detection separates moving intruders from stationary background items, allowing for higher level processing taking place while reducing computation time. In order to accurately follow and analyse the actions of the detected intruders, they are further categorised in order to set them apart from one another. The algorithm for linear discriminative analysis is used to perform the classification. This algorithm is used to separate photographs from distinct image classes and group images belonging to the same class. Reference identifiers are stored with the identified targets. The segmented targets are temporally identified by the object tracking, which also produces cohesive data on those targets. Occlusion could happen occasionally while following the targets. During target tracking, occlusion is exploited to get around this.

VI. Intruder Detection

The ability to recognize intruder faces automatically based on dynamic facial captures is important in security and surveillance domains [18, 19]. The input for the detection is recorded videos of type's mp4, avi, etc., the video undergoes Pre- processing technique so that the video is converted into frames. Following the selection of these frames, the Viola-Jones algorithm is used. The Haar basis feature filters used in this technique. Some characteristics are common to all human faces. Haar characteristics can be used to match these regularities. By creating an integral image first, the Viola-Jones algorithm's efficiency can be greatly improved. This is given by

$$II(y, x) = p = 0yq = 0xY(p, q) \qquad (1)$$

From the above equation 1, explores the prevention of attacks over the intruder. Within a detection window, detection takes place. Each of the so-called "N" filters for face recognition is claimed to have a group of cascade-connected classifiers. Each classifier searches the supplied detection window for a rectangle subset and decides whether it resembles a face. If so, the subsequent classifier is used. The face is recognised if every classifier returns a positive result, which is followed by a positive response from the filter. If not, the subsequent filter in the group of "N" filters is made to run. The speed at which the traits can be assessed is insufficient to make up for their quantity. It would be extremely expensive to assess each of the $M = 162,336$ potential features in a typical 24x24 pixel sub-window when examining an image. Each classifier is made up of weak classifiers called Haar feature extractors. A linear combination of weighted, straightforward weak classifiers is how the Viola-Jones algorithm creates a strong classifier. The classifiers with the fewest characteristics are positioned at the start of the cascade, reducing the amount of computation that must be done in the overall cascade design.

VII. Intruder Classification

The detected faces are then classified based on Linear Discriminant Analysis by identifying faces with samples. Linear Discriminant Analysis is a dimensionality reduction technique which is commonly used for the supervised classification problems [20]. According to the field of Computer Vision, face recognition is said to be a popular application because each face is represented by a very large number of pixel values. Linear Discriminant Analysis is used in order to reduce the number of features to a more manageable numbers before the process of classification. It is used for modelling differences in groups i.e. separating two or more classes. It is used to present the features in the higher dimension space into a lower dimension space. Each of the new dimensions that is been generated is a linear combination of pixel values, which form a template. Here the detected faces are classified based on the samples. The identified samples are recognised and cropped based on the facial properties.

VIII. Intruder Tracking

A. Tracking

The main focus of this system is the intruder tracking without occlusion. There are many target tracking techniques with same approach but the drawback is that when two targets cross each other or overlaps, the target becomes misguided or their tracking identifiers gets exchanged and sometimes even gets lost. Here we use Background Subtraction algorithm for detecting the intruder. Background Subtraction usually separates the foreground objects i.e., the intruders from the background. This technique is considered to be important for tracking as it is used for detecting dynamically moving objects. Here we go through a loop in which the background frame is compared with the next foreground frames to detect the foreground elements. This is then checked for further noise using trimmed median filter with edge detection in order to track the intruder. This iterative trimmed median filter works in such a way that selected window of noise pixel which is seemed to be completely noisy will be left without any changes in the current iteration and it will be processed in the next iteration, also the selected window size increases till an image pixel is found.

B. Occlusion Handling

Here the occlusion is handled by using Kalman Filter and Edge Histogram. Usually,

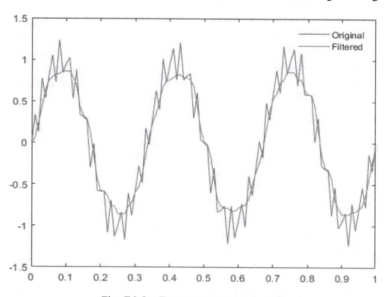

Fig. 54.2 Trimmed median filtering

Source: MATLAB

these algorithms are proposed to overcome the occlusion partially. In order to overcome completely, trimmed median filter is applied initially to remove any distorted noise in the detected image and then the stated algorithms are applied. Using this practice, a better result can be provided. The Kalman filter is specifically used for tracking. It tracks the objects based on their velocity and acceleration which can be derived from the measurement of its location. In order to face the occlusion, we apply edge histogram. Since we fed the de-noised frame, here we can completely overcome occlusion using edge histogram.

IX. Conclusion

Here the occlusion is handled by using Kalman Filter and Edge Histogram. Usually, these algorithms are proposed to overcome the occlusion partially. In order to overcome completely, trimmed median filter is applied initially to remove any distorted noise in the detected image and then the stated algorithms are applied. Using this practice, a better result can be provided. The Kalman filter is specifically used for tracking. It tracks the objects based on their velocity and acceleration. In order to face the occlusion, we apply edge histogram. Since we fed the de-noised frame, here we can completely overcome occlusion using edge histogram.

References

1. "Smart Video Surveillance: Object Detection, Tracking and Classification", International Journal of Innovations and Advancement in Computer Science (IJIACS) ISSN-8616 Volume 7, Issue 3 - March 2018.
2. "A Study on Face Detection Using Viola-Jones Algorithm for Various Background, Angles and Distance", Conference on Soft Computing - May 2018.
3. "A Study on Video Surveillance System for Object Detection and Tracking", IEEE 3rd International Conference on Computing for Sustained Global Development – 2016.
4. "Intelligent Video Surveillance System Based on Moving Object Detection and Tracking", International Conference on Information Engineering and Communication Technology (IECT 2016) - November 2016.
5. "Moving Object Detection for Video Surveillance", Hindawi Publishing Corporation The Scientific World Journal Volume 2015, Article ID 9047469 – 2015.
6. "Object Tracking with Occlusion Handling using Mean Shift, Kalman Filter and Edge Histogram", Conference Paper – March 2015.
7. V. R, R. S, V. K and S. P, "Integrated Communal Attentive & Warning System via Cellular Systems," 2022 International Conference on Electronic Systems and Intelligent Computing (ICESIC), 2022, pp. 370–375, doi: 10.1109/ICESIC53714.2022.9783481.
8. "Adaptive Visual Tracking System using Artificial Intelligence", Proceedings of the IEEE/OSA/IAPR International Conference on Informatics, Electronics and Vision (ICIEV'12), pp. 954–957, IEEE – May 2012
9. V. R, R. S, V. K and S. P, "Integrated Communal Attentive & Warning System via Cellular Systems," 2022 International Conference on Electronic Systems and Intelligent Computing (ICESIC), 2022, pp. 370–375, doi: 10.1109/ICESIC53714.2022.9783481.
10. "A Convolutional Neural Network Based on Tensor Flow for Face Recognition", 978-1-8972-2/17/$31.00 ©2017 IEEE.
11. Priya Vijay, S.Bharatwaaj S.Bharadwaj, Multi tier security system using ULK kit for secured door operation, International Journal of Pure and Applied Mathematics, Academic Publications Ltd., 111(18),2401-2411,2018,
12. Arockia Raj, Y & Alli, P 2019, 'Turtle edge encoding and flood fill based image compression scheme', Springer. Cluster Computing. Volume 22, Supplement 01, pp. 361–377.
13. "Moving Object Detection, Tracking and Classification for Smart Video Surveillance" Thesis Paper – August 2014.
14. Shanthalakshmi M, Lakshmipriya K,Raveena A, "Multi-Functional Surveillance System", International Journal of Creative Research Thoughts – Volume No. 06, Apr 2018.
15. Arockia Raj Y, V. K. K. S. J. G. K. P. V. (2021). Enhancing the Security of Data Using Digital

Stemage Technique. Annals of the Romanian Society for Cell Biology, 25(6), 9138–9143.

16. Nilaiswariya, R., J. Manikandan, and P. Hemalatha. "Improving Scalability And Security Medical Dataset Using Recurrent Neural Network And Blockchain Technology." In 2021 International Conference on System, Computation, Automation and Networking (ICSCAN), pp. 1–6. IEEE, 2021.

17. Nilaiswariya, R., J. Manikandan, and P. Hemalatha. "Improving Scalability And Security Medical Dataset Using Recurrent Neural Network And Blockchain Technology." In 2021 International Conference on System, Computation, Automation and Networking (ICSCAN), pp. 1–6. IEEE, 2021.

18. Sriram, S., J. Manikandan, P. Hemalatha, and G. Leema Roselin. "A Chatbot Mobile Quarantine App for Stress Relief." In 2021 International Conference on System, Computation, Automation and Networking (ICSCAN), pp. 1–5. IEEE, 2021.

19. Leema, Roselin G., K. Kiruba, P. Hemalatha, J. Manikandan, M. Madhin, and Raj S. Mohan. "Revolutionizing Secure Commercialization In Agriculture Using Blockchain Technology." In 2021 International Conference on System, Computation, Automation and Networking (ICSCAN), pp. 1–6. IEEE, 2021.

20. Hemalatha, P., J. Manikandan, G. Leemarosilin, and P. Kanimozhi. "Sea Food Processing Using Internet of Things and Cloud Technologies." PalArch's Journal of Archaeology of Egypt/Egyptology 17, no. 9 (2020): 5877–5885.

Recent Trends in Computational Intelligence and its Application – Sugumaran D. et al. (eds)
© 2023 Taylor & Francis Group, London, ISBN 978-1-032-48410-5

55

Dynamic Request Redirection and Workload Balancing of Cloud in Business Applications

Vijay K.[1]
Assistant Professor, Department of Computer Science and Engineering,
Rajalakshmi Engineering College, Chennai

K. Jayashree[2]
Professor, Department of Artificial Intelligence and Data Science,
Panimalar Engineering College, Chennai

Abstract—Video service providers (VSPs) can now take advantage of cloud computing to manage their video application infrastructure efficiently and cheaply. Virtual machines (VM) are rented from different clouds that are close geographically to the requester to use this method. Cost-effectiveness for VSPs depends on optimising the set of possible VMs decided to rent from public cloud in various locations over a given period due to the unpredictability of user requests and the time- and location-dependent pricing of VMs. The rented VMs must also be designed to ensure a high quality of experience (QoE) for the end user. In order to solve this issue, we can use a method called Interactive Request Redirection and Resource Provisioning (D3RP). To address this issue, we propose framing it as a conceptual optimization model and developing an optimization algorithm template that can be executed in an online environment. The theoretical investigation shows that the online algorithm provides a solution that is superior to the optimal solution obtained by using offline computing.

Keyword—Virtual machines, Quality of experience, Optimization algorithm

I. Introduction

Hosts of video-based services (VSPs) use public cloud providers (PCPs) and content delivery networks (CDNs) to continue serving users in various geographical locations. Upwards of one CSP per geographic area is required for this approach, and users' Quality of Experience (QOE) is compromised so that the VSP can save money on CSP and CDN rental [1] [2]. The resource scheduling algorithm solves this issue by moving the CDN towards the cloud, where a variety of virtual machines can serve the user's video needs according to his QoE preferences [3]; if a given VM is unavailable in a given geographic region, the CDN maximally tries to search for a further data centre and diverts the incoming requests here already [4] [7].

[1]vijayk.btech@gmail.com, [2]k.jayashri@gmail.com

DOI: 10.1201/9781003388913-55

Existing system is the sufficient QoE delivered to consumers in a cost-effective sense by Video Service Suppliers who lease computing infrastructure resources using cloud [6]. Both the needs of the consumers and the arrival rates of requests are difficult to forecast. Because of the wide variety of QoE needs and individual demands [5], it can be challenging to determine how best to map them towards the numerous resource types available in the cloud.

Cloud services, as demonstrated by prior research, are rapidly becoming the de facto standard for both business and personal computing [8] [10]. We consider a mathematical framework of a data center cluster, wherein jobs arrive per a theoretical aspect and demand VMs, that are depicted by computational, memory, and storage resources.

It also demonstrated that IPTV, as an application interface, has the possibilities to overwhelm existing ISP infrastructure with an influx of brand-new users and data. We found that 200,000 IPTV users were consuming 100 GB/s during a single broadcast [9] [12]. A number of point-to-point configurations have been successfully implemented in the Internet, while other frameworks are possible. We conduct a thorough detailed survey of PPLive to learn more about point-to-point IPTV systems and ISP traffic loads [11]. To simulate incoming request in a cloud-based media network, we enable the arriving look to switch among low- and high-crowd states (CCMN).

We also look into the energy-efficient job routing and scheduling methodology for transcoding services (TaaS) in an audio-visual cloud, as was shown in a previous study [4]. To reduce the amount of power needed to run cloud-based service engines, we focus on lowering the latency of TaaS. Within the foundation of global optimization, we cast the job routing and scheduling problem as an optimization problem [12]. We suggest a computation that deploys the encoding tasks to support engines,

including a goal to minimize the Usage while accomplishing the Queue consistency.

It also included from the viewpoint of tournament coders; player's characteristic is perhaps the most significant considerations they just had to consider when developing a game. To accomplish a foundational knowledge of the rules, play actions of gamer available on the internet, investigating user's tournament game time gives a useful point of reference.

The authors present a scheme for optimally allocating customer request to resources rented from numerous CSPs [14]. In specific, the structure can accommodate a wide variety of user needs, workloads, and quality-of-experience specifications. Video services are hosted in datacentres across the world, and this is made possible by the emergence of CDNs [15] [16]. To solve this issue, we propose the D3RP algorithm for dynamic request diversion and bandwidth utilization. The video network operator can serve an unlimited number of customers with high-quality, cost-effective offering using this model.

II. Materials and Methods

A. Architecture

Architecture diagram is shown in Fig. 55.1, which contains the following components in it.

B. CDN and Data Centres

Web applications or other forms of Online content can be delivered to users via a distribution platform, which is a collection of interconnected servers that takes into account the user's location as well as the arrival and departure of the requested content (CDN). With these credentials, the CDN administrator can access the backend of the CDN [17]. He is able to create, delete, and alter virtual machines across multiple data centres. If a user needs to reroute their requests on the fly, a policy folder is created for them.

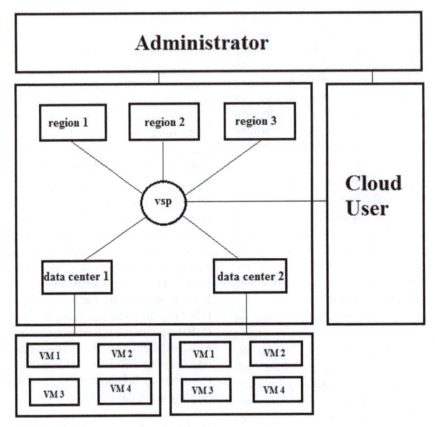

Fig. 55.1 Architecture diagram

Source: Authors

C. Video Service Provider and CDN Request

A video service provider has asked a content delivery network to host their platform online. High-quality, standard-quality, and low-quality videos are all available in the video streaming service provider app. When a video service provider wants to host it's own services, it contacts a CDN and tells it which virtual occurrences to use on which clouds [18, 19]. When a video service provider signs up for a CDN's services, the CDN will determine the cost of using the cloud and present that cost to the provider

D. Banking and Application Deployment

Video streaming service application deployment occurs in this phase. If the VSP is pleased with the invoice it has generated, it can move on to the next step of the banking procedure. After a transaction is initiated, a one-time password (OTP) is engendered and sent to the VSP's email address for use in subsequent steps. After a successful exchange, the customer can use any number of cloud services or virtual machines at their disposal. The operator can now package up his own streaming platform application and send it to the CDN or other clouds for deployment. The services will be activated immediately and made available to everyone via CDN.

E. User Request and Dynamic Redirection

Classifications of cloud-based request management and resource distribution can be made from both the cloud service provider's or the cloud customer's points of view. Many people are working to develop System apply for cloud service providers. Enhancing resource utilisation or being fair are often prioritised when dealing with a single cloud provider [20].

Scheduling approaches are presented to minimise electricity costs for numerous cloud computing by distributing workloads across multiple datacenters in different locations. As soon as the VSP receives an order, it will adaptively reroute it to the most suitable datacenter in terms of Quality of Experience (QoE), position, and computation complexity.

F. Experimental Setup

This method employs a system with three data centres located in two different regions to serve users in five different regions. The experiment employs three distinct types of virtual servers: low, intermediate, and high. When using our Automated system, you can save money when acquiring a virtual server from a Cloud Providers. When first booting up, a virtual machine requires only a few seconds to fully load.

III. Result Discussion

A. Implementation Screenshots

See Fig. 55.2.

B. Experimental Result and Inference

To ensure that effective algorithm can handle videos of varying lengths and resolutions, a comparative experiment is carried out.

Impact of Length L on Time T

The loading time for a video grows exponentially as Length L rises. To rephrase, the time required to load the video grows proportionally with the length of the video. Time and length character traits are graphed below.

Impact of Resolution R on Time T

Time needed to pile the video grows exponentially as Resolution R is increased. Therefore, the time required to complete a task is proportional to its resolution and vice versa. Here is a graph illustrating some of Resolution on Time's key features:

IV. Conclusions

In this article, we introduce a new method named D3RP for dynamic demand diversion and demand configuration management, and we demonstrate that it can be used to efficiently host videos in the cloud with minimal outlay of resources and time. One of our future goals is to better leverage VM sharing practices by accounting for the social media network team's typical habits regarding the consumption of video. Instead of considering job match, this article suggests instead to solve the issue at the stage of job/virtual machine match. Third, redefining the Quality of Experience feature and expanding the scope of the modelling objectives to include resource utilization, holding cost, virtual machine migration cost, and so on.

References

1. JianhuaTang,Wee Peng Tay, Yonggang Wen, Dynamic Request Redirection and Elastic Service Scaling in Cloud-Centric Media Networks
2. Ravikumar, S. and Kannan, E., A swift unrest horde system for curtail SDO hit in cloud computing (2018), International Journal of Engineering and Technology(UAE) Volume 7, Issue 1.7 Special Issue 7, pp. 156–160.

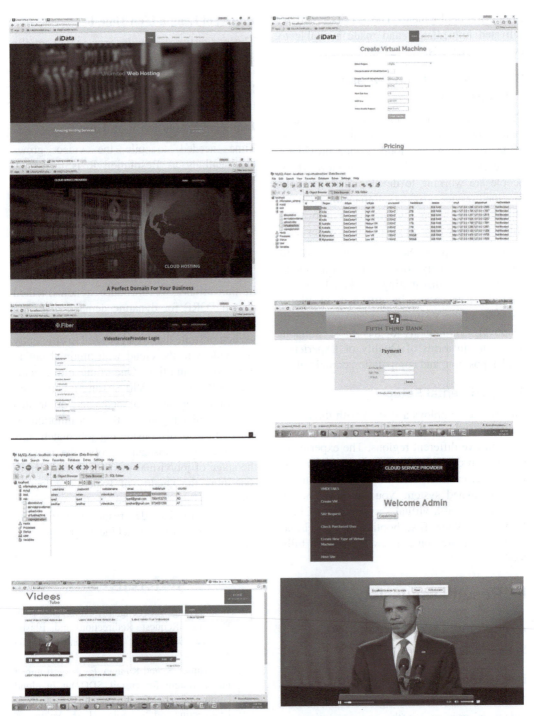

Fig. 55.2 Implementation screenshots

Source: Authors

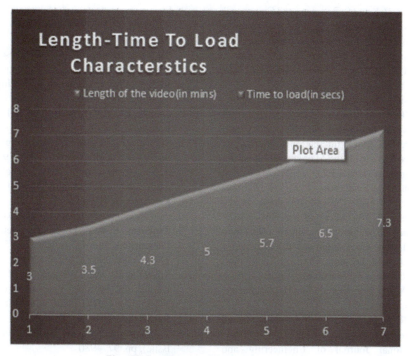

Fig. 55.3   Impact of length L on time T

Source: Authors

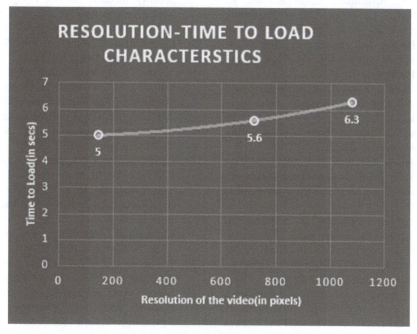

Fig. 55.4   Impact of resolution R on time T

Source: Authors

3. V. R, R. S, V. K and S. P, "Integrated Communal Attentive & Warning System via Cellular Systems," 2022 International Conference on Electronic Systems and Intelligent Computing (ICESIC), 2022, pp. 370–375, doi: 10.1109/ICESIC53714.2022.9783481.

4. Vijay, K., Sivaranjani, P., & Sowmia, K. R. (2016). Optimised Online Algorithm Framework for Video Services in Cloud with Perfect QoE. International Journal of Advanced Research in Computer Engineering & Technology (IJARCET), 5(12).

5. Arockia Raj, Y & Alli, P 2018, 'Pattern-based chain code for bi-level shape image compression', TAGA Journal 14: 3064–3080.

6. Wenhua Xiao, Weidong Bao, Xiaomin Zhu, Chen Wang, Lidong Chen, Laurence T. Yang, Senior Member, IEEE, Dynamic Request Redirection and Resource Provisioning for Cloud-based Video Services under Heterogeneous Environment.

7. Babu, R, & Jayashree K, "Prioritizing Cloud Infrastructure Using MCDM Algorithms" International Journal of Engineering and Advanced Technology (IJEAT), vol. 8, no. 3S, 2019, pp. 696–698.

8. Vijay, K., Sivaranjani, P., & Sowmia, K. R. (2016). Optimised Online Algorithm Framework for Video Services in Cloud with Perfect QoE. International Journal of Advanced Research in Computer Engineering & Technology (IJARCET), 5(12).

9. Narayana K.E & Jayashree, K "A Overview on Cloud Computing Platforms and Issues" International Journal of Advanced Research in Computer Science and Software Engineering Volume 7 pp. 238–22, 2017.

10. R. Babu, K. Jayashree & R. Abirami "Fog Computing QoS Review and Open Challenges" International journal of Fog Computing, vol.1, no.2, 2018, pp. 109–118.

11. Priya Vijay, S.Bharatwaaj S.Bharadwaj, Multi tier security system using ULK kit for secured door operation, International Journal of Pure and Applied Mathematics, Academic Publications Ltd., 111(18), 2401–2411, 2018,

12. Chahat Agarwal, Akshay S, Vijay P., Intelligent System based on pollution intensity, Journal of Advanced Research in Dynamical and Control Systems, Institute of Advanced Scientific Research, 11(5), 2152–2161, 2019

13. Arockia Raj, Y & Alli, P 2017, 'Improving the performance for edge-based image compression using various techniques', Asian Journal of Information and Technology, pp. 16, no. 1, pp. 148–152.

14. Arockia Raj Y, V. K. K. S. J. G. K. P. V. (2021). Enhancing the Security of Data Using Digital Stemage Technique. Annals of the Romanian Society for Cell Biology, 25(6), 9138–9143. Retrieved from https://www.annalsofrscb.ro/index.php/journal/article/view/7160

15. Kiruthika S, Kirubasri G. (2020). Improving the Efficiency of a Dual Corpus Text To Speech Synthesis System Using a Prefetch Buffer. International Journal of Advanced Science and Technology, 29(06), 4253–4258.

16. Mariprabhu, M.; Niranjana, R.; Praveen Kumar, R.; Jananee, V, Stock prediction by ensembling LSTM using adaboost algorithm, Journal of Advanced Research in Dynamical and Control Systems, 2020, 12(5 Special Issue), pp. 977–980

17. Nilaiswariya, R., J. Manikandan, and P. Hemalatha. "Improving Scalability And Security Medical Dataset Using Recurrent Neural Network And Blockchain Technology." In 2021 International Conference on System, Computation, Automation and Networking (ICSCAN), pp. 1–6. IEEE, 2021.

18. Sriram, S., J. Manikandan, P. Hemalatha, and G. Leema Roselin. "A Chatbot Mobile Quarantine App for Stress Relief." In 2021 International Conference on System, Computation, Automation and Networking (ICSCAN), pp. 1–5. IEEE, 2021.

19. Leema, Roselin G., K. Kiruba, P. Hemalatha, J. Manikandan, M. Madhin, and Raj S. Mohan. "Revolutionizing Secure Commercialization In Agriculture Using Blockchain Technology." In 2021 International Conference on System, Computation, Automation and Networking (ICSCAN), pp. 1–6. IEEE, 2021.

20. Hemalatha, P., J. Manikandan, G. Leemarosilin, and P. Kanimozhi. "Sea Food Processing Using Internet of Things and Cloud Technologies." PalArch's Journal of Archaeology of Egypt/ Egyptology 17, no. 9 (2020): 5877–5885.

Recent Trends in Computational Intelligence and its Application – Sugumaran D. et al. (eds)
© 2023 Taylor & Francis Group, London, ISBN 978-1-032-48410-5

56

Deep Dive on Oversampling and Under Sampling Techniques in Machine Learning

Vijay K.[1], Manikandan J.[2]
Assistant Professor, Department of Computer Science and Engineering,
Rajalakshmi Engineering College, Chennai

Babu Rajendiran[3]
Assistant Professor, Department of Computational Intelligence,
School of Computing, SRM Institute of Science and Technology, Chennai

Sowmia K. R.[4], Eugene Berna I.[5]
Assistant Professor, Department of Artificial Intelligence and Machine Learning,
Rajalakshmi Engineering College, Chennai, India

Abstract—Data imbalance is a term used in Computer Vision to refer to an unbalanced distribution between classes throughout a dataset. In classification tasks, where the dispersion of classes or labels in a given data is not regular, this problem is most frequently encountered. The resampling approach, which involves adding data to the minority class and removing entries from majority class, is the most feasible approach for solving this problem. In this study, we conducted experiments with two frequently used resampling techniques: oversampling and under sampling, both of which have been widely accepted. In order to investigate both resampling methodologies, we used a publicly available imbalanced dataset from the Kaggle website and a collective of very well ML algorithms with distinct hyperparameters that produced the best outcomes for both resampling methodologies. One of the most important discoveries of this study is the observation that oversampling outperforms under sampling for various classifiers and results in increased scores in many evaluation metrices.

Keywords—Computer vision, Imbalanced distribution, ML algorithm, Hyperparameters

I. Introduction

In computer vision and image processing, classification is characterized as having trained an approach to determine a previously undiscovered dataset in what class it pertains.

As data has proliferated, it is difficult to find properly labelled information. The probabilities of the predefined classes were hypothesized to be the same in conventional machine learning approaches. There are a few instances where this presumption is inaccurate, such as when

[1]vijayk.btech@gmail.com, [2]jmanekandan@gmail.com, [3]babu.rajen17@gmail.com, [4]sowmiakr@gmail.com, [5]eugeneberna31@gmail.com

DOI: 10.1201/9781003388913-56

predicting the weather, diagnosing illness, or detecting fraud. In these instances, almost everything cases are classified in one category whereas only just few illustrations are classified in the opposite category. Therefore, the models are in the best interests of the majority while excluding the minority. Data imbalance has an adverse effect on model performance, as poorly balanced sets of data will negatively affect model performance. To put it another way, this is known as a problem of social inequity. When it comes to other assessment methods like precision, recall, F1- score [4,], ROC score, and other parameters—even though we can achieve better accurateness in this situation—we fall short. There's been a lot of discussion recently about class inequality. According to several research studies [5] [6], it is a complicated problem that needs more attention. Resampling techniques were commonly used to balance the dataset.

Under-sampling or oversampling the dataset can be used in resampling techniques. As a general term, under sampling refers to decrease the quantity of majority target occurrences or samples. We draw a representative sample out from large percentage class in order to equalize the size of the minority group. As a result, there is a risk of obliterating useful data from the dataset. cluster centroids [7], Tomeks' links [8], and other approaches are examples of frequent under sampling strategies. Oversampling can be accomplished by doubling the frequency of samples from the minority class whereas simultaneously creating new illustrations of reiterating existing instances. Borderline-SMOTE [9] is an example of an oversampling method. The contrast between the two strategies, oversampling and under sampling, is depicted in Figure 1. Oversampling and under sampling methodologies were tested with various machine learning concepts in this work using the extremely unbalanced dataset of 'Santander Customer Transaction Prediction' from a Kaggle contest (released in February 2019). On Github, you can find the code for this

experiment [10]. For several machine learning classifier models, the results suggest that oversampling performs well than the under-sampling strategies.

An outline of the paper's organisation is given below. Section II features examples of relevant work. Detailed information about the dataset used in this study can be found in Section III. Section IV explains our research methods and assessment metrics. Experiments and their results are discussed in Section V. This concludes the paper's discussion in Section VI.

II. Literature Review

Experts have only recently begun to pay attention to the issue of data imbalance. This well-known method for under sampling was presented by the researchers in [11]. Removes data points from the KNN that aren't representative of the target class as a whole. A wide range of issues regarding learning with class imbalance scatterings were investigated by the authors of [12]. Class scatterings and valuation information, for example, and use of erroneous regularity and accurateness to evaluate predictive accuracy. These approaches to gaining knowledge from unequal classes were examined by [13]'s authors. Class sprinkling, accuracy, and error costs are the three major causes of the poor performance of model derived using machine learning techniques on imbalanced classes. The authors of [14] provided resampling strategies because trying to find the minority goal is complicated. For the new re - sampling scheme, they used category sub-clustering to create an artificial situation and then oversampled the positives while sampling the quasi majority. They claim that their new resampling alternative assessments traditional random resampling in terms of accuracy. The authors of [15] analysed an advertising range of data in three main forms. There are three ways to detect interrelations: logical regression, Chi-squared automatic interaction detection, and neural networks. In order to generate the results of the selected techniques, accuracy, AUC, and precision were just used together. This was done by comparing various imbalanced data generated from the original data. conclusion: For all authors, specificity is a valid measure for inferring from an unbalanced dataset.

When compared to previous research, our study is unique in that we will use a Kaggle competition set of data to investigate oversampling and under sampling techniques (published in February 2019). In [8], k-means grouping was suggested as a way to balance the extremely unbalanced occurrences by decrease the quantity of majority instances. Under-sampling was used to completely remove data points from vast majority of instances by the authors in [16] as well.

III. Datasets

Kaggle's "Santander Detailed Customer Prediction" contest dataset was utilised in our research. To predict which consumers will offer a specific purchase in the long term, terms of the quantity of money at stake, this information was released on December 1, 2019. We was using this data to address and evaluate the unbalancing data issue because we knew that the set of data was imbalanced. With both techniques, we evaluated the classifier performance using a variety of performance indicators, such as k-means clustering.

Figure 56.1 shows that Oversampling outperforms under sampling for a variety of classification algorithms and analysis methods. We plan to use a variety of deep learning methods including both resampling methods in the long term to compare their performance.

IV. Methodology

As a first step, we examine the data set in detail in order to gain a thorough understanding of the imbalanced data issue. Following the download of the training sample from Kaggle, we splitted the information into training and target datasets. After that, the data set was shrunk. To make things easier, we've ranked all the features and selected only the most essential for this investigation. Consequently, the frequency analysis of the attributes was examined, and the correlation analysis was determined. Because of this, we found there is only a small connection between characteristics, which suggests that they are largely unrelated. That is why features were selectively culled using a feature selection technique. After analysing the set of data and trying to prepare it for use with machine learning techniques, we used two re - sampling strategies based on changing the classification performance. We also glanced at some other different classifiers as part of this investigative process. Table 56.1 lists the various classifiers employed in the two

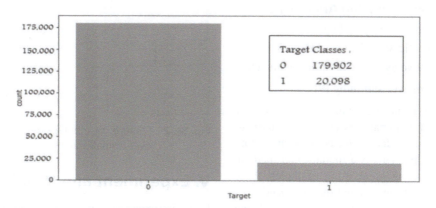

Fig. 56.1 Distribution of target classes [7]

resampling methods. We compared classifiers using a variety of metrics (Accuracy, Precision, Recall, F1- Score [17], and ROC [18]) . The most common assessment metric for traditional models is precision [20-22]. However, accuracy is not really an useful statistic to use when interacting with poorly balanced datasets [21]. When the target dispersion is highly skewed, numerous experts have started to notice that recall for such minority goal is sometimes zero, implying no categorisation guidelines were produced specifically for this group of individuals or organisms. The minority target has lower precision and recall than huge percentage target, which can be described using terms from the field of information retrieval. It's complicated for a classification algorithm to perform well on the minority goal since accuracy places more importance on the majority target than minority target. Because it reduces the need for human intervention and creates more accurate predictions, machine learning is already widely used in the medical field. Once the dataset has been properly analysed for data analysis, the dependent and independent variables can be found by performing proper variable identification. It then uses an appropriate algorithm to analyse the data and discover patterns in it. Predicting the outcome is made easier by using different algorithms.

A. Different Sampling Approaches

(a) Random Over Sampling

(b) Synthetic Minority Oversampling

(c) Adaptive Synthetic sampling approach

(a) Random over sampling [9]

There are numerous duplicates of the tiny segment of minimal classes that are used to complement the data collected through random over sampling. This is a fundamental method. Instead of using pre-existing examples, these are selected at random and then replaced. In a nutshell, New samples are used to resample the originals.

(b) Synthetic minority oversampling (SMOTE) [9]

On the other hand, SMOTE uses a particular heuristic method, with size n and f features on the other side of the data. In this method, the percentage of closest neighbours is used. For the purpose of generating synthetic data points, we need to boost the variable by an integer between 0 and 1, which is equivalent to adding x to the vector.

(c) Adaptive synthetic sampling approach (ADASYN) [9]

Smote-based heuristics are also used. Figure 56.2, It moves from the majority to the minority side of the spectrum. Weighted dissemination is used for minority class samples. When we apply our methods to the poorly balanced datasets, we'll see an increase in accuracy and a decrease in recall rate.

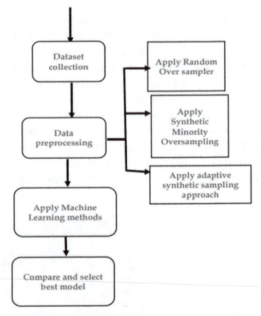

Fig. 56.2 Feature distribution representation over a small sample

V. Experiment and Results

In our first experiment, we used the oversampling strategy. Random sampling is a non-heuristic process of gathering data. Through the use of random repetition, it attempts to achieve a more even distribution of social classes. This approach, however, has two drawbacks. In the first place, it raises the risk of over-fitting because it duplicates the minority class cases. We have a large but unbalanced dataset that makes this process even more difficult to learn from. This strategy is ideal information to collaborate with. Multi - class classification features and random selection were used to create the forecasting. Several parameters of the model were tested for every model. SMOTE and Adaptive synthetic sampling were used in our two experiments as well. The best simple under sample selection method is random under sampling [17]. It's a non-heuristic technique that sporadically removes occurrences from the classification model in order to balance target distributions. Classifier models involve

a lot of data, and this methodology can help reduce the amount of data they need without sacrificing accuracy. After adopting this model to our dataset, and in this particular instance is 14,000 individuals, the target class is shown in Fig. 56.6.

(a) Experimentation results of various classifiers using random over sampling

In this method Random Forest classifier gives accuracy of 96% and recall rate of 97.23%. And the respective comparison table and graph is given in Table 56.1 and Fig. 56.3.

(b) Experimentation results of various classifiers using synthetic minority oversampling technique

With this method, random forest gives 96.34% accuracy and recall rate 98.38%, but it gives less accuracy than previous method, it is

Table 56.1 Comparisons of classification algorithms using random over sampling technique

Algorithm	Precision (%)	Recall (%)	Accuracy (%)
Logistic Regression [17]	68.23	71.34	73.13
Decision Tree [7]	90.24	89.32	91.25
Naïve Bayes [7]	76.23	79.32	80.21
Random Forest [9]	94.32	97.23	96.34

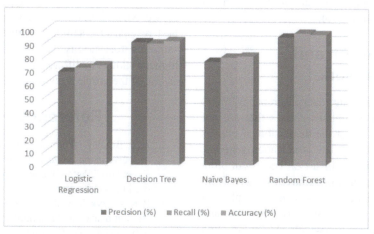

Fig. 56.3 Comparisons of classification algorithms using synthetic minority oversampling technique

Table 56.2 Comparisons of classification algorithms using synthetic minority oversampling technique

Algorithm	Precision (%)	Recall (%)	Accuracy (%)
Logistic Regression [17]	72.32	75.53	75.32
Decision Tree [7]	91.65	90.33	93.97
Naïve Bayes [7]	77.38	80.32	83.74
Random Forest [9]	95.37	98.38	96.34

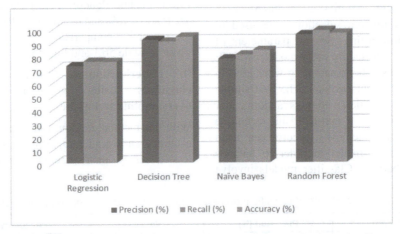

Fig. 56.4 Comparisons of classification algorithms using synthetic minority oversampling technique

considered as real time data sample because it uses heuristic method. And the respective comparison table and graph is given in Table 56.2 and Fig. 56.4.

VI. Conclusion and Future Work

In Machine Learning, data discrepancy refers to a dataset that contains classes that are not evenly distributed. If the dispersion of classes or labels in a set of data is not homogeneous, then this problem will arise. The resampling method, which adds or removes records from majority or minority classes, is a simple and direct solution to this issue. Oversampling and under sampling are two widely used resampling techniques that have been tested in this paper. Using Kaggle's Santander Detailed Customer

Prognostication set of data, we managed to run a number of well-known supervised learning models with a variety of hyper parameters to see which resampling method yielded the best results. Oversampling outperforms under sampling for various classifiers and achieves higher rankings in different assessment metrics, as one of the article's main insights exposes.

References

1. A. P. Bradley, "The use of the area under the roc curve in the evaluation of machine learning algorithms," Pattern recognition, vol. 30, no. 7, pp. 1145–1159, 1997.
2. B. Krawczyk, M. Galar, Ł. Jelen, and F. Herrera, "Evolutionary undersampling boosting for imbalanced classification of breast cancer malignancy," Applied Soft Computing, vol. 38, pp. 714–726, 2016.

3. C. J. Van Rijsbergen, The geometry of information retrieval. Cambridge University Press, 2004.

4. D. L. Wilson, "Asymptotic properties of nearest neighbor rules using edited data," IEEE Transactions on Systems, Man, and Cybernetics, no. 3, pp. 408–421, 1972.

5. E. Duman, Y. Ekinci, and A. Tanrıverdi, "Comparing alternative classifiers for database marketing: The case of imbalanced datasets," Expert Systems with Applications, vol. 39, no. 1, pp. 48–53, 2012.

6. G. Cohen, M. Hilario, H. Sax, S. Hugonnet, and A. Geissbuhler, "Learning from imbalanced data in surveillance of nosocomial infection," Artificial intelligence in medicine, vol. 37, no. 1, pp. 7–18, 2006.

7. G. Lemaˆıtre, F. Nogueira, and C. K. Aridas, "Imbalanced-learn: A python toolbox to tackle the curse of imbalanced datasets in machine learning," The Journal of Machine Learning Research, vol. 18, no. 1, pp. 559–563, 2017.

8. H. Han, W.-Y. Wang, and B.-H. Mao, "Borderline-smote: a new oversampling method in imbalanced data sets learning," in International conference on intelligent computing. Springer, 2005, pp. 878–887.

9. H. He and E. A. Garcia, "Learning from imbalanced data," IEEE Transactions on knowledge and data engineering, vol. 21, no. 9, pp. 1263–1284, 2009.

10. https://github.com/Roweida-Mohammed/ Code for Santander Customer Transaction Prediction.

11. I. Mani and I. Zhang, "knn approach to unbalanced data distributions: a case study involving information extraction," in Proceedings of workshop on learning from imbalanced datasets, vol. 126, 2003. [17] A. Estabrooks and N. Japkowicz, "A mixture-of-experts framework for learning from imbalanced data sets," in International Symposium on Intelligent Data Analysis. Springer, 2001, pp. 34–43.

12. Sriram, S., J. Manikandan, P. Hemalatha, and G. Leema Roselin. "A Chatbot Mobile Quarantine App for Stress Relief." In 2021 International Conference on System, Computation, Automation and Networking (ICSCAN), pp. 1–5. IEEE, 2021.

13. Kannan, E., Ravikumar, S., Anitha, A. et al. Analyzing uncertainty in cardiotocogram data for the prediction of fetal risks based on machine learning techniques using rough set. J Ambient Intell Human Comput (2021). https:// doi.org/10.1007/s12652-020-02803-4

14. M. C. Monard and G. E. Batista, "Learmng with skewed class distrihutions," Advances in Logic, Artificial Intelligence, and Robotics: LAPTEC, vol. 85, no. 2002, p. 173, 2002.

15. Leema, Roselin G., K. Kiruba, P. Hemalatha, J. Manikandan, M. Madhin, and Raj S. Mohan. "Revolutionizing Secure Commercialization In Agriculture Using Blockchain Technology." In 2021 International Conference on System, Computation, Automation and Networking (ICSCAN), pp. 1–6. IEEE, 2021.

16. N. V. Chawla, N. Japkowicz, and A. Kotcz, "Special issue on learning from imbalanced data sets," ACM SIGKDD explorations newsletter, vol. 6, no. 1, pp. 1–6, 2004.

17. Q. Gu, L. Zhu, and Z. Cai, "Evaluation measures of the classification performance of imbalanced data sets," in international symposium on intelligence computation and applications. Springer, 2009, pp. 461–471.

18. Ravikumar, S. and Kannan, E., 2021. Machine Learning Techniques for Identifying Fetal Risk During Pregnancy. International Journal of Image and Graphics, p. 2250045.

19. S. Choi, Y. J. Kim, S. Briceno, and D. Mavris, "Prediction of weatherinduced airline delays based on machine learning algorithms," in 2016 IEEE/AIAA 35th Digital Avionics Systems Conference (DASC). IEEE, 2016, pp. 1–6.

20. V. K, K. Jayashree, V. R and B. Rajendiran, "Forecasting Methods and Computational Complexity for the Sport Result Prediction," 2022 International Conference on Electronic Systems and Intelligent Computing (ICESIC), 2022, pp. 364–369, doi: 10.1109/ ICESIC53714.2022.9783514.

21. S. Visa and A. Ralescu, "Issues in mining imbalanced data sets-a review paper," in Proceedings of the sixteen midwest artificial intelligence and cognitive science conference, vol. 2005. sn, 2005, pp. 67–73.

22. W. Wei, J. Li, L. Cao, Y. Ou, and J. Chen, "Effective detection of sophisticated online

banking fraud on extremely imbalanced data," World Wide Web, vol. 16, no. 4, pp. 449–475, 2013.

23. Y. Freund and R. E. Schapire, "A decision-theoretic generalization of on-line learning and an application to boosting," Journal of computer and system sciences, vol. 55, no. 1, pp. 119–139, 1997.

24. Y. Sun, A. K. Wong, and M. S. Kamel, "Classification of imbalanced data: A review," International Journal of Pattern Recognition and Artificial Intelligence, vol. 23, no. 04, pp. 687–719, 2009.

Recent Trends in Computational Intelligence and its Application – Sugumaran D. et al. (eds)
© 2023 Taylor & Francis Group, London, ISBN 978-1-032-48410-5

57

Water Quality Monitoring and Notification System Based on Arduino, MATLAB and Visualization Techniques

Ravikumar S.[1]

Assistant Professor, Department of CSE,
Vel Tech Rangarajan Dr. Sagunthala R&D Institute of Science and Technology, Chennai, India

Vijay K.[2], Bhuvaneswaran B.[3]

Assistant Professor, Department of Computer Science and Engineering,
Rajalakshmi Engineering College, Chennai

Babu Rajendiran[4]

Assistant Professor Department of Computational Intelligence,
SRM Institute of Science and Technology, Chennai

Vidhya S.[5]

Assistant Professor, Department of Information Technology,
Saveetha Engineering College, Chennai

Abstract—Water quality measurement is vital now days to identify the useful resource present in the globe and utilize the resource properly. In the existing system, a hardware-based approach is evaluated in which only a part of the implementation is developed using embedded sensors. The sensor will detect the water quality and convey the same into the communication channels which is transmitted into the IOT platform. Our new system focused mainly on improvising the existing system by including high quality PH sensor, turbidity sensor and conductivity sensor. As an advancement of the system, a MATLAB based analysis model is extended. With MATLAB, you can integrate results into your existing applications. The live data which is measured is analyzed with the dataset of various locations and their resources. Different techniques like Neural Network, Deep Learning also Machine Learning techniques that learn directly from input data. The compared analysis provides the useful resource available locations in the globe. The proposed system detects the location and the live data of PH value of the water and other conductivity, Turbidity is displayed in the Thing Speak platform.

Keywords—IoT, pH sensor, Turbidity sensor, Conductivity sensor, Arduino, Location mapping, ThingSpeak, MATLAB

[1]ravikumars@veltech.edu.in, [2]vijayk.btech@gmail.com, [3]bhuvaneswaran@rajalakshmi.edu.in, [4]babu.rajen17@gmail.com, [e]vidhyas_1983@yahoo.com

DOI: 10.1201/9781003388913-57

I. Introduction

Water is one of the essential necessities and vital for supporting the personal satisfaction. In India its importance is more than normal because of the agrarian idea of the economy. Increasing urbanization and industrialization patterns suggest that the quality of water is deteriorating [5]. In this way, the nature of the drinking water should be estimated progressively while it is provided to purchasers. Need for water resources is one of the major issues that we are facing nowadays as many sectors such as Industries, Agriculture etc. consume large amount of water [6] [7]. "Water quality" is a term utilized here to communicate the appropriateness of water to support different uses or cycles. A specific use will have certain prerequisites [8] for the physical, chemical, or organic qualities of water; for instance, limits on the convergences of poisonous substances for drinking water use, or limitations on temperature and pH ranges for water supporting networks [9].

Subsequently, water quality can be characterized by a scope of factors which breaking point water use. While numerous utilizations have some basic prerequisites for specific factors, each utilization will have its own requests and effects on water quality [10]. Quality requests of various clients won't generally be viable, and the exercises of one client may limit the exercises of another, either by requesting water of a quality external the reach needed by the other client or by bringing down quality during utilization of the water [11]. Here in this paper, we propose a current framework by extemporizing with cutting edge quality sensors and MATLAB examination calculations [12].

The main focus of this paper is developing some prototype to record the water quality parameters in the cloud using Wi-Fi a real time. The data which is processed by these methods can be analyzed in MATLAB and is connected to Thing Speak which can create instant visualization of live data, and send alerts [13]. Also, the objective is to improvise

or to maintain a certain level of water quality in order to compromise the quality demands of the users. Based on the quality of the live locations water, the system suggests us for the areas which have best quality of water [14].

Water quality is being estimated by gathering water tests for examination or by utilizing tests which can record information at a solitary point on schedule, or logged at standard stretches over an all-encompassing period. Dataset got from live area regions are examined utilizing MATLAB calculations. To guarantee precise and solid examination in observing water quality we need an enormous number of tests, where, IoT resolves such issues of information assortment, investigation and correspondence. The boundaries needed for observing the nature of water, incorporate turbidity and pH levels [15] [16]. World Health Organization (WHO) has characterized safe reaches for every one of the water quality boundaries like pH - 6.5 to 8.5, Conductivity – 2000 µS/cm, Turbidity – 0 to 5 NTU.

The remainder of this essay is structured as follows. The extant literature is described in Section 2 in depth. Section 3 presents the suggested architecture, and Section 4 details its execution and outcomes. The conclusion and upcoming work are presented in Section 5.

II. Literature Survey

Brinda Das and P.C. Jain [1] have carried out a water quality estimation to check the nature of water progressively through different sensors. The ZigBee module transfers data collected by the sensors remotely to the microcontroller, and the GSM module transfers data further from the microcontroller to the high-tech mobile phone or PC. The nearness sensor makes the authorities by communicating something specific aware of them by means of the GSM module.

Monira Mukta, et al., [2] [17] has introduced a water quality framework where the sensors are associated with arduino-uno. Extricated information from the sensors is communicated to a work area application created in NET stage and contrasted and the WHO (World Health Organization) standard qualities. Quick backwoods paired classifier is utilized to characterize if the test water test is drinkable.

Zhu Wang et al., [3] [19] introduce the Wireless Sensor Network (WSN) for Remote Data Center and Water Quality Overseeing Network. The WSN evaluates the water's quality. The sensor network and Zigbee remote correspondence understanding are in implicit solidarity.

A framework for evaluating water quality has been developed by Mo Deqing et al., [4] [24]; it includes a data communication unit, several sensors for measuring water quality, a module for information acquisition with a single-chip microcontroller unit, and an observation focus. Water quality is naturally recognized heavily influenced by single chip miniature regulator. The information is immediately shipped off observing focus by GSM network as SMS [18].

In the existing system, a hardware-based approach is evaluated in which only a part of the implementation is developed using embedded sensors. The sensor will detect the water quality and convey the same into the communication channels which is transmitted into the internet of things platform [20, 21].

III. Proposed System

The proposed system focused on improvising the existing system by including high quality PH sensor, turbidity sensor and conductivity sensor. As an advancement of the system, a MATLAB based analysis model is extended. The data measured live is analyzed with the dataset of various locations and their resources. The compared analysis provides the similar resource available across locations in the globe. The proposed system detects the location and the live data of PH value of the water and other conductivity, Turbidity is displayed in the Thing Speak platform.

The proposed Architecture is shown in Fig. 57.1. And it is divided into four modules,

A. Senso0072

In this module, sample water for our project is taken from different locations in Chennai. Quality of Sample water is measured using pH sensor, Conductivity sensor and Turbidity sensor which is connected to Arduino and LCD to display the quality. Quality Analysis of

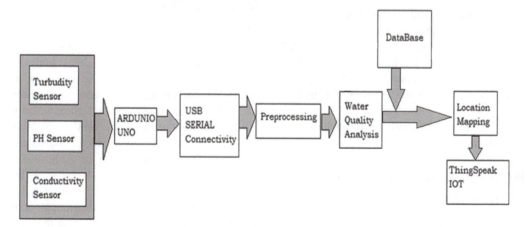

Fig. 57.1 Architecture of proposed water monitoring system

Source: Authors

sample water is done on pH and conductivity values. Turbidity values are used to visualize the quality levels [22].

B. Database

Information esteems estimated from IoT sensors are store in a cloud called "ThingSpeak". The water quality boundaries are put away in the cloud utilizing Wi-Fi for constant stockpiling of information. Sensors distinguish and measure data on a wide range of things like Acidity, Electrons and shadiness from test water and they convey that information in some structure, like a number or a signal.

C. MATLAB Analysis

ThingSpeak gives admittance to MATLAB to help you sort out information through which you can change over, consolidate, and ascertain new information. We can Schedule computations to run at specific stretches with the goal that examination result is refreshed on current estimation [23]. Consolidate information from numerous channels to assemble a more modern examination. Examination is finished utilizing Machine Learning Algorithm. For more exact outcomes more elevated level calculations are utilized.

D. Visualization

This module manages the Implementation consequence of Project. Yield of the task is shown in ThingSpeak stage. The MATLAB Analysis and MATLAB Visualizations applications furnish code layouts to help you with essential procedure on your information base or live information. ThingSpeak gives diagrams and outlines to imagine the consequence of the Project.

IV. Implementation and Results

The Internet of Things is an environment made up of physically connected objects that can be accessed online. The objects in the Internet of Things (IoT) could be a person wearing a heart monitor or a car with built-in sensors; in other words, they could be anything that has been given an IP address and is capable of assembling and transmitting data over a network without human intervention or input. The substance's embedded innovation forces them to collaborate with internal or external environmental elements, which affect the actions conducted.

ThingSpeak: ThingSpeak is IoT investigation platform administrations which permit us to aggregate, envision, and dissect the live information streams in the cloud. As a result, moving the data from our device to Thing Speak is simple. The purposeful information stored in the cloud can be presented by Thing Speak. Since the information got are sensor esteems, they can be effectively envisioned utilizing ThingSpeak as outlines. For sensor information Spline Charts are best. Along these lines, the immediate perceptions of continuous live information and cautions will be given to the specialists utilizing web administrations.

Figures 57.2, 57.3, and 57.4 demonstrate, respectively, the proposed model's performance outcomes, training states, and error plots.

V. Conclusion

The Arduino serves as the system's core processing unit during design and implementation, MATLAB for analysis, visualization app to display the live data to the end users and the water detection sensor to monitor the water quality automatically without the help of the people on duty. So, this water quality model is economical, convenient and fast. The framework can be additionally extended by expanding the boundaries by adding different sensors and can be additionally utilized for observing air contamination, modern and horticulture creation, etc. Further the trial arrangement can be improved by consolidating

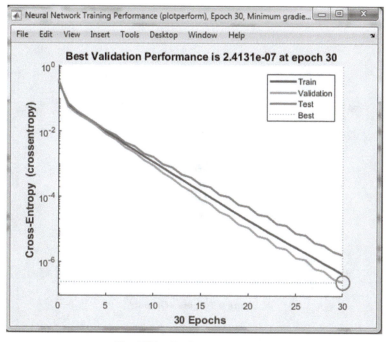

Fig. 57.2 Performance results

Source: MATLAB

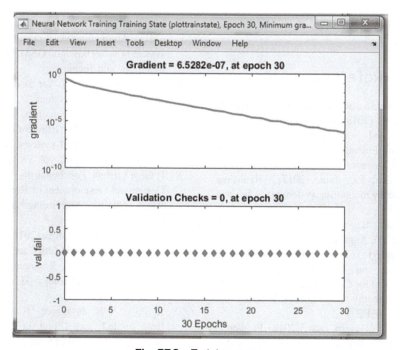

Fig. 57.3 Training states

Source: MATLAB

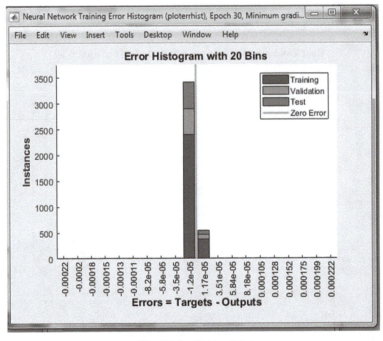

Fig. 57.4 Error plots

Source: MATLAB

calculations for peculiarity identifications in water quality.

References

1. Alex David, S., Ravikumar, S. and Rizwana Parveen, A., (2018). Raspberry Pi in computer science and engineering education. In Intelligent Embedded Systems (pp. 11–16). Springer, Singapore.
2. Brinda Das, P.C. Jain, (2017), "Real-time water quality monitoring system using Internet of Things".
3. Kamlaesan, B., Kumar, K.A. and David, S.A., (2017), August. Analysis of transformer faults using IoT. In 2017 IEEE International Conference on Smart Technologies and Management for Computing, Communication, Controls, Energy and Materials (ICSTM) (pp. 239–241). IEEE
4. Karthick. T, Gayatri Dutt, (2018) Prediction of Water Quality and Smart Water Quality Monitoring System in IoT Environment&quot.
5. Mo Dequing, Zhao Ying, Chen Shangsong, (2011). Automatic measurement and reporting system of water based on GSM, Department of Electronic and Technology.
6. Monira Mukta (2019), "Iot based Smart Water Quality Monitoring System"
7. N. Natheeswari, P. Sivaranjani, K. Vijay, and R. Vijayakumar (2020), "Efficient data migration method in distributed systems environment," Adv. Parallel Comput., vol. 37, pp. 533–537.
8. P. Gopi krishna, Panchumarthi Harish, (2019), Design and Development of Remote Location Water Quality Monitoring System using IoT&quot.
9. R. Babu & Dr. K. Jayashree (2016) "Prominence of IoT and Cloud in Health Care" International Journal of Advanced Research in Computer Engineering & Technology (IJARCET), vol. 5 no. 2.
10. R. Babu & K. Jayashree "A Survey on the Role of IoT and Cloud in Health Care", International Journal of Scientific, Engineering and Technology Research (IJSETR) ISSN

2319-8885 Vol. 04, Issue. 12, May-2015, Pages: 2217–2219. ISSN 2319-8885.

11. Dr.R. Renugadevi, Jeya Prakash, Dr B. Sakthivel, Arockia Raj Y, "AN IOT-BASED SYSTEM FOR EFFECTIVE COVID PATIENT HEALTH MONITORING WITH SVM DECISION MAKING", Turkish Journal of Physiotherapy and Rehabilitation, Vol. 32, Issue 3, Pages 3649–3653

12. S. Harini, K. Jothika & K. Jayashree (2017), "A Survey on Privacy and Security in Internet of Things" International Journal of Innovations in Engineering and Technology, vol.8, pp 129–134.

13. Daigavane, V. V. and Gaikwad, M. A., (2017). Water quality monitoring system based on IoT. *Advances in wireless and mobile communications*, *10*(5), pp.1107–1116.

14. Vivek, J., Kumar, K. A., Sahtihi, R., Sravani, R.S. and Sravani, V., (2018). Automated Extraterritorial Averting System. Journal of Computational and Theoretical Nanoscience, 15(11-12), pp. 3538–3541

15. V Priya, D. Prabakar, K. Vijay, "The Healthy Skin Framework-Iot Based", Psychology and Education Journal 58 (2), 7731–7734

16. Zhanwei Sun, Chi Harold Liu, Chatschik Bisdikia_, Joel W. Branch and Bo Yang, (2012) 9th Annual IEEE Communications Society Conference on Sensor, Mesh and Ad Hoc Communications and Networks.

17. Zhu Wang, Qi Wang, Xiaoqiang Hao, (2009), "The Design of the Remote Water Quality Monitoring System Based on WSN"

18. Shanthalakshmi M, Puvvada Siva Naga Lakshmi Sai Prithvi Raj, RahulGuda, "IOT Enhanced Car Control System with Voice Commands using Arduino" was published in Journal of Network Communications and Emerging Technologies – Volume No. 08, April 2018.

19. Priya Vijay, S.Bharatwaaj S.Bharadwaj, Multi tier security system using ULK kit for secured door operation, International Journal of Pure and Applied Mathematics, Academic Publications Ltd., 111(18), 2401–2411, 2018

20. Arockia Raj, Y & Alli, P 2019, 'Turtle edge encoding and flood fill based image compression scheme', Springer. Cluster Computing. Volume 22, Supplement 01, pp. 361–377.

21. Nilaiswariya, R., J. Manikandan, and P. Hemalatha. "Improving Scalability And Security Medical Dataset Using Recurrent Neural Network And Blockchain Technology." In 2021 International Conference on System, Computation, Automation and Networking (ICSCAN), pp. 1–6. IEEE, 2021.

22. Sriram, S., J. Manikandan, P. Hemalatha, and G. Leema Roselin. "A Chatbot Mobile Quarantine App for Stress Relief." In 2021 International Conference on System, Computation, Automation and Networking (ICSCAN), pp. 1–5. IEEE, 2021.

23. Leema, Roselin G., K. Kiruba, P. Hemalatha, J. Manikandan, M. Madhin, and Raj S. Mohan. "Revolutionizing Secure Commercialization In Agriculture Using Blockchain Technology." In 2021 International Conference on System, Computation, Automation and Networking (ICSCAN), pp. 1–6. IEEE, 2021.

24. Hemalatha, P., J. Manikandan, G. Leemarosilin, and P. Kanimozhi. "Sea Food Processing Using Internet of Things and Cloud Technologies." PalArch's Journal of Archaeology of Egypt/ Egyptology 17, no. 9 (2020): 5877–5885.

Recent Trends in Computational Intelligence and its Application – Sugumaran D. et al. (eds)
© 2023 Taylor & Francis Group, London, ISBN 978-1-032-48410-5

58

Computational Drug Discovery Using QSAR Modeling

Deepa J.*, R. Naveen[1], Kambam Aneka[2]
Vel Tech Rangarajan Dr. Sagunthala R&D Institute of Science and Technology

J. Velmurugan[3]
Saveetha Institute of Medical and Technical Sciences

Uma Rani V.[4]
Saveetha Engineering College

Merry Ida[5]
Loyola Institute of Science and Technology

Abstract—The biological activity of a compound refers to the way it interacts with the target protein/enzyme/virus either to inhibit or catalyze its actions. This bioactivity largely depends on the molecular structure along with some chemical properties. The proposed technique provides a computational way to analyze the potency of biological compounds in the treatment of bacterial infection caused by staphylococcus aureus. The ChEMBL-29 database (a collection of curated bioactivity of more than 2 million compounds) is used to acquire a dataset for the target disease. Further, this dataset is cleaned and probed to acquire bioactivity data of the drug compounds of the target and prepared using molecular descriptors (Lipinski and Padel descriptors). A statistical interpretation of the efficiency of compounds is obtained via a Quantitative structure-activity relationship model (a regression model) developed using the prepared dataset. The regression model of Staphylococcus aureus inhibitors used was compared with a set of various other models based on metric R-squared error as well as time taken measure using python's Lazy predict library, and the comparisons were visualized based on two Lipinski descriptors, Logp and MW that analyses active and inactive classes of chemicals and is observed that the two bioactivity classes occupy similar chemical regions.

Keywords—Computational drug discovery, QSAR modeling, Anti-MRSA, Drug prediction, Lipinski and Padel molecular descriptors, CheMBL-29

---

*Corresponding Author: vasdeepa03@gmail.com
[1]naveenvelur4@gmail.com, [2]anekareddy8@gmail.com, [3]velmuruganj.sse@saveetha.com, [4]umaranibharathy@gmail.com, [5]idamerry@gmail.com

DOI: 10.1201/9781003388913-58

I. Introduction

In the biological sciences, bioinformatics has risen to prominence as a discipline that allows scientists to grasp and handle the huge volumes of data that are now available. Drug discovery is one such discipline. The identification of new molecular entities that may prove to be of some value in diseases that qualify as unmet medical needs is the primary goal here. Quantitative structure-activity relationship is a paradigm that uses statistical or machine learning techniques to predict the biological activity of a compound of interest as a function of its molecular fingerprint [6]. Fundamental idea is that differences in structural characteristics result in various biological activity. Inspired by the above paradigm the proposed model aims to apply this technique [9] to biodata retrieved from the ChEMBL-29 database and help towards treatment difficulties of mrsa-caused infections by helping predict compounds' potential of acting as promising tools. The most common cause of human bacterial infections is Staphylococcus aureus. Despite the fact that these infections are generally asymptomatic, they can have devastating consequences in immunocompromised people, such as pneumonia, meningitis, or septicemias. Furthermore, in clinical infections, Staphylococcus aureus generates a high risk of morbidity, particularly in youngsters. In developed countries, getting medicine is tough. The advent of methicillin-resistant bacteria has aggravated the situation. MRSA strains with a high resistance rate are discovered every year all around the world.

The process includes the use of python libraries like pandas, matplotlib, etc., for dataset preparation and some standard molecular descriptors to finetune data. Finally, the Random forest algorithm, which proves to be a highly robust ensemble technique in the field of QSAR, was used in the model for prediction of useful compounds. Questions that can be answered using these computational models:

- What target proteins could a specified compound bind to and modulate?
- Would this compound bind unspecifically to other off target activities?
- Are there similar compound to a specific query compound that exert similar binding behavior?

The research investigates the vast field of bioinformatics using some data analysis and machine learning methodologies to gain useful insights. The bioinformatics field is mostly concerned with developing various new algorithms and analyzing biological systems for advances in the field of biology. The objective of the proposed work is Exploratory Data Analysis to summarize the main characteristics of drug compounds of the target, to understand the origins of biological activity data, to develop a QSAR model for Computational Drug Discovery, Interpretation of the specified model to design better therapeutic agents. A potential approach that concerns current needs of advancement in bioinformatics. Uses fact-checked and FDA approved characteristics, also financially viable due to minimum system software requirements.

II. Related Work

Cortes, E et al. [3] analyzed 24-active cannabinoids against MRSA by making use of various mathematical models obtained by multiple regression methods. Affinity to penicillin binding proteins divided the compounds to low and high affinity groups. Both types of compounds had showed similar interactions with penicillin binding proteins and hydroxyl group in the chemical structure played important role. Further examination with robust models and QSAR interpretation helped propose 3 new compounds that were ready for possible testing.

Liu, y, [8] explored the machine learning algorithms when applied to Quantitative structure-activity relationship modeling technique. No single strategy can be said to be consistently better than others. Ensemble Machine Learning, developed in this work, is a collection of classifiers whose individual conclusions are somehow merged to improve the overall system performance. Bagging and AdaBoost, two popular ensemble learning techniques for QSAR modelling, were compared. Inhibition of dihydrofolate reductase from E. coli by pyrimidines and inhibition of dihydrofolate reductase from rat or mouse tumors by triazines were investigated as test case issues. We found that the Bagging and AdaBoost ensemble learning algorithms significantly improve the performance of C4.5 and 1-R decision trees (p 0.05). but 1-N-N and naive Bayes did not. Adaboost was found to perform better.

Kwon, S et al. [6] Concerned about the limitations of prevalant RF and other ensemble techniques being limited to a single subject, Researchers have developed an exhaustive ensemble method that can build multi-subject models and combine them using second-level meta-learning An end-to-end neural network is also presented and focuses on an automatic sequential feature extractor from SMILES. When the separate models were integrated, they produced outstanding results.

Alsenan, S.A et al. [4] explain that in the field of huge boom in chemical data has been caused by medication research and computational methods for creating new chemical compounds. In this work, we investigate the impact of dimensionality reduction techniques on high-dimensional QSAR datasets. Since QSAR is multidimensional, dimensionality reduction approaches have become an important part of the modeling process. Principal Component Analysis is a extraction method along numerous uses in dimensionality reduction, visualization, and exploratory data analysis. Anyhow, due to the complicated structure of QSAR data, linear PCA is insufficient. Given the wide range of current feature extraction strategies,

a review was conducted to investigate different extraction techniques.

Likitha, S. et al. [7] propose a possible method for determining a compound's permeability across a certain layer. In this study, a method for predicting Caco-2 cell pharmacological permeability using QSAR regression is proposed. Chemical descriptors are calculated using the compounds' formation and structural properties to represent them. The chemical compounds' structures were used to generate a set of descriptors. Linear regression, nonlinear regression, and nonlinear artificial neural network models were all used to relate the reported permeability values. Each different machine learning and neural network model is used to train each of the two sets of chemical descriptors. Comparing the results with different regression models, we found that the boosting-based ML model returned the lowest values for standard metrics such as mean squared error and r-squared error.

Ishola, A. A et al. [5] employs QSAR modeling technique with genetic multiple regression to discover inhibitor compounds capable of treating Sars coronavirus infection. The model used had an R2 value of 0.50 for predictive correlation on the test set. The paper could successfully suggest 3 compounds worthy of further exploration.

Saw Simeon et al. [9] used QSAR modelling and molecular docking to look into the causes of human acetylcholinesterase inhibition. To learn more about the origins of the bioactivity of the two thousand five seventy compounds with described values IC50 against AChE, a sizable non-redundant data set was gathered from ChEMBL and used in the QSAR investigation. A set of 12 fingerprint descriptors were used to describe AChE inhibitors and used the random forest to build prediction models from 100 different data splits. The best model built using substructure numbers was selected according to OECD guidelines and provided R2, Q2CV, and Q2Ext values of 0.92 0.01, 0.78 0.06, and 0.78 0.05 respectively.

III. Methodology

Vast and diverse datasets with structurally distinct chemotypes are used in QSAR studies. Multiple linear regression and support vector machines have been used in numerous research to forecast target protein inhibition for a variety of medicines. The existing system [5] used a genetic algorithm multiple regression method applied to QSAR for the prediction of possible inhibiting compounds for the key drug target protein 3-C like protease. Several inhibitors were suggested that could be further investigated for the treatment of Covid-19.The developed QSAR model had an R2 value of 0.907, and the test data indicated an R2 pred value of 0.517 for the correlation coefficient.

In a nutshell, our model incorporates Modeling and evaluating MRSA inhibition using a large-scale QSAR. It is carried out according to the OECD guidelines:

1. A collection of data with a defined endpoint;
2. A clear learning algorithm;
3. The QSAR model's application domain;
4. Utilizing acceptable goodness-of-fit and predictivity measurements;
5. The QSAR model's mechanistic interpretation.

The proposed methodology applies Random Forest regression to QSAR modeling for the prediction of possible inhibitors against the target protein Methicillin Resistant Staphylococcus aureus(MRSA).The process used a similar Molecular descriptor for fingerprint generation but made a different approach for model building and finally analyzed the robustness of the model by using python's Lazypredict Library. The developed model had an improved R2 pred value of 0.55 and an R2 error value of 0.88 for the test set predicted correlation coefficient.

Figure 58.1 shows the architecture diagram that presents a collective briefing of the entire process followed to achieve the aim of the proposed work. Starting with the retrieval of bioactivity data using MRSA as a search keyword for pre-processing data and applying molecular descriptors. Finally building the QSAR model using the RF algorithm and using it to predict pIC50 values.

The data flow diagram which is stepwise guidance of the methodology followed throughout the proposed model is shown in Fig. 58.2. The oval boxes describe the actions are taken and arrows show the flow order.

A. Random Forest Algorithm with QSAR

A supervised learning method that internally makes use of various decision trees to get atomic results and finally integrates them all to give final output by majority vote in case of classification and averaging atomic results in case of regression. Due to this combined result of various individual trees, it also wipes out the problem of overfitting problem to much extent despite being time taking. The developing of dataset and working algorithm of RF (bagging ensemble method) model with QSAR used in this work is as follows:

1. Extract data for MRSA targets from CheMBL-29 database.
2. Retrieve data only for ORGANISM as target and preprocess data using pandas.
3. Check drug-likeness of compounds using Lipinski descriptor and use canonical smiles to obtain PubChem fingerprint dataset.
4. Build a Model using a predefined RF model from sklearn and keras.
5. The model takes x random samples from PubChem prints and assesses each one with a separate decision tree.
6. Finally predicts pIC50 values by averaging all x outputs.

B. QSAR Modeling

Following the principle that the biological activity of a compound varies with its structural

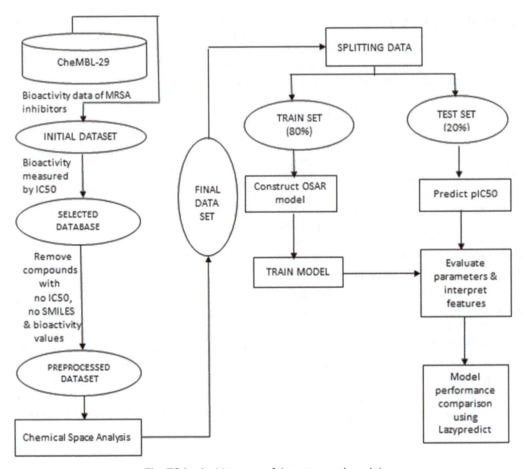

Fig. 58.1 Architecture of the proposed model

Source: Author

changes, quantitative structure-activity relationship is a mathematical approach that helps rank a large number of compounds according to their desired biological activity. QSAR analysis can be performed only when a training set of ligands with known biological activity is available. These training sets are used to build statistical models that link biological activities to molecular characteristics. Calculation of molecular descriptors (here Padel descriptor is used to obtain PubChem fingerprint data) that describe the topological or physicochemical features of molecules is required in QSAR analysis. Following the calculation of descriptors for the entire

dataset, statistical techniques such as several linear regression are acclimated and used to investigate the relationship between descriptors and experimental activities (MLR). RF is quite commonly applied machine learning algorithm to QSAR prediction in order to prevent its constraints as RF proves to be quite robust in terms of other.

IV. Experimental Results

The fingerprints dataset obtained after data preprocessing and applying Lipinski and Padel molecular descriptors is next used as input for the prediction model. The RF model used was

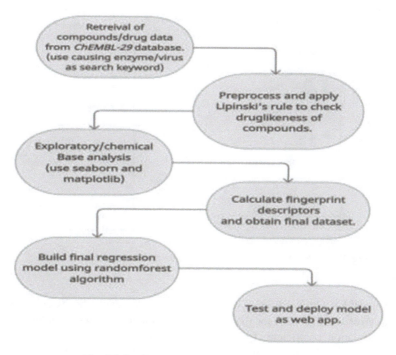

Fig. 58.2 Data Flow of the proposed model

Source: Author

found to be among top six performing models. The statistical values of the top six performing models from the lazypredict library are shown in Table 1. R-squared error and time taken are the measures employed. The R-squared error indicates how close the regression line (that is, the predicted values) is to the actual data values. The R-squared multiplier is a number between 0 and 1, with 0 indicating that the model does not fit the data and 1 indicating that the model fits the data set perfectly.

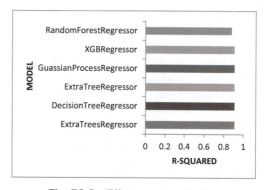

Fig. 58.3 Efficiency visualization

Source: Author

Figure 58.3 shows the graphical representation of efficiency of models when compared with used model. A better predictive correlation coefficient R2pred value of 0.55 and an R2 error value of 0.88 were achieved by the constructed model. The model lies among top 6 performing models when analyzed using Lazypredict python library as visualized above.

Table 58.1 Model performance statistics

Model	R-Squared	Time Taken
ExtraTreesRegressor	0.91	1.33
DecisionTreeRegressor	0.91	0.06
ExtraTreeRegressor	0.91	0.05
GuassianProcessRegressor	0.91	0.39
XGBRegressor	0.91	0.59
RandomForestRegressor	0.88	1.03

Source: Author

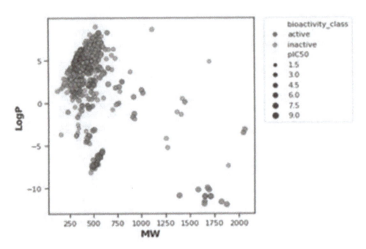

Fig. 58.4 Scatterplot comparing active and inactive compounds

Source: Author

The generated model has a higher R2pred value of 0.55 and R2 error value of 0.88 for the test set predicted correlation coefficient. Existing systems suggested several inhibitors that could be further investigated for the treatment of Covid-19. The QSAR model developed had an R2 value of 0.907 and the test set predicted a correlation coefficient R2 pred value of 0.517. Fig. 58.4 illustrates a scatterplot based on two Lipinski descriptors, Logp and MW that analyses active and inactive classes of chemicals and is observed that the two bioactivity classes occupy similar chemical regions.

V. Conclusion and Future Enhancements

The proposed model not only helps in the identification of potential bio compounds/drugs that could possibly help in the development of a better therapeutic agent for the serious bacterial infections caused by staphylococcus aureus but the descriptors used while building models also enable understanding of the origin of the biological activities. Therefore such a model could prove to be of great help in the field of drug discovery. Performance comparison with various other models from Lazypredict also proves the robustness of the prediction model.

This model could be extended further via a deployment as a web application that could make it open for researchers throughout the world and prove to be a great time-saver in proper chemical identification although limited to a single subject and useful only within its applicability domain. Of course, the methodology could be made a lot more robust and efficient with new algorithms existing or being proposed ahead.

References

1. Alsenan, S. A., Al-Turaiki, I. M., Hafez, A. M. (2020). Feature extraction methods in quantitative structure–activity relationship modeling: A comparative study. IEEE Access, 8, 78737–78752.
2. Bosc, N., Atkinson, F., Felix, E., Gaulton, A., Hersey, A., Leach, A. R. (2019). Large scale comparison of QSAR and conformal prediction methods and their applications in drug discovery. Journal of cheminformatics, 11(1), 1–16.
3. Cortes, E., Mora, J. and Marquez, E., 2020. Modelling the anti-methicillin-resistant Staphylococcus aureus (MRSA) activity of cannabinoids: A QSAR and docking study. Crystals, 10(8), p. 692.
4. Hu, S., Chen, P., Gu, P. and Wang, B., 2020. A deep learning-based chemical system for

QSAR prediction. IEEE Journal of Biomedical and Health Informatics, 24(10), pp. 3020–3028.

5. Ishola, A. A., Adedirin, O., Joshi, T., Chandra, S. (2021). QSAR modeling and Pharmacoinformatics of SARS coronavirus 3C-like protease inhibitors. Computers in biology and medicine, 134, 104483.

6. Kwon, S., Bae, H., Jo, J., Yoon, S. (2019). Comprehensive ensemble in QSAR prediction for drug discovery. BMC bioinformatics, 20(1), 1–12.

7. Likitha, S., Kamath, S. (2021, September). ML based QSAR Models for Prediction of Pharma-cological Permeability of Caco-2 Cell. In 2021 IEEE 4th International Conference on Computing, Power and Communication Technologies (GUCON) (pp. 1–6). IEEE.

8. Liu, Y. (2005, December). Drug design by machine learning: ensemble learning for qsar modeling. In Fourth International Conference on Machine Learning and Applications (ICMLA'05) (pp.7-pp). IEEE.

9. Saw Simeon, Nuttapat Anuwongcharoen, Watshara Shoombuatong, Aijaz Ahmad Malik, Virapong Prachayasittikul, Jarl E.S. Wikberg and Chanin Nantasenamat1, 2016. Probing the origins of human acetylcholinesterase inhibition via QSAR modeling and molecular docking. DOI 10.7717/peerj.2322.

10. Zhou, Y., Li, S., Zhao, Y., Guo, M., Liu, Y., Li, M., Wen, Z. (2021). Quantitative Structure–Activity Relationship (QSAR) Model for the Severity Prediction of Drug-Induced Rhabdomyolysis by Using Random Forest. Chemical Research in Toxicology, 34(2), 514–521.

Recent Trends in Computational Intelligence and its Application – Sugumaran D. et al. (eds)
© 2023 Taylor & Francis Group, London, ISBN 978-1-032-48410-5

59

Biomedical Image Segmentation Using Enhanced U-NET

Hariharan R.*, Mohan R.[1], Krunal Randive[2]

Department of Computer science and Engineering,
National Institute of Technology, Tiruchirappalli, India

Toonesan T. A.[3]

Department of Computer science and Engineering,
Government College of Technology, Coimbatore, India

Abi Sundar S.[4]

Department of Information Technology,
Vel Tech Rangarajan Dr. Sagunthala R&D Institute of Science and Technology, Chennai, India

Abstract—The research about tumor segmentation is a major topic infield of biomedical. Nowadays many research has developed in biomedical image segmentation in multimodal MRI to diagnose and monitoring of sickness progression. The world is developing in a rapid phase and we need to keep up and develop methods that can be sustained for a longer period of time as biomedical keeps demanding for new and innovative ideas for the benefits of human beings. From the study the gliomas are maligant and heterogeneous its very tedious task to find. As of considered many algorithmic activities are performed for the segmentation pre friendly environment and to perform the exact segmentation strategies which are used for the effectual delineation of tumor into intra tumor activities. In recent years we have seen a variety of methods for brain tumor segmentation and tumor detection based on the concept of deep learning and most of the approaches are still based on U-Net or CNN which limits processing speed and accuracy. This article propose the Enhanced U-Net for brain tumor segmentation is used to identify tumors more accurately and precisely and get the proper shapes and sizes of the tumor in output widely used for biomedical image segmentation, Convolutional Neural Networks have significant performance in multiple image processing field include brain tumor segmentation.

Keywords—Unet, Brats, Segmentation, CNN

I. Introduction

We planed to develop a model that helps radiologist to detect different types of brain tumor segmentation in each and every stage. Our model will help doctors to work in a more precise manner and form a more accurate view. Early stage of Identification of brain tumor

*Corresponding Author: hharanbtech@gmail.com

[1]rmohan@nitt.edu, [2]krunalrandive@gmail.com, [3]iamtoonesen@gmail.com, [4]asundar2002@gmail.com

DOI: 10.1201/9781003388913-59

will improve the life time of patients and also a chance of speed recovery. Processing and identification of tumors in MRI is much needed technique nowadays and it became a part in a treatment. As per WHO identification brain tumor is major deadly disease. Glioma is most normal tumor which occurs because of carcinogenesis of glial cells in brain as well as spinal. Magnetic Resonance Imaging (MRI), Is a general non damaging imaging technique which is captures medical image with more clear and detailed information about each pixel, so MRI is world wide used biomedical imaging technique.

Even though MRI imaging giving detail information about each pixel manual segmentation and analysis is very difficult task and time consuming process for radiologist. And also chance of getting different result in human segmentation. So avoid the deviation in result many researchers proposed automatic and semi automatic segmentation techniques using computer vision. In this article proposed a enhanced U net model. and we tested our model with Brats 2020 dataset which contains MRI images There are five type of images in the datatset including T1, T1ce, T2, Flair and mask. We tend to ignore T1 because T1ce is basically the same image as T1 but more detailed and hence provides, more information. Then we convert the image into numPy arrays for future tasks. Then training model has created and feed the data that we preprocessed to get results.

Our article focuses on the task of determining the segmenation of 3D-models of the brain to find the tumour growth and development using U-Net. In comparison to the original U-Net template, we made some changes to the U-Net architecture for the BraTS task.

II. Related Work

There have been many researches conducted onmedical image processing and segmentation across many fields. Many authors have published their research on MRI images segmentation by imagesusing CNN. Some of them include follows. RaminRanjbarzadeh, et al.proposed a attention deep learning model for multi model MRI brain images tested with 2018 BraTS dataset using a C-ConvNet/C-CNN algorithm where they are performing the computation on a small portion of the dataset to save computational time and then replicating them throughout the entire dataset [1].

Sourodip Ghosh, et al., they come up with an enhanced U –net using model for segmentation with VGG-16 for Brain MRI images which recognize 4 different region of tumor cell[2]. Automated analysis and segmentation of brain MRI images facilitates detection of neurological diseases and disorders. By analyzing his TCGA-LGG dataset (3929 images) from the TCI archive, the author designed improved U-Net results using a custom U-Net architecture, achieve the results from 0.994 to 0.9975. His U-Net pixel accuracy of range has achieves better. The results of this paper are relatively better than other CNN-based models. Mahnoor Ali, et al. Present a hybrid model with the combination of 3D CNN and a U-Net which give more accurate predictions[3]. BraTS-19 dataset has used to train the model separately to assess the terms segmentation tumor sub-regions which differs from each other in the segmentation maps and to clearly identify the difference in range of the pixel in final prediction. The propose hybrid model achieve dice score of 0.750for enhancing tumor, 0.906 for whole tumor, 0.846 for enhancing tumor in validation.

Hatamizadeh, et al. put forward a technique for 3D medical image segmentation, UNETR is the first successful transformer design [4]. They tried a concept of transformer as encoder which train the input sequence which observe the data in multi scaling information. The structure of UNERT follow similar as U net which having encoder and decoder part. Transformer encoder is directly connected Decoders over skip connections with different resolutions Compute the final output of the semantic segmentation. Tsung-Yi Lin, et al., put forward the Retina Net model based on two stages[5]. first is the one stage detector and second is the two stage detector. The concept for the retina net model arose with the problem of forward and backward image imbalance. One stage detector is technique is applied on the regular basis which has dense sampling of location of the possible item it is also faster and simpler way to implement but lags behind the accuracy of two stage detectors. Single unified network is a basic principle for retina detector model which composed of backbone network and two task-specific subnetworks.

Sergio Pereira et al., came up with a idea of preprogrammed segmentation method based on Convolutional Neural Networks (CNN), exploring small three cross three kernels [6]. The manipulation of less kernels approves designing a deeper architecture, except using a fantastic impact in opposition to overloading, given the network with the fewer variety of weights in the network. the notion was validated in the database BRATS 2013 dataset, acquiring simultaneously the first function for the complete, core, and Dice Similarity Coefficient metric (0.88, 0.83, 0.77) for the Challenge facts set.

III. Proposed System

Our proposed model uses UNET algorithm to segment brain tumor in MRI images. We have collected the BraTS 2020dataset from BRATS official website with legal access from the authorities. The images in the dataset are too large and uses a lot of computational resource and time to compute the result. So, we trimmed the images as per the requirement in the preprocessing part and filtered through the samples to get the bundle of images that can actually contribute significantly in the training part. We made the dataset into a training folder and validation folder in a ratio of 0. 75

and 0.25. We have adopted a preprocessing approach to obtain a flexible and effectual brain tumor segmentation system which work only on sample images, rather than the whole part of the data set. This method decrease the computing time 10 and overcomes the over fitting problems. Basically, we do all our computations in one image at first and then repeat that across the dataset. Then we create a custom data generator since keras image generator does not work for numpy images. Then we build a 3D UNET model. Either we can import the model or we can build the model from scratch. We have chose to build the model from scratch. Then we use the custom made data generator to feed the pre processed images into the UNET model. The advantage of using our proposed model is Faster and more precise results. Efficient system for diagnosis. And also Processing of modalities would lead to significant increase in sensitivity [8,9].

In this article, we have used the Brats 2020 dataset to test and train a model to segment brain tumor MRI images. The images in the dataset are too large and uses a lot of computational resource and time to compute the result. So, we trimmed the images as per the requirement in the preprocessing part and filtered through the samples to get the bundle of images that can actually contribute significantly in the training part. There are five type of images in the datatset including T1, T1ce, T2, Flair and mask. We tend to ignore T1 because T1ce is basically the same image as T1 but more detailed and hence provides, more information. Then we merge the image into numPy arrays for future tasks. Then we create the training model and feed the data that we pre processed to get results [10].

A. Architecture

Model architecture comprises the design of the proposed model and how the algorithm has been used in the model. Figure 59.1 The picture is a representation of how actually Enhanced UNET architecture models are present and stacked.

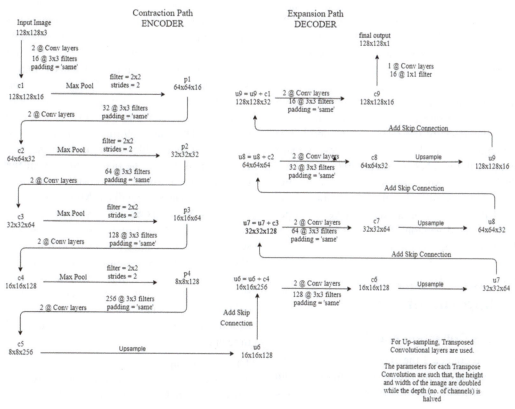

Fig. 59.1 Path of enhanced U Net architecture diagram

Source: Author

Convolution layer

The convolution layer is the first layer of our model which is responsible for extracting the features from an image. The convolution layer contains many hidden layer inside it and each of which is responsible to detect an unique feature of the image. The convolution layer uses filters to scan through the images. The size of the filter can be specified by the user and this is moved across the image from top right corner to bottom left corner. A value is calculated for each point in the image that the filter passes through by a convolution operation. A feature map is generated for each filter after the filters passes through the image. These feature maps are then passed through an activation function which decides if the respective feature is present or missing. The output of a convolution layer is always downsampled.

Padding layer

Padding layer basically helps in increasing the area of the image which is processed by CNN by adding rows and columns to the outer dimension of the images. This way,it helps save the information which are present at the corner of the images from getting chopped off by the convolution process. This helps in maintaining the output size of the data. It helps with a better and more accurate analysis of the image. We have used 'same' padding in our proposed method which basically adds zero values in the outer frame of the image

Pooling layer

The Pooling layer works on each feature map individually and it gives out a new set of pooled feature maps. Pooling basically is like applying a filter to a feature map. The size of the filter is smaller than the size of the image. In most of the cases,the size of pooling filter is usually 2x2 with a stride of 2 pixels. There are mainly two types of pooling are max Pooling and average pooling. We are using max pooling in our proposed method. The Max Pooling layer is used to take the maximum pixel value from an adjacent group of pixels thereby reducing the dimensions of the feature maps and hence reducing the number of parameters to learn and also the number of computations required. Basically, the whole point of using a pooling layer and the minimize the dimension of the input image by using a pooling layer is to get a summarized version of important features of the input image.

IV. Steps Involved in Implementation

A. Data Collection and Preprocessing

We have collected the BraTS 2020 dataset for multi-modal brain tumor segmentation from Brats official website. After collecting the

dataset pre-processing to be done in the dataset. The dataset contains images in nii format. For each volume of brain, there are four channels of information in terms of T1,T1ce,T2 and FLAIR. T1ce(T1 contrast enhanced) image is the enhanced version of T1 image and basically,contains the same data but more detailed. So, we can ignore the T1 Fig. 59.2 and Fig. 59.3. There are 369 images/- folder in the dataset and each folder contains 4 images and 1 segmented mask. The pixel values of all the images is not uniform. So, to bring uniformity among the images, we will use Min Max Scaler. But to use Min Max Scaler, we have to convert the 3D images to 1D image and then use the scaler and then we have to transform it back to 3D. The next part is to combine all the volumes of the image into a numpy array to be used for testing. A lot of the area inside the image contains no data and is blank and we have to trim down on those part to save computational time and resources.

Then we compare all the numPy arrays we got against the mask and if the array con tains less then 1 percent of the mask, then we

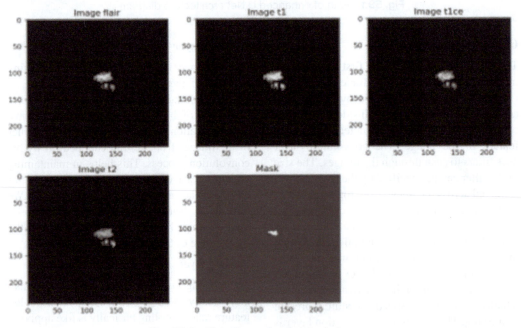

Fig. 59.2 Five channels of a single image

Source: Author

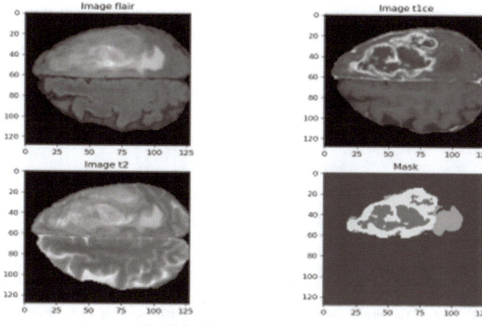

Fig. 59.3 After cropping of the channels

Source: Author

tend to ignore it because it contribute nothing to the training of the model and is a waste to the training resources. Pre-processing of data involves transforming the image samples in the dataset in a manner which can be easily readable by the model including augmenting the images to help the model train better.

B. Building The Custom Data Generator

The Keras Image Data Generator does not work for. npy image. So, we have to build a custom data generator that loads our images into the training model that we build. After building the generator, we test it before using it.

C. Building and Training the Model

Building the UNET model is about importing all the necessary packages and the CNN layers that are required for the building of the model. A CNN model is made using keras and all the layers like convolution layers, ReLU and max pooling layer. The UNET architecture consists

of a contracting path and an expansive path. The contracting path follows the typical architecture of a convolutional network. Training the UNET model is about using and feeding the image samples in the dataset to the UNET model for it to learn continuously over and over again. Target size determines the final size of the image that is being fed into the model. The dataset is way too large for the model to pass all the images in a single epoch. So, batch size determines the number of images to be passed through the model in a single epoch. Increasing the number of epochs helps the model to train better. The trained model is then again tested against the testing dataset. Optimizer is used to increase the accuracy metric. We are using Keras Adam optimizer. The other metrics that we are using is dice loss, focal loss and total loss in equation 1.

Total loss = dice loss + (1 * focal loss)   (1)

We can assign weights to the four classes as we want but not assigning the weights gives a message to the model for all the weights to be

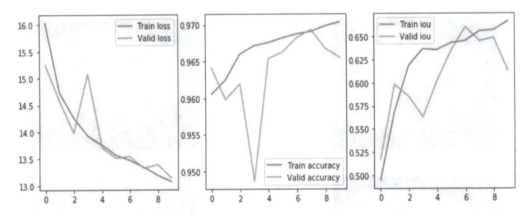

Fig. 59.4 Graph for loss, accuracy and IoU for enhanced U-Net

Source: Author

taken equal. Another metric that we are using is IOU score. The input shape of our model is (imglength, imgwidth, imgheight, channel) i.e. (128, 128, 128, 3). The input shape of our model is (imglength, imgwidth, imgheight, channel) i.e. (128, 128, 128, 4). Fig. 59.4 is graphical representation of training and validation loss, accuracy and IOU.

V. Result and Discussion

In Table 59.1 we have compared proposed model with various Standard u net and hybrid models with three different mean values of enhanced tumour (ET), whole tumour (WT) and tumour core (TC). Among this three mean values proposed method gives best results in mean ET and Mean TC.

Table 59.1 Comparison with other models

Model	Mean (ET)	Mean (WT)	Mean (TC)
nn U-Net	0.74	0.90	0.83
nn U-Net+Ranger	0.77	0.90	0.83
nn U-Net+GWDL	0.77	0.91	0.83
nn U-Net+DRO	0.76	0.91	0.83
Proposed Model (Enhanced UNet)	0.81	0.88	0.84

Source: Author

The Output is a segmented image fig 5 which is basically the brain tumor that we aim to find in the project. The output we have achieved has got an IOU score of 83.4 percent. IOU score is basically the amount of image that coincides with the mask. We have achieved the model which provides efficient and robust segmentation compared to manual delineated ground truth.

VI. Conclusion

There are a lot of CNN networks available and there have been a lot of research papers that actually use other CNN networks for MRI image segmentation but none of them actually come close to UNET when it comes to image segmentation. The dataset that we are using is BraTS 2020 multimodal image segmentation dataset which contains 369 images and 4 different channels and 1 mask each for each image. We have divided it into a ratio of 0.25 to 0.75 for validation to training image folder. We apply the preprocessing to one image first, and then apply the same to all the images. Then we create a custom data generator since keras image generator does not work for npy images. Then we build a 3D UNET model. Either we can import the model or we can build the model

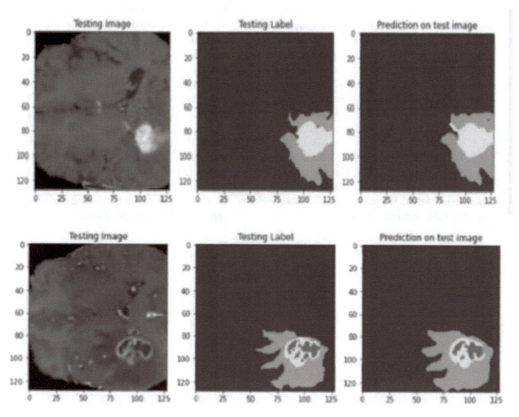

Fig. 59.5 Sample output image

Source: Author

from scratch. We have chose to build the model from scratch. Then we use the custom made data generator to feed the pre processed images into the UNET model. We have used IOU score as a metric to track the progress record. We have achieved a mean IOU score of 83.4 percent. There are many papers using UNET architecture for brain tumor segmentation but they are ignoring the T1 channel of the image while pre-processing of the dataset just because T1ce is available. But using T1 can be proven beneficial.

Reference

1. Ranjbarzadeh, R., Bagherian Kasgari, A., Jafarzadeh Ghoushchi, S., Anari, S., Naseri, M. and Bendechache, M., 2021. Brain tumor segmentation based on deep learning and an attention mechanism using MRI multi-modalities brain images. Scientific Reports, 11(1), pp. 1–17.

2. Ghosh, S., Chaki, A. and Santosh, K.C., 2021. Improved U-Net architecture with VGG-16 for brain tumor segmentation. Physical and Engineering Sciences in Medicine, 44(3), pp. 703–712.

3. Ali, M., Gilani, S.O., Waris, A., Zafar, K. and Jamil, M., 2020. Brain tumour image segmentation using deep networks. IEEE Access, 8, pp. 153589–153598.

4. Hatamizadeh, A., Nath, V., Tang, Y., Yang, D., Roth, H.R. and Xu, D., 2022. Swin unetr: Swin transformers for semantic segmentation of brain tumors in mri images. In International MICCAI Brainlesion Workshop (pp. 272–284). Springer, Cham.

5. Lin, T.Y., Goyal, P., Girshick, R., He, K. and Dollár, P., 2017. Focal loss for dense object detection. In Proceedings of the IEEE international conference on computer vision (pp. 2980–2988).

6. Pereira, S., Pinto, A., Alves, V. and Silva, C.A., 2016. Brain tumor segmentation using convolutional neural networks in MRI images. IEEE transactions on medical imaging, 35(5), pp. 1240–1251.

7. Chen, W., Liu, B., Peng, S., Sun, J. and Qiao, X., 2018, September. S3D-UNet: separable 3D U-Net for brain tumor segmentation. In International MICCAI Brainlesion Workshop (pp. 358–368). Springer, Cham.

8. Mehta, R. and Arbel, T., 2018, September. 3D U-Net for brain tumour segmentation. In International MICCAI Brainlesion Workshop (pp. 254–266). Springer, Cham.

9. Futrega, M., Milesi, A., Marcinkiewicz, M. and Ribalta, P., 2022. Optimized U-Net for brain tumor segmentation. In International MICCAI Brainlesion Workshop (pp. 15–29). Springer, Cham.

10. Dong, H., Yang, G., Liu, F., Mo, Y. and Guo, Y., 2017, July. Automatic brain tumor detection and segmentation using U-Net based fully convolutional networks. In annual conference on medical image understanding and analysis (pp. 506–517). Springer, Cham.

Recent Trends in Computational Intelligence and its Application – Sugumaran D. et al. (eds)
© 2023 Taylor & Francis Group, London, ISBN 978-1-032-48410-5

60

A Survey of Relevance Prediction Based Focused Web Crawlers on the Web

Sakunthala Prabha K. S.[1], Mahesh C.[2]
Department of Information Technology,
Veltech Rangarajan Dr. Sagunthala R & D Institute of Science and Technology, Chennai

Abstract—The demand for efficient and economical crawling strategies has skyrocketed together with the rapid expansion of dynamic web contents. As a corollary, multiple novel ideas have been put forth including focused crawling, which emerged as the most relevant. The focused crawlers are useful to search for web pages that satisfy the already established notions. Because to its excellent filtering and minimal memory and processing time demands, focused crawler caught the eye of several search engines. This paper offers a survey on web crawling that is centered on relevance computation. From the literature that is currently accessible, 63 focused crawlers are grouped into three categories as classic, semantic and learning focused crawler. Each metric's importance and impact on precision, recall, and harvest rate are examined. For the advantage of the users, future trends, bottlenecks, and solutions are also discussed.

Keywords—Web crawler, Focused crawler, Semantic crawler, Learning crawler, Machine learning

I. Introduction

The World Wide Web (WWW) now includes more than 1.9 billion web pages in total, and that number is still increasing daily. Web content like statistics, multimedia and schedules also grows dynamically over a specific period of time. The colossal data formation on the internet, searching a valuable information within stipulated time becomes challenging. Web crawler in search engines are the key facing the aforementioned challenges. Web crawlers also known as Internet robots, bots, or spiders is a prime part of search engine. The programmed bots or scripts behind a search engine which retrieve Web sites by repeatedly accessing Uniform Resource Locator (URL) links. The dynamic change in web contents creates hard challenges in maintaining the current indices. Traditional crawlers also consume large storage and bandwidth resources. The focused crawlers only download the most relevant web pages rather than downloading URLs randomly they come across. Beginning its visits to websites

[1]prabha1414@gmail.com, [2]chimahesh@gmail.com

DOI: 10.1201/9781003388913-60

from the seed URLs, the focused web crawler calculates the similarity score of the unexplored outgoing links in the seed website pages. Priority will be assigned and preserved in a web page repository depending on the relevance score. Until a considerable number of web pages have been downloaded, this procedure will be continued.

A comparative study focusing category, algorithms used, actions, dataset used, metrics used for evaluation, and evaluation results, Target variables as six different features for comparison are discussed. The comparison gives a overview and weight of individual inputs as given in Table 60.1, 60.2, 60.3.

Table 60.1 Classic focused crawlers

Category	Algorithms used	Actions	Evaluation Metrics and results	Target Variables
Focused Crawler [1]	Reinforcement learning	To prioritize document requests	Average relevance score - Not Provided	Contents of the web page
Focused Crawler [2]	Fish search crawling algorithm	To find the "optimal" order in which to retrieve nodes	The average measure	Keyword, Regular Expression, URL
Focused Crawler [3]	Shark Search crawling algorithm	To enter regions where a higher populace of relevant pages	Improvement Ratio	Contents of the web page, Inherited Score, Neighborhood Score
Hybrid focused crawler [4]	Vivisimo clustering algorithm	To compute a number of links relevant to the given query	Harvest rate - 85.5%	In-link, out-link and contents of the web page
Focused Crawler [5]	Relevancy-Context-Graph	To find the visit order of URLs by calculating the relevancy of a given topic	Average Relevance	Links, Rank of the web page
SmartCrawler[6]	Adaptive Learning	To prioritize most relevant web pages	Harvest Rate	Textual contents of the web page, Anchor text, URL
Focused crawler for digital repository [7]	Popularity of crawled pages	To download most relevant URLs based on popularity	Precision	Full Text and Anchor Text
Topic-driven focused crawler [8]	Hypertext Content Latent Analysis	Discover the information without having the previous knowledge of link structure	Harvest Rate - 95%	Contents of the web page, URL context
Focused Crawler [9]	Directed Acyclic Graph (DAG)	To download most relevant URLs based on search criteria	Crawling Time	Textual contents of the web page, URL context
Focused Crawler [10]	Dynamic Threshold Adjustment	Calculates the relationship between web page and some particular topic by the contents in the web page and the crawler topic	Precision and Recall	Textual contents of the web page, Title of the web page

Source: Author

A. Comparative Study of the Classic Focused Web Crawlers

The greater part of classic focused crawlers are integrated into extravagant systems like web search engines, while some are designed as standalone tools. Classic focused crawlers find the relevance on the basis of text contents in the web, anchor texts, and URL links. The performance evaluation of most of the classic focused crawlers done using the following metrics precision, recall, harvest rate and F-Measure, average relevance score in some cases. The avoidance of evaluation metrics and outcome measures a low efficient crawling of URLs in classic focused crawlers. The categorization of classic focused crawler with the relevance of dataset used, evaluation metrics and results and target variables used are given in Table 60.2.

Table 60.2 Semantic focused crawlers

Category	Algorithms used	Actions	Ontology used	Evaluation Metrics and results	Target Variables
LSCrawler [11]	Link Extractor and Hypertext Analyzer	Assign priority to the URL based on the relevancy	Ontology developed using ARP of Jena API	Harvest Rate and Recall	Textual contents of the web page and the domain ontology
Onto crawler [12]	fuzzy inference system	To find the web pages which are semantically relevant	Ontology developed using protégé	Precision - 0.87	Extracted terms from the web page and terms in domain ontology
ALVIS [13]	Subject Classification	To retrieve highly relevant web pages by ranking	Ontology developed manually	Relevance Score	Textual contents of the web page and the domain ontology
THESUS [14]	DBSCAN	Collect all the relevant web pages and split into smaller sets	DMOZ Web Directory	F-Measure - 0.519	Textual contents of the web page and the domain ontology
Semantic Focused Crawler [15]	Simulated Annealing	Minimize the semantic conflicts by gathering relevant information	Ontology developed using Web Services Modeling Language	Average Matching Ratio - 12%	User context and semantic annotations
Semantic Focused Crawler [16]	SSVSM Algorithm	The priority of the unvisited hyperlink	WordNet	Harvest Rate- 0.393, Average Similarity, Average Error	Textual contents of the web page and the domain ontology
Topic-Specific Focused Crawler [17]	Semantic Ranking	To retrieve relevant web pages	Google Web API	Precision, Recall, Harvest Rate	Core and non-core concepts of the web page
OntoPortal Crawler [18]	Semantic Webpage Generator	Web pages that are searched and retrieved semantically	Ontology developed using protégé	Precision - 21.93% Recall - 38.51%	Textual contents of the web page and the domain ontology

Category	Algorithms used	Actions	Ontology used	Evaluation Metrics and results	Target Variables
Swoogle [19]	N-Gram	To index and retrieve web documents semantically	FOAF ontology	Mean of rank, Harvest Rate	Textual contents of the web page and the domain ontology
BioSpider [20]	Genetic algorithms	To collect semantic data	Knowledge Base	Energy, Content exploration rate, Unknown state rate - 0.615	Textual contents of the web page and the domain ontology
Bio-Crawler [21]	BioTope	To retrieve web pages semantically	Manually developed OWL files	Harvest Rate	Textual contents of web page and domain ontology
Semantic focused Crawler [22]	RDF-based relational metadata	To retrieve document and metadata in parallel	Karlsruhe Ontology	Harvest Rate	Textual contents of web page and domain ontology

Source: Author

B. Comparative Study of the Semantic Focused Crawlers

The main component of a semantic focused crawler is ontology. They determine the degree to which URLs and ontologies are relevant. Precision, recall, harvest rate, and F-Measure are employed in most semantic focused crawlers' performance evaluations, along with average matching ratio in some instances. The categorization of semantic focused crawler with the relevance of dataset used, evaluation metrics and results and target variables used are given in Table 60.2.

C. Comparison of the Learnable Focused Crawlers

Training examples is the prime part of the learning focused crawlers. The user preferences for the topic are learned by learnable focused crawlers from a set of training pages known as a training set. The performance evaluation of most of the Learning focused crawlers done using the following metrics precision, recall, harvest rate and F-Measure, Search length in some cases. But some of the classic focused crawlers do not convey the evaluation metrics and evaluation outcomes. The categorization of learning focused crawler with the relevance of dataset used, evaluation metrics and results and target variables used are given in Table 60.3.

II. Conclusion

An overview of the current focused web crawlers is provided in this paper. The present focused web crawlers are categorized into 3 classes based on how they perform as classic, semantic and learning focused web crawlers. Precision, recall, and harvest rate are the three major areas of improvement that are common for every group. For possible new options, readers were also given a relationship between each input and output.

References

1. A. Patel and N. Schmidt, "Application of structured document parsing to focused web crawling," *Comput. Stand. Interfaces*, vol. 33, no. 3, pp. 325–331, 2011.
2. P. De Bra, G.-J. Houben, Y. Kornatzky, and R. Post, "Information Retrieval in Distributed

Table 60.3 Learning focused crawlers

Category	Algorithms used	Actions	Evaluation Metrics and results	Target Variables
Learnable focused crawler [23]	Hidden Markov Model (HMM)	To predict the relevant web pages	Recall - 61.8	Textual contents of the web page
Learnable focused crawler [24]	Decision Tree	To prioritize the unvisited URL's	Recall - 50%	Anchor texts of the web page
Learnable focused crawler [25]	Naive Bayes classifier	To find the off-topic documents which are highly relevant to the search	Harvest Rate	Textual contents of the web page and the Labels
Context Focused Crawler (CFC) [26]	Modified Naive Bayes Classifier	To find most valuable pages	the fraction of pages	Textual contents of the web page and the Labels
Learnable focused crawler [27]	Sequential covering algorithms	To calculate the relevance score of unvisited links.	Harvest Rate - 91.36%	Textual contents of the web page and the Labels
Panorama [28]	Naive Bayes, Spherical k-means clustering algorithm	To gather a huge collection of related Web pages	Harvest Rate - Not Provided	Contents of the web page
Graph-Based Sentiment (GBS) Crawler [29]	Graph-based tunneling mechanism	To identify the pages which contains opinions	Recall - 0.7888 Precision - 0.3199 F-Measure - 0.3915	Retrieved words, Retrieved relevant words
Learnable focused crawler [30]	Markov chain	To find the nearest relevant web page	Recall, Harvest Rate	Relevant and Irrelevant web pages
Intelligent crawler [31]	Q-Value Approximation algorithm	To select a query from the experience	Harvest Rate	Reward function of state and action
Learnable focused crawler [32]	ANN classifier	To find the most relevant web pages	Harvest Rate - 0.7924	Crawled topics and the topics in the domain ontology
Learnable focused crawler [33]	Support Vector Machine	To find the relevancy of unvisited web pages	Target Recall - 85.23%	Unvisited hyperlink, URL
Learnable focused crawler [34]	Apprentice Learner	To prioritize the URL's	Recall	Semantic annotated web page contents and the given topic
History crawler [35]	History classifier	Predict unvisited website segment	Harvest Rate	Link based feature, Anchor text based feature
InfoSpiders [36]	Local selection algorithm	Find relevant web pages to user's query	Precision and recall	Compact document, Relevance feedback

Category	Algorithms used	Actions	Evaluation Metrics and results	Target Variables
MySpiders [37]	Neural Network	Relevant information discovery	Precision and recall	Compact document, Relevance feedback
Learning Focused Crawler [38]	O-SVM	Download most relevant web pages for a given topic	Harvest rate	Textual contents of the web page, URL, Meta data, Parent page
Learning Focused Crawler [39]	Q-learning Based Link Prediction (QBLP) algorithm	Download most relevant web pages for a given topic	Precision, Recall	Textual contents of the web page, Anchor text, Link type, Document type, Link relevance, Document relevance

Source: Author

Hypertexts," *RIAO '94 Intell. Multimed. Inf. Retr. Syst. Manag.*, pp. 481–491, 1994.

3. M. Hersovici, M. Jacovi, Y. S. Maarek, D. Pelleg, M. Shtalhaim, and S. Ur, "The shark-search algorithm. An application: tailored Web site mapping," *Comput. Networks ISDN Syst.*, vol. 30, no. 1–7, pp. 317–326, 1998.

4. M. Jamali, H. Sayyadi, B. B. Hariri, and H. Abolhassani, "A method for focused crawling using combination of link structure and content similarity," *Proc. - 2006 IEEE/WIC/ACM Int. Conf. Web Intell. (WI 2006 Main Conf. Proceedings), WI'06*, pp. 753–756, 2007.

5. C. C. Hsu and F. Wu, "Topic-specific crawling on the Web with the measurements of the relevancy context graph," *Inf. Syst.*, vol. 31, no. 4–5, pp. 232–246, 2006.

6. F. Zhao, J. Zhou, C. Nie, H. Huang, and H. Jin, "SmartCrawler: A two-stage crawler for efficiently harvesting deep-web interfaces," *IEEE Trans. Serv. Comput.*, vol. 9, no. 4, pp. 608–620, 2016.

7. J. R. Park, C. Yang, Y. Tosaka, Q. Ping, and H. El Mimouni, "Developing an automatic crawling system for populating a digital repository of professional development resources: A pilot study," *J. Electron. Resour. Librariansh.*, vol. 28, no. 2, pp. 63–72, 2016.

8. G. Almpanidis, C. Kotropoulos, and I. Pitas, "Combining text and link analysis for focused crawling-An application for vertical search engines," *Inf. Syst.*, vol. 32, no. 6, pp. 886–908, 2007.

9. G. H. Agre and N. V. Mahajan, "Keyword focused web crawler," *2nd Int. Conf. Electron. Commun. Syst. ICECS 2015*, pp. 1089–1092, 2015.

10. Z. Geng, D. Shang, Q. Zhu, Q. Wu, and Y. Han, "Research on improved focused crawler and its application in food safety public opinion analysis," *2017 Chinese Autom. Congr.*, pp. 2847–2852, 2017.

11. M. Yuvarani, N. C. S. N. Iyengar, and A. Kannan, "LSCrawler: A framework for an enhanced focused web crawler based on link semantics," *Proc. - 2006 IEEE/WIC/ACM Int. Conf. Web Intell. (WI 2006 Main Conf. Proceedings), WI'06*, pp. 794–797, 2007.

12. O. Jalilian and H. Khotanlou, "A new fuzzy-based method to weigh the related concepts in semantic focused web crawlers," *ICCRD2011 - 2011 3rd Int. Conf. Comput. Res. Dev.*, vol. 3, pp. 23–27, 2011.

13. A. Ardö, "Focused crawling in the ALVIS semantic search engine," *Proceedings, 2nd Annu. Eur. Semant. Web ...*, 2005.

14. M. Halkidi, B. Nguyen, I. Varlamis, and M. Vazirgiannis, "THESUS : Organizing Web Document Collections Based on Semantics and Clustering."

15. J. J. Jung, "Towards open decision support systems based on semantic focused crawling," *Expert Syst. Appl.*, vol. 36, no. 2 PART 2, pp. 3914–3922, 2009.

16. Y. Du, W. Liu, X. Lv, and G. Peng, "An improved focused crawler based on Semantic

Similarity Vector Space Model," *Appl. Soft Comput. J.*, vol. 36, pp. 392–407, 2015.

17. Y. Du, Q. Pen, and Z. Gao, "Data & Knowledge Engineering A topic-speci fi c crawling strategy based on semantics similarity," *Datak*, vol. 88, pp. 75–93, 2013.

18. S. Y. Yang, "OntoPortal: An ontology-supported portal architecture with linguistically enhanced and focused crawler technologies," *Expert Syst. Appl.*, vol. 36, no. 6, pp. 10148–10157, 2009.

19. L. Ding, T. Finin, P. Reddivari, R. S. Cost, and J. Sachs, "Swoogle : A Search and Metadata Engine for the Semantic Web," *ACM Conf. Inf. Knowl. Manag.*, pp. 652–659, 2004.

20. C. Dimou, A. Batzios, A. L. Symeonidis, and P. A. Mitkas, "A multi-agent simulation framework for spiders traversing the semantic web," *Proc. - 2006 IEEE/WIC/ACM Int. Conf. Web Intell. (WI 2006 Main Conf. Proceedings), WI'06*, pp. 736–739, 2007.

21. A. Batzios, C. Dimou, A. L. Symeonidis, and P. A. Mitkas, "BioCrawler: An intelligent crawler for the semantic web," *Expert Syst. Appl.*, vol. 35, no. 1–2, pp. 524–530, 2008.

22. M. Ehrig and A. Maedche, "Ontology-focused crawling of Web documents," *Proc. 2003 ACM Symp. Appl. Comput. - SAC '03*, pp. 1174–1178, 2003.

23. H. Liu, J. Janssen, and E. Milios, "Using HMM to learn user browsing patterns for focused Web crawling," *Data Knowl. Eng.*, vol. 59, no. 2, pp. 270–291, 2006.

24. J. Li, K. Furuse, and K. Yamaguchi, "Focused crawling by exploiting anchor text using decision tree," *Spec. Interes. tracks posters 14th Int. Conf. World Wide Web - WWW '05*, p. 1190, 2005.

25. W. Wang, X. Chen, Y. Zou, H. Wang, and Z. Dai, "A focused crawler based on naive Bayes classifier," *3rd Int. Symp. Intell. Inf. Technol. Secur. Informatics, IITSI 2010*, pp. 517–521, 2010.

26. M. Diligenti, F. Coetzee, S. Lawrence, C. L. Giles, and M. Gori, "Focused crawling using context graphs," *Proc. 26th ...*, pp. 527–534, 2000.

27. Q. Xu and W. Zuo, "First-order focused crawling," *Proc. 16th Int. Conf. World Wide Web - WWW '07*, p. 1159, 2007.

28. G. P. G. Pant, K. Tsioutsiouliklis, J. Johnson, and C. L. Giles, "Panorama: extending digital libraries with topical crawlers," *Proc. 2004 Jt. ACM/IEEE Conf. Digit. Libr. 2004.*, no. Section 4, pp. 142–150, 2004.

29. T. Fu, A. Abbasi, D. Zeng, and H. Chen, "Sentimental Spidering," *ACM Trans. Inf. Syst.*, vol. 30, no. 4, pp. 1–30, 2012.

30. K. H. K. Hu and W. S. W. W. S. Wong, "A probabilistic model for intelligent Web crawlers," *Proc. 27th Annu. Int. Comput. Softw. Appl. Conf. COMPAC 2003*, pp. 0–4, 2003.

31. Q. Zheng, Z. Wu, X. Cheng, L. Jiang, and J. Liu, "Learning to crawl deep web," *Inf. Syst.*, vol. 38, no. 6, pp. 801–819, 2013.

32. H. T. Zheng, B. Y. Kang, and H. G. Kim, "An ontology-based approach to learnable focused crawling," *Inf. Sci. (Ny).*, vol. 178, no. 23, pp. 4512–4522, 2008.

33. G. Pant and P. Srinivasan, "Link contexts in classifier-guided topical crawlers," *IEEE Trans. Knowl. Data Eng.*, vol. 18, no. 1, pp. 107–122, 2006.

34. S. Chakrabarti, K. Punera, and M. Subramanyam, "Accelerated focused crawling through online relevance feedback," *Proc. Elev. Int. Conf. World Wide Web - WWW '02*, p. 148, 2002.

35. T. Suebchua, B. Manaskasemsak, A. Rungsawang, and H. Yamana, "History-enhanced focused website segment crawler," *Int. Conf. Inf. Netw.*, vol. 2018-Janua, pp. 80–85, 2018.

36. M. A. E. Menczer F., "Scalable Web Search by Adaptive Online Agents: An InfoSpiders Case Study," *Klusch M. Intell. Inf. Agents, Springer, Berlin, Heidelb.*, 1999.

37. F. Menczer, "Complementing search engines with online web mining agents," *Decis. Support Syst.*, vol. 35, no. 2, pp. 195–212, 2003.

38. M. Kovačevič and C. H. Davidson, "Crawling the construction web - A machine-learning approach without negative examples," *Appl. Artif. Intell.*, vol. 22, no. 5, pp. 459–482, 2008.

39. C. Dong, L. Fang, J. Yan, and B. Shi, "Semantic focused crawler based on Q-learning and Bayes classifier," *Proc. - 2010 3rd IEEE Int. Conf. Comput. Sci. Inf. Technol. ICCSIT 2010*, vol. 8, pp. 420–423, 2010.

Recent Trends in Computational Intelligence and its Application – Sugumaran D. et al. (eds)
© 2023 Taylor & Francis Group, London, ISBN 978-1-032-48410-5

61

IndexedByText

Arthi A.[1]

Professor, Department of Artificial Intelligence and Data Science,
RIT Institute of Technology, Chennai

Mahesh[2]

Professor,Department of Information Technology,
Veltech Rangarajan Dr Sagunthala R&D Institute of Science and Technology, Chennai

Abu M.[3]

Senior Transaction Processing Facility Developer, Tata Consultancy Services, Chennai

Muthuraj M.[4]

Program Analyst Trainee, Cognizant Technology Solutions, Chennai

Sanjay Gnanam U.[5]

Software Engineer, MC Consulting, Chennai

Abstract—In an era where information has become wealth, we are used to handle high quantity of data in the form of images. Lots of images containing text is regularly circulated among students and business individuals either through personal chats or emails. This excessive distribution leads to a disorganized setup where it takes a significant amount of time to find the image resource that we need from the huge image collection on our physical storage. With the help of optical character recognition algorithms, it is possible to extract the text from an image, which will be paired with the respective image. The images can be searched with the help of keywords given by the user. By comparing the keyword with the extracted text of each image, the resultant matching images will get displayed. We will also provide constraints based on image metadata, which includes parameters such as date. Each individual image will be provided with options to share it to various communication platforms. Along with it, if multiple images are shown as result, we offer a PDF generation function which would compile the respective images into a single PDF file.

Keywords—Indexing, Optical character recognition, Text search, PDF, Unorganized images

I. Introduction

Images have become a key part of information circulation. In this decade data generation has been an indispensable component of everyday life, it's no surprise that images [1] constitute a major part of it. In social media, millions of images are being circulated between friends,

[1]arthi179@gmail.com, [2]chimahesh@gmail.com, [3]abumurugesan@gmail.com [4]mugeshrajmd37900@gmail.com, [5]cdmsanjayrajan@gmail.com

DOI: 10.1201/9781003388913-61

posted on the timelines, shared on Whatsapp, posted as status messages, etc. Therefore, it is natural to assume that an average human would have stored a major chunk of the data on his mobile in the form of images.

Due to the frenzy lifestyle followed by many, not everyone will spend their precious time on organizing the content on their phones for simplifying access. Some of the images they save will practically get lost in the mountain load of images that will get saved as time passes. On such cases, when the user suddenly has a reminiscence about the image they need at that moment's notice, it would be a labored task for them to go through each image to find the one they need and waste their time over such simple issues. This is where we aim to make a difference through IndexedByText.

Being an "Image Gallery" application as it's core, it saves the user time to needlessly scroll through other file formats. The main component of this application is a search module, which performs keyword search on the images which are now identified by the text present in them instead of the filename.

The text extracted with the Google Machine Learning Kit's Text Recognition Application Programming Interface is used for the search. Through this, the average search time for the desired image is reduced to a significant level. The user can either directly share the searched image to the desired receiver (or) to his social media accounts. They are provided with the option of selecting individual images and share them, or select multiple images from the obtained result into a PDF to reduce the file clutter as well.

By the implementation of the above methods, the user can significantly reduce the time being wasted in tedious search through unorganized files and also provide an option to organize them neatly into PDF documents.

In this paper, we will be discussing about the objective of developing this application with all the mentioned features, along with the scope of the application. We will also explain about the various research papers being referred to have a detailed understanding about the technology that would be used for the extraction of text from the images [2-8]. Along with that, an account of how the application works from start to the end is also described in detail, supported with the necessary information about the architecture of the system and the supporting components which constitute the application as a whole.

A. Objective

The main objective of developing this application is to reduce the time involved in searching for the desired image by manual means of individually looking through the text content in each image to find the desired one.

The application also aims to provide options for instant sharing of the resultant image and enables PDF document generation on multiple image selections which lead to neat organization and reduces the clutter.

B. Scope

The application's target audience is the general public, which includes youngsters who tend to handle a lot of image data on social media. It is also targeted towards businesspeople who might handle images containing sensitive information. Since the application is offline, it provides a layer of privacy when it comes to extract sensitive information.

II. Literature Survey

The paper "Efficient Search in Document Image Collections" by Anand Kumar et al. discussed about the efficient hashing method which perform text-based search on document images. This paper explains an alternate method of searching text in document images. Adapting the usage of context sensitive hashing, a method where existing word images are used to compare the words in the document images to get a more accurate representation. The words which would be used are hashed with a key describing the features of the word, such as the characters present in it, and the language, etc. Through that they claim that it speeds up image search considerably, with 20000 images searched within milliseconds. [9].

Jawahar et al. published "Matching word images for content-based retrieval from printed document images." Which talked about the various processes that can be performed on document image text which gets distorted due to scanning issues? This explains the method mentioned in the previous paper with respect to printed document images. The difference is that, here the author discusses more in detail about the challenges faced while performing this process on those images. Factors such as character or print degradation, either due to age or the printer, are discussed. The performance of their model on such images is also explained in elaborately with statistics. [10].

A Detailed Analysis of Optical Character Recognition Technology, written by Karez Abdulwahhab Hamad et al., explains about the various processes and challenges involved in OCR technology. This paper gives an introduction to the OCR technology. It explains about the various challenges faced on extracting text from images which are unclear, skewed, badly aligned, etc. In addition to that, it also explains about the phases involved in OCR - preprocessing, segmentation, normalization, feature extraction, classification, post processing. It then gives an overview of the various fields where OCR is being used. [11].

Scene Text Recognition in Mobile Applications by Character Descriptor and Structure Configuration, written by Chucai Yi and Yingli Tan, describes the challenges faced by OCR algorithms while processing images taken from a scenery. The various environmental factors such as lighting, skew, focus issues are highlighted to be fundamental for extracting the text correctly. To rectify them, image processing methods such as Edge detection of objects and image sharpening are employed to improve image readability. [12].

A Text Extraction Approach towards Document Image Organization for the Android Platform, written by M. Maduram and Aruna Parameshwaran, explains the methodology used for document search using the MLKit Technology incorporated through usage of Firebase, a cloud based Artificial Intelligence service hosted by Google. Besides the brief explanation on the usage of it, emphasis is also given on the search algorithm used by them for increasing the speed to search on large sized image collections. [13].

Convolutional Neural Networks for Document Image Classification, written by Kang et al., goes in depth about adopting the neural network way of recognizing text. By employing various image processing techniques such as greyscale, sharpening and posterization, a sharp image is obtained. From that image, a 2D array of bits representing the pixels is generated. These are trained to compare with pre-existing datasets which contain array representations of images with text. From that, it allows us to speed up the whole process and ensures a stable performance for the classification process. [14].

III. Proposed Methodology

The proposed system reduces the average time required to retrieve the images the user desires to acquire. By storing the extracted text in the database and using the image name as key, the time to find the image is reduced considerably. In addition to it, setting certain constraints based on image metadata also helps to filter the results.

A. Flow Diagram

Through the flow diagram, we can understand that the functioning of the application depends upon the user input. Before the user inputs something, the images on the local storage are processed for extracting the text. They are stored along with the image properties such as Date taken, Path, Last modified and Size. The Search module manages to obtain the information from there and using that the search operation is performed. The results are displayed, with the option to share them individually or to combine multiple results into a single PDF document. The process flow of our proposed methodology is given below.

Through the flow diagram, we can understand that the functioning of the application depends upon the user input. Before the user inputs something, the images on the local storage are processed for extracting the text. They are stored along with the image properties such as Date taken, Path, Last modified and Size. The Search module manages to obtain the information from there. Using that, search operation is performed. The results are displayed, with the option to share them individually or to combine multiple results into a single PDF document.

IV. Working of the System

A. Starting the Application

The application, once installed, will be accessible from the App drawer by pressing the dotted button (on older devices) or by swiping up.

B. Background Process

Once the application is started, the Text Recognition API powered by Google MLKit will scan the image for the text. This API can be operated completely offline, as a security measure to ensure that the sensitive data of the user is not leaked through internet.

To use these functionalities, dependencies for mlkit-vision should be added to the Gradle file. We would need to initialize a "Text Recognizer" instance. This instance will accept images through the "Input Image" object. To load the image onto the object, we would require the URI (Uniform Resource Identifier) of the

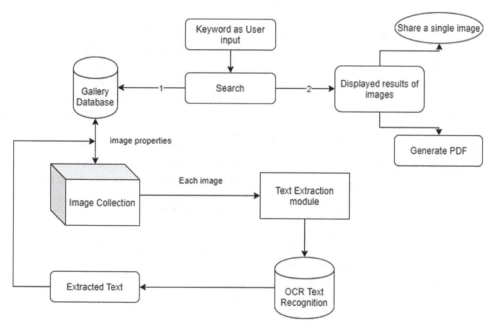

Fig. 61.1   Process flow diagram

Source: Author

image. The URI is unique for each image and can be obtained through accessing the Media Store object of Android.

Once the image gets passed to the Text Recognizer instance, text will be generated by using the Text Recognizer Process () function. This also handles exceptions through on Success/ on Failure functions. Once text gets generated, it will get added to the details of the image on the Gallery Database, along with the image properties such as Name, Date taken, Last modified, path and Random. This process happens every time when the app is started.

These details are stored in gallery.db, an SQLite Database which is standardly used on Android devices. This database is constructed by Room API, which simplifies the queries into programming syntax. Each of the database queries will be represented by a program instruction. These queries will be written on a Dao (Data access object) file. The purpose of using Room is to reduce boilerplate code. Boilerplate Code is a term used to describe code which is repeated on multiple locations. The instance of the database will be passed as a parameter to the module in concern. It will get skipped for previously processed images. These are all the activities that happen in the background once the application is launched.

C. Visible Activity

The home screen will display all the folders present in the device. The app is an image gallery, so only the folders containing images on the device will be displayed. The user can choose the folder they would like to access. Inside the folder, they can make use of the search bar, along with the images displayed in a grid format. Once a keyword is typed, it will filter the results based on the keyword and the image parameters that we can choose on the "Sort" menu. The user can select the multiple image results, and press Share button. They will be presented with an option to Generate PDF.

D. PDF Generation

The PDF generation is powered by iText API. It consists of various functions enabling us to seamlessly handle PDF documents. An instance of Document () will be initialized first.

By using PdfWriter object, we will initialize an empty PDF document at a predetermined path. Then an instance of Image() will be initialized. The images shared from the previous step will be loaded onto it. The scaling and alignment of the images is then specified.

The Image() instance will be loaded onto Document() instance. Once the images are loaded, the Document() instance will be closed. The PDF will be successfully generated on the IndexedPDFs folder.

V. Architecture of the System

IndexedByText stores the information about the image on the local database. No internet connection will be used, which means there will not be any online database connections.

A. Home Screen

No login information has to be provided to use this application. The application's design will resemble a Gallery application. Image Gallery shows a collection of all the folders containing image files present in the user's device. Search module performs search on all the available directory names. Sort module is used to sort the folders based on various parameters such as Date, Size, Path, Last modified and Random.

B. Folder View

Search module performs search against the extracted search specific to the image and display results. Image Gallery displays all the images on the folder in a Grid format. Select module provides the option to select multiple images on search results as well as the default view. Share module allows us to share the selected

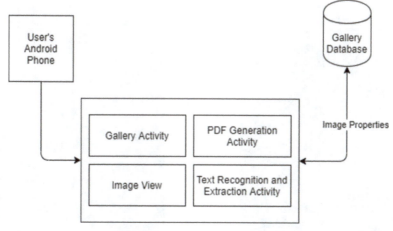

Fig. 61.2 Architecture Diagram

Source: Author

image individually. On multiple selections, they can select "Generate PDF" button. This will generate PDF on the "IndexedPDFs" folder present in the user's device.

C. Image View

View module displays the selected image in full size. Share module consists of options which are given to share the image to any of the communication apps installed on the device.

D. Generate PDF

Success message mentions the folder where the PDF is saved. In general The PDFs are saved with the Date and time as unique filename.

VI. Implementation

All the described features were implemented using Android Studio as the development environment. Tests were carried out on the Android Virtual Device, which acts as a virtual machine running the version 7 of the Android operating system.

User interface was designed based on the Material Guidelines published by Android. Usage of a common accent colour, along with

flatter icon designs consistent throughout the application is the hallmark of this design standard. The home screen greets the user with a very simple user interface, consisting of the name of the application, along with the name of the folder which contains the image. For the demonstration purpose, a single folder was used with a given set of images. Once we click and open the folder, we are presented with a grid view of the images stored in that folder. The grid format used here is 3x3. Along with the image grid, the user is also provided with the option to search the keyword on the Action Bar. The action bar houses the Sort button which allows the user to arrange the images based on the various image properties. Once the user chooses the option and press OK button, the images will be reordered based on the property. Here we are organizing the image based on the file size. The text box can be accessed by tapping the "Search" button. Then we can type the keyword. In the above figure, we have used a simple keyword "not", which displayed the resultant images based on the Sort property chosen earlier. It is also to be noted that, even while searching, the user can access the Sort menu.

Fig. 61.3 Home screen

Source: Author

Fig. 61.5 Sort menu

Source: Author

Fig. 61.4 Folder view

Source: Author

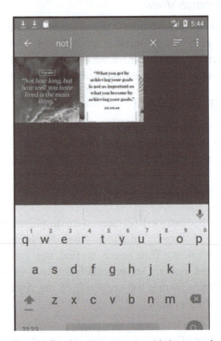

Fig. 61.6 Results screen with keyword

Source: Author

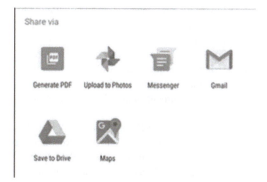

Fig. 61.7 Share sheet

Source: Author

Share sheet is the default share menu for Android devices [15]. It displays all the applications to which data can be shared. Then the user can share them to the available applications, or can make use of the Generate PDF button. This button will combine the selected results into a PDF document, which will be saved in a folder named "IndexedPDFs".

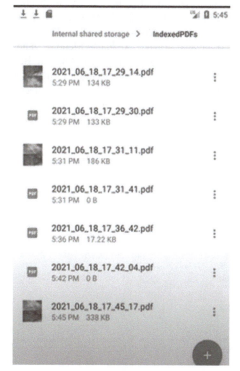

Fig. 61.8 IndexedPDFs folder

Source: Author

Fig. 61.9 Image view

Source: Author

Each of these PDF files are named with the Date and Time of file creation, thus creating a unique filename. This demonstrates the neat organization of the documents as explained before. Conversely, the user is also provided with the option to view each individual image and share them. The Image view screen has a simple design which displays the image on full size. The availability of two quick access buttons allows the user to either share the image instantly or delete them. This concludes the overall implementation of the application.

References

1. Nisha Pawar, Zainab Shaikh, Poonam Shinde and Prof. Y.P. Warke, "Image to Text Conversion Using Tesseract ", International Research Journal of Engineering and Technology (IRJET)., vol. 06, issue 02, (2019), pp. 516–519.
2. Pratik Madhukar Manwatkar, Dr. Kavita R. Singh," A Technical Review on Text Recognition from Images", IEEE Sponsored

9th International Conference on Intelligent Systems and Control (ISCO), (2015).

3. Chwan-Yi Shiah, "Content-Based Document Image Retrieval Based on Document Modeling", Journal of Intelligent Information Systems (2020) 55, Springer Science+Business Media, (2020), pp. 287–306.

4. Ondrej Krejcar, "Smart Implementation of Text Recognition (OCR) for Smart Mobile Devices", The First International Conference on Intelligent Systems and Applications., (2012), pp. 19–24.

5. K. Karthick, K.B.Ravindrakumar, R.Francis and S.Ilankannan, "Steps Involved in Text Recognition and Recent Research in OCR; A Study", International Journal of Recent Technology and Engineering (IJRTE)., vol.8, issue 1, (2019), pp. 3095–3100.

6. Qixiang Ye, and David Doermann, "Text Detection and Recognition in Imagery: A Survey", IEEE Transactions on Pattern Analysis and Machine Intelligence., vol. 37, no. 7, (2015).

7. Pratik Madhukar Manwatkar, Dr. Kavita and R. Singh "Text Recognition from Images: A Review", International Journal of Advanced Research in Computer Science and Software Engineering., vol. 4, issue 11, (2014), pp. 390–394.

8. Ravneet Kaur, "TEXT RECOGNITION APPLICATIONS FOR MOBILE DEVICES", Journal of Global Research in Computer Science., vol. 9, no.4, (2018), pp. 16–20.

9. Kumar Anand, Jawahar C and Manmatha R, "Efficient Search in Document Image Collections", Asian Conference on Computer Vision, Springer link, (2007), pp. 586–595.

10. Meshesha Million and Jawahar C, "Matching word images for content-based retrieval from printed document images", IJDAR, (2008), pp. 29–38.

11. Hamad Karez and Kaya Mehmet, "A Detailed Analysis of Optical Character Recognition Technology", International Journal of Applied Mathematics., Electronics and Computers, (2016), pp. 244–244.

12. C. Yi and Y. Tian, "Scene Text Recognition in Mobile Applications by Character Descriptor and Structure Configuration," IEEE Transactions on Image Processing., vol. 23, no. 7, (2014), pp. 2972–2982.

13. M. Maduram and Aruna Parameshwaran, "A Text Extraction Approach towards Document Image Organisation for the Android Platform", IRJET., vol. 5, issue 9, (2018), pp. 1560–1565.

14. Kang, Le, J. Kumar, Peng Ye, Y. Li and D. Doermann, "Convolutional Neural Networks for Document Image Classification", 22nd International Conference on Pattern Recognition, (2014), pp. 3168–3172.

15. Bhavana Rajak, Archana Menon, Janhavi Patel, Vivek Solavande, "An Android Based Detection of Text Extraction from Image", IOSR Journal of Computer Engineering, vol. 22, issue 2, (2020), pp. 16–19.

Recent Trends in Computational Intelligence and its Application – Sugumaran D. et al. (eds)
© 2023 Taylor & Francis Group, London, ISBN 978-1-032-48410-5

62

IOT Based Non Invasive Blood Glucose Level Monitoring

Vidhya S.[1], Swetha S.[2]
Assistant Professor, Department of Electronics and Communication Engineering,
Er. Perumal Manimekalai College of Engineering, Hosur, India

Pradeep R.[3], Vijay K.[4]
Assistant Professor, Department of Computer Science and Engineering,
Rajalakshmi Engineering College, Chennai, India

Sowmia K. R.[5]
Assistant Professor, Department of Artificial Intelligence and Machine Learning,
Rajalakshmi Engineering College, Chennai, India

Abstract—One of the biggest global problems is Diabetes which is rapidly increasing day by day. On the other side measurement of blood glucose level constantly is a challenging task. The invasive procedure is one of the conventional methods which involve pricking of blood with lancet which is uncomfortable. In order to overcome this problem a solution is proposed using IOT based architecture. In this paper a design of a wearable device for glucose monitoring system is done. This IOT system architecture design involves a sensor device which is connected to the backend system and this represents real time glucose level of a patient. This monitoring of glucose level can of two types: Continuous glucose monitoring system (CGMS) and Non-invasive glucose monitoring (NGM). This proposed method involves the detection of glucose level using GSR sensor which gives galvanic skin response and this system is more efficient for blood glucose monitoring from patient sweat. This glucose level obtained from patient sweat will be furthermore communicated to the Wi-Fi controller for storing in the cloud and further analysing the data and to monitor it at regular intervals. The insulin injector pump will be activated if the glucose level reaches critical stage and this insulin will be injected to the patient automatically. The insulin stock can be monitored by using ultrasound sensor and information can be sent and vice versa. The need of usage of the recent technologies is for perpetual observance of patient health against chronic disease.

Keywords—GSR Sensor, Insulin control unit, Non-invasive measurement of glucose, Insulin injector, Zig-bee

[1]vidhyasukumar11@gmail.com, [2]swethamani7777@gmail.com, [3]pradeeprajame@gmail.com, [4]vijayk.btech@gmail.com, [5]swomiakr@gmail.com

DOI: 10.1201/9781003388913-62

I. Introduction

One of the most common lifelong chronic disorders among human beings is Diabetes. Diabetes occurs when the pancreas is not able to make insulin or the body is no longer capable of using the insulin which it has generated. This disease involves many reasons like immune disorders, lack of physical exercise, genetic factors and many others due to which imbalance of blood glucose level occurs in human body [1]. The international diabetes federation has stated that diabetes is spiraling out of control and also 1 out of 10 people are living with diabetes and half of the people are undiagnosed. It has also estimated that the count of world's population afflicted with diabetes will. The people affected with diabetes should check their blood glucose level two times a day [2]. This procedure is complicated to do every time. There is also a danger of infection during pricking the finger for testing the blood and the cost of Lancet and strips are the factors to be considered as each test requires a new test-strip. There are two categories of diabetes which are Type 1 and Type 2.

Type 1 Diabetes

Type 1 diabetes normally occurs when the defense system of the body attacks the cells which produce insulin and this is an autoimmune reaction. Hence the result is body will not be able to produce sufficient amount of insulin [3]. This type can occur for people at any age but children and young adults are mostly to get affected. Some of the symptoms of this type include frequent urination, blurred vision, constant hunger etc. People affected by this type diabetes should take insulin injection to keep their blood glucose level in normal range or it leads to adverse effect [4].

Type 2 Diabetes

Type 2 diabetes is the most common type which contributes for 90% of all the diabetes records. The characteristic of this diabetes is it is insulin resistance [4]. This happens because the blood glucose levels will be increasing so that insulin cannot work properly. This type normally exists in older adults. The frequent symptoms of this diabetes are lack of energy, excessive thirst, dry mouth etc. This diabetes can be overcome by maintaining healthy life style along with oral medication.

Impaired fasting glycaemia and impaired glucose tolerance

The above two conditions are at high risk of advancing towards type 2 diabetes hence an intermediate condition which is the transition between diabetic and normality. The health impact of diabetes includes damaging the body organs, damaging the blood vessels which eventually lead to problems in major organs [5]. In this proposed system we design an IoT based system architecture that includes a sensor device connected to a back end system which monitors the real time glucose level periodically from patient sweat using GSR sensor [6]. This monitored data is transmitted through the Wi-Fi controller and further analyzed and this data is stored in cloud. The GSR sensor probe will measure the blood glucose level from the sweat and the LCD displays the blood glucose range [12]. If patient has normal glucose level range, it displays normal in the LCD. For further analysis the obtained glucose level is transmitted to WiFi controller and it is also stored in IOT cloud platform online webpage. Hence the glucose level is monitored at regular interval. If glucose level exceeds the normal level the insulin injector pumps the insulin to the patient from the insulin tank [5]. The ultrasonic sensor monitors the insulin stock, if the insulin level is low an online notification is sent to the patient. The buzzer indication is also used when the glucose level exceeds the normal level.

II. Literature Survey

This method explains about the most traditional method which diabetic patients do as their daily routine. Normally for a diabetic patient the blood glucose level will be changing constantly [6]. Hence regular monitoring of blood glucose level should be performed. This observance can be done by pricking the skin, by taking a drop of blood. This is most excruciating task and hence many patients don't prefer this method to do it on a regular basis. After doing this method necessary medicines will be prescribed by the doctor [6].

This second method is also invasive which uses elf monitoring glucose meter. But the major drawback of this method is it uses test strips and blood should be extracted from the forearm by pricking the finger and further chemical analysis will be performed [7]. The entire process involves discomfort to the patient. On the other hand non invasive techniques are more users friendly and they are also useful. It overcomes the drawbacks of normal conventional methods [7].

In this paper blood glucose monitoring using a real time wireless system using electrochemical sensor is discussed[8]. This sensor tag contains a potentiate, an on chip temperature sensor, a 10 bit sigma delta ADC, a 13.56 MHz RFID front end, a winding ferrite core antenna and also a digital baseband for protocol [8]. The ISO15693 is observed for air interface. The entire system involves a sensitivity of 0.4 dBm. The implanted glucose monitoring application contains a simple sensor tag which is used for low power and also used for low cost applications [11]. When used in real time applications this system gives accurate result. The improvement of system on chip performance is the major disadvantage of this method [9].

In this paper the device for design and possible development of blood glucose monitoring system invasively is discussed. The proposed technique involves a near infrared sensor which does the work of transmission and reception of rays from forearm [10]. The intensity variation can be analyzed from the received signal which uses a photo detector at other side of fore arm

and from this prediction of blood glucose level will be done [10]. Then for further analysis the data will be transmitted to a remote android device. Subsequent section describes basic principle for measurement of glucose level in blood

III. Methodology and Working

The whole method is incorporated with the sensor along with microprocessor to transmit the signals to operate to the functions given. The GSR sensor measures the glucose level from the sweat. If glucose level exceeds the threshold value insulin control unit will inject the insulin to the patient. The insulin stock is maintained by ultrasonic sensor and updated to the patient.

A. Architecture Diagram

See diagram (Fig. 62.1) on next page.

B. ESP8266/Node MCU

Node MCU is an open source platform based on ESP8266 that is used to connect objects and allows data transfer using the Wi-Fi protocol. All the elements like RAM, CPU, networking, SDK and a modern operating system are present in ESP8266. It is an open source and provides a easy programming environment which is based on eLua. The developer community can use it conveniently as it has got a fast scripting language. This is an adaptable IOT Controller which creates and open source hardware design and software SDK. Also to operate properly the components like crystal, capacitor and resistor, a 4 MB SPI flash memory for programming ESP 12 E are included in ESP8266. Further an antenna with a range of (-70 to -80dBm at 50 feet) is included.

C. GSR Sensor

The other name of Galvanic skin response is Electrodermal activity and Skin Conductance. It is defined as the measure of electrical characteristics by continuously monitoring

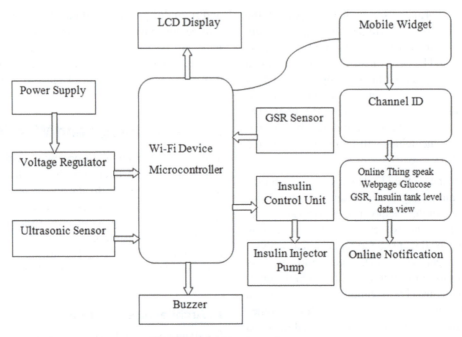

Fig. 62.1 Architecture diagram

Source: Author

the variations of the skin. In other words it measures the variation in human body sweating for any instance. The moisture level of the skin will be directly proportional to the electric conductance. The sweat from the human body is constrained by autonomic nervous system. The sweat gland activity increases if the autonomic nervous system activity due to sympathetic branch increases. As a result Skin conductance increases and this is a regenerative process.

D. Insulin Control Unit

A 12 V Single pole double throw PCB Mount Sugar Cube Relay Switch is an electromechanically operated switch. The 12 V 10 A SPDT is a single pole double throw (SPDT) relay, having a common terminal that connects to either of two others, including two for the coil such a relay has five terminals. With a compatible voltage of 12 V, it can pass a max 10 A current. And for the max power rating as it is written on the product 30 VDC, with 30 VDC it can pass max current 10 A.

Specification

- Voltage: 12
- Rated with:
  - 7 A/240 VAC
  - 7 A/14 VAC
  - 7.5 A/250 A

E. Insulin Pump

It is a small, wearable device used to deliver insulin for the body. This insulin pump can be used as a flipside for insulin injection. This insulin pump continuously provides insulin to the body whenever needed. The insulin pump is more efficient and comfortable than insulin injections. The insulin pump delivers insulin in a small and continuous dose. This insulin pump overcomes the use of daily injections as this pump supplies to the body according to the varying glucose level.

F. Ultrasonic Sensor

In this product HC-SR04 ultrasonic sensor which uses SONAR is used. This is an electronic device which measures the distance of an object as this sensor emits ultrasonic sound waves and this reflected sound wave is converted into an electrical signal. The features of this sensor are high accuracy and stable reading along with good non contact range detection. The operation of ultrasonic sensor will not be affected when it is exposed to sun light or when it comes across black material but when soft materials are used it is difficult to detect. The ultrasonic sensor consists of an ultrasonic transmitter for transmitting the signal and a receiver module for receiving the signals.

IV. Results and Discussion

The result curve shows it continuously monitors the glucose in real time and it is more efficient.

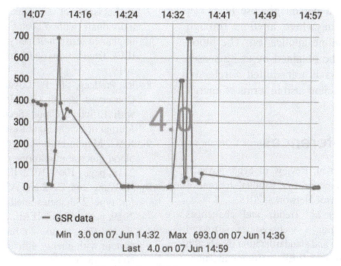

Fig. 62.2 GSR sensor value

Source: Author

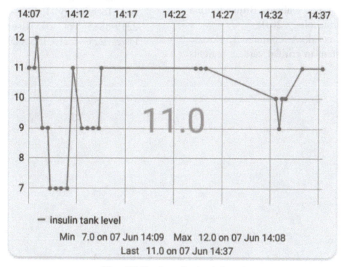

Fig. 62.3 Insulin tank level

Source: Author

The first curve shows the tracking by using GSR sensor and the second shows insulin tank level. This curve also provides more services to users by providing online notification and this can be still extended to many applications.

V. Conclusion

This is an absolute system which starts from sensor node connected to back end server. By using this system a person glucose level can be monitored at anywhere anytime through a mobile or an online application. The blood glucose level can be obtained by using sensor nodes and this data is transmitted wirelessly. This method is efficient also in terms of energy consumption.

References

1. S. A. Haque et al. Review of cyber-physical system in healthcare. International Journal of Distributed Sensor Networks, 2019.

2. A. Aragues et al. Trends and challenges of theemerging technologies toward interoperability and standardization in e-health communications. IEEE Communications Magazine, 2019.

3. P. King et al. The UK prospective diabetes study (UKPDS): clinical and therapeutic implications for type 2 diabetes. British Journal of Clinical Pharmacology, 2019.

4. A. Murakami et al. A continuous glucose monitoring system in critical cardiac patients in the intensive care unit. In 2016 Computers in Cardiology, pages 233–236. IEEE, 2020.

5. Dr. N. Shunmugakarpagam, P. Deepika, R.Pradeep, Blockchain as the Answer to IoT Challenges and their use cases. International Journal of All Research Education and Scientific Methods Issue 1, Vol. 10, pages 1875–1881, 2022.

6. M. Ali et al. A bluetooth low energy implantable glucose monitoring system. In EuMC 2011, pages 1265–1268. IEEE, 2020.

7. J. Lucisano et al. Glucose monitoring in individuals with diabetes using a long-term implanted sensor/telemetry system and model. IEEE Transactions on Biomedical Engineering, 2019.

8. R. Pradeep, T. P. Dayana Peter Privacy-Conserved Health Care Information System with Efficient Cloud Security. International Journal of Science and Research 2015.

9. KAU. Menon et al. A survey on non-invasive blood glucose monitoring using nir. In ICCSP 2018, pages 1069–1072. IEEE, 2018.

10. N. Wang and G. Kang. A monitoring system for type 2 diabetes mellitus. In Healthcom 2019, pages 62–67. IEEE, 2019.

11. T. N. Gia et al. Iot-based fall detection system with energy efficient sensor nodes. In NORCAS 2016, pages 1–6. IEEE, 2016.

12. R. Babu & K. Jayashree "A Survey on the Role of IoT and Cloud in Health Care", International Journal of Scientific, Engineering and Technology Research (IJSETR) ISSN 2319-8885 Vol. 04, Issue. 12, May-2015, Pages: 2217–2219. ISSN 2319–8885.

Recent Trends in Computational Intelligence and its Application – Sugumaran D. et al. (eds)
© 2023 Taylor & Francis Group, London, ISBN 978-1-032-48410-5

63

Bio Bank–An Innovation Authentication System for B2B Transaction

S. Vidhya[1]
Associate Professor, Department of Information Technology,
Saveetha Engineering College, Thandalam, Chennai

S. Amudha[2]
Assistant. Professor, Department of Information Technology, VIT, Chennai

S. Ravi kumar[3]
Assistant Professor, Department of Computer Science and Engineering,
Vel Tech Deemed to be University

Abstract—The number of people using online banking has steadily expanded during the last ten years. Due to this, many developers are looking into more user-friendly ways for users to do remote banking transactions. A new practical method for clients to conduct transactions, Bio Bank is anticipated to boost the number of mobile phone users Activities at Bio Bank are covered within the definition of the banking industry. This essay's main topic is using a bio bank. This paper is application based and examines various obstacles, such as security, and the technologies needed in bio banks. Through the suggested system, one can get the essential information about his account, such as balance details, inquire about check status, share price, etc. In addition to this, a customer can execute transactions through this service, such as money transfers between accounts or loan applications. The Bio Bank service can be advantageous for fostering unique connections between the bank and its clients.

Keywords—Bio bank, One time password, Personal identification number, Internet information service, Personal digital assistant

I. Introduction

Banking is the practise of doing balance checks, account transactions, payments, credit applications, and other financial activities using a mobile device, such as a phone or personal digital assistant (pda). Given the amount of mobile phone users and the quick growth of the mobile industry, the mobile sector is one of the market's most dynamic segments, and one should note the probable requirement for a new standard in banking services. Mobile devices

[1]vidhyas_1983@yahoo.com, [2]amuthabenziker@gmail.com, [3]ravikaurs.86@gmail.com

DOI: 10.1201/9781003388913-63

are becoming increasingly prevalent in daily life and are also being used more frequently for financial services, such as the m-banking system, which is based on short message service. Normally the customer would have to make a visit to the bank to do transactions. Mobile banking is the option by which the bank customers can make transactions from their homes conveniently. Customers will be able to check their account summary, move money across accounts, and receive notifications for significant payments or transactions that surpass a predetermined threshold, giving them immediate and total control over their funds. Next-generation mobile banking services will provide significant improvements at a lower session cost with user-friendly icon-driven instruction, quicker access, enhanced security, and quick payment services. By offering everywhere, anytime banking, banks will boost both client pleasure and loyalty. They will also profit from reduced administrative expenses, fewer branches, decreased headcount, streamlined call centres, and decreased handling fees, savings that should, ideally, be passed on to clients.

II. Background

A. Business Models for Mobile Banking

Numerous branchless/mobile banking concepts are now being developed. The main difference between these models is who will establish the relationship with the end customer, the bank or the non-bank/telecommunication company (Telco). The three basic categories of branchless banking models are bank focused, bank led, and non bank led.

B. Model

The bank-focused model is created and shown in Fig. 63.1 when a typical bank offers banking services to its present customers through unorthodox, affordable distribution channels.

Examples include providing bank customers with a limited selection of financial services through automated teller machines (ATMs), internet banking, or mobile phone banking. This paradigm, which is additive in nature, could be considered a slight improvement over traditional branch-based banking.

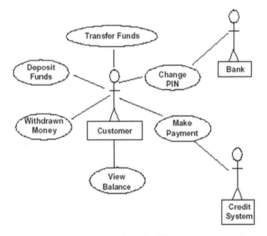

Fig. 63.1 Model of bio bank

Source: Author

C. The Bank-Led Model

In contrast to traditional branch-based banking, the bank-led approach which is shown in Fig. 63.2. and allows clients to complete financial transactions at a range of retail agents (or by mobile phone) as opposed to bank branches or with bank employees. This strategy provides the potential to greatly expand the financial services outreach by employing a different delivery channel (retailers/mobile phones), a different trade partner (Telco/Chain Store), and a target market that is different from typical banks. Additionally, it might be considerably less expensive than the bank-based options. The bank-led concept may be put into practise by establishing a joint venture (JV) between the bank and a telco or non-bank or by using correspondent agreements. In this approach, the bank controls the customer account connection.

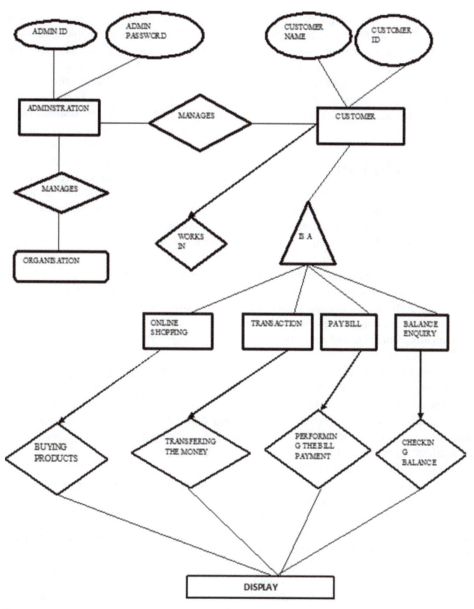

Fig. 63.2 Bank LED model

Source: Author

D. Non Bank-Led Model

In the non-bank-led approach, a non-bank (such a telco) takes on all the responsibilities, with the exception of maybe acting as a safe-keeper of surplus funds.

III. Existing System

If an application operates on a portable computing device and is regularly or constantly linked to a network, it is considered mobile. This definition covers software that runs on many devices, such as mobile phones, PDAs, and notebook PCs. Additionally, it suggests a client-server architecture in which the programme executing on the device is a user of a service made available over the network. The main categories of current mobile banking applications are introduced in the section that follows

IV. Proposed System

The proposed system and architecture is shown in Fig. 63.3. is a proper SMS authentication for each and every banking transaction. A new authentication called Biometric Iris concept have been included for the banking operations and online Shopping. While performing the registration in banking the customer have to give their Iris. While performing other operations the user must use their Iris image. The ideal method for automatically identifying and authenticating people is iris recognition. Iris patterns are one of the biometrics used in the planned study to identify people. This requires the gathering of iris images, picture pre-processing, feature extraction, and ultimately the conversion of this data into encrypted code. Each person's iris has a different pattern, and even the iris in one's left and right eyes is different from one another. The properties of the iris can seldom ever be altered without endangering vision. The iris identification systems can even identify a blind person who has an iris in his or her blind eye. Age has little effect on the properties of the iris. Using this system's secure authentication will improve the user's productivity. Individual person's Security and Authentication is essential necessary for our daily lives in net Banking [17].

V. Project Description

A. Problem Definition

Bio Bank is a developing area of m-commerce where customers can communicate with service providers via a wireless mobile network while using mobile devices to complete transactions. Utilizing a variety of wireless and mobile networks and mobile devices, users can access Bio Bank services and applications. However, limitations of mobile networks and devices have an impact on how well they operate, thus it is imperative to take these limitations into account when designing and developing m-banking services and applications. Identifying the needs of mobile users is another crucial issue. In order to derive a set of requirements to mobile banking applications we pursue two steps: firstly we identify general characteristics of the mobile use which are relevant. Secondly we closely watch the user and his context when wanting to use mobile banking. This will allow us to evaluate the existing solutions according to their fulfilment to these requirements.

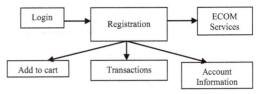

Fig. 63.3 Architecture of the proposed approach

Source: Author

B. View Transaction Records

Table 63.1

Use Case	UC1 View Transaction Records
Main Flow of Events	The use case starts when the customer selects an account from a drop down menu. The application then lists last ten days transaction records of selected account to the customer.
Exceptional Flow of Events	If the database server or database connection string is unavailable or wrong that will cause an exception.

Source: Author

C. Transfer Money

See Table 63.2 on next page.

Table 63.2

Use Case	UC 2: Transfer Money
Main Flow of Events	The use case starts when the application shows a Transfer Money Form. The customer can select a source account and a destination account from two different drop-down menus. The customer then can enter an amount and transaction date. If everything is correct, the application shows the transaction detail to customer. The customer can submit the request by pressing Submit button if he accepts the transaction detail, thus ending the use case.
Exceptional Flow of Events	The customer can cancel a transaction at any time by pressing Cancel button. No changes made to customer's accounts.
Exceptional Flow of Events	By clicking the Cancel button, the consumer can end a transaction at any time. The accounts of the customers have not changed.
Exceptional Flow of Events	If the customer enters an incorrect amount, an error message will display and the use case stays.
Exceptional Flow of Events	If the database server or database connection string is unavailable or wrong that will cause an exception.

Source: Author

VI. Overview of the Proposed System

With proper authentication, this solution enables the user to conduct all banking transactions including online shopping. A mobile application is what a mobile banking app is first and foremost. Additional informational added values must be addressed utilising mobile added values in order to conceptualise a mobile application. In other words, simply transferring an existing Internet application to a mobile device is far from sufficient. Mobile apps must be particularly tailored to fit the demands and expectations of the mobile user on the one hand, and the limitations imposed by mobile communication methods and mobile

devices on the other. ASP.NET is used in this project to do Mobile Banking.

VII. Module Description

A. Banking Module

This module involves Customer, Service, Bank forms. If the user want to create new account, he must provide all the necessary details so that the bank will provide the account number, pin number. With that necessary account number the user can perform further transactions. If two user details are same then the registration is not performed further. One Time Password (OTP) will be generated automatically.

In the service form the fund transactions can be performed from one account number to another account number. If the account given account number is typed wrongly then the transactions cannot be performed further. If the necessary amount is not there in the account number then it will display the error message that the given amount is not in the account so that the amount cannot be transferred. The amount will be transferred with the proper account number and necessary amount in the account. The amount will be transferred with the current date and time. The statement will be appeared with the proper transaction details.

In the bank form all the necessary banking operations can be performed. They can able to perform deposit, withdraw from the necessary account. If the user want to deposit the amount he must give the account number and it will show the balance. Then amount to be deposited must be entered .After entering all the details it will display the total amount in account. . If the user want to withdraw the amount he must give the account number and it will display the amount in the account. If the amount is not there in the account then withdraw cannot be performed. The statement will display the current operations done on the specific account. For performing all the banking operations it must check with the database. After performing all the operations the user can logout.

Login Module:

To access this module, go to the login page. The system will show a notice asking the user to type the username and password correctly if they have entered them incorrectly. This will carry on until the user enters the proper username and password. The bank administrator enters all the customer information in this module. This would be the first step to enter in to this application. Every user given a specific id and password. If user name and password found to be correct then the user will allowed entering in to the site.

Fig. 63.4 Login module

Source: Author

Registration

If the user is new to this application then the user should submit the registration form to get the user name and the password.

B. Ecom Module

Product Management

The management of items, offers, and bundles for businesses and mass-market clients is supported by product management. Offering cross-product discounts, reasonable pricing, and consumer loyalty programmes are all common components of product management.

Fig. 63.5 Registration module

Source: Author

Product Information

This page gives the product information that is being manufactured by them. The client to whom this project designed was a medicine related company. They are preparing the necessary items based computer related medical products. Hence this page gives all the details regarding the products that are manufactured in their head office and their branches.

Shop Cart

In shopping cart management the site should give provisions for adding, modifying and deleting the contents of the shopping cart, after confirmation of the contents of the shopping cart, the personal information of the purchaser are accepted like Name, Address, credit card number or Bank cheque and demand Draft, Expiry date, password etc. On submission a new Order number is generated, with display of the order.

Add to Cart

Add to Cart, enables the user to buy a product. Whenever Add to Cart is clicked the quantity needed by the user is entered. The product

Fig. 63.6   Ecom module

Source: Author

name, its price, and quantity are stored in the add cart table.

View Cart

View cart is used, when the user wants to view the product he has selected so far to buy. The items that he has selected are retrieved from the table and listed in the browser.

Biometric Authentication

The study of automated systems for distinguishing individuals from each other based on one or more intrinsic physical or behavioural attributes is known as biometrics. Features of the body include. In this System the left and right eye of all whose are registered in e-commerce portal, If any user want login this system their Biometric data will captured and validated. If both matches present will be stored otherwise the user would not be allowed to the site.

VIII. System Implementation and Testing

A. System Implementation

The stage of implementation is when a new system is designed. Planning, training, and system testing are the primary phases of implementation. A new or updated system is implemented when it is made operational. The primary component of implementation is conversion. It is the transition from the previous system to the new one. After the system is put into use, the user evaluates it. It is employed to collect data for system maintenance. The fundamental review technique involves data collection through questionnaires, interviews, etc.

B. Feasibility Study

Internal expenses, existing spending (for example, on communication resources and information technology), and a thorough cost scenario of what would occur with the implementation of the online strategy should all be specifically examined in this research.

C. Cost Justification

Understanding the financial and business milestones towards which a company strives should be done before developing the project. This step's objective is to calculate the cost of devoting priceless resources to the project. One of the most crucial factors in internally justifying the project will be this. This project will be less expensive than comparable banking applications.

D. System Maintenance

This maintenance work's goal is to ensure sure the system boots up every time without a glitch. Environmental changes that could impact the computer system or software system must be planned for. This is referred to as system maintenance. The world of software is changing quickly today. The system should be able to respond to this change due to its quick nature. The procedure in this project can be added without affecting other system components.

Maintenance is essential. After it is implemented, the system should be obligated to accept any adjustments. This mechanism is set up to favour any fresh developments. Both the system's accuracy and performance won't be impacted by doing this.

IX. System Testing

A. Testing

The system's success depends on testing. System testing makes the logical premise that the objective will be effectively attained if every component of the system is correct. System testing is the process of "measuring" a component's performance when it is placed in a computer system that is (reasonably close to or identical to) functional. This criteria cannot be met by 'System Testing' configurations. This doesn't imply that there isn't a test result value; it just means that it's unclear how 'good' it might be. Unit testing is testing changes made in an existing (or) a new program. System testing is executing a program to check logic changes made in it and with the information of finding errors – making the program fail.

There are various types of testing all the below testing are done in this project.

- Unit testing
- Integration testing
- Performance testing

X. Result and Discussion

Thus in this system the customer can do their banking and shopping in secured manner using their iris concept and OTP method. The results obtained with the proposed method are efficient one because user can provide the account details in a safe platform with full confidence of secured version. Compared with the previous approaches the security provided to the bio bank approach gives the 10 % good results while seeing the output produced by the proposed approach. In future they can implement this security method in all application

XI. Conclusion

The project demonstrates a way for providing banking services and enables online shopping utilising a banking account and an iris image for security. Due to the rapid rise of mobile phones, a wide spectrum of users can be served by this system. This technology not only reduces visits to the interior of the banks, but it can also reduce some visits to ATMs (Automatic Teller Machines) for handling tasks like checking our account balance and inspecting them, as well as resolve issues with these devices. The ability to transfer balances is one of our suggested financial system's most alluring features.

Reference

1. Anderson, J. M-banking in developing markets. Info: The journal of policy, regulation and strategy for telecommunications, 2010. 12(1): pp. 18–25.
2. Weber, R. & Darbellay, Legal issues in mobile banking. Journal of Banking Regulation, 2010. 11(2).
2. Azim, A. F. and Zibran, M. F. "Alternative frameworks of E-Commerce and electronic payment systems especially suitable for the developing countries like Bangladesh", ICCIT 2005.
3. Niina Mallat, Matti Rossi and Virpi Kristiina Tuunainen (2010), Mobile Banking Services Communications of the ACM, 47, 5 pp. 42–46.
4. Shirali-Shahreza, M and Shirali-shahera, M.H. "Mobile banking Services in the bank area", SICE , 2007.
5. Tiwari, R and Herstatt, C."Customer on the Move: Strategic Implications of Mobile Banking for Banks and Financial Enterprises," Proceedings of the 8th IEEE International Conference on E-Commerce Technology, San Francisco, California, pp. 81–88, 26–29 June, 2006
6. https://www.ibm.com/developerworks/mydeveloperworks/wikis/home/wiki/W1302f61f2e98_46e8_8f4b_649337b014b6/page/Project%20Scenario
7. http://www.mobilecomms-technology.com
8. http://www.wikipedia.org/wiki/Short_message_service/projects/sms
9. Niina Mallat, Matti Rossi and Virpi Kristiina Tuunainen (2010), Mobile Banking Services Communications of the ACM, 47, 5 pp. 42–46.

10. Shirali-Shahreza, M and Shirali-shahera, M.H. "Mobile banking Services in the bank area", SICE, 2007.

11. Tiwari, R and Herstatt, C. "Customer on the Move: Strategic Implications of Mobile Banking for Banks and Financial Enterprises," Proceedings of the 8th IEEE International Conference on E-Commerce Technology, San Francisco, California, pp. 81–88, 26–29 June, 2006

12. https://www.ibm.com/developerworks/mydeveloperworks/wikis/home/wiki/W1302f61f2e98_46e8_8f4b_649337b014b6/page/Project%20Scenario

13. http://www.mobilecomms-technology.com

14. Azim, A. F. and Zibran, M. F. "Alternative frameworks of E-Commerce and electronic payment systems especially suitable for the developing countries like Bangladesh", ICCIT 2005.

15. Rachid Aloaui and Ahmed alloui (2022) "Secure internet banking authentication based on multimodal biometric in distributed technology"

Recent Trends in Computational Intelligence and its Application – Sugumaran D. et al. (eds)
© 2023 Taylor & Francis Group, London, ISBN 978-1-032-48410-5

64

A Cloud Based Healthcare Data Storage System Using Encryption Algorithm

M. Jaeyalakshmi[1], Vijay K.[2]
Assistant Professor, Department of Computer Science,
Rajalakshmi Engineering College, Thandalam, Chennai, India

Jayashree K.[3]
Professor, Department of Artificial Intelligence and Data Science,
Panimalar Engineering College, Chennai, India

Priya Vijay[4]
Assistant Professor, Department of Computer Science,
Rajalakshmi Engineering College, Thandalam, Chennai, India

Abstract—Cloud computing is a commercial and financial model allowing the users to utilize all the storage virtually with minimal infrastructure and make the proper services. The healthcare system plays the major role in the current technology in that electronic medical records (EMRs) are modernized records which contain a list of details regarding patients' health. Due to their centralised data storage, which creates a single point of failure for patients, electronic medical records are efficient in comparison to currently used conventional storage systems. One of the key components of the medical management system that enables users to access patient health information in a secure manner is the encryption algorithms. The proposed work is to figure an encrypted access controller outline, comparing the different encryption algorithms available and use the best suited algorithm to protect the sharing of EMRs between dissimilar gears to elaborate in the clever health care organization and make the life to live in the secured and healthy manner.

Keywords—Cloud computing, Encryption algorithm, Health care

I. Introduction

Health is the very important for all the human beings. But the awareness about the health and people careless takes the more disadvantage. Even the people are very careful in their health they but they are afraid that the data is not be saved secured. Electronic medical records (EMRs) are modernized records which contain a list of details regarding patients' health. Electronic medical records are effective compared to the existing conventional storage

[1]jaeyalakshmi.m@rajalakshmi.edu.in, [2]vijayk.btech@gmail.com, [3]k.jayashri@gmail.com, [4]priyavijay03@gmail.com

DOI: 10.1201/9781003388913-64

approaches due to its centralized storage of data, it primes to a solitary point of disaster. An appearance of the novel healthiness disaster has occupy yourself the part of an accelerator pedal, creation is possible to cross in a rare days phases that usually mark out the cycle of adoption of an innovation. It was essential to decrease the experience of professionals and their patients to the virus. So, a remote access to the EMRs was a necessary action. EMRs that include sensitive, discrete patient data that might be readily exploited must be genuine. Therefore, the concept is to create a framework for encrypted access control to safeguard the exchange of EMRs between various actors in the smart healthcare system. One of the key challenges in the medical management system is cloud computing, which uses encryption methods to allow users to access patient health information remotely in a secure manner.

II. Related Work

Jinyuan Sun & Yuguang Fang [3], is proposed to design of delegation device a structure block of cross-domain collaboration, includes swapping and allocation of pertinent patient information that are measured highly private and confidential. A safe EHR scheme, founded on cryptographical buildings, to permit safe distribution of complex patient information throughout collaboration and preserve patient information is kept privacy. EHR structure is established to fulfill ideas precise to the cross-domain allocation situation of attention.

Smart materials, according to research by Dina Hussein et al. [10], may operate independently to sense/actuate, retain, and comprehend data that was either generated by them or was present in the environment. Human rights are to be assessed based on societal standards by the community of intelligent objects sharing a shared task, on behalf of those who have store boundaries.

The rational and intelligent framework provided by Syed Umar Amin et al. [11]

depicts the patient's current role in real time and offers accurate, cost-effective medical care of the highest calibre. The patient's EEG signals can be sent to the cloud using clever IoT techniques, where they can be processed and sent to a cognitive system. The system controls who has access to the persistent by monitoring sensor readings including makeover words, voice, EEG, movement, and gestures. The best deep learning system outperforms cutting-edge methods in terms of accuracy.

Debajyoti Pal et al., [5], proposed to evolve and examine a theoretic outline aimed at defining the important issues that distress the ageing people to accept clever home services for healthcare. Incomplete least square structural equation modeling. User insights slows the happening theoretical equal somewhat than the real custom meaning in the direction of a precise facility delivers the basis to discover the procedure of the definite acceptance in clever homebased service area for health care by the mature people with possible future investigate zones.

A strong wireless and mobile communication infrastructure is required to connect and access Smart healthcare services, people, and sensors seamlessly, wherever they are and whenever they need to, according to MD. Mofijul Islam et al. [11]. By shifting Big Health care Data-related functions, such as data exchange, dispensing, and analysis to the cloud, mobile cloud computing (MCC) can play a significant role in this context by alleviating pressure on local mobile app resources. access smart healthcare data in an effort to improve inhabitants' quality of life A new joint virtual machine migration approach for mobile cloud computing that uses ant colony optimization.

III. Table

Table 64.1 shows the objectives discussed and the different encryption algorithms used with the disadvantages of the system.

Table 64.1 Literature survey

Title of the paperand author	Objectives	Methodology used	Limitation of the system
"Cross-Domain DataSharing in DistributedElectronic Health Record Systems", Jinyuan Sun & Yuguang Fang, [3]	To allow safe distribution of enduring data during collaboration and maintain the confidentiality of the patient data.	EHR system, basedon cryptographic constructions.	Classifying such occurrences to safeguard flexibility of our scheme under thefears of malicious assailants.
"A Community-Driven Access Control Approach inDistributed IoT Environments", DinaHussein et al, [10]	To establish the idea of admission of human rights in a dispersed IoT Atmosphere.	Community-basedstructure	The different Stages of civic formation and growth with ACcivil rights related with each segment. Extend prototype to significant an IoT systems, mainly in an innovativeness
"Cognitive Smart Healthcare for Pathology Detection and Monitoring", Syed Umar Amin et al, [11]	To use the EEG pathology classification technique to keep track of a patient's health.	Deep learning, EEG smart sensor.	Examine the patient and then it just used to monitor record.. It uses to detect and just monitor the healthcare process.
Debajyoti Pal et al, [5] , "Internet-of-Things and Smart Homes for Elderly Healthcare: An End-User Perspective"	To evolve and examine a hypothetical summary for decisive the important aspects that move mature people to accept clever home facilities.	Partial a smallest amount square operational equation modeling.	The environmental distribution of the elderly subjects. The people can't able to protect themselves with their data. When they lost the device, the data will not be stored in the secured way.
MD. Mofijul Islam et al, [5], "Mobile Cloud-Based Big Healthcare Data Processing in Smart Cities	To access Smart healthcare data to have a peaceful life without any in secured data storing.	Mobile Cloud computing, The techniques used here is based on the optimization algorithm for the migration puropose.	The developing advantage is only based on the execution of the time. The awareness about the health is very important. People are not making that awareness about the work.

IV. Proposed System

The proposed system explains, the users registered their account and they need to make the request to authorize person information and including doctors to access the report of specific users. The report of the user can only be accessed by the doctor when the approval of the management is been given. The management have the full authority to have the doctor and patient detail. The data should be stored inside the cloud in which the AES algorithm and hybrid cloud model used to supply the data in the private cloud. The information now has to be kept in the secure encryption key and it will go with the secured data in cloud management.

Figure 64.1, the flow structure of the working model about the proposed work. The management plays the major role in the stream of the working. Patient have the full rights to know about the doctor and then the doctor cannot access the patient report without the permission of the management. To have the most secured data the working of the study goes with the advanced algorithm AES and DES algorithm. Based on several factors and

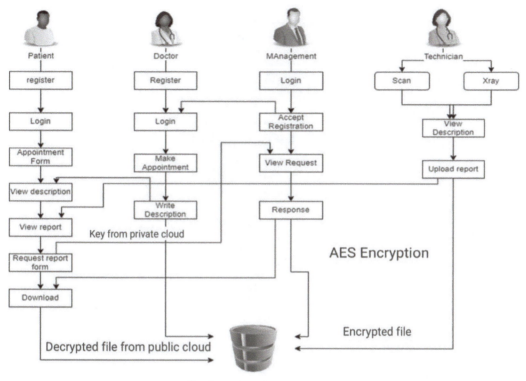

Fig. 64.1 Architecture diagram

Source: Authors

as a conclusion to the study, the proposed model will use AES encryption with SHA-256 hashing.

The model depicts an application interface of the proposed model which can be accessed by a single or multiple medical organizations and it shows the flow of functions from login of different types of users to how the data is encrypted and decrypted, how the key is stored and retrieved from private cloud and how it is used to decrypt and retrieve the data from the public cloud.

The cloud storage plays the major role to save the data in the secured way.

V. Implementation and Result Discussion

Figures 64.2–64.5 explains about the result and implementation of the proposed work.

The model explains about the full structure implementation. The view of all the doctors and then the patient login page and the consultant view. Each and every stage of the employed is clarified with the above diagram. The confidentiality matters of the data have been replied and obtained the solution to keep the data in the secured way. The persistent data and the doctor data are taken care by the management. The patient can have the secured access of the data and the management used have the secured data for the future use.

VI. Conclusion

Although many people feel their lives are not secure, modern technology makes enormous profits by using knowledge. Although health is a crucial factor in a higher quality of life for humans, personal information about them is still not securely protected. With the help

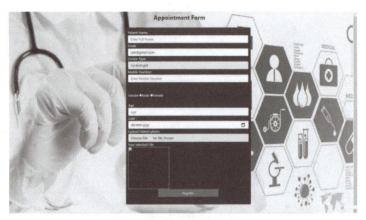

Fig. 64.2 Patient appointment form

Source: Authors

Fig. 64.3 Doctor home page

Source: Authors

Fig. 64.4 Consultant view

Source: Authors

Fig. 64.5 Management home page

Source: Authors

of a security algorithm, the cloud assumes responsibility for helping people live secure and safe lives. With the key, the encrypted data is saved in the cloud. The data must be preserved for usage in the future. Management plays a crucial role in maintaining patient record security. Information processing and subsequent fencing are both crucial. The record and all the details will be kept secure as a result of the encryption key's hard work. Therefore, more research is needed to discuss this problem. Another future attempt is to advance a motivational tool for EMR owners in the suggested scheme in order to further improve the liveliness and fairness of the system.

References

1. A. Ouaddah, H. Mousannif, A. A. Elkalam, and A. A. Ouahman, "Access control in the Internet of Things: Big challenges and new opportunities," Comput. Netw., vol. 112, pp. 237–262, Jan. 2017.

2. N. Fatema and R. Brad, "Security requirements, counterattacks and projects in healthcare applications using WSNs—A review," 2014. [Online]. Available: arXiv:1406.1795.

3. J. Sun and Y. Fang, "Cross-domain data sharing in distributed electronic health record systems," IEEE Trans. Parallel Distrib. Syst., vol. 21, no. 6, pp. 754–764, Jun. 2010.L. J. Kish and E. J. Topol, "Unpatients-why patients should own their medical data," Nat. Biotechnol., vol. 33, no. 9, p. 921, 2015.

4. S. Ravidas, A. Lekidis, F. Paci, and N. Zannone, "Access control in Internet-of Things: A survey," J. Netw. Comput. Appl., vol. 144, pp. 79–101, Oct. 2019.

5. S. Osborn, R. Sandhu, and Q. Munawer, "Configuring role-based access control to enforce mandatory and discretionary access control policies," ACM Trans. Inf. Syst. Security, vol. 3, no. 2, pp. 85–106, 2000.

6. R. S. Sandhu, "Role-based access control," in Advances in Computers, vol. 46. London, U.K.: Elsevier, 1998, pp. 237–286.

7. V. C. Hu, D. R. Kuhn, D. F. Ferraiolo, and J. Voas, "Attribute-based access control," Computer, vol. 48, no. 2, pp. 85–88, 2015.

8. R. S. Sandhu and P. Samarati, "Access control: Principle and practice," IEEE Commun. Mag., vol. 32, no. 9, pp. 40–48, Sep. 1994.

9. E. B. D. Hussein and V. Frey, "A community-driven access control approach in distributed IoT environments," IEEE Commun. Mag., vol. 55, no. 3, pp. 146–153, Mar. 2017.

10. Syed Umar Amin, M. Shamim Hossain; Ghulam Muhammad; Musaed Alhussein; Md. Abdur Rahman, "Cognitive Smart Healthcare for Pathology Detection and Monitoring", IEEE, vol 7, 2019.

11. Mofijul Islam et al," Mobile Cloud-Based Big Healthcare Data Processing in Smart Cities", IEEE, 2017.

Recent Trends in Computational Intelligence and its Application – Sugumaran D. et al. (eds)
© 2023 Taylor & Francis Group, London, ISBN 978-1-032-48410-5

65

Deduplication in Cloud Computing for Heterogeneous Data Storage Management

Vijay K.[1], Pradeep R.[2]
Assistant Professor, Department of Computer Science and Engineering,
Rajalakshmi Engineering College, Chennai, India

Sorna Shanthi D.[3]
Associate Professor, Department of Computer Science and Engineering,
Rajalakshmi Engineering College, Chennai, India

Sowmia K. R.[4], Eugene Berna I.[5]
Assistant Professor, Department of Artificial Intelligence and Machine Learning,
Rajalakshmi Engineering College, Chennai, India

Abstract—With cloud storage, users can expand their storage capacity without upgrading their hardware, removing a major resource bottleneck that can otherwise prevent them from taking full advantage of the benefits of cloud computing. Cloud data is always transmitted in an encrypted format to safeguard sensitive information and user privacy. However, encrypted data may require a great deal of extra space, and gaining quick access to shared data in the cloud may prove challenging for authorised users. We are still having issues with replica as a storage and management mechanism for encrypted data. Historically, deduplication techniques have been built around a single application architecture, giving data owners or cloud servers full authority over the process. However, they are limited in their ability to meet the changing needs of data owners by the volume of sensitive information they store. We introduce a strategy for managing data storage that is heterogeneous, spanning many Cloud Service Providers, in a way that is both flexible and synchronous in terms of access control and deduplication (CSPs).

Keywords—Cloud, Privacy, Deduplication, Storage, Resources

I. Introduction

It's safe to say that cloud storage is the most well-known service offered by cloud computing[3]. Proof of data ownership is a crucial part of the evidence of data replication process for encrypted data in the cloud. However, this method precludes individualised

[1]vijayk.btech@gmail.com, [2]pradeeprajame@gmail.com, [3]sornashanthi.d@rajalakshmi.edu.in, [4]sowmiakr@gmail.com, [5]eugeneberna31@gmail.com

DOI: 10.1201/9781003388913-65

deduplication preferences across multiple CSPs (CSPs). In this paper, we present a multi-cloud service provider (CSP) architecture for cloud data storage, in which files are uploaded by their owners and data replication is checked with the MD5 hash method. It has several layers of security and data replication protection. Likewise, a protocol known as "Provable Ownership of the File" has been set up (POF) [1][2]. The end result is a reliable system for managing data storage that is both easy to use and highly secure.

II. Related Work

In order: Dr. S. Masood Ahamed1, N. Mounika2, N. Vasavi3, M. Vinitha Reddy4, Using a deduplication strategy, cloud storage providers can reduce their space and bandwidth needs by keeping only a single duplicate of each user's data. A user can prove ownership of data stored in the cloud in a variety of ways[4][8]. Several deduplication solutions have recently been proposed as a means of dealing with this problem, with the idea being that the owner can re-use their encryption key for all of their duplicated data[6]. However, most of the systems are insecure because they fail to consider data ownership shifts that occur often in an actual cloud storage service[5]. In this way, neither a nosy cloud storage server nor a user whose access has been revoked but who was previously in possession of the data may gain access to it. In addition to protecting against tag inconsistency attacks, the proposed approach guarantees the authenticity of all stored information[7]. Therefore, the proposed solution is more secure. An efficiency analysis shows that the proposed method adds just a minimal amount of time and effort to the computations compared to the previous methods.

Steffi Researchers Miriam Philip, Shahana M. S. A., R. K. Saranya, R. Sanjana, If you choose to use a cloud service, another company will be in charge of making sure your data is secure

and up to date. Integration of files allows for their centralization and control, making them available to users[9]. Administrators have no means of knowing how well the storage node is being used because of the vast number of linked users and devices[8][10]. As a result, resources are misplaced and it becomes more challenging to maintain one's files.

Harikesh Pandy, and Shubham When it comes to data integrity and efficiency, distributed storage is a must. Respectability of data in distributed storage is ensured by technologies like proof of retrievability (POR) and confirmation of information ownership (PDP). Proof of Ownership (POW) improves skill accumulation by safely erasing data duplicates from the capacity server[12][15]. The sites of POW can be rendered invalid by the use of a mixture of the two, functioning performances, which also achieves information credibility and capacity proficiency through the use of non-minor metadata duplication (i.e., validation labels). Recent solutions to this issue have been demonstrated to be unsecure and to impose substantial computational and communication overhead[13]. Another solution is required to allow efficient and secure information trustworthiness assessment with capacity deduplication for dispersed storage[11].

In this study, we employ a novel approach based on homomorphic straight authenticators and polynomial-based validation labels to solve this unresolved issue. Both duplicate papers and confirmation labels can be removed using our method[14]. Resources are deduplicated while data integrity is verified simultaneously. The client's side computational expense and the need for constant, continuous contact are also used to highlight our proposed strategy. We recommend both individual and collective inspection. As a result, our proposed method improves upon both the POR and PDP strategies now in use.

Chiu C. Tan, Guojun Wang, Jie Wu, and Qin Liu Inevitably, the expansion of "cloud computing" will affect future IT trends. This article discusses two major concerns in a cloud setting: privacy and efficiency. We begin by talking about Ostrovsky et al., private keyword-based .'s method of file retrieval. Then, to further lessen cloud querying costs, we provide the efficient information retrieval for ranked query (EIRQ) method, which is based on an aggregation and distribution layer (ADL). Considering that searches are ranked, the ones that are higher up have a better opportunity of getting relevant results. We have thoroughly analysed a mathematical model to determine the efficacy of our technique.

The cast includes Madhuri Kavade, A.C. Lomte, presents that, As opposed to keeping several copies of the same data, de-duplication keeps only one copy and makes references to the superfluous parts of that one copy. Convergence encryption, also known as content hash keying, is a cryptosystem that can generate identical ciphertext from identical plaintext files [15]. Eliminating unnecessary data copies through a process called "data de-duplication" is a common practise in cloud storage that helps save money and resources. Although convergent encryption has seen widespread application for safe deduplication, a major component to making convergent encryption feasible is the effective and consistent management of a large number of convergent keys. Typically, we get things off by presenting a rudimentary technique that operates solely at the file level and does not involve an internet component. Current environments often use a formal approach to resolving the challenge of delivering cost-effective and trustworthy key management in safe deduplication. The proposed layout for the Dup key eliminates the need for users to physically handle any keys. It implements de-duplication at the block level and provides high levels of security. Dup key is a web-based de-duplication and cloud DB space reduction service.

Dr. Rashmi Rachh, Sneha Chandrashekhar Parit, The introduction of cloud computing has had far-reaching effects on the way data is stored and accessed. However, concerns about safety are the biggest obstacle to wider adoption. Several solutions are proposed and discussed in the canonical literature to combat this problem. For example, "fine grained access control" can be achieved via attribute-based encryption. Encrypting data based on a set of attributes termed policy is made possible by the "ciphertext policy Attribute based Encryption" system implemented in this project, which also provides cloud security and fine grained access control. The person who uploads the file is the one who gets to decide on the attributes. Those who are allowed access to the data are those hand-picked by the owner. The cipher text produced will be accessible only if the policy is satisfied.

In addition to Jin Li and Xiaofeng Chen, Mingqiang Li, Jingwei Li, and Patrick Data deduplication, the process of removing duplicate data, is widely used in cloud storage to lessen the quantity of data uploaded and the amount of storage space needed. While safe deduplication of cloud-based data has clear advantages, it is becoming increasingly challenging to implement. The efficient and trustworthy handling of a large number of convergent keys is a fundamental difficulty in making convergent encryption successful, despite its widespread use for secure deduplication. In the first part of this paper, we provide a baseline method in which users encrypt their convergent keys with their own private master keys before storing them in the cloud. The number of keys generated by this easy method grows exponentially with the number of users, therefore protecting the master key is critical. Instead of depending on a single server to manage all of the keys, we offer the unique structure of de-key to safely spread the convergent key shares over several servers. Using the proposed security model's definitions, security analysis concludes that de-key is secure. We show that the Ramp secret sharing approach may be used to implement De-key and that the scheme only suffers a small performance cost in real-world circumstances.

Cloud computing is a relatively new phenomenon in the IT sector, and it has already attracted the attention of three authors: Zhiguo Wan, Jun'e Liu, and Robert H. Deng. Many methods have been proposed for use in regulating who has access to cloud-based data; however, most of these solutions rely on attribute-based encryption (ABE), which has proven to be inadequate for enforcing policies with a high degree of complexity. We propose hierarchical attribute-set-based encryption (HASBE), which builds on ciphertext-policy attribute-set-based encryption (ASBE) with a hierarchical user structure, to achieve scalable, flexible, and fine-grained access control of outsourced data in cloud computing. Moreover, HASBE assigns a range of values to the time remaining before access expires, making it more adaptable than current systems in dealing with user revocation. To explicitly establish the security of HASBE, we leverage the performance and computational complexity studies of Bethencourt et alciphertext-policyattribute-based .'s encryption (CP-ABE) technique. We have demonstrated the efficacy of our method through a series of experiments.

As explained in a paper by Mihir Bellare, Sriram Keelveedhi, and Thomas Ristenpart, deduplication is used by cloud storage providers like Dropbox, Mozy, and others to save space by keeping only a single copy of each uploaded file. Customers can save money if they decide against using conventional data encryption. Most commonly used message-locked encryption, convergent encryption, aids in relieving this stress. Our Dup-LESS architecture allows us to provide deduplicated storage that is immune to brute-force attacks. Message-based keys are used by Dup-LESS clients to encrypt data after being requested from a key-server via a secret PRF protocol. A client's privacy is safeguarded when their data is encrypted and stored with a reputable service that offers deduplication services. We show that deduplicated storage using encryption yields similar benefits to those obtained by storing data in its native, unencrypted form.

III. Problem Definition

The phrase "cloud computing" describes the on-demand delivery of IT resources, such as data management (cloud storage) and processing power that does not compromise the user's ability to exercise centralised control. Analysis of cloud performance and security isn't always reliable. They are unable to address the issue of redundant data storage on the cloud. It's possible that storing data twice is a waste of space. The issue of access control remains unresolved. Administrators and the end users who rely on them for their services both have a vested interest in limiting who can view and change any given set of data.

Table 65.1 Experimentation findings

S.No	Number of Users involved in Communication	Number of Social Workers	Average of attack Finding Time in Existing system (ms)	Average of attack Finding Time in Proposed system (ms)
1	5	25	33	29
2	10	30	35	31
3	15	35	42	39
4	20	40	54	48
5	25	45	63	57
6	30	50	72	68
7	35	55	81	75

Source: MATLAB

A extensive comparison between both the existing system and the proposed system is provided in the Table 65.1.

In the above Fig. 65.1, the effectiveness in finding attacker is done for existing and proposed system.

COMPARISON FOR EXISTING AND PROPOSED FINDING ATTACKER EFFECTIVENESS (Time)

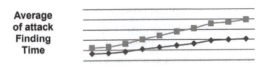

Average of attack Finding Time

Number of User Working

—■— Average of attack Finding Time in Proposed system (ms)

Fig. 65.1 Comparing the efficiency of the proposed and existing systems for finding intruders

Source: MATLAB

IV. Proposed System

As a result of our investigation, we offer a strategy for storing data across many CSPs without endangering its security by employing deduplication management. In addition, we set up a system we're calling "Provable Ownership of the File" (POF). They make users more secure in their personal information and increase the effectiveness of implementation. When using the suggested system, customers will feel secure, assured, and efficient. Users can rest easy knowing their data is secure while taking advantage of the steep discounts provided by many CSPs, all because to the encryption supplied by these companies. They provide a set of characteristics that can be used to confirm a user's identity before encrypting any sensitive data. Using Attribute based encryption, as proposed, greatly improves cloud security. The current policy ensures that users' identities

are genuine by verifying their registration information raises overall deduplication system effectiveness.

V. Architecture Diagram

Figure 65.2 represents the architectural internal structure of the system which shows all the sub algorithms at work.

In the above diagram, digital signature and deduplication check is done to ensure security in cloud storage.

VI. Implementation

A. Cloud User Authentication

The Owner has completed the Basic Cloud Service Provider Registration (CSP). Users' personal information is used in this analysis. This data is then added to the server's database. The next thing to do is sign in to your cloud storage account.

B. File Upload and Comparison

Here, the data owner signs up for a public cloud service and stores their data there. To address this issue, we propose the Provable Ownership of the File (POF) system. The file's hash key is generated using the MD5 algorithm at the time of upload by the data owner. Each and every one of your uploads will have its own unique hash key. If another data owner attempts to upload the same file, however, the system will prevent the upload and instead update the reference id with a new index mapping. As an added security measure, it verifies whether or not both data Owners are physically present when a file is accessed.

C. Set Access Policy for File

The HDFS file system is used in this section to upload the selected file. The system will create a signature and divide it into multiple chunks within a designated file. Each block will have a key-generated signature. To prevent duplicate

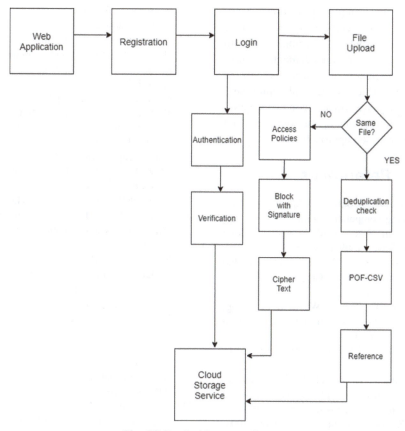

Fig. 65.2 Architecture diagram

Source: Authors

files from being created, the signature uses the MD5 message-digest technique, which produces a cryptographic hash function that generates a 128-bit hash value, which is commonly written in text format as a 32-digit hex number. Create convergent keys for each block split after that in order to save the CSV file. Like block keys, block paths, blocks, usernames, and filenames.

D. Requesting and Managing Downloads of Files

Here, the data's rightful owner accesses the cloud storage service and downloads the file. If it doesn't work, they'll ask to get it from a different Cloud storage service, compare it to what they have, and report back.

VII. Conclusion

When it comes to massive amounts of data, Data deduplication is an absolute must for cloud storage. Based on our findings, we propose a cloud-based data deduplication and access-management framework suitable for storing disparate types of data in a unified format. The flexibility of our solution allows it to meet the needs of a wide variety of applications and ensures that enormous amounts of data are stored efficiently across multiple CSPs. Data deduplication and access control may be possible, depending on the level of security implemented. Our system's safety, modernity, and efficacy have all been shown by analyses, comparisons, and empirical testing. Since cloud storage uses encryption to protect user information, our solution is compatible with cloud storage. Further enhancements to user privacy and efficiency optimizations will be made as we get closer to practical deployment of our technology. We will also conduct a game-theoretic research to back up the soundness and safety of the proposed approach.

References

1. R. Chow, et al., "Controlling data in the cloud: outsourcing computation without outsourcing control," in Proc. ACM Workshop Cloud Comput. Secur., 2009, pp. 85–90.

2. S. Kamara, and K. Lauter, "Cryptographic cloud storage," Financ. Crypto. Data Secur., 2010, pp. 136–149.

3. Q. Liu, C. C. Tan, J. Wu, and G. Wang, "Efficient information retrieval for ranked queries in cost-effective cloud environments," in Proc. IEEE INFOCOM, 2012, pp. 2581–2585.

4. M. Kallahalla, E. Riedel, R. Swaminathan, Q. Wang, and K. Fu, "Plutus: scalable secure file sharing on untrusted storage," in Proc. USENIX Conf. File Storage Technol., 2003, pp. 29–42.

5. R. Babu & K. Jayashree "A Survey on the Role of IoT and Cloud in Health Care", International Journal of Scientific, Engineering and Technology Research (IJSETR) ISSN 2319-8885 Vol.04, Issue.12, May-2015, Pages: 2217-2219. ISSN 2319-8885.

6. E.-J. Goh, H. Shacham, N. Modadugu, and D. Boneh, "SiRiUS: securing remote untrusted storage," in Proc. Netw. Distrib. Syst. Secur. Symp., 2003, pp. 131–145.

7. R. Babu & Dr. K. Jayashree "Prominence of IoT and Cloud in Health Care" International Journal of Advanced Research in Computer Engineering & Technology (IJARCET), vol.5 no.2, February 2016.

8. J. Bethencourt, A. Sahai, and B. Waters, "Ciphertext-policy attribute-based encryption," in Proc. IEEE Symp. Secur. Privacy, 2007, pp. 321–334.

9. Malathi S, Arockia Raj Y, Abhishek Kumar, V D Ashok Kumar, Ankit Kumar, Elangovan D, V D Ambeth Kumar, Chitra B & a Abirami (2021): Prediction of cardiovascular disease using deep learning algorithms to prevent COVID 19, Journal of Experimental & Theoretical Artificial Intelligence, DOI: 10.1080/0952813X.2021.1966842

10. V. Goyal, O. Pandey, A. Sahai, and B. Waters, "Attribute-based encryption for fine-grained access.

11. Narayana K.E & Jayashree, K "A Overview on Cloud Computing Platforms and Issues" International Journal of Advanced Research in Computer Science and Software Engineering Volume 7, pp. 238-22, 2017.

12. Babu, R, & Jayashree K, "Prioritizing Cloud Infrastructure Using MCDM Algorithms" International Journal of Engineering and Advanced Technology (IJEAT), vol. 8, no. 3S, 2019, pp. 696–698

13. R. Vijayakumar, K. Vijay, P. Sivaranjani, and V. Priya, "Detection of network attacks based on multiprocessing and trace back methods," Adv. Parallel Comput., vol. 38, pp. 608–613, 2021.

14. Bhavani, M., Shrijeeth, S., Rohit, M., Sanjeev Krishnan, R., Sharveshwaran, R., "A detailed study on sentimental analysis using Twitter data with an Improved deep learning model", Proceedings of the 5th International Conference on I-SMAC (IoT in Social, Mobile, Analytics and Cloud), 2021.

15. Chahat Agarwal, Akshay S, Vijay P., Intelligent System based on pollution intensity, Journal of Advanced Research in Dynamical and Control Systems, Institute of Advanced Scientific Research, 11(5), 2152–2161, 2019

Recent Trends in Computational Intelligence and its Application – Sugumaran D. et al. (eds)
© 2023 Taylor & Francis Group, London, ISBN 978-1-032-48410-5

66

Application of Neutrosophic Network Using Efficient Domination

J. Senbagamalar*, A. Meenakshi[1], T. Kujani[2]
Department of Mathematics, Vel Tech Rangarajan Dr. Sagunthala R&D Institute of
Science and Technology, Chennai, India

A. Kanchana[3]
Department of Mathematics,
Rajalakshmi Institute of Science and Technology, Chennai, India

Abstract—The most secured computational technique is to find the secret information using encryption and decryption defined in this paper. Mathematical modeling of single valued neutrosophic network is defined and constructed to elude the burgeoning intruder. The studies of efficient domination of single valued neutrosophic graph is initiated and this domination parameter plays a nuance technique to decrypt the framed network. The algorithm is framed to encrypt and decrypt the given secret number.

Keywords—Efficient domination, Neutrosophic network, Single valued, Mathematical modeling

I. Introduction

L.A. Zadeh proposed a mathematical framework to characterize the phenomenon of uncertainty in real-world situations in 1965 [18]. The idea of fuzzy graphs and various fuzzy analogues of graph theory ideas with connectedness were first presented by Rosenfeld [14]. Ore [13] and Berge started researching graphs' domination sets. Paired domination studies begun by Teresa et.al [17]. Efficient domination was initiated by Biggs [2], V.R. Kulli [5], begun the study of split domination of graph and also he[6] wrote the theory of domination in graphs. The independent domination number was first used in graphs by Cockayne [3]. Equitable domination was introduced by Swaminathan and Dharmalingam [16]. Paired equitable domination was introduced and studied by A. Meenakshi [7] and it continued by in inflated graph and its complement of a graph[8,9]. Intuitionistic fuzzy relations and Intuitionistic fuzzy graphs(IFGS) were developed by K.T. Atanassov[1]. IFG was defined by M.G. Karunambigai et. al [4] which is a special case of IFGS defined by A.Shannon and Atanassov

---
*Corresponding Author: senbagamalar2005@yahoo.com, [1]meenakshiannamalai1@gmail.com, [2]kujani@veltech.edu.in, [3]kanchana.anbazhagan@gmail.com

DOI: 10.1201/9781003388913-66

of [15]. The terms "order," "degree," and "size" of IFG were defined by A. Nagoor Gani and Shajitha Begum [12]. Split domination in Intuitionistic fuzzy graph was introduced by A. Nagoor Gani and S. Anu priya [11]. Encryption and decryption of crisp graph was studied by Meenakshi et. al [10].

II. Efficient Domination in Neutrosophic Graph

Definition 2.1: A neutrosophic graph (NG) is of the form $H_{NG} = (V_s, E_s)$ where

(i) $V_s = \{o_1, o_2,..., o_n\}$ such that $T_{V_s}:V_s \to [0,1]$; $I_{V_s}:V_s \to [0,1]$ and $F_{V_s}:V_s \to [0,1]$ denote the degree of truth membership value, degree of indeterminacy membership value, and degree of falsity membership value respectively and $0 \le T_{V_s}(v_s) + I_{V_s}(v_s) + F_{V_s}(v_s) \le 3$ for every $v_s \in V$

(ii) $E \subseteq V \times V$ where $T_{E_s}:V \times V \to [0,1]$; $I_{E_s}:V \times V \to [0,1]$; $F_{E_s}:V \times V \to [0,1]$ are defined by $T_{E_s}\{a_i, a_j)\} \le \min\{T_{V_s}(a_i), T_{V_s}(a_j)\}$; $I_{E_s}\{a_i, a_j)\} \le \min\{I_{V_s}(a_i), I_{V_s}(a_j)\}$; $F_{E_s}\{a_i, a_j)\} \ge \max\{F_{V_s}(a_i), F_{V_s}(a_j)\}$ denote the degree of truth membership value, degree of indeterminacy membership value, and degree of falsity membership value of the edge $(a_i, a_j) \in E_s$ respectively where $0 \le T_{E_s}\{(a_i, a_j)\} + I_{E_s}\{(a_i, a_j)\} + F_{E_s}\{(a_i, a_j)\} \le 3 \ \forall \ (a_i, a_j) \in E_s$.

Definition 2.2: A subset T of V^1 is said to be dominating set of a single valued neutrosophic graph if for every vertex in V^1-T is dominated by at least one vertex of V^1. The dominating set T is said to be minimal if no proper subset of T is a dominating set.

Definition 2.3: An arc $(u$-$v)$ is said to be strong arc if its degree of edge membership value is is equal to strength of connectedness between u and v.

Definition 2.4: Let $e = (a, b)$ be an edge of a single valued neutrosophic graph. We say that a dominates b if there exists a strong arc between them.

Definition 2.5: Single Valued Neutrosophic Network(SVNN) is defined as a group of same category peoples (a set of nodes) they interact with each other and work together (link is a relation which represents sharing work or sharing information) such that every node (person) has true degree membership value (T), indeterminacy degree membership value (I) and falsity degree membership (F). The relation (information, knowledge sharing, etc.) between any two persons is represented by link. The link also has true degree membership value (T), indeterminacy degree membership value (I) and falsity degree membership (F).

Definition 2.6: A SVNN is said to be strong if it satisfies the following

$$T_{E_s}\{(a_i, a_j)\} = \min\{T_{V_s}(a_i), T_{V_s}(a_j)\};$$
$$I_{E_s}\{(a_i, a_j)\} = \min\{I_{V_s}(a_i), I_{V_s}(a_j)\};$$
$$F_{E_s}\{(a_i, a_j)\} = \min\{F_{V_s}(a_i), F_{V_s}(a_j)\}$$
$$\forall \ (a_i, a_j) \in E_s \text{ and } a_i \ \& \ a_j \in V_s.$$

Definition 2.7: A dominating set T of V_s is said to be efficient dominating set of a single valued neutrosophic graph $T \cap N[v] = 1$ if for every vertex v in $V_s - T$, where $N[v]$ represents the closed neighborhood of v. The dominating set T is said to be minimal efficient dominating if no proper subset of T is a efficient dominating. The following SVN graph is strong.

III. Main Frame Work

This main frame work consists of this paper is

- Construction of SVNN from sub SVNN
- Secret key
- Encryption Algorithm
- Decryption Algorithm

A. Construction of SVNN from Sub SVNN

The secret number (numerical value) to be encrypted is non zero integer. Select the suitable numerical value ($NV \neq 0$) (as we have to split this under modulo r, $r \neq 0$). Now NV is sub divided in to 'r' values say NV_1, NV_2, ..., NV_r such that $NV_1 \equiv R_1 \pmod{r}$ (where $R_1 = 0$), $NV_2 \equiv R_2 \pmod{r}$ (where $R_2 = 1$), $NV_3 \equiv R_3 \pmod{r}$ (where $R_3 = 2$), ..., $NV_r \equiv R_r \pmod{r}$ (where $R_r = r - 1$). Since we have 'r' subdivision values, have to frame 'r' sub network and planned to assign 'r' efficient domination nodes in the constructed network. Let the efficient dominating nodes (EDN) be o_1, o_2, o_3, ..., o_r. These nodes are the center of the sub networks say SN_1, SN_2, SN_3, ..., SN_r respectively. Let the neighbors of o_1, o_2, o_3, ..., o_r be o_{11}, o_{12}, ..., o_{1l_1}; o_{21}, o_{22}, ..., o_{2l_2}; o_{31}, o_{32}, ..., o_{3l_3}, ..., o_{r1}, o_{r2}, ..., o_{rl_r} respectively. first SVN sub network is SN_1 whose center is o_1 and its neighbors are o_{11}, o_{12}, ..., o_{1l_1}. o_{11}, o_{12}, ..., o_{1l_1} First subdivision value $NV_1 \equiv R_1 \pmod{r}$. Set $V_1 = \dfrac{NV_1}{r}$ and $D_1 = D_{v1}/V_1$ (where D_{v1} is the numerical value 1 followed by the number of 0's digits of integral part of V_1) partitioned into sum of l_1 values say d_{11}, d_{12}, ..., d_{1l_1} respectively and assign these values are minimum value of either o_1 or o_{11}, o_{12}, ..., o_{1l_1}, degree of truth membership value. By the definition of SVNN, the degree of membership values of the edges o_1, o_{11}, o_1, o_{12}, ..., $o_1 o_{1l_1}$ are $(\min\{o_1(t_1), o_{11}(t_{11})\}$, $\min\{o_1(i_1), o_{11}(i_{11})\}$, $\max\{o_1(f_1), o_{11}(f_{11})\})$ $(\min\{o_1(t_1), o_{12}(t_{12})\}$, $\min\{o_1(i_1), o_{12}(t_{12})\}$, $\max\{o_1(f_1), o_{12}(f_{12})\})$,...,$(\min\{o_1(t_1), o_{1l_1}(t_{1l_1})\}$, $\min\{o_1\{i_1\}$, $o_{1l_1}\{i_{1l_1}\}\}$, $\max\{o_1\{f_1\}, o_{1l_1}\{f_{1l_1}\}\})$ respectively. Repeat the process till to frame the sub network SN_r and by the definition of SVNN, the rest of the edge's degree membership values will be defined, the readers may refer encryption algorithm for further explanation.

B. Secret Key

The key is to break the encrypted SVNN is the efficient dominating set of the SVN encrypted network. Once we find the efficient dominating set of this SVF network, we can decrypt it.

C. Encryption Algorithm

Input: $NV \geq r$, $r \neq 0$ is the secret number; Output: Encrypted SVN Network begin

Step 1: Sub divide the secret number NV into "r" values NV_1, NV_2, ..., NV_r such that $NV_1 \equiv R_1 \pmod{r}$ (where $R_1 = 0$), $NV_2 \equiv R_2 \pmod{r}$ (where $R_2 = 1$), $NV_3 \equiv R_3 \pmod{r}$ (where $R_3 = 2$), ..., $NV_r \equiv R_r \pmod{r}$ (where $R_r = r - 1$)

Step 2: Frame 'r' sub network and planned to assign 'r' efficient domination nodes in the constructed network. Hence the minimum number of edges present in the network is $(r - 1) + l_1 + l_2 + ... + l_r$.

Step 4: Define (Maximum no. of edges present in the constructed network is denoted by Max $E = \{(a, b); 1 \leq a, b \leq l_1 + l_2 + ... + l_r + r; a \neq b\}$

$$-\begin{cases} o_1 o_{k_1} \text{ where } 2 \leq k_1 \leq r, \\ o_2 o_{k_2} \text{ where } 3 \leq k_2 \leq r, \\ o_3 o_{k_3} \text{ where } 4 \leq k_3 \leq r, ..., o_{r-1} o_r, \\ o_1 o_{2j_2}, o_1 o_{3j_3}, ..., o_1 o_{rj}; \\ o_2 o_{3j3}, o_2 o_{4j4}, ... o_2 o_{rjr}; \\ o_3 o_{4j4}, o_3 o_{5j5}, ... o_3 o_{rjr}; ...; o_{r-1} o_{rjr} \\ \text{where } 1 \leq j_2 \leq l_2, 1 \leq j_3 \leq l_3..., 1 \leq j_r \leq l_r. \end{cases},$$

Step 5: $V_1 = \dfrac{NV_1}{r}$, $V_2 = \dfrac{NV_2}{r}$, $V_3 = \dfrac{NV_3}{r}$, ...,

$V_r = \dfrac{NV_r}{r}$ $D_1 = D_{v1}/V_1$

(where D_{v1}-the numerical value 1 followed by the number of 0's digits of integral part of V_1).$_,$ $D_2 = D_{v2}/V_2..., D_r = D_{vr}/V_r$

Step 6: Split $D_1 = d_{11} + d_{12} + ... + d_{1l_1}$; $D_2 = d_{21} + d_{22} + ... + d_{2l_2}$; ...; $D_r = d_{r1} + d_{r2} + ... + d_{rl_r}$;

Assign min $\{o_1\}(t_1)$, $o_{11}(t_{11})\} = d_{11}$ min $\{o_1\}$ (t_1), $o_{12}(t_{12})\} = d_{12}$, ..., min $\{o_1\}(t_1)$, $o_{1l_1}(t_{1l_1})\}$ $= d_{1l_1}$ in the first sub network. continuing the process till to assign min $\{o_r\}(t_1)$, $o_{r1}(t_{r1})\} = d_{r1}$, min $\{o_r\}(t_1)$, $o_{r2}(t_{r2})\} = d_{r2}$, ..., min $\{o_r\}$ (t_1), $o_{rl_1}(t_{rl_r})\} = d_{rl_r}$, in the r th sub network.

Step7: Rest of the edge's membership values will be followed by the definition of SVNN

End.

The Output, encrypted SVN network with minimum number of links is shown in Fig. 66.1

D. Decryption Algorithm

Input: Encrypted SVNN; Output: NV, the secret number

Begin

Step 1: Find the Efficient dominating members of SVNN o_1, o_2, o_3, ..., o_r such that $N[o_1] \cap N[o_2] \cap ... N[o_r] = \phi$ where $N[o_i]$ represent the neighbours of the vertex o_i

Step 2: $V_1 = D_{v1}\left(\sum_{j_1=1}^{l_1} d_{1j_1}\right)$; $V_2 = D_{v2}\left(\sum_{j_2=1}^{l_2} d_{1j_2}\right)$,

..., $V_r = D_{vr}\left(\sum_{j_r=1}^{l_r} d_{1j_r}\right)$;

Step 3: $NV = r\left(\sum_{i=1}^{r} V_i\right)$

End

IV. Illustration

B. Construction of SVNN from Sub SVNN

The secret number is $NV = 10810$. The suitable numerical value NV. we have to split this NV under modulo $r = 5$. Now NV is sub divided in to '5' values say NV_1, NV_2, NV_3, NV_4, NV_5 such that $NV_1 = 3000 \equiv 0(\text{mod } r)$, $NV_2 = 3001$ $\equiv 1(\text{mod } r)$, $NV_3 = 1502 \equiv 2(\text{mod } r)$, $NV_4 = 1503 \equiv 3(\text{mod } r)$ $NV_5 = 1804 \equiv 4(\text{mod } r)$. First sub network is SN_1 whose center is o_1 and its neighbors are o_{11}, o_{12}, o_{13}, o_{14}, o_{15}, o_{16}. First subdivision value $NV_1 = 3000 \equiv 0(\text{mod } r)$ Set

$V_1 = \dfrac{NV_1}{r} = \dfrac{3000}{5} = 600$ and $D_1 = D_{v1}/V_1 =$ 1000/600 = 0.600, the rest of the edge's degree membership values will be defined, the readers may refer encryption algorithm for further explanation. Degree membership values of the edges are given in Table 66.1.

B. Secret Key

The key is to break the encrypted SVNN is the efficient dominating members of the SVN encrypted network which is shown in fig 13. Once we find the efficient dominating set of this network, we can decrypt it. The efficient dominating members of this SVNN is o_1, o_2, o_3, o_4, and o_5.

C. Encryption Algorithm

Input: $NV = 10810$ is the secret number

Output: Encrypted SVN Network

begin

Step 1: Sub divide the secret number NV into '5' values NV_1, NV_2, NV_3, NV_4, NV_5 such that $NV_1 = 3000 \equiv 0(\text{mod } r)$ $NV_2 = 3001 \equiv 1(\text{mod } r)$, $NV_3 = 1502 \equiv 2(\text{mod } r)$, $NV_4 = 1503 \equiv 3(\text{mod } r)$, $NV_5 = 1804 \equiv 4(\text{mod } r)$

Step 2: Frame '5' sub network and planned to assign '5' efficient domination nodes in the constructed network.

Step 3: The number of nodes present in the SVN network is 33.

Minimum no. of edges present in the constructed network is denoted by Min E = $\{o_1o_{1j_1}, o_2o_{2j_2}, ..., o_5o_{5j_5}/1 \leq j_1 \leq 6, 1 \leq j_2 \leq 6, 1 \leq j_3 \leq 5, 1 \leq j_4 \leq 5, 1 \leq j_5 \leq 6\} \cup \{o_{1j_1}o_{2j_2}, o_{2j_2}o_{3j_3}, o_{4j_4}o_{5j_5}\}$ for only one $j_1, j_2, ..., j_5$ where $1 \leq j_1 \leq 6, 1 \leq j_2 \leq 6$, $1 \leq j_3 \leq 5, 1 \leq j_4 \leq 5, 1 \leq j_5 \leq 6$.

Hence the minimum number of edges present in the network is 32

Step 4: Define (Maximum no. of edges present in the constructed network is denoted by

Table 66.1   Edge membership values of Fig. 66.2

Edges	Degree of Membership values	Edges	Degree of Membership values	Edges	Degree of Membership values
o_1o_{11}	(0.1,0.02,0.33)	o_3o_{31}	(0.05,0.03,0.36)	o_5o_{51}	(0.05,0.03,0.4)
o_1o_{12}	(0.1,0.02,0.35)	o_3o_{32}	(0.05,0.03,0.4)	o_5o_{52}	(0.05,0.03,0.36)
o_1o_{13}	(0.1,0.22,0.32)	o_3o_{33}	(0.05,0.03,0.4)	o_5o_{53}	(0.05,0.03,0.4)
o_1o_{14}	(0.1,0.02,0.3)	o_3o_{34}	(0.05,0.03,0.4)	o_5o_{54}	(0.05,0.02,0.4)
o_1o_{15}	(0.1,0.02,0.33)	o_3o_{35}	(0.050,0.03,0.4)	o_5o_{55}	(0.05,0.03,0.4)
o_1o_{16}	(0.1,0.02,0.3)	$o_{31}o_{35}$	(0.0504,0.03,0.36)	o_5o_{56}	(0.0608,0.03,0.4)
$o_{11}o_{21}$	(0.1,0.02,0.36)	$o_{32}o_{33}$	(0.05,0.15,0.35)	$o_{14}o_{52}$	(0.05,0.1,0.25)
$o_{12}o_{26}$	(0.1,0.12,0.35)	$o_{33}o_{34}$	(0.05,0.04,0.3)	$o_{14}o_{51}$	(0.05,0.03,0.36)
$o_{13}o_{26}$	(0.1,0.12,0.32)	$o_{34}o_{35}$	(0.05,0.04,0.25)	$o_{15}o_{56}$	(0.0608,0.02,0.33)
$o_{11}o_{16}$	(0.1,0.02,0.33)	o_4o_{41}	(0.1,0.03,0.4)	$o_{51}o_{56}$	(0.05,0.03,0.36)
$o_{11}o_{12}$	(0.1,0.02,0.36)	o_4o_{42}	(0.05,0.03,0.4)	$o_{51}o_{52}$	(0.05,0.03,0.361)
$o_{14}o_{15}$	(0.1,0.02,0.25)	o_4o_{43}	(0.05,0.03,0.4)	$o_{53}o_{54}$	(0.05,0.02,0.35)
$o_{15}o_{16}$	(0.1,0.02,0.33)	o_4o_{44}	(0.05,0.02,0.4)	$o_{54}o_{55}$	(0.05,0.02,0.25)
o_2o_{21}	(0.1,0.02,0.36)	o_4o_{45}	(0.0506,0.02,0.4)	$o_{24}o_{53}$	(0.05,0.02,0.35)
o_2o_{22}	(0.1,0.02,0.3)	$o_{41}o_{45}$	(0.0506,0.02,0.36)	$o_{34}o_{53}$	(0.05,0.22,0.35)
o_2o_{23}	(0.1,0.02,0.3)	$o_{41}o_{42}$	(0.05,0.03,0.36)	$o_{45}o_{42}$	(0.05,0.04,0.25)
o_2o_{24}	(0.1,0.02,0.3)	$o_{43}o_{44}$	(0.05,0.02,0.3)	$o_{21}o_{26}$	(0.1,0.03,0.36)
o_2o_{25}	(0.1,0.02,0.3)	$o_{41}o_{53}$	(0.05,0.03,0.36)	$o_{25}o_{26}$	(0.1,0.02,0.2)
o_2o_{26}	(0.1002,0.02,0.3)	$o_{45}o_{54}$	(0.05,0.02,0.2)	$o_{23}o_{24}$	(0.1,0.02,0.3)
		$o_{41}o_{54}$	(0.05,0.02,0.2)	$o_{22}o_{23}$	(0.1,0.15,0.35)

Max E = $\{(a, b); 1 \le a, b \le 33; a \ne b\}$

$$=\begin{cases} o_1o_{k_1} \text{ where } 2 \le k_1 \le 6, \\ o_2o_{k_2} \text{ where } 3 \le k_2 \le 6, \\ o_3o_{k_3} \text{ where } 4 \le k_3 \le 5,. \\ o_4o_5, o_1o_{2j_2}, o_1o_{3j_3}, ..., o_1o_{5j_5}; \\ o_2o_{3j3}, o_2o_{4j4}, ...o_2o_{5rj5}; \\ o_3o_{4j4}, o_3O_{o5j5}; o_4o_{5j5} \\ \text{where } 1 \le j_2 \le 6, 1 \le j_3 \le 5, \\ 1 \le j_4 \le 5, 1 \le j_5 \le 6. \end{cases}$$

Step 5: $V_1 = \dfrac{NV_1}{r} = \dfrac{3000}{5}, \quad V_2 = \dfrac{3001}{5},$

$V_3 = \dfrac{1502}{5}, \quad V_4 = \dfrac{1503}{5} \quad V_5 = \dfrac{1804}{5}$

$D_1 = D_{v1}/V_1 = 1000/600 = 0.600; D_2 = 0.6002;$
$D_3 = 0.3004; D_4 = 0.3006; D_5 = 0.3608$

Step 6: Split $D_1 = d_{11} + d_{12} + d_{13} + d_{14} + d_{15} =$
$0.1 + 0.1 + 0.1 + 0.1 + 0.1 + 0.1;$
$D_2 = 0.1 + 0.1 + 0.1 + 0.1 + 0.1 + 0.1002$
$D_3 = 0.05 + 0.05 + 0.05 + 0.1 + 0.0504$
$D_4 = 0.05 + 0.05 + 0.05 + 0.1 + 0.0506$
$D_5 = 0.05 + 0.05 + 0.05 + 0.05 + 0.05 + 0.0608$

Step7: Rest of the edge's membership values will be followed by the definition of SVNN

End. The output of the encrypted network is shown in Fig. 66.2.

D. Decryption Algorithm

Input: Encrypted SVNN ;Output: *NV*, the secret number

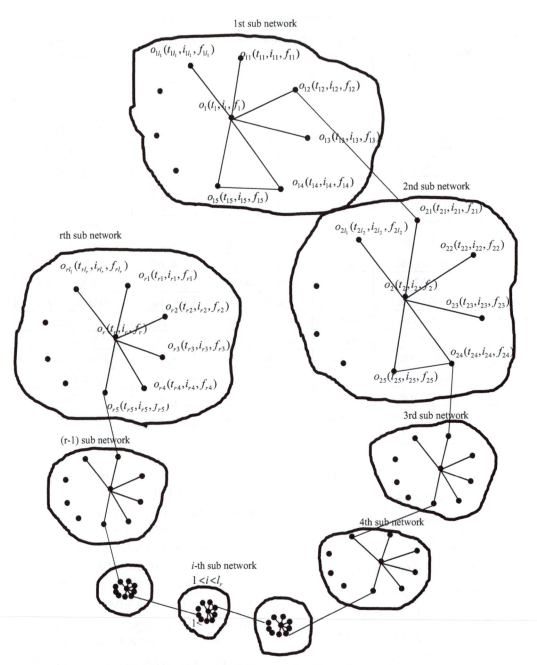

Fig. 66.1 Encrypted SVN network with minimum edges

Begin

Step 1: Find the Efficient dominating members of SVNN o_1, o_2, o_3, o_4, o_5 such that

$$N[o_1] \cap N[o_2] \cap N[o_3] \cap N[o_4] \cap N[o_5] = \phi$$

Step 2: $V_1 = D_{v1}\left(\sum_{j_1=1}^{6} d_{1j_1} \right) = 600,\ V_2 = 600.4,$

$V_3 = 300.4,\ V_4 = 300.6,\ V_5 = 360.8$

Step 3: $NV = 10810.$

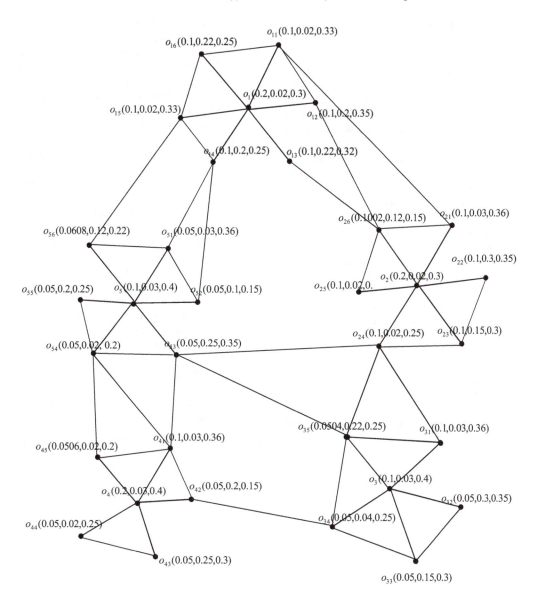

Fig. 66.2 Illustration encrypted SVNN with secret number -10810

V. Conclusion

The secret key is to find the efficient dominating set of SVNN. Repetition of assigning the d_{ij} values and maximum number of edges presented in the constructed SVNN then it becomes more complicated. Moreover, according to the construction of SVNN, the efficient dominating set of the constructed network is unique and also it is strong.

References

1. K. T. Atanassov, (1999) Intutionistic fuzzy set theory and applications physica, New York.
2. Biggs, N, (1973) Perfect codes in graphs, J. Combin, Theory Ser. B, 15(1) 289–296.
3. E. J. Cockayne & S. T. Hedeniemi, (1977) Townards a theory of Domination in Graph Networks. 7(1) 247–261.
4. M. G. Karunambigai, R. Parvathi & R.Bhuvaneswari, (2011) Constant Intutionistic Fuzzy Graphs, NIFS 17 37–47.
5. V. R. Kulli & Janakiram B. (1997) The split domination number of a graph, Graph theory notes of Newyork, New York Academy of sciences XXXXII, 16–19.
6. V. R. Kulli, (2012), Theory of domination in Graphs, Vishwa International Publications.
7. A. Meenakshi & J. Baskar Babujee, (2016) Paired equitable Domination in Graphs, International Journal of Pure and Applied Mathematics, 109(7), 75–81.
8. A. Meenakshi, (2019), Paired Equitable domination in inflated graph, International Journal of innovative Technology and exploring engineering, 8(7), 1117–1120.
9. A. Meenakshi, (2020), Equitable Domination in inflated graphs and its complements, AIP Conference Proceedings, 2277, 100006 -18.
10. A. Meenakshi, J. Senbagamala & A. Neel Armstrong, (2021) Advances in Intelligent systems and Computing, 1422(1) 427–433.
11. A. Nagoor Gani & S. Anupriya, (2012) Split domination in Intuitionistic Fuzzy Graphs, International Journal of algorithms, Computing and Mathematics, 3(3) 11–16.
12. A. Nagoor Gani & Shajitha Begum (2010) Degree, Order and Size in Intuitionistic Fuzzy Graphs, International Journal of Algorithms, Computing and Mathematics, 3(3) 11–16.
13. Ore, O. (1962), Theory of Graphs, Amer. Math, Soc, 38(1)
14. A. Rosenfield, L. A. Zadeh, K. S. Fu & M. Shimura,(1975), Fuzzy sets and their applications, Academic press, Newyork. (1975)
15. A. Shannon & K. Atanassov (2006), On a Generalization of Intuitionistic Fuzzy Graphs, NIFS, 12(1), 24–29.
16. V. Swaminathan & K. M. Dharmalingam, (2011) Degree equitable domination on graphs, Kragujevac Journal of Mathematics, 35(1), 191–197.
17. Teresa W. Haynes & Slater P. J. (1998) Paired Domination in Graphs, 32(1) 199–206.
18. L.A. Zadeh, (1965) Fuzzy Sets, Information and Control 8(1).

Note: All of the figures and table in this chapter were made by the author.

Recent Trends in Computational Intelligence and its Application – Sugumaran D. et al. (eds)
© 2023 Taylor & Francis Group, London, ISBN 978-1-032-48410-5

67

A Comparative Study of CPM Analysis in the Fuzzy and Neutrosophic Environment

T. Nagalakshmi*

Assistant Professor, Veltech Rangarajan Dr. Sagunthala R&D Institute of Science and Technology, Chennai, Tamilnadu, India

J. Shivangi Mishra[2]

Student (M.Sc.), Veltech Rangarajan Dr. Sagunthala R&D Institute of Science and Technology, Chennai, Tamilnadu, India

Abstract—Network Scheduling is a method used to plan and schedule significant projects in the sectors of fabrication, maintenance, and production, among others. A project is characterized as a collection of connected tasks that must be carried out in a specific sequence in order to be completed. In order to build a project network, E.I. Du Pont de Nemours and Company created the critical path method (CPM), which was later expanded by Mauchly Associates. The CPM approach starts off with static activity times. Nowadays, fuzziness prevails in every activity. A project involving nine activities is considered in this proposed approach. For the first case, the project duration is taken as L-R Generalized Trapezoidal Fuzzy Numbers (L-RGTrFNs). For the second case, the project duration is taken as Neutrosophic Triangular Fuzzy Numbers (NTFNs). The crisp, fuzzy solution along with its critical path are obtained for both the fuzzy numbers. The crisp project completion time is obtained by applying the existing ranking functions of both L-RGTrFNs and NTFNs. Additionally, by using the appropriate techniques, the fuzzy solution for the project completion time is also obtained. The results are compared with a numerical example and conclusion is drawn. This proposed approach paves the way for the fuzzy and neutrosophic analysis of a critical path network problem.

Keywords—Fuzzy optimization, Fuzzy critical path network problem, Defuzzification technique, L-R type generalized trapezoidal fuzzy numbers, Neutrosophic triangular fuzzy numbers, Trapezoidal fuzzy numbers

I. Introduction

A project is modelled as a network for analytical processing to provide solutions for organising and managing its activities. A network is made up of a collection of arcs that are meaningfully connected by a collection of nodes. A network is a useful way to show the order of importance

---
*Corresponding Author: drnagalakshmit@veltech.edu.in, nagalakshmi.1979@gmail.com;
[1]shivangimishra1427@gmail.com

DOI: 10.1201/9781003388913-67

between different project tasks. Thus, the project network is the collection of precedence relationships between the various project tasks. CPMs are essential resources for the effective management of all kinds of projects. CPM gives the absolute minimum project duration and encapsulates the critical path when projects are deterministic and known.

In 1965, Lotfi A. Zadeh proposed the concept of fuzzy set theory. Several fuzzy mathematics problems have since been constructed and developed. The information of uncertainty emerges as the point of interest for concern. The issues start with the focus on the problem of complexity. The concept of a fuzzy set emerges as an essence of a generalization of the classical or crisp set. Numerous scholars have conducted extensive study in the area of fuzzy optimization. In 1975, the Fuzzy Ranking Technique was proposed for dealing different kinds of fuzzy numbers. There are many suggestions for ordering L-R type generalised trapezoidal fuzzy numbers to overcome this issue with fuzzy numbers. To determine the crisp and fuzzy length of the activities and the project network built, which provides a critical path, the ranking method, and the operation of L-RGTrFNs were utilised.

To represent ambiguous, contradictory, and incomplete facts as well as real-world issues, neutrosophic sets have been presented as a generalisation of crisp sets, fuzzy sets, and intuitionistic fuzzy sets. Truth-membership, falsity-membership, and indeterminacy membership functions are used to describe the characteristics of neutrosophic set factors. The ranking technique and the operations of the NTFNs have been used to find the crisp and fuzzy duration of the activity and project network constructed.

Much research has been carried out in the field of network analysis. Zhou et al. (2016) have used some special L-R fuzzy numbers to deal with vague information. An operational law for independent regular L-R fuzzy was proposed.

Uthra et al. (2018) proposed a novel technique for ranking L-RGTrFNs. The study presented an average approach for the ranking technique, and it was compared with other fuzzy ranking techniques. The ranking technique suggested in this research is simple to understand, adaptable, and real number translation invariant. Dorfeshan and Mousavi (2018) dealt with soft computing, which is founded on an interval type - ii fuzzy decision modelling technique. This model is applied to solve a Project-Critical Path Selection Problem. Begum et al. (2019) described a method to determine a critical path to decrease the time duration for completing the jobs by applying interval-valued hexagonal fuzzy numbers. The method to determine a fuzzy critical path developed by Rameshan and Dinagar (2020) taken into account Octagonal Intuitionistic Fuzzy Numbers for the activity time.

The arithmetic operations of Neutrosophic single-valued numbers have been defined by Smarandache (2016). The limitations for these procedures for neutrosophic single-valued numbers are provided. Ye (2017) made use of the Simplified Neutrosophic Sets (SNSs) and corresponding constraint conditions for the subtraction and division operations. In the meantime, numerical examples were given to demonstrate the division and subtraction operations over SNSs.

In this proposed model, a novel approach is suggested for determining a project network's critical path in a fuzzy and neutrosophic environment. The second section defines L-RGTrFNs and NTFNs. Section 3 discusses its ranking techniques, and Section 4 gives its arithmetic operations. Section 5 states the mathematical formulation of a fuzzy critical path method. Sections 6 and 7 discusses the illustration of numerical examples in fuzzy and neutrosophic environments. Sections 8 and 9 give the comparison of results and the conclusion. To handle project analysis using the critical path method (CPM), network techniques were employed.

II. Terminology

A. L-R type Generalized Trapezoidal Fuzzy Number (L-RGTrFN)

$\tilde{A} = (m, n, a, b, \omega)_{LR}$ is considered as L-RGTrFN, if its function of membership is given as

$$\mu_{\tilde{A}}(x) = \begin{cases} \omega L\left(\dfrac{m-x}{a}\right), x \leq m, a > 0 \\ \omega R\left(\dfrac{x-n}{b}\right), x \geq n, b > 0 \\ \omega, \text{ otherwise} \end{cases}$$

In the above function, L and R stands for reference functions.

B. Neutrosophic Triangular Fuzzy Number (NTFN)

$\tilde{A} = (a_1, a_2, a_3; b_1, b_2, b_3; c_1, c_2, c_3)$ is considered as NTFN, if its truth, indeterminacy, and falsity functions of membership are given as

$$T_{\tilde{A}_N}(x) = \begin{cases} \dfrac{x-a_1}{a_2-a_1}, a_1 \leq x \leq a_2 \\ 1, x = a_2 \\ \dfrac{a_3-x}{a_3-a_2}, a_2 \leq x \leq a_3 \\ 0, \text{ otherwise} \end{cases},$$

$$I_{\tilde{A}_N}(x) = \begin{cases} \dfrac{b_2-x}{b_2-b_1}, b_1 \leq x \leq b_2 \\ 0, x = b_2 \\ \dfrac{x-b_2}{b_3-b_2}, b_2 \leq x \leq b_3 \\ 1, \text{ otherwise} \end{cases},$$

$$F_{\tilde{A}_N}(x) = \begin{cases} \dfrac{c_2-x}{c_2-c_1}, c_1 \leq x \leq c_2 \\ 0, x = c_2 \\ \dfrac{x-c_2}{c_3-c_2}, c_2 \leq x \leq c_3 \\ 1, \text{ otherwise} \end{cases}$$

III. Ranking Technique

A. Ranking Technique of L-RGTrFNs

The ranking technique used to defuzzify a L-RGTrFN taken as $\tilde{A} = (m, n, a, b, \omega)_{LR}$ is given as follows:

$$R(\tilde{A}) = \frac{\omega}{2} \frac{(m-a+m+n+n+b)}{6}$$
$$= \frac{\omega(2m+2n-a+b)}{12}$$

B. Ranking Technique of NTFNs

The ranking technique used to defuzzify a NTFN taken as $\tilde{B} = (q_1, q_2, q_3; t_1, t_2, t_3; s_1, s_2, s_3)$ is given as follows:

$$R(\tilde{B}) = \frac{(q_1 + 2q_2 + q_3 + t_1 + 2t_2 + t_3 + s_1 + 2s_2 + s_3)}{12}$$

IV. Arithmetic Operations

A. Arithmetic Operations on L-RGTrFNs:

Let $\tilde{A}_1 = (m_1, n_1, \alpha_1, \beta_1, \omega)_{LR}$, $\tilde{A}_2 = (m_2, n_2, \alpha_2, \beta_2, \omega)_{LR}$ be any two L-RGTrFNs, then the addition and subtraction of 2 L-RGTrFNs is given by

1. $\tilde{A}_1 \oplus \tilde{A}_2 = (m_1 + m_2, n_1 + n_2, \alpha_1 + \alpha_2, \beta_1 + \beta_2, \omega)_{LR}$
2. $\tilde{A}_1 \ominus \tilde{A}_2 = (m_1 - m_2, n_1 - n_2, \alpha_1 + \beta_2, \beta_1 + \alpha_2, \omega)_{LR}$

Arithmetic Operations of NTFNs:

Let $\tilde{A} = \{T_{\tilde{A}_{N1}}(x); I_{\tilde{A}_{N1}}(x); F_{\tilde{A}_{N1}}(x)\}$ & $\tilde{B} = \{T_{\tilde{A}_{N2}}(x); I_{\tilde{A}_{N2}}(x); F_{\tilde{A}_{N2}}(x)\}$ be two NTFNs, then the operations such as addition, subtraction & multiplication are defined as follows:

1. $\tilde{A} + \tilde{B} = \{x, \ T_{\tilde{A}_{N1}}(x) + T_{\tilde{A}_{N2}}(x) - T_{\tilde{A}_{N1}}(x) T_{\tilde{A}_{N2}}(x); \ I_{\tilde{A}_{N1}}(x) \ I_{\tilde{A}_{N2}}(x); \ F_{\tilde{A}_{N1}}(x) T_{\tilde{A}_{N2}}(x)\}$

2. $\tilde{A} - \tilde{B} = \left\{ x, \dfrac{T_{\tilde{A}_{N1}}(x) - T_{\tilde{A}_{N2}}(x)}{1 - T_{\tilde{A}_{N2}}(x)}; \dfrac{T_{\tilde{A}_{N1}}(x)}{I_{\tilde{A}_{N2}}(x)}; \dfrac{F_{\tilde{A}_{N1}}(x)}{F_{\tilde{A}_{N2}}(x)} \right\}$

3. $\tilde{A} \times \tilde{B} = \left\{ \left\langle \begin{array}{c} x, T_{\tilde{A}_{N1}}(x) T_{\tilde{A}_{N2}}(x); I_{\tilde{A}_{N1}}(x) + I_{\tilde{A}_{N2}}(x) - I_{\tilde{A}_{N1}}(x) I_{\tilde{A}_{N2}}(x); \\ F_{\tilde{A}_{N1}}(x) + F_{\tilde{A}_{N2}}(x) - F_{\tilde{A}_{N1}}(x) T_{\tilde{A}_{N2}}(x) \end{array} \middle| x \in X \right\rangle \right\}$

V. FCPM—A Mathematical Formulation

In this proposed approach, a project is considered with nine activities in which its duration is taken as L-RGTrFNs and NTFNs. The crisp optimal solution is obtained by applying the ranking technique of L-RGTrFNs as well as NTFNs. The proposed approach is used to obtain the fuzzy optimal solution using fuzzy forward slack and fuzzy backward slack methods. By comparing both the slacks, the fuzzy optimal solution, and its corresponding critical path is obtained. This approach is carried out under the Fuzzy Critical Path Method.

VI. Numerical Example with L-RGTrFNs

The following project is considered with nine activities. The activities which are to

be preceded before every activity are also considered. The duration of all the activities is considered as L-RGTrFNs. It is given in Table 67.1.

By using the ranking technique of L-RGTrFNs, the fuzzy durations of the above problem are converted to crisp durations as given in Table 67.2.

Critical Path:

By comparing the optimal crisp solutions of forward slack methods and the backward slack methods, the critical path network is framed and its crisp optimal solution is obtained in Fig. 67.1. It is observed that the critical path is 1-2-5-7 (A-D-H). The critical path A-D-H along with its optimal crisp solution 3.333 is obtained.

Fuzzy Critical Path:

By comparing the optimal fuzzy solutions of fuzzy forward slack methods and fuzzy

Table 67.1 Project with fuzzy duration (L-RGTrFNs)

Activity	Precedence	Duration	Activity	Precedence	Duration
A	-	(3,4,1,1,0.5)	F	B	(4,5,1,1,0.5)
B	-	(4, 5, 2, 2, 0.5)	G	C	(6, 8, 2, 2, 0.5)
C	-	(4, 6, 2, 2, 0.5)	H	D	(10,12, 2, 2, 0.5)
D	A	(5,6, 1, 1, 0.5)	I	E,F	(8, 9, 1, 1, 0.5)
E	A	(5, 7, 2, 2, 0.5)			

Source: Authors

Table 67.2 Project with crisp duration (L-RGTrFNs)

Activity	Precedence	Duration	Activity	Precedence	Duration	Activity	Precedence	Duration
A	-	0.583	D	A	0.917	G	C	1.167
B	-	0.750	E	A	1.000	H	D	1.833
C	-	0.833	F	B	0.750	I	E, F	1.417

Source: Authors

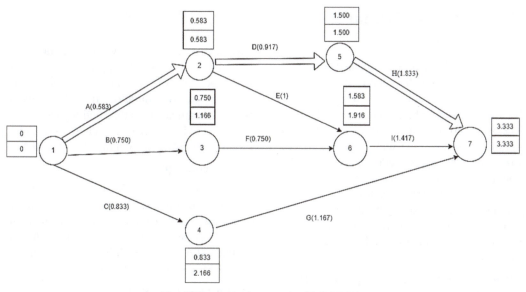

Fig. 67.1 Critical network of L-RGTrFNs

Source: Authors

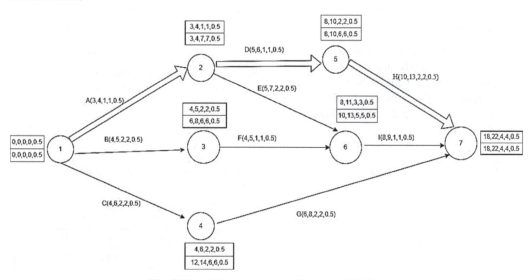

Fig. 67.2 Critical network of fuzzy L-RGTrFNs

Source: Authors

backward slack methods, the critical path network is framed and its optimal fuzzy solution is obtained in Fig. 67.2. It is observed that the critical path is 1-2-5-7 (A-D-H).

Results and Discussions:

In this proposed approach, a project with nine activities is considered, whose fuzzy duration is taken as L-RGTrFNs. It is then solved by the fuzzy forward slack method and the fuzzy backward slack method. It is observed that the critical path is 1-2-5-7 (A-D-H). The critical path A-D-H along with its optimal fuzzy solution (18, 22, 4, 4,0.5) is obtained.

VII. Numerical Example with NTFNs

In this example, a project is considered with nine activities in which its durations are taken as NTFNs. The optimal crisp solution is obtained by applying the ranking technique of NTFNs. The proposed approach is used to obtain the optimal fuzzy solution using fuzzy forward and fuzzy backward slack methods. By comparing both the slacks, the optimal fuzzy solution, and its corresponding critical path is obtained. The activities which are to be preceded before every activity are also considered. The duration of all the activities is considered as NTFNs. It is given in Table 67.3.

By using the ranking technique of NTFNs, the fuzzy durations of the above problems are converted to crisp durations as given in Table 67.4.

Critical Path:

By comparing the optimal crisp solutions of forward slack methods and backward slack methods, the critical path network is framed and its optimal crisp solution is obtained in Fig. 67.3. It is observed that the critical path is 1-2-5-7 (A-D-H).

Results and Discussions:

In this proposed approach, a project with nine activities is considered where the duration was taken as NTFNs. By applying the ranking technique of NTFNs, fuzzy duration was converted to crisp duration. It is then solved by the forward slack method and the backward slack method. The critical path A-D-H along with its optimal crisp solution 22 is obtained.

Fuzzy Critical path:

By comparing the optimal fuzzy solutions of fuzzy forward slack methods and fuzzy backward slack methods, the critical path network is framed and its optimal fuzzy solution is obtained in Fig. 67.4. It is observed that the critical path is 1-2-5-7 (A-D-H).

Results and Discussions:

In this proposed approach, a project with nine activities is considered where the fuzzy duration was taken as NTFNs. It is then solved by the fuzzy forward slack method and the fuzzy backward slack method. The critical path A-D-H along with its optimal fuzzy solution (61, 166, 337; 0, 0, 0; 0, 0, 0) is obtained.

Table 67.3 Project with fuzzy duration (NTFNs)

Activity	Precedence	Fuzzy Duration	Activity	Precedence	Fuzzy Duration
A	-	(3,4,5;2,3.5,4.5;3.5,4.5,6)	F	B	(4,5,6;3,4.5,5.5;4.5,5.5,7)
B	-	(4,5,6;3,4.5,5.5;4.5,5.5,7)	G	C	(7,8,9;6,7.5,8.5;7.5,8.5,10)
C	-	(5,6,7;4,5.5,6.5;5.5,6.5,8)	H	D	(11,12,13;10,11.5,12.5;11.5,12.5,14)
D	A	(4,6,8;3,5,7;5,7,9)	I	E, F	(8,9,10;7,8.5,9.5;8.5,9.5,11)
E	A	(6,7,8;5,6.5,7.5;6.5,7.5,9)			

Source: Authors

Table 67.4 Project with crisp duration (NTFNs)

Activity	Precedence	Crisp Duration	Activity	Precedence	Crisp Duration	Activity	Precedence	Crisp Duration
A	-	4	D	A	6	G	C	8
B	-	5	E	A	7	H	D	12
C	-	6	F	B	5	I	E, F	9

Source: Authors

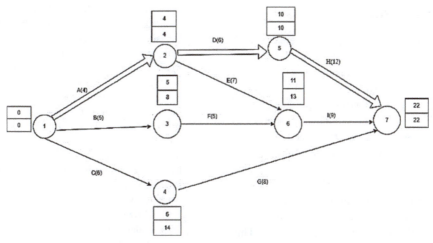

Fig. 67.3 Critical network of NTFNs

Source: Authors

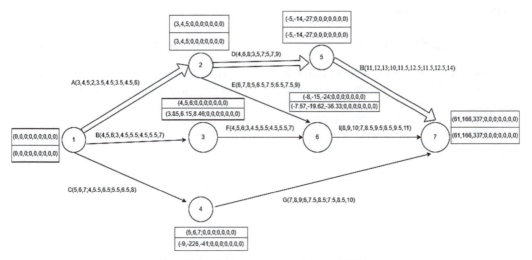

Fig. 67.4 Critical network of fuzzy NTFNs

Source: Authors

VIII. Comparison of Results

Table 67.5 Comparison of crisp and fuzzy duration of critical path network using L-RGTrFNs and NTFNs

Operation	CRISP/FUZZY	Optimal Critical Solution	Critical Path	Conclusion
L-RGTrFNs	Optimal Crisp solution	3.333	A-D-H	L-RGTrFNs play a major role in finding the critical path and it yields the best optimal solution to the problem
	Optimal Fuzzy solution	(18,22,4,4,0.5)		
NTFNs	Optimal Crisp Solution	22	A-D-H	
	Optimal Fuzzy solution	(61, 166, 337; 0, 0, 0; 0,0,0)		

Source: Authors

IX. Conclusion

In this proposed work, a project network with nine activities is considered. The duration of all the activities was taken as L-RGTrFNs and NTFNs. It was defuzzified into crisp duration using the ranking technique of both L-RGTrFNs and NTFNs. The optimal crisp and fuzzy solutions were framed for both L-RGTrFNs and NTFNs. While comparing the optimal crisp and fuzzy solutions, it is observed from Table 67.5 that L-RGTrFNs yield the best solution, which is the minimum. Additionally, it is noted that the critical path network is the same throughout all analyses carried out. This proposed approach will pave the way for the researchers to apply this concept to other fuzzy numbers with more parameters.

References

1. Begum, S. G., Praveena, J. P. N., Rajkumar, A. (2019) 'Critical path through interval valued hexagonal fuzzy number', *Int. J. Innovative Technol. Expl. Eng*. Vol. 8, No. 11, pp. 1190–1193.

2. Dorfeshan, Y. and Mousavi, S. M. (2018) 'Soft Computing Based on an Interval Type-2 Fuzzy Decision Model for Project-Critical Path Selection Problem', *International Journal of Applied Industrial Engineering*, Vol. 5, No.1, pp. 1–24.

3. Rameshan, N., Dinagar, D. S. (2020) 'Solving fuzzy critical path with octagonal intuitionistic fuzzy number', *AIP Conf. Proc.* 2277(090022), pp. 1–8.

4. Smarandache F., (2016) Subtraction and Division of Neutrosophic Numbers, Critical Review, Volume XIII.

5. Uthra, G. Thangavelu, K. and Umamageswari, R.M. (2018) 'Ranking L-R Type Generalized Trapezoidal Fuzzy Numbers', *International Journal of Mathematics and its applications*, Vol. 6, No. 1-C, pp. 411–416.

6. Ye, J. (2017) 'Subtraction and Division Operations of Simplified Neutrosophic Sets', *Information*, Vol. 8, No. 2, 51.

7. Zhou, J., Yang, F. and Wang, K, (2016) 'Fuzzy Arithmetic on L-R Fuzzy Numbers with Applications to Fuzzy Programming', *Journal of Intelligent & Fuzzy Systems*, Vol. 30, No. 1, pp. 71–87.

Recent Trends in Computational Intelligence and its Application – Sugumaran D. et al. (eds)
© 2023 Taylor & Francis Group, London, ISBN 978-1-032-48410-5

68

Secret Sharing of Information using First Order ODE and Variable Separable Method

Nagadevi Bala Nagaram*, Kala Raja Mohan[1], Sathish Kumar Kumaravel[2]
Department of Mathematics, Vel Tech Rangarajan Dr. Sagunthala R&D Institute of
Science and Technology, Avadi, Chennai – 600 062, Tamil Nadu, India

Suresh Rasappan[3]
Information Technology Department,
University of Technology and Applied Sciences - Ibri, Sultanate of Oman

Abstract—Secret Information sharing is all time requirement in this internet world. It is widely used in system security, transaction etc. The mathematical technique is adapted to maintain the secrecy of the information. This paper proposes solving first order ordinary differential equations by variable separable method. The conversion of the given information into secret message is known as encryption. It is kept the original information secretly when has been reached unauthorized persons. With the help of secret key one can redeem the original information. For this secrecy the mathematical technique formation of first order ordinary differential equation is considered. To reveal the original information the variable separable method is applied.

Keywords—Cryptography, Data encryption, Decryption, First order ordinary differential equations, Variable separable method

I. Introduction

In today's life mobile communications and netsource plays an vital role of our society[1]. Providing information security is obviously needed to protect from hackers. Cryptography is one of the most important approaches for information security [2],[3],[4]. The goal of cryptography is to assure the secret communication between two persons [5],[6],[7],[8]. Encryption hide the original information. Also it makes unreadable without secret key [9],[10],[11]. This paper aims at developing a cryptographic algorithm using first order ordinary differential equation and variable separable method.

In this paper, section 2 explains the basics of crypto analysis. Section 3 presents the algorithm for encryption. Section 4 demonstrates the algorithm explained in section 3 with an

*Corresponding Author: nagaramnagadevibala@gmail.com,
[1]kalamohan24@yahoo.co.in, [2]k.sathi89@gmail.com, [3]mrpsuresh83@gmail.com

DOI: 10.1201/9781003388913-68

example. Section 5 depicts the algorithm applied for decryption. With the help of the cipher text obtained in Section 4, decryption process is explained in section 6. The conclusion is given in section 7 followed by references.

II. Preliminaries

The following basic concepts are applied in this article.

- *Plain Text:* It is the original information can read by anyone.
- *Cipher Text:* The secret information send it to the receiver.
- *Encryption:* The plain text can be converted by using algorithm.
- *Decryption:* The original text is retained by secret key.
- *Ordinary Differential Equations:* It is an equation consisting of one or more functions of one independent variable and its derivatives.
- *Variable Separable Method:* It is a method to find the solution of the ordinary differential equation. As the name indicates, find the given equation is isolated into separate variables. Then by integrating both sides the solution is obtained.

III. Algorithm for Encryption:

The procedure to encrypt the information are as follows:

Step 1: Assuming 1 to A, 2 to B and so on., Space is assumed as 27 and NULL character is assumed as 28. The numerical equivalent of the given plain text is generated.

Step 2: These numerical characters are seperated into groups of 4 elements each. To the last group NULL charaters are included at the end to bring out 4 number count in the group.

Step 3: For each group the elements are assumed to be 'a','b','c','d'.

Step 4: The first order ordinary differential equation is obtained as $\dfrac{dy}{dx} = \dfrac{ax+b}{cy+d}$

Step 5: The solution of the equation obtained in step 4 is found by variable separable method, omitting the integral constant.

Step 6: The combination of the solutions obtained in each group is taken as the coded message.

IV. Encryption Process

In order to illustrate the encryption process, the word MATHS is taken and its corresponding numerical values are assigned in Table 68.1.

Table 68.1 Numerical values of plain text

M	A	T	H	S
13	1	20	8	19

With a motive to form groups of four elements, three NULL characters are included at the end which is presented in Table 68.2.

Table 68.2 Numerical values of plain text with NULL character

M	A	T	H	S	NULL	NULL	NULL
13	1	20	8	19	28	28	28

Group 1:

The elements in this group are 13, 1, 20, 8.

Assume $a = 13, b = 1, c = 20, d = 8$

The first order Ordinry Differentil Eqution obtined is $\dfrac{dy}{dx} = \dfrac{13x+1}{20y+8}$

Solving by applying variable separable method, the solution obtained is $10y^2 + 8y = \dfrac{13}{2}x^2 + x$ by omitting the integral constant.

Group 2:

Repeating the procedure to Group 2 the following solution is obtined.

$$a = 19, b = 28, c = 28, d = 28$$

$$14y^2 + 28y = \frac{19}{2}x^2 + 28x$$

From the solutions obtained for each group, the coded message is formed by inserting the solution in blocks as given in Table 68.3.

Table 68.3 Cipher text

$10y^2 + 8y = \frac{13}{2}x^2 + x$	$14y^2 + 28y = \frac{19}{2}x^2 + 28x$

V. Algorithm for Decryption

Decryption is done to retrieve the original information. The cipher text gets converted as a plain text in this step. The steps are as follow:

Step 1: Blocks of solution of the first order differential equation is given as cipher text.

Step 2: The solution in each block is taken and differentiated to find $\frac{dy}{dx}$

Step 3: From $\frac{dy}{dx}$, the character values of 'a', 'b', 'c', 'd' are obtained.

Step 4: Combining the character value of 'a', 'b', 'c', 'd' corresponding to each blocks, the corresponding plain text character is marked.

Step 5: Omitting the NULL character at the end, the plain text is obtained.

VI. Decryption Process

The decryption process with respect to the example specified in step 4 is in Table 68.4. The coded message obtained is

Table 68.4

$10y^2 + 8y = \frac{13}{2}x^2 + x$	$14y^2 + 28y = \frac{19}{2}x^2 + 28x$

From Block 1,

$$10y^2 + 8y = \frac{13}{2}x^2 + x$$

Differentiating both sides, the value of $\frac{dy}{dx}$ is obtained.

$$\frac{dy}{dx} = \frac{13x + 1}{20y + 8}$$

From the ordinary differential equation obtained,

$$a = 13, b = 1, c = 20, d = 8$$

The same procedure is repeated for Block 2 also.

$$14y^2 + 28y = \frac{19}{2}x^2 + 28x$$

$$\frac{dy}{dx} = \frac{19x + 28}{28y + 28}$$

$$a = 19, b = 28, c = 28, d = 28$$

Combining the values of 'a', 'b', 'c', 'd' obtained from each group and decoding

Table 68.5 Decryption plain text

13	1	20	8	19	28	28	28
M	A	T	H	S	NULL	NULL	NULL

By omitting the NULL character, the plain text is obtained as MATHS from Table 68.5.

VII. Conclusion

A novel cryptographic algorithm which applying first order ordinary differential equation and variable separable method has been proposed. The plain text MATHS has been taken and converted as a secret information using the proposed algorithm. Also its reverse process using Decryption algorithm has been emerged the plain text.

References

1. A. P. Hiwarekar (2014), "New Mathematical Modeling for Cryptography", Journal of Information Assurance and Security, Vol. 9 (2014) pp. 027–033

2. M. Tuncay Gencoglu, "Cryptanalaysis of a New Method of Cryptography using Laplace Transform Hyperbolic Functions", Communications in Mathematics and Applications, Vol. 8 (2017), pp. 183–189.

3. A.P. Hiwarekar (2013), "A new method of Cryptography using Laplace transform of Hyperbolic functions", International Journal of Mathematical Archive, Vol. 4(2), pp. 208–213.

4. Dr. Hemant K. Undegaonkar (2019), Security in Communication By Using Laplace Transform and Cryptography", International Journal of Scientific & Technology Research, Vol. 8, pp. 3207–3209.

5. S. Sujatha (2013), "Application of Laplace Transforms in Cryptography", International Journal of Mathematical Archive, Vol. 4, pp. 67–71.

6. CH. Jayanthi and V. Srinivas (2019), "Mathematical Modelling for Cryptography using Laplace Transform", International Journal of Mathematics Trends and Technology, Vol. 65, pp. 10–15.

7. G. Nagalakshmi, A. Chandra Sekhar and D. Ravi Sankar (2020), "Asymmetric key Cryptography using Laplace Transform", International Journal of Innovative Technology and Exploring Engineering, Vol. 9, pp. 3083–3087.

8. Swati Dhingra, Archana A. Savalgi and Swati Jain (2016) "Laplace Transformation based Cryptographic Technique in Network Security", International Journal of Computer Applications, Vol. 136, pp. 6–10.

9. Mampi Saha (2017), "Application of Laplace – Mellin Transform for Cryptography", Raj Journal of Technology Research & Innovation, Vol. V, pp. 12–17.

10. Abdelilah K. Hassan Sedeeg, Mohand M. Abdelrahim Mahgoub and Muneer A. Saif Saeed (2016), "An application of the New Integral "Abooth Transform" in Cryptography", Pure and Applied Mathematics Journal, Vol. 5, pp. 151–154.

11. Kala Raja Mohan, Suresh Rasappan, Regan Murugesan, Sathish Kumar Kumaravel andAhamed A. Elngar (2022), "Secret Information Sharing Using Probability and Biliear Transformation", Proceedings of 2nd International Conference on Mathematical Modeling and Computational Science, pp. 115–122.

Note: The author created all of the tables in this chapter.

Recent Trends in Computational Intelligence and its Application – Sugumaran D. et al. (eds)
© 2023 Taylor & Francis Group, London, ISBN 978-1-032-48410-5

69

A Cryptographic Technique Using Conformal Mapping

Kala Raja Mohan*, Nagadevi Bala Nagaram[1], Regan Murugesan[2]
Department of Mathematics, Vel Tech Rangarajan Dr. Sagunthala R&D Institute of
Science and Technology, Avadi, Chennai – 600 062, Tamil Nadu, India

Suresh Rasappan[3]
Information Technology Department,
University of Technology and Applied Sciences - Ibri, Sultanate of Oman

Abstract—Secret Information sharing is all time requirement in this internet world, especially in Electronic communications such as system security, smart card, mobile communications etc. Cryptography is based on transformation of multiple rounds of transformation of messages in the form of plain text as input into encrypted text message. Through suitable mathematical technique, secrecy of the information is maintained. Many researchers have shown their interest in making use of mathematical techniques in Cryptography. This paper proposes a cryptographic technique applying conformal mapping.

Keywords—Cryptography, Encryption, Decryption, Conformal mapping, Graph, Secret information

I. Introduction

Nowadays an inevitable part of our life is mobile communications, computer networks which uses internet for its base. Still, the internet provides option for many hackers to steel the data stored in the system such as mobile or computer. It is very much essential to provide safety measure while sharing the information. Cryptography plays a major role in providing such information security. Cryptography ensure communication between any two individuals in a secured way.

A. P. Hiwarekar in 2014 propoed two cryptographic technique applying Laplace Transform and Hyperbolic functions [1,3]. M. Tuncay Gencoglu in 2017 introduced a crytographic process involing Laplace Transform with Hyperbolic functions [2]. Dr. K. Hemant K. Undegaonkar introduced a secured communication method involving Laplace Transform [4]. S. Sujatha in 2013 made use of the application of Laplace Transform in the field of cryptography [5]. C. H. Jayanthi and V. Srinivas in 2019 framed a new mathematical modelling involving Laplace Transform [6].

*Corresponding Author: kalamohan24@yahoo.co.in,
[1]nagadevibalaarun@gmail.com, [2]mreganprof@gmail.com, [3]mrpsuresh83@gmail.com

DOI: 10.1201/9781003388913-69

G. Nagalakshmi et al., in 2020 involved Laplace Transform Laplace Tranform using Asymmetric key for secured communication [7]. S. Dhingra et al., proposed a network security method involving Laplace Transform [8]. M. Saha in 2017 utilized Laplace Mellin Transform in forming a cryptographic method for secured information sharing [9]. A. K. H. Sedeeg et al., in 2016 formulated a new cryptographic algorithm applying Abooth Tranform [10]. Kala Raja Mohan et al., in 2022 applied Bilinear Transform with Probability in identifying a secured information sharing algorithm [11]. A. Meenakshi et al., applied graph network in designing a cryptographic algorithm [12]. This paper aims at developing a cryptographic algorithm using conformal mapping.

In section 2, the standard definitions made use of, in this crypto analysis are described. Section 3 represents the algorithm which is applied for encryption. The encryption algorithm which is explained in section 3 is demonstrated with an example in section 4. Section 5 depicts the algorithm applied for decryption. With the help of the cipher text obtained in Section 4, decryption process is explained in section 6. Section 7 is about the conclusion followed by references.

II. Standard Definitions

The cryptographic analysis make use of the following standard position in this paper.

- The information which is to be shared to the other person secretly is the plain text.
- The encrypted message making use of the key specified for the process is the cipher text.
- The process by which the plain text gets transformed into the cipher text is the cipher.
- The process involved in converting plain text into secret message is encryption.
- The reverse process of encryption is decryption, which convert secret message to plain text.

- *Conformal Mapping:* Angle preserving transformation at a non-zero derivative point is conformal mapping. It is a function which preserves with the angles but not the with the length defined with it.

III. Algorithm for Encryption

The procedure to be followed in the process of encryption is as given below.

Step 1: Assuming 1 to A, 2 to B and so on upto 26 to Z. In addition to this, 27 is assigned to 'space' and 28 is assigned to 'NULL' character. The numerical equivalent of the given plain text is generated.

Step 2: These numerical characters are seperated into groups of 4 elements each. To the last group NULL characters are included at the end to bring out 4 number count in the group.

Step 3: The actual number of character is assumed as 'a' and the number of characters after including null character is assumed as 'b'. Using the values of 'a' and 'b', the mapping

$$w = az + b$$

is framed

Step 4: To each of the group, first two number are taken to be 'x' co-ordinate values and last two numbers are taken to be 'y' co-ordinate values. Using the values of 'x' and 'y', four pair of elements are formed. While pairing the procedure mentioned below is followed.

Two values of 'x' are assumed as and two values of 'y' are   The pair of elements are taken as

Step 5: To each pair of element, the corresponding pair of values of 'u' and 'v' are found. Also, the graph relating to the newly order pair is drawn.

Step 6: Using the pair of elements of 'u' and 'v', coded message is formed by inserting the graphs inside blocks. This block of images together with the values of 'a' and 'b' to be shared to the receiver.

IV. Encryption Process

In this section, the encryption process discussed here is illustrated using the sentence "ALL IS WELL"

The word ALL IS WELL and its numerical equivalent assigned are as follows.

A	L	L		I	S		W	E	L	L
1	12	12	27	9	19	27	23	5	12	12

This sentence consists of 11 characters. Hence assume $a = 11$.

A	L	L		I	S		W	E	L	L	NULL
1	12	12	27	9	19	27	23	5	12	12	28

In order to make the number of characters as a multiple of 4, one NULL character is included at the end. Now the number of characters is changed to 12. Now, assume $b = 12$.

$$w = 11z + 12$$

Let $z = x + iy$ and $w = u + iv$

We get $u = 11x + 12$; $v = 11y$

Group 1

A	L	L	
1	12	12	27

Consider $x = 1, 12$ & $y = 12, 27$

Forming pairs we get A_1: (1, 12) A_2: (1, 27) A_3: (12, 12) A_4: (12, 27)

To each of the pairs, find the corresponding values of u and v, which are as follows.

B_1: (23, 132) B_2: (23, 297) B_3: (144, 132) B_4: (144, 297)

To these values, the following graph is obtained, which to be used in the cipher text message.

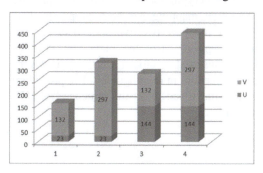

Group 2

I	S		W
9	19	27	23

Repeating the procedure to group 2, the graph is obtained.

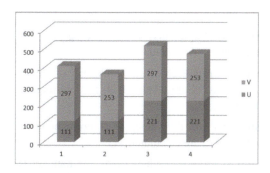

Group 3

E	L	L	NULL
5	12	12	28

The graph obtained for group 3 is as follows.

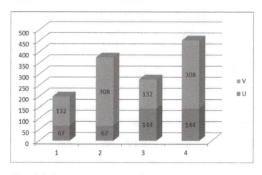

Combining the graphs of all the three groups, coded message obtained is given in blocks as mentioned below (for combined graph see next page).

This block of graphs together with the values of '*a*' and '*b*' to be shared to the receiver.

V. Algorithm for Decryption

The decryption process involves the steps as specified below.

Step 1: Using the values of '*a*' and '*b*', the mapping $w = az + b$ is framed.

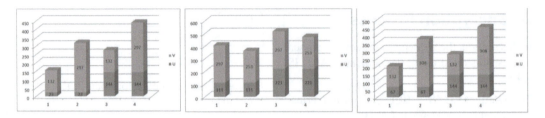

Step 2: Equations corresponding to the mapping is also found.

Step 3: From each of the block, pair of values of 'u' and 'v' are taken which are assumed as B_1, B_2, B_3, B_4

Step 4: From the values of B_i corresponding values of A_i are calculated.

Step 5: From the values of A_i, the data set is formed and from the data set decoded message is formed.

VI. Decryption Process

This section describes the decryption process involved with the example coted in section 4. Before proceeding with the decryption process, using the values of 'a' and 'b', the conformal mapping

$$w = 11z + 12$$

Using this mapping, the equations related to the mapping are obtained as

$$u = 11x + 12; v = 11y$$

Block 1

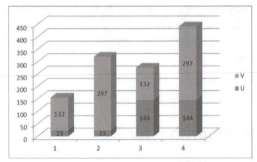

From the graph, the points obtained are $B_1(23, 132), B_2(23, 297), B_3(144, 132), B_4(144, 297)$

To each of the points, its corresponding x, y values are obtained using the equations

$$(23, 132) \rightarrow 23 = 11x + 12 \;\&\; 132 = 11y$$
$$x = 1; y = 12$$
$$(23, 297) \rightarrow 23 = 11x + 12 \;\&\; 297 = 11y$$
$$x = 1; y = 27$$
$$(144, 132) \rightarrow 144 = 11x + 12 \;\&\; 132 = 11y$$
$$x = 12; y = 12$$
$$(144, 297) \rightarrow 144 = 11x + 12 \;\&\; 297 = 11y$$
$$x = 12; y = 27$$
$$A_1(1, 12), A_2(1, 27), A_3(12, 12), A_4(12, 27)$$

Repeating the procedure to blocks 2 and 3, the points obtained are as follows.

$$A_1(9, 27), A_2(9, 23), A_3(19, 27), A_4(19, 23)$$
$$A_1(5, 12), A_2(5, 28), A_3(12, 12), A_4(12, 28)$$

By combining the points obtained from each block, the decryption table is given below.

1	12	12	27	9	19	27	23	5	12	12	28
A	L	L		I	S		W	E	L	L	NULL

By omitting the NULL characters at the end, the plain text is obtained as ALL IS WELL.

VII. Conclusion

A new cryptographic algorithm applying conformal mapping has been proposed. This process involves two types of data sharing in decryption. Also, the plain text involves graphs inserted in blocks Thus, this is a very safe procedure in cryptography. The procedure is also illustrated using the sentence "ALL IS WELL".

References

1. A. P. Hiwarekar, "New mathematical modeling for cryptography," *Journal of Information Assurance and Security*, **9**, 027–033 (2014).

2. M. Tuncay Gencoglu, "Cryptanalaysis of a New Method of Cryptography using Laplace Transform Hyperbolic Functions." *Communications in Mathematics and Applications*, **8**, 183–189 (2017).

3. A. P. Hiwarekar, "A new method of Cryptography ussing Laplace transform of Hyperbolic functions," *International Journal of Mathematical Archive*, **4**, 208–213 (2013).

4. Dr. K. Hemant K. Undegaonkar, "Security in Communication By Using Laplace Transform and Cryptography," *International Journal of Scientific & Technology Research*, **8**, 3207–3209 (2019).

5. S. Sujatha, "Application of Laplace Transforms in Cryptography," *International Journal of Mathematical Archive*, **4**, 67–71 (2013).

6. C. H. Jayanthi and V. Srinivas, "Mathematical Modelling for Cryptography using Laplace Transform," *International Journal of Mathematics Trends and Technology*, **65**, 10–15 (2019).

7. G. Nagalakshmi, A. Chandra Sekhar and D. Ravi Sankar, "Asymmetric key Cryptography using Laplace Transform," *International Journal of Innovative Technology and Exploring Engineering*, **9**, 3083–3087 (2020).

8. S. Dhingra, A. A. Savalgi and S. Jain, "Laplace Transformation based Cryptographic Technique in Network Security," *International Journal of Computer Applications*, **136**, 6–10 (2016).

9. M. Saha, "Application of Laplace – Mellin Transform for Cryptography," *Raj Journal of Technology Research & Innovation*, **5**, 12–17 (2017).

10. A. K. H. Sedeeg, M. M. Abdelrahim Mahgoub, and M. A. Saif Saeed, "An Application of the New Integral "Aboodh Transform" in Cryptography," *Pure and Applied Mathematics Journal*, **5**, 151–154 (2016).

11. Kala Raja Mohan, Suresh Rasappan, Regan Murugesan, Sathish Kumar Kumaravel and Ahamed A. Elngar, "Secret Information Sharing Using Probability and Biliear Transformation", Proceedings of 2nd International Conference on Mathematical Modeling and Computational Science, pp. 115–122 (2022).

12. A. Meenakshi, J. Senbagamalar, and A. Neel Armstrong, "Encryption on Graph Networks", Proceedings of 2nd International Conference on Mathematical Modeling and Computational Science, pp. 123–130 (2022).

Recent Trends in Computational Intelligence and its Application – Sugumaran D. et al. (eds)
© 2023 Taylor & Francis Group, London, ISBN 978-1-032-48410-5

70

Average Resolving Number of a Fuzzy Graph and its Application in Network Theory

R. Shanmugapriya[1], M. Vasuki[2], P. K. Hemalatha[3]

Department of Mathematics,
Vel Tech Rangarajan Dr. Sagunthala R&D Institute of Science and Technology, Chennai

Abstract—Consider an ordered fuzzy subset $H = \{(u_1, \sigma(u_1)), (u_2, \sigma(u_2)), \dots (u_k, \sigma(u_k))\}$, $|H| \geq 2$ of a FG G, then $(z, \sigma(z)) \in \sigma - H = \{(u_{k+1}, \sigma(u_{k+1})), (u_{k+2}, \sigma(u_{k+2})), \dots (u_n, \sigma(u_n))\}$ whereas $w(z, u_1)$, $w(z, u_2), \dots w(z, u_k)$ is an ordered k-tuple with regard to H. If any two elements of $\sigma - H$ does not have similar representations with regard to H, then the fuzzy resolving set (FRS) is then described as having the fuzzy subset H. As seen, the fuzzy resolving number (FRN) is $Fr(G)$ which has the minimum cardinality of the fuzzy resolving set (FRS). Also, we introduce the average resolving number (AFRN) $(\sigma_{av}(G))$ of G, the average super fuzzy resolving number(FSRN) of a FG and fuzzy resolving excellent. We have also defined their properties and an application related to social network connection. The application helps to resolve all the social network storage units to a few signal storage units using the concept of strong arcs in FRS.

Keywords—Fuzzy graph, Fuzzy resolving set, Fuzzy resolving number, Fuzzy super resolving number, Average resolving number, Fuzzy resolving excellent

I. Introduction

In the 21st century, Graph theory has been put forward by the Fuzzy set. It was evaluated by Zadeh [9] in 1965 whereas Graph theory was proclaimed by Euler in 1936. The fuzzy graph (FG) theory was derived from the fuzzy set by Kauffman in 1973 and Rosenfled extended the fuzzy graph theory in 1975. Moderson, Sunil Mathew, et al., [5][6] developed a Fuzzy graph with a wide range of applications in it. A Fuzzy set is more likely the uncertainty and vagueness in the set. Later, many real-life applications are developed using this concept. In 2000, Chartrand [3] introduced resolvability in Graphs [2] and an upper dimension of graphs[4] which has been later extended to resolvability in Fuzzy graphs by Shanmugapriya and Mary Jiny [8]. Also, the Fuzzy graph was lengthened to an Intuitionistic fuzzy graph(IFG) by Atanassov [1]. In recent years, the average resolving number of graphs [7] was developed by Saravanan, Sujatha, and Sundareswaran in 2019. A Fuzzy system was developed in the case of uncertainty and vagueness in many real-life situations. Later, it has been developed by

[1]spriyasathish11@gmail.com, [2]vasukimani1997@gmail.com, [3]pkhemalathamsc@gmail.com

DOI: 10.1201/9781003388913-70

many researchers. A Fuzzy Graph is represented by giving membership value to all the vertices and edges of a graph. An application of Fuzzy sets includes traffic light problems, network theory, mathematical biology, identification of drugs, cancer detection, examination schedules, and image capturing. Many applications are developed using the parameter of clustering the sets.

In this paper, we define a new fuzzy resolving parameter namely the average fuzzy resolving number $\sigma_{av}(G)$ as the average of all the FRS of G's minimal cardinality. Also, we have developed FR excellent and FSR excellent of a provided fuzzy graph.

II. Preliminaries

Definition 2.1

A FG $G(V, \sigma, \mu)$ where μ is a symmetric fuzzy relation on σ: $V \to [0, 1]$ and μ: $V \times V \to [0, 1]$ such that

$$\mu(u, v) \le \sigma(u) \wedge \sigma(v), \forall u, v \in V.$$

Definition 2.2

A FG G is regarded as a complete fuzzy graph (CFG) if $\mu(u, v) = \sigma(u) \wedge \sigma(v) \forall u, v \in V$ and $\mu^{\infty}(u, v) = \mu(u, v), \forall u, v \in V$.

Definition 2.3

The strength or membership value of the path's weakest edge is referred to as the path's weight. The greatest weight among all possible pathways between v_1 and v_n is given by $\mu^{\infty}(v_1, v_n)$ and it represents the weight of the connectedness between v_1 and v_n.

Definition 2.4

An $n \times n$ matrix which is regarded as $X_{ij} = \mu(v_i, v_j)$ for $i \ne j$ and $X_{ij} = \sigma(v_i)$ when $i = j$ is called an adjacency matrix A of G.

Definition 2.5

In a fuzzy graph, an arc (u, y) is regarded as a strong arc if $\mu^{\infty}(u, y) = \mu(u, y)$.

Definition 2.6

The representation of $(z, \sigma(z)) \epsilon \sigma - H = \{(u_{k+1}, \sigma(u_{k+1})), (u_{k+2}, \sigma(u_{k+2})), \dots (u_n, \sigma(u_n))\}$ with

regards to H is $\{w(z_j, u_1), w(z_j, u_2), \dots w(z_j, u_k)\}$, where $j = k + 1, k + 2, \dots n$ are written in a row form. This matrix is called the Fuzzy Resolving Matrix of order $n - k \times k$ and it is represented as $R_{n-k \times k}$.

Definition 2.7

Consider an ordered fuzzy subset $H = \{(u_1, \sigma(u_1)), (u_2, \sigma(u_2)), \dots (u_k, \sigma(u_k))\}$, $|H| \ge 2$, the representation of $(z, \sigma(z)) \epsilon \sigma - H = \{(u_{k+1}, \sigma(u_{k+1})), (u_{k+2}, \sigma(u_{k+2})), \dots (u_n, \sigma(u_n))\}$ and $\{w(z, u_1), w(z, u_2), \dots w(z, u_k)\}$, $w(z, y)$ is the importance of the relationship between z and y. If each pair of elements in the fuzzy subset H has a unique representation with regard to H of G, the set is referred as a FRS. The FRN is represented as $Fr(G)$, the minimal cardinality of FRS.

III. Average FRN of a FG

We apply graph theory's average fuzzy resolving number(AFRN) to a fuzzy graph. We also discuss a simple connected graph with the vertices $n \ge 3$ in this paper.

The AFRN of a fuzzy graph denoted by $\sigma_{av}(G)$ is written as

$$\sigma_{av(G)} = \frac{1}{|V(G)|} \sum_{v_i \in V(G)} \sigma_{v_i}(G)$$

where $V(G)$, the vertex set and $\sigma_{v_i}(G)$ is the lowest cardinality of FRS of G containing v_i.

Also, the fuzzy graph is called fuzzy resolving excellent if $\sigma_{v_i}(G) = Fr(G) \forall v_i \in V(G)$.

Example 3.1

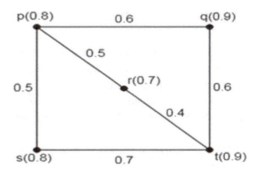

Fig. 70.1 Fuzzy graph

Let $V = \{p, q, r, s, t\}$ and $\sigma = \{(p, 0.8), (q, 0.9),$ $(r, 0.7), (s, 0.8), (t, 0.9)\}$

Let $\mu(pq) = 0.6, \mu(qt) = 0.6, \mu(st) = 0.7, \mu(sp) = 0.5, \mu(pr) = 0.5, \mu(rt) = 0.4$

The adjacency matrix is

$$\begin{array}{c} \\ p \\ q \\ r \\ s \\ t \end{array} \begin{bmatrix} p & q & r & s & t \\ 0.8 & 0.6 & 0.5 & 0.5 & 0 \\ 0.6 & 0.9 & 0 & 0 & 0.6 \\ 0.5 & 0 & 0.7 & 0 & 0.4 \\ 0.5 & 0 & 0 & 0.8 & 0.7 \\ 0 & 0.6 & 0.4 & 0.7 & 0.9 \end{bmatrix}$$

Let $H_1 = \{\sigma_1, \sigma_5\}, \sigma - H_1 = \{\sigma_2, \sigma_3, \sigma_4\}$

Then $\sigma_2/H_1 = \{\mu^\infty (\sigma_1, \sigma_2), \mu^\infty (\sigma_5, \sigma_2)\}$

$$= (0.6, 0.6)$$

Similarly, $\sigma_3/H_1 = (0.5, 0.5), \sigma_4/(H_1 = (0.6, 0.7)$

The matrix representation of the above $\sigma - H$ representation is

$$R_{3\times2} = \begin{bmatrix} 0.6 & 0.6 \\ 0.5 & 0.5 \\ 0.6 & 0.7 \end{bmatrix}$$

Since the values are unique, then H_1 is the FRS of G.

Likewise, $H_2 = \{\sigma_1, \sigma_4\}, H_3 = \{\sigma_2, \sigma_4\}, H_4 = \{\sigma_2, \sigma_3, \sigma_4\}$ are all having distinct representation with respect to σ.

For H_2, $\sigma_2/(H_2 = (0.6, 0.6), \sigma_3/H_2 = (0.5, 0.6),$ $\sigma_5/H_2 = (0.6, 0.7)$

For H_3, $\sigma_1/H_3 = (0.6, 0.6), \sigma_3/H_3 = (0.5, 0.5),$ $\sigma_5/H_3 = (0.6, 0.7)$

For H_4, $\sigma_1/H_4 = (0.5, 0.6, 0.6),$
$\sigma_5/H_4 = (0.5, 0.6, 0.7)$

Therefore, H_1, H_2, H_3 and H_4 are the minimum FRS containing all the vertices of G whereas $H_5 = (\sigma_1, \sigma_2), H_6 = (\sigma_1, \sigma_3)$ and $H_7 = (\sigma_2, \sigma_3)$ are all not fuzzy resolving subsets of a given Fuzzy graph since the representations $\sigma - H$ are not unique.

(i.e) $\sigma_p(G) = 2, \sigma_q(G) = 2, \sigma_r(G) = 3, \sigma_s(G) = 2,$ $\sigma_t (G) = 2$.

FRN is $Fr(G) = 2$.

Hence $\sigma_{av}(G) = \dfrac{2 + 2 + 3 + 2 + 2}{5} = \dfrac{11}{5}$

Also, G is not fuzzy resolving excellent since $\sigma_r(G) = Fr(G)$.

THEOREM 3.1

The average FRN of any connected graph is 'n' if all the subsets of G have n-number of vertices.

Proof: Let G needs to be a plain linked fuzzy network with vertices. Let $H = \{\sigma_1, \sigma_2, ..., \sigma_m\}$, $n > m$ be a FRS of G, then the representations are all distinct of σ_i/H for $i = m + 1, m + 2,$ ..., n. However, for all $i = 1$ to n, σ_i/H may or may not be distinct. Using this concept, we can find $\sigma_{v_i} \forall i = 1, 2, ..., m$. Let us consider that $\sigma_{v_i} = n \forall i = 1, 2, ..., m$ then the average fuzzy resolving number is the average of all the σ_{v_i}'s with regards to H of the set V. This implies that the average fuzzy resolving number is n if $\sigma_{v_i} = n \forall v_i \in V$ which satisfies the theorem.

THEOREM 3.2

$Fr(G) \leq \sigma_{av}(G)$.

Proof: Let G needs to be a plain linked fuzzy network with n vertices. Let $H = \{\sigma_1, \sigma_2, ..., \sigma_m\}$ be a FRS of G, then all the representation of $\sigma - H$ are all different. $Fr(G)$ is the minimum cardinality of all the FRS of G. Similarly, we can find the representation with regards to each vertex to the FRS H. The AFRN has been calculated using the cardinality of each vertex value. It is obvious to see that $Fr(G)$ must be equal or smaller than the average fuzzy resolving number $\sigma_{av}(G)$ which proves the statement.

IV. Average SRN of a Fuzzy Graph

We now develop an average super resolving number (ASRN) of a fuzzy graph G.

If any two elements of a fuzzy graph G have unique representations with regard to H, the FRS of G is treated as a FSRS, and we also write the SRN which is indicated by $Sr(G)$ and the super resolving matrix by $S_{n \times k}$, is the lowest cardinality of a SRS of a FG G.

An ASRN of a FG is denoted by $\sigma_{sav}(G)$ and defined as,

$$\sigma_{sav}(G) = \frac{1}{|V(G)|} \sum_{v_i \in V(G)} \sigma_{v_i}(G)$$

A fuzzy graph G is said to be super resolving excellent if $\sigma_{v_i}(G) = Sr(G) \ \forall \ v_i \in V(G)$.

EXAMPLE 4.1:

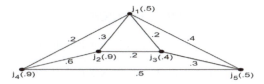

Fig. 70.2 Fuzzy graph

Here $V = \{j_1, j_2, j_3, j_4, j_5\}$ and $\sigma = \{(j_1, 0.5),$ $(j_2, 0.9), (j_3, 0.4), (j_4, 0.9), (j_5, 0.5)\}$

$\mu(j_1, j_2) = 0.3, \ \mu(j_1, j_3) = 0.2, \ \mu(j_1, j_4) = 0.2,$ $\mu(j_1, j_5) = 0.4$

$\mu(j_2, j_3) = 0.2, \ \mu(j_2, j_4) = 0.6, \ \mu(j_3, j_5) = 0.3,$ $\mu(j_4, j_5) = 0.5$

Adjacency Matrix is

$$\begin{array}{c c c c c c} & j_1 & j_2 & j_3 & j_4 & j_5 \\ j_1 & \begin{bmatrix} 0.5 & 0.3 & 0.2 & 0.2 & 0.4 \\ j_2 & 0.3 & 0.9 & 0.2 & 0.6 & 0 \\ j_3 & 0.2 & 0.2 & 0.4 & 0 & 0.3 \\ j_4 & 0.2 & 0.6 & 0 & 0.9 & 0.5 \\ j_5 & 0.4 & 0 & 0.3 & 0.5 & 0.5 \end{bmatrix} \end{array}$$

The connectedness matrix between the vertices of G is

$$\begin{array}{c c c c c c} & j_1 & j_2 & j_3 & j_4 & j_5 \\ j_1 & \begin{bmatrix} 0.5 & 0.4 & 0.3 & 0.4 & 0.4 \\ j_2 & 0.4 & 0.9 & 0.3 & 0.6 & 0.5 \\ j_3 & 0.3 & 0.3 & 0.4 & 0.3 & 0.3 \\ j_4 & 0.4 & 0.6 & 0.3 & 0.9 & 0.5 \\ j_5 & 0.4 & 0.5 & 0.3 & 0.5 & 0.5 \end{bmatrix} \end{array}$$

We can manually search the minimum subset of σ which has the unique representation in order with all the vertices. Hence, $H_1 = \{\sigma_1, \sigma_2\}, H_2 = \{\sigma_1, \sigma_4\}, H_3 = \{\sigma_2, \sigma_5\}$ and $H_4 = \{\sigma_2, \sigma_3\}$ are all the minimum SRS of a fuzzy graph.

The $\sigma - H$ representation of these subsets are

$$\begin{bmatrix} 0.5 & 0.4 \\ 0.4 & 0.9 \\ 0.3 & 0.3 \\ 0.4 & 0.6 \\ 0.4 & 0.5 \end{bmatrix}, \begin{bmatrix} 0.5 & 0.4 \\ 0.4 & 0.6 \\ 0.3 & 0.3 \\ 0.4 & 0.9 \\ 0.4 & 0.5 \end{bmatrix}, \begin{bmatrix} 0.4 & 0.4 \\ 0.9 & 0.5 \\ 0.3 & 0.3 \\ 0.6 & 0.5 \\ 0.5 & 0.5 \end{bmatrix} \text{ and}$$

$$\begin{bmatrix} 0.4 & 0.3 \\ 0.9 & 0.3 \\ 0.3 & 0.4 \\ 0.6 & 0.3 \\ 0.5 & 0.3 \end{bmatrix} \text{ respectively.}$$

Also, $H_5 = \{\sigma_1, \sigma_3\}$ and $H_6 = \{\sigma_1, \sigma_5\}$ is not SRS since it does not have unique representation in order with all the vertices.

That is, $Sr(G) = 2$.

Also $\sigma_{j_1} = 2, \ \sigma_{j_2} = 2, \ \sigma_{j_3} = 2, \ \sigma_{j_4} = 2$ and $\sigma_{j_5} = 2$

Hence $\sigma_{sav}(G) = \dfrac{2+2+2+2+2}{5} = 2$

The above given fuzzy graph is super resolving excellent since $\sigma_{j_i} = Sr(G)$ for all j_i.

Theorem 4.1

The AFRN of a fuzzy graph G does not need to be an ASRN of G. But the converse is true.

Proof: Let $H = \{\sigma_1, \sigma_2, \dots \sigma_m\}$ be a resolving set and here m is the FRN of a fuzzy graph G since m is the minimum cardinality of the given set H Similarly, there will be many resolving subsets containing all the vertices of G and they will have different resolving numbers according to G. The average of all these resolving subsets may or may not be the same for super resolving subsets of the same graph G. Hence, AFRN need not be an ASRN of G.

Now, let H be a super resolving subset of G. It is obvious to see that the same resolving subsets will be true for the resolving subset of a fuzzy graph. Hence, an ASRN and AFRN of a fuzzy graph are the same. Therefore, the converse of this statement is also true.

Theorem 4.2

1. For a connected fuzzy graph, $Sr(G) \leq \sigma_{sav}(G)$.
2. A fuzzy graph is super resolving excellent if and only if $\sigma_{sav}(G) = Sr(G)$.

V. Application

Consider a social network connection where the vertices are represented by different social network units and the edges represent the strength of the signal passing through them which we have demonstrated in Fig. 70.3. We now find many FRS in the graph. The vertices are chosen randomly and checked for FRS using a fuzzy strong arc. This is just an example to understand the nature of the problem. The FRS is marked as a triangle in the diagram and the obtained triangle vertices can have the capacity to provide signals to all the other signal storage units. The obtained fuzzy resolving set is enough to provide signals to all the other vertices. This application is true for -vertices.

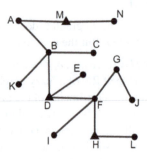

Fig. 70.3 Network graph

VI. Conclusion

We have deeply explained about an AFRN and ASRN of a FG . We have also defined resolving excellent and super resolving excellent corresponding to the FRN and FSRN of a fuzzy graph respectively and we have derived an example to understand the concept of FRS. We have also written an application based on FRS. In the future, we will try to find an algorithm based on finding the average FRS.

References

1. Atanassov, K. T. (1986). Intuitionistic Fuzzy sets, *Fuzzy sets and systems*, 20(1), 87–96.
2. Chartrand, G. Poisson, C. & Zhang, P. (2000). Resolvability and the upper dimension of graphs, *Comput. Math. Appl.* 39, 19–28.
3. Chartrand, G. Eroh, L. Johnson, M. A. & Oellermann, O. R. (2000). Resolvability in graphs and the metric dimension of a graph, *Discrete Applied Mathematics*, 105, 99–113.
4. Chartrand, G. Zhang, P. (2003). The theory and application of Resolvability in Graphs: A Survey, *Congr.Numer.*, 160, 47–68.
5. Mordeson, J. N. Mathew, S. & Malik, D. S., (2018), Fuzzy Graph Theory with applications to Human Trafficking, *Springer International Publishing*, 14–41.
6. Mordeson, J. N. Mathew, S., (2019), Advanced topics in Fuzzy Graph Theory, *Springer science and Business media*.
7. Saravanan, P., Sujatha, R. & Sundareswaran, R. (2019). The Average Resolving Number of Graphs, *International Journal of Future Generation Communication and Networking*, 12, 27–38.
8. Shanmugapriya, R. Mary jiny, D., (2021). Fuzzy super resolving number and resolving number of some special graphs,. *TWMS J.App. and Eng Math*, 11, 459–468.
9. Zadeh, L. A. (1965). Fuzzy sets, *Information and Control*, 8, 338–353.

Note: The author created all of the figures in this chapter.